Lecture Notes in Civil Engineering

Volume 208

Lecture Notes in Civil Engineering (LNCE) publishes the latest developments in Civil Engineering - quickly, informally and in top quality. Though original research reported in proceedings and post-proceedings represents the core of LNCE, edited volumes of exceptionally high quality and interest may also be considered for publication. Volumes published in LNCE embrace all aspects and subfields of, as well as new challenges in, Civil Engineering. Topics in the series include:

- Construction and Structural Mechanics
- Building Materials
- Concrete, Steel and Timber Structures
- Geotechnical Engineering
- Earthquake Engineering
- Coastal Engineering
- Ocean and Offshore Engineering; Ships and Floating Structures
- Hydraulics, Hydrology and Water Resources Engineering
- Environmental Engineering and Sustainability
- Structural Health and Monitoring
- Surveying and Geographical Information Systems
- Indoor Environments
- Transportation and Traffic
- Risk Analysis
- Safety and Security

To submit a proposal or request further information, please contact the appropriate Springer Editor:

- Pierpaolo Riva at pierpaolo.riva@springer.com (Europe and Americas);
- Swati Meherishi at swati.meherishi@springer.com (Asia - except China, and Australia, New Zealand);
- Wayne Hu at wayne.hu@springer.com (China).

All books in the series now indexed by Scopus and EI Compendex database!

More information about this series at https://link.springer.com/bookseries/15087

Dat Vu Khoa Huynh · Anh Minh Tang ·
Dinh Hong Doan · Phil Watson
Editors

Proceedings of the 2nd Vietnam Symposium on Advances in Offshore Engineering

Sustainable Energy and Marine Planning

Springer

Editors
Dat Vu Khoa Huynh
Norwegian Geotechnical Institute
Oslo, Oslo, Norway

Dinh Hong Doan
Subsea7
Suresnes CEDEX, France

Anh Minh Tang
École des ponts ParisTech
Marne-la-Vallée CEDEX 2, France

Phil Watson
University of Western Australia
Perth, WA, Australia

ISSN 2366-2557 ISSN 2366-2565 (electronic)
Lecture Notes in Civil Engineering
ISBN 978-981-16-7734-2 ISBN 978-981-16-7735-9 (eBook)
https://doi.org/10.1007/978-981-16-7735-9

This Springer imprint is published by the registered company Springer Nature Singapore Pte Ltd.
The registered company address is: 152 Beach Road, #21-01/04 Gateway East, Singapore 189721, Singapore

Preface

Global demand for energy is rising rapidly, because of the world population and economic growth, especially in emerging market economies. Marine energy has been gaining attention as a potential contributor to the diversification of the world renewable portfolio as well as to sustainable energy generation. The growth of marine energy can result in the increasing use of sea space and potential for conflict with existing ocean uses, both of which can be resolved by marine spatial planning. In this context, the second Vietnam Symposium on Advances in Offshore Engineering (VSOE2021) focused on the theme of "Sustainable Energy and Marine Planning", aiming at enhancing the sustainable use of our marine resources while ensuring the health of the ecosystem as well as the effective management of marine activities including energy production and infrastructures.

This publication covers the second edition of VSOE, which was originally planned to be held in Ho Chi Minh City, Vietnam, in 2021, but then postponed to 2022 due to the global Covid-19 situation. The second VSOE event is organized by the Association of Vietnamese Scientists and Experts (AVSE Global) in collaboration with Ho Chi Minh City University of Technology (HCMUT) and the Vietnam Administration of Seas and Islands (VASI). VSOE2021 is also held under the auspices of both the International Society for Soil Mechanics and Geotechnical Engineering (ISSMGE) and the Vietnam Society for Soil Mechanics and Geotechnical Engineering (VSSMGE).

Following the success of the first symposium (VSOE2018), VSOE2021 continues providing a platform for all participants to exchange knowledge and experience gained recently in offshore engineering, technology innovations, and marine spatial plan to achieve the goal of economic, reliable, and sustainable solutions for offshore energy development and ecosystem-based management of the marine environment. VSOE2021 brings together researchers, practitioners, policymakers, and entrepreneurs to discuss and promote technology and policy changes as well as to generate business opportunities in marine energy, both domestically in Vietnam and globally.

The papers collected in the book comprise 4 keynote lectures and 56 technical papers from 15 different countries which cover six topics including: site characterization and ground modeling, design of offshore foundations and structures, performance of offshore foundations and structures, fossil energy: technology opportunities and challenges, offshore renewable energy systems, and marine spatial planning. Each paper has been peer-reviewed by at least two experts in the field, drawn from academic institutions and industry from around the world.

We would like to thank the two co-organizers, all members of the scientific and organizing committees, and reviewers for their important contributions to VSOE2021. We wish to thank the keynote speakers and authors for their efforts in preparing papers in such a challenging Covid-19's context.

Last but not least, we are very grateful for the wonderful support of the two symposium patrons: Viet Nam National University Ho Chi Minh City (VNU-HCM) and The Ministry of Natural Resources and Environment (MONRE).

Dat Vu Khoa Huynh
Dinh Hong Doan
Anh Minh Tang
Phil Watson

Organisation

Organising Committee

Chairmen

Doan, Dinh Hong	Subsea7, France
Le, Minh Phuong	Ho Chi Minh City University of Technology, Vietnam
Ta, Dinh Thi	Vietnam Administration of Seas and Islands, Vietnam

International Members

Bui, Man	Geotechnical Testing Consultants, Dubai, UAE
Do, Tuan Anh	University Caen, France
Dinh, Van Nguyen	ANTS & MaREI Centre, University College Cork, Ireland
Dong, Hoang Yen	Lyon, France
Huynh, Dat Vu Khoa	Norwegian Geotechnical Institute, Norway
Le, Thi Xiu	Ecole des Ponts ParisTech, France
Nguyen, Binh	Ecole Internationale des sciences du traitement de l'information, France
Nguyen, Tien Dung	Vietnam Japan University, Hanoi, Vietnam
Nguyen, Tran Dieu My	Ho Chi Minh City Institute for Development Studies, Vietnam
Tang, Anh Minh	Ecole des Ponts ParisTech, France
Tran, Xuan Nghiem	Seoul National University, South Korea

Local Members

Bui, Trong Vinh	Ho Chi Minh City University of Technology, Vietnam
Dao, Hong Hai	Ho Chi Minh City University of Technology, Vietnam
Luu, Anh Duc	Vietnam Administration of Seas and Islands, Ministry of Natural Resources and Environment, Vietnam
Phan, Tan Thi	Ho Chi Minh City University of Technology, Vietnam
Ta, Quoc Dung	Ho Chi Minh City University of Technology, Vietnam
Vu, Dinh Hieu	Vietnam Administration of Seas and Islands, Ministry of Natural Resources and Environment, Vietnam

International Scientific Committee

Coordinators

Dinh, Van Nguyen	ANTS & MaREI Centre, University College Cork, Ireland
Huynh, Dat Vu Khoa	Norwegian Geotechnical Institute, Norway

Members

Ansong, Joseph	Belfast School of Architecture and Built Environment, Ulster University, UK
Baniotopoulos, Charalampos	University of Birmingham, UK
Basu, Biswajit	Trinity College Dublin, Ireland
Bui, Man	GTC Soil Analysis Services, United Arab Emirates
Bui, Trong Vinh	Ho Chi Minh City University of Technology, Vietnam
Collin, Frédéric	Université de Liège, Belgium
Du, Van Toan	Vietnam Administration of Seas and Islands, Vietnam
Elkadi, Ahmed	Deltares, the Netherlands
Georgiev, Valentin	GeoMarine Ltd., Bulgaria
Kim, Ki-Du	Konkuk University and Korean Society of Civil Engineers, South Korea
Kim, Sung-Ryul	Seoul National University, South Korea
Lambrughi, Angelo	Salini Impregilo, Italy

Le, Ba Vinh	Ho Chi Minh City University of Technology, Vietnam
Lehane, Barry	University of Western Australia, Australia
Lunne, Tom	Norwegian Geotechnical Institute, Norway
Mai, Hong Quan	National University of Civil Engineering, Vietnam
Nguyen, Danh Thao	Ho Chi Minh City University of Technology, Vietnam
Nguyen, Giang	The University of Adelaide, Australia
Nguyen, Hoang	Benthic, USA
Nguyen, Minh	ODE, UK
Nguyen, Tien Dung	Vietnam Japan University, Vietnam
Nicolini, Emilio	Cathie Associates, France
O'hagan, Anne Marie	MaREI Centre, ERI, University College Cork, Ireland
Paredes, Guilherme Moura	Universidade Lusófona do Porto, Portugal
Pavan, Gianni	Università di Pavia, Italy
Pham, Huy Giao	Asian Institute of Technology, Thailand
Ramsey, Nick	Fugro, Australia
Randolph, Mark	University of Western Australia, Australia
Robertson, Amy	National Renewable Energy Laboratory, USA
Samui, Pijush	National Institute of Technology Patna, India
Sánchez, Marcelo	Texas A&M University, USA
Shin, Hyunkyoung	University of Ulsan, South Korea
Ta, Quoc Dung	Ho Chi Minh City University of Technology, Vietnam
Taborda, David M. G.	Imperial College London, UK
Tang, Anh Minh	Ecole des Ponts ParisTech, France
Thorel, Luc	French Geotech. Soc., France
Tran, Anh Tu	Ho Chi Minh City University of Technology, Vietnam
Tran, Quoc Tuan	INSTN-Paris Saclay University and Grenoble Alpes University, France
Tran, Thanh Toan	National Renewable Energy Laboratory (NREL), USA
Vanneste, Maarten	Norwegian Geotechnical Institute, Norway
Villard, Laurent	Seaway 7, France
Vu, Tuyen	Clarkson University, USA
Wang, Dong	Ocean University of China, China
Wang, Lizhong	Zhejiang University, China
Watson, Phil	University of Western Australia, Australia

Invited Reviewers

Algren, Loui, Denmark
Ballard, Jean-Christophe, Belgium
Barwise, Andy, UK
Bohloli, Bahman, Norway
Bretelle, Sylvie, France
Bui, Ha, Australia
Carrington, Tim, UK
Cauquil, Eric, France
Choi, Jung Chan, Norway
Delimi Zine Labidine, United Arab Emirates
Gavin, Ken, the Netherlands
Govoni, Laura, Italy
Harun, Kürşat Engin, Norway
Ho Karkov, Knud, Denmark
Huvaj, Nejan, Turkey
Johansson, Jörgen, Norway
Klinkvort, Rasmus Tofte, Norway
Le, Thanh Vinh, Vietnam
Le, Van Sy, Vietnam
Le, Viet Hung, Germany
Li, Zheming, UK
Liu, Lele, China
Liu, Zhongqiang, Norway
Mai, Tri Cao, Vietnam
Mandyam, Ram Babu, Australia
Mentani, Alessio, Italy
Muhammed, Rawaz-Dlawar, France
Mukherjee, Kaushik, Singapore
Newson, Tim, Canada
Nguyen, Khanh Duc, Norway
Nguyen, Minh Quy, Vietnam
Nguyen, Quoc Khanh, Vietnam
Nguyen, Van Hung, Vietnam
Nguyen, Viet, France
Pham Hien Hau, Vietnam
Prakasha Kuppalli, India
Serra, Edy, France
Skomedal, Eiliv, Norway
Sturm, Hendrik, Norway
Stuyts, Bruno, Belgium
Terente, Victor, UK
Thumann, Vladimir, the Netherlands
Tran, Quang Huy, Australia

Valjarevic, Aleksandar, Vietnam
van Dijk, Bas, the Netherlands
Vu, Xuan Hong, France
Xiang Yiqiang, China
Yang, Shaoli, Norway
Yu, Long, China
Zhang, Youhu, China

Organisers

Patronages

http://www9.monre.gov.vn/wps/portal/english

https://vnuhcm.edu.vn/

Under the Auspices of

SIMSG ISSMGE

https://www.issmge.org/

VSSMGE
VIETNAMESE SOCIETY FOR SOIL MECHANICS AND GEOTECHNICAL ENGINEERING

https://vssmge.org/

Contents

Design of Offshore Foundations and Structures

Keynote Lectures

Data-Driven Ground Models: The Road to Fully-Integrated Site Characterization and Design

Maarten Vanneste(✉), Guillaume Sauvin, Jean-Rémi Dujardin, Carl Fredrik Forsberg, Rasmus T. Klinkvort, Cathinka S. Forsberg, and Ragnhild C. Hansen

NGI (Norwegian Geotechnical Institute), Sognsveien 72, Oslo, Norway
maarten.vanneste@ngi.no

Abstract. This keynote paper describes recent developments in optimizing the data-driven ground model approach, from acquisition to the full site characterization part, applicable for all types of offshore energy developments, with particular focus to offshore wind applications. For offshore renewables in shallow water depths, the sub-surface is often highly complex, due to the various large-scale geological processes that have affected these areas over the last millions of years. The result is that spatial variability is a key aspect to address over the entire workflow, from geophysical and geotechnical data acquisition, processing, analysis, geostatistics and ultimately engineering design. The direct consequence is that the site has to be assessed in full 3D, which is rarely adopted for engineering applications on large scale. The data-driven approach exploits the benefits of the geological, geophysical and geotechnical data and integrates those into a consistent ground model. In this paper, we illustrate a number of ground model approaches – with variable levels of complexity – and present a few novel methods that will further enhance these models in the future.

Keywords: Data-driven ground models · Data integration · Site characterization · Machine learning · Acoustic impedance inversion · Geology-geophysics-geotechnics

1 Introduction

Engineering applications for offshore energy developments rely on a proper and detailed quantitative characterization of the sub-surface. The most important elements are seismo-acoustic data to characterize the seafloor morphology and sub-seafloor structure and stratigraphy and a combination of geotechnical data from either in situ continuous measurements (like cone penetration testing) or boreholes from which samples are taken and tested in advanced laboratories. The main purpose of these data is to obtain an in-depth understanding of the active and static processes on the site in a consistent geological framework, which sheds light on the site's spatial variability, the key geotechnical parameters and their uncertainty for engineering design, and information on geohazard

D. V. K. Huynh et al. (Eds.): VSOE 2021, LNCE 208, pp. 3–21, 2022.
https://doi.org/10.1007/978-981-16-7735-9_1

and risk assessment. This integration is commonly labelled under the term of a ground model.

However, ground models come in different flavors, depending on the data available, the purpose of the study, but also the experience of the developers and end-users or stakeholders. Traditionally, there has been little integration – or only qualitative integration – of the different data sets. As the seismic data are often collected ahead of the geotechnical data, the seismic interpretation is typically conducted as a separate activity, with minimum calibration or ground-truthing once the geotechnical data become available. As the seismo-acoustic and geotechnical data do not necessarily respond to the same sub-surface properties, their correlation is not always straightforward.

As part of the green transition, the offshore industry is shifting focus from conventional oil and gas developments towards offshore renewables in shallow water depth, for example in formerly glaciated margins. These areas are typically characterized by a complex sub-surface stratigraphy in response to glacio-eustatic sealevel changes and glacial-interglacial processes, having a distinct influence on the subsurface, causing significant spatial variability in these shallow water environments. This spatial complexity needs to be properly captured in the development, as part of the ground model.

As of late, there is a trend of developing data-driven ground models, which aim at a full integration and quantitative assessment of soil properties, their spatial variability and their uncertainty in 3D covering the development area. Such models imply an iterative approach from establishing the preliminary ground model in the early stages of the project, systematically update the model when new data become available, maximize the information from the ground model to optimize additional surveys (and hence optimize the additional survey costs) to fill in the gaps and deliver results in 3D which can be further used for engineering applications and early design. In addition, technological developments like the advent of ultra-high-resolution 2D and 3D seismic systems and machine learning approaches linking the physical and geotechnical properties allow for a new generation of quantitative ground models. These become important as the data collected is not necessarily perfectly suitable for the final layouts, which often changes during the development phase.

This paper presents the current status of data-driven ground models, and starts with a brief summary of the critical data that need to be incorporated, i.e., geology, geophysics and geotechnics, which all have their strengths and weaknesses. Then, we present some examples of ground models with various levels of complexity, using different types of integration algorithms. Before summarizing the paper, we highlight a number of new methods that can be adopted to further improve the quality of the ground models, by either obtaining better data or performing additional analyses.

2 Data Integration Aka Ground Models: Trilogy of Geology, Geophysics and Geotechnics

2.1 Ground Models – Definition and Semantics

A ground model (GM) is a 3D representation of the surface and sub-surface, which provides a spatial and temporal understanding of the static and dynamic conditions affecting the site. It is a dynamic tool, such that it should be updated whenever more data become available throughout the lifetime of the project, from initial conception to decommissioning. As such, the GM can be a decision-making tool for geohazard and risk assessment, optimizing layout plans and collecting complementary or additional data. Ultimately, the GM provides specific input for foundation design. Hence, it combines the information from various disciplines, essentially geology, geophysics and geotechnics into a consistent model (Fig. 1; Fig. 2). There are different flavors of ground models, largely classified as qualitative vs. quantitative ground models (Fig. 2), with the latter being the focus here.

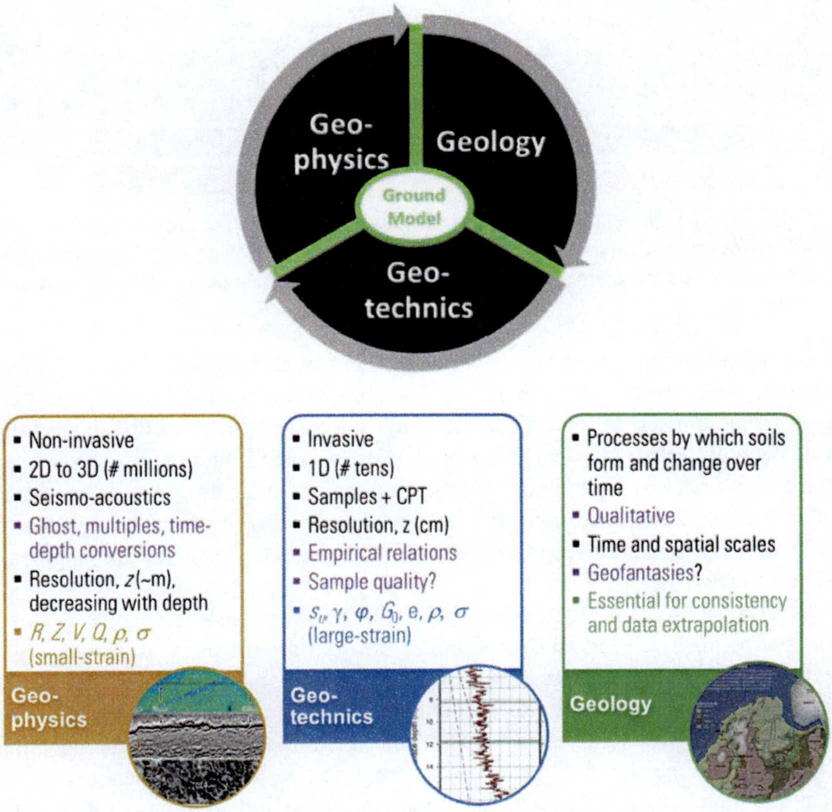

Fig. 1. (Top) The "ground model" is the link between the geophysics, geology and geotechnical engineering into a consistent and reliable 3D framework that can be used for survey planning, geohazard and risk assessment, uncertainty assessment and (preliminary) engineering design. (Bottom) The three building blocks of the ground model each have their specific deliverables, characteristics, advantages and limitations (purple text), highlighted in this figure. # stands for the typical number of data points.

The quantitative GM uses the strengths of the different methods to better link the observations and understanding of the site into a coherent model that respects each

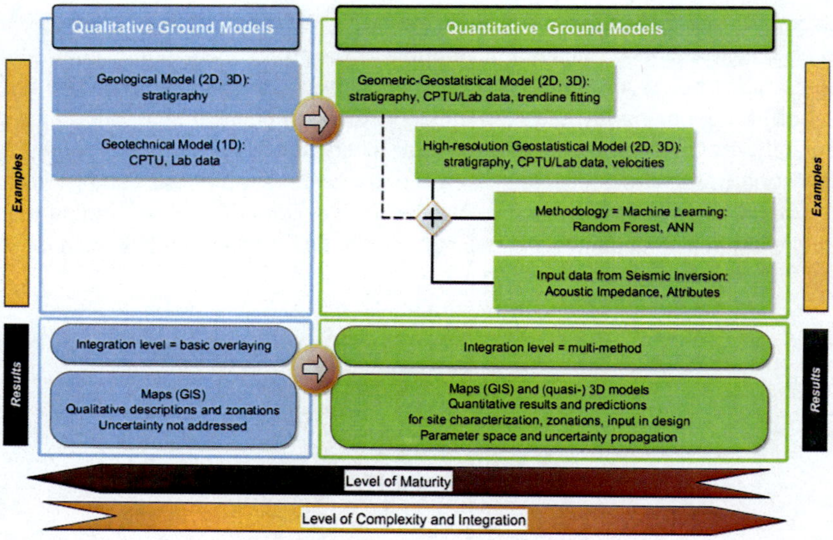

Fig. 2. Ground models are largely split into two categories, qualitative and quantitative, with the qualitative or descriptive models more abundant, but less detailed. However, the level of complexity and integration – and therefore added value – are larger for the quantitative ground models.

method but quantifies the predictions of the geotechnical parameters. The implications are first and foremost that the ground model should be constructed by a group of people with expertise in each of these fields, and that – ideally – the workflow is performed jointly. Each discipline is essential but often uses specific terminology (see Vardy et al. 2017), and not necessarily the same physical or engineering parameters are obtained.

This section describes briefly the strength and weakness of the three disciplines used to create a quantitative ground model, with focus on near-surface engineering applications (Fig. 1).

Geophysical methods are non-invasive and cover extensive areas relatively quickly. In the marine environments, seismo-acoustic methods are the prime tools (see seismic reflection profile and swath bathymetry image in Fig. 3 as an example, as well as a single data trace for comparison with other techniques). The data comes in multi-dimensions (2D to 4D) with millions of data points. There is a trade-off between the penetration and resolution of the signal. For shallow site characterization, the vertical resolution is typically in meter-scale. The main weaknesses are noise and artefacts[1] and time-depth conversions. Key information from seismic surveys includes stratigraphy, small-strain

[1] For example: ghosts, i.e., interference of up- and downgoing energy at source and receiver side related to the spherical wavefronts and the towing depth below sealevel, or seafloor multiples, i.e., energy reflected back and forth between the seafloor or subsurface horizons and the top of the water column.

P-wave velocity (and sometimes S-wave velocity and Poisson ratio), density, quality factors[2], acoustic impedance[3] and reflectivity[4] (Fig. 1).

Geotechnical soil parameters come from in situ testing tools and invasive 1D boreholes with sampling. For the ground modelling work, we focus largely on Cone Penetration Testing (CPT, see Fig. 3). The vertical resolution is in the order of centimeters. The main weaknesses are related to sample quality (e.g. disturbance, sands are typically heavily disturbed) and uncertainty of empirical relations between measured CPT tip resistance q_c and geotechnical parameters. Key parameters are large-strain stiffness, Poisson ratio, unit weight, void ratio, shear strength and friction angle (Fig. 1).

A proper understanding of the geology provides important information on the soil conditions and will put the site into a realistic and consistent regional context. A large-scale understanding of how climate and environmental changes have evolved through time and the various geological processes (like sedimentology, erosion, diagenesis) that have taken place, is key in the data interpretation, linking the geophysical and geotechnical data, particularly where discrepancies or discontinuities may occur. Geological interpretations are often qualitative (Fig. 1). Additional geological analysis tools, like multi-sensor core logging (MSCL) may shed further light on the data interpretation and understanding (see Fig. 3) (see also Vanneste et al. 2013; Vanneste et al. 2015; Vardy et al. 2012).

2.2 Generic Workflow

The generic approach for the data-driven quantitative ground models is shown in Fig. 4. The essential part is to perform a multi-disciplinary integration, which starts with a proper understanding of the geological processes that have shaped the site, particularly with respect to sealevel changes, erosional and depositional processes, faulting, etc. It is, ultimately, the geology of the site which will explains the geophysical and geotechnical data. The site-specific geophysical and geotechnical data – after detailed quality-control and processing – provide the details to create the structure and characterization of the site, but the geology provides the means for a consistent interpretation and explanation of the spatial variability.

In situ geotechnical data allow establishing a detailed 1D soil stratigraphy and unit definition. This interpretation shall be tied with the interpretation from the seismic data, such that the key soil units are properly captured in the seismic data set, i.e., the seismic data are calibrated by the boreholes and in situ tests. Once the seismic and geotechnical units are matched with confidence, one can start using the subtle details from either data set, combined with the structural framework from the seismic data, to interpolate the properties spatially across the site, using either geostatistical or machine learning algorithms. Recently, also advanced data analysis and ultra-high-resolution seismic inversion techniques, which map the measured data into acoustic impedance and/or attenuation values, can further be used in trace-by-trace prediction of geotechnical properties

2 The quality factor is a soil property indicative of the energy that dissipates during wave propagation, providing information on lithology, internal friction and velocity dispersion.

3 Acoustic impedance equals bulk density times P-wave velocity.

4 Reflectivity is the acoustic impedance contrast across an interface.

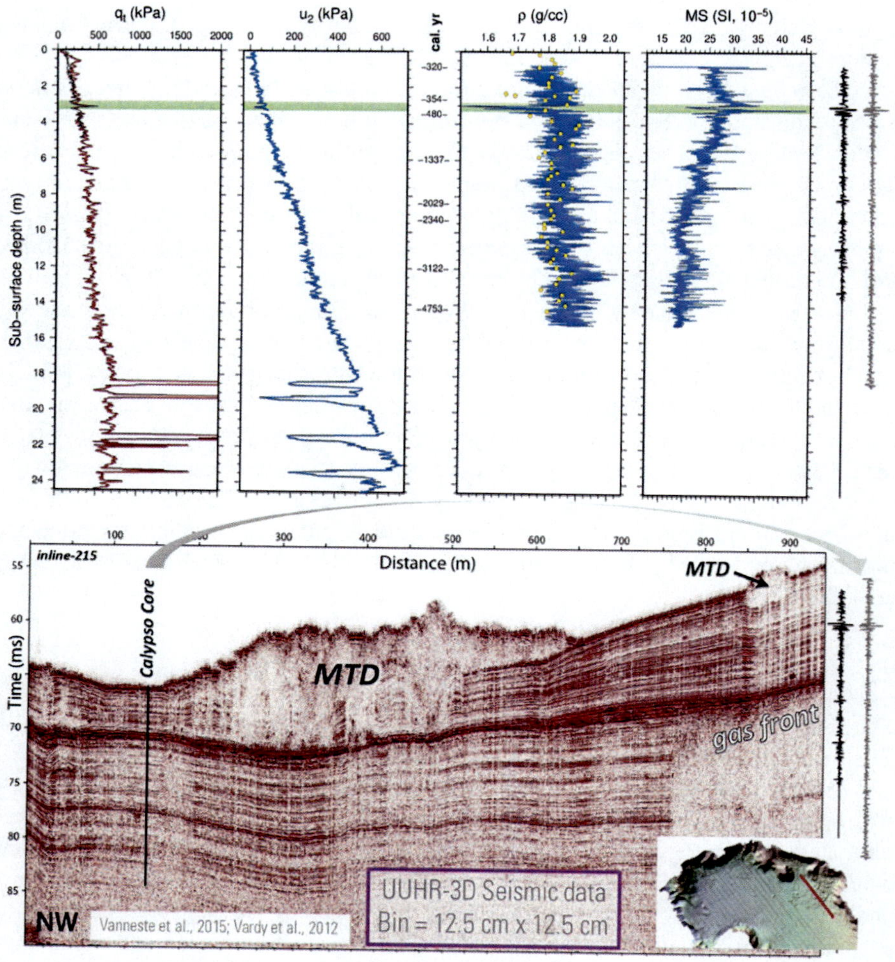

Fig. 3. (Top) Example of 1D geotechnical soil profiles from cone penetration tests (q_t and u_2) used for geotechnical site characterization and slope stability studies. The bulk density (ρ) and magnetic susceptibility (MS) come from Multi-sensor Core Logging (MSCL) data, with bulk density validated by laboratory testing (golden dots). Furthest to the right, we display a synthetic seismic trace derived from the measured bulk density (and P-wave velocity, not shown) from the MSCL data (black trace) as well as the recorded trace at the nearest location from an ultra-high-resolution 3D chirp system. (Bottom) In-line from the 3D UHR seismic chirp data showing a distinct landslide deposit, close to the core location, from where the seismic traces were derived/taken. The inset shows the bathymetry of the 1996 Finneidfjord landslide event.

(Vardy 2015; Vardy et al. 2015, 2017, 2018; Sauvin et al. 2019). The result of this approach is actually the construction of a (pseudo-)3D cube of predicted geotechnical properties or synthetic CPT properties, with an uncertainty assessment attached to it. In other words, these models provide the engineers with a tool to perform design calculations in 3D.

The advantages are multifold, as it allows for (1) quantitative interpretation over qualitative interpretation alone; (2) improved site characterization, geohazard and risk assessment; (3) establishing the baseline for time-lapse analysis and monitoring/intervention; (4) input into engineering applications and design; and (5) a data-driven assessment on where additional data will add most value to further reduce uncertainties in the models.

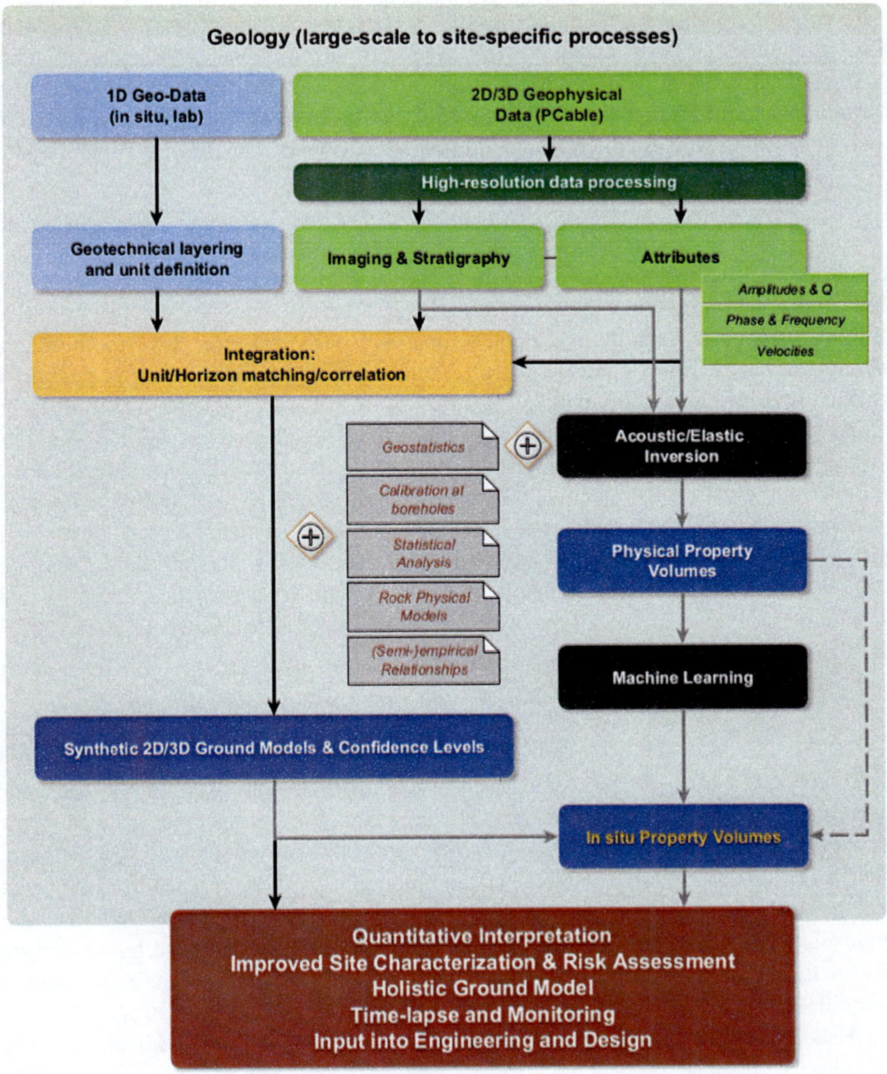

Fig. 4. Generic workflow to generate data-driven, quantitative ground models, with the aim of performing quantitative interpretation, improved site characterization and risk assessment, as well as a key element for time-lapse surveying and monitoring. Ultimately, the ground model should also serve to optimize surveying as well as a key driver in early-phase engineering and design activities.

This generic workflow is not too dissimilar from reservoir characterization approaches developed for conventional oil and gas developments.

3 Some Examples of Data-Driven Ground Models

In this section, we summarize some of the data-driven GMs. While the approach is different, the end results are similar, such that we create a 3D cube of predicted geotechnical parameters and their uncertainties. These attribute cubes can then be incorporated in standard 3D seismic data interpretation softwares, like the freeware *Opendtect*, or commercial solutions like *Petrel*, *Kingdom Suite* or *GVerse* (or other), and interpreted or used in the same way as conventional 3D seismic attributes.

3.1 synCPT

The simplest – but most intuitive – approach of a data-driven ground model uses robust linear trendline fitting of the measured or derived geotechnical properties. This trendline fitting is applied to each defined geotechnical soil units and at each of the borehole locations. This approach yields a sparse 1D intercept and gradient at each geotechnical location and for each soil unit (Fig. 5). In order to construct a pseudo-3D model, we adopt spatial interpolation techniques on the intercept and gradient. While doing so, we use the interpretation of key seismic horizons as the structural framework across the development area. The models take into account whether or not soil units occur at given locations or if the boreholes or CPTs may not have reached the target depths. As one obtains gradients, intercepts and the horizon depths (both in depth below seafloor and below sealevel), one can reconstruct the given parameter across the entire area, and obtain an estimate of the goodness of fit. For further details, see Forsberg et al. (2017) and Sauvin et al. (2019).

3.2 Co-kriging

Co-kriging is one of the main geostatistical approaches used in the oil and gas industry for reservoir modelling and characterization. Kriging and co-kriging are generalized forms of multivariate regression models, with the weight not only depending on distance, but also on direction and orientation of the neighbouring data. It takes the advantage of the co-variance between two or more regionalized variables that are related and are appropriate when the main attribute of interest (here, CPTU data) is sparse, but secondary information (here, seismics) is abundant, hence using the best of both methods (Sauvin et al. 2019).

Also for this approach, we use the structural model as defined by the seismic horizons picked across the development area, and bin those into a suitable spatial grid (x,y,z). The geotechnical data are upscaled to match then the vertical bins, in say typically 50 cm to 1 m steps. These upscaled data are then interpolated using the structural model, with additional input from the *P*-wave velocities. The reasons for the latter was that, at this site, there appeared to be a relationship between the *P*-wave velocity and the tip resistance (Fig. 6) (see also Sauvin et al. 2019).

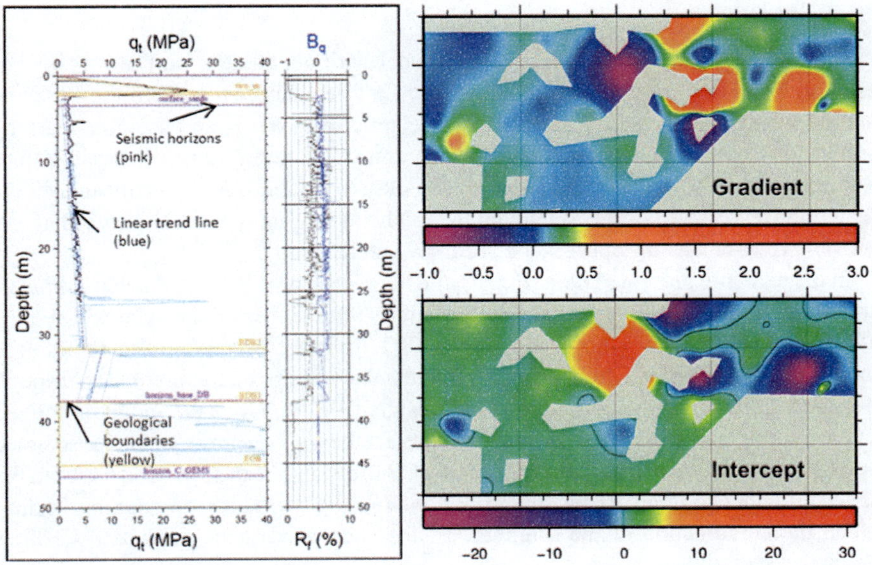

Fig. 5. Example of the trendline fitting per soil unit (left) and spatial gridding of the resulting intercept and gradient for a given soil unit, in order to construct pseudo-3D volumes of geotechnical properties (for more details, see Forsberg et al. 2017; Sauvin et al. 2019).

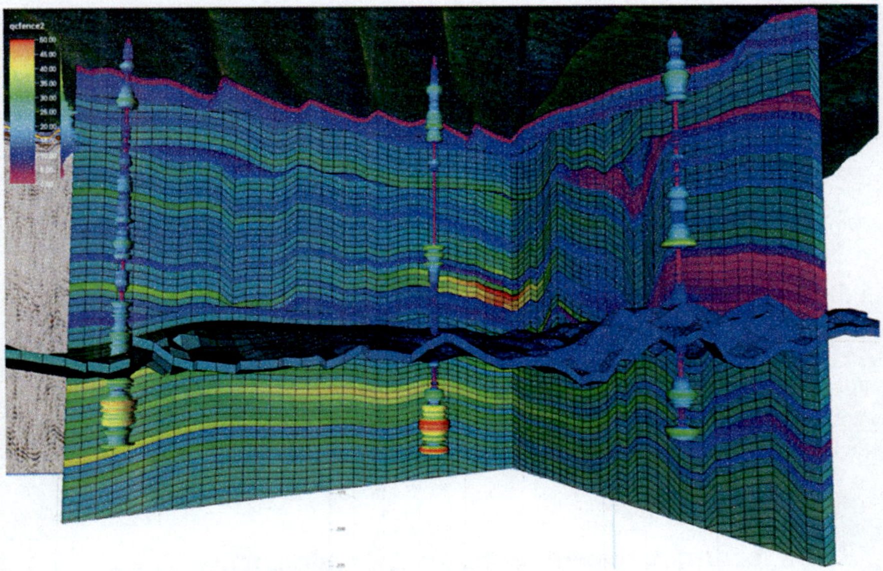

Fig. 6. Fence-diagram illustrating the prediction of tip resistance, qc, from the co-kriging approach linking the seismic data and measured tip resistance data (colour-coded cones), using the *P*-wave velocity as part of the co-kriging.

3.3 Random Forest

Another approach to populate the ground model with quantitative soil information is based on Random Forests (RF) to predict the CPT response away from the borehole location while making the most out of the seismic data. RF is an ensemble-learning method for classification or regression which constructs a multitude of decision trees at training time and outputs the mode of the classes (classification) or mean/average prediction (regression) of the individual trees. The averaging makes RF better than the Decision Tree; hence, it improves the accuracy and reduces overfitting.

In the example presented here (Fig. 7), the RF regressor was trained on coordinates, sub-surface depth, geological model, seismic amplitude, its envelope, acoustic energy, and *P*-wave velocity, in order to predict the CPT tip resistance. We estimate both the random forest model prediction accuracy and the uncertainty on the prediction through *k*-fold cross-validation, yielding promising results (Fig. 7). Once the model is trained and the hyper-parameters tuned, we then can predict the CPT response and the associated uncertainty from the seismic data. The derived CPT response is then gridded between the 2D seismic lines to populate the 3D ground model (Fig. 8). The estimated uncertainty is then the combination of the kriging variance and the variance of the RF CPT tip resistance prediction.

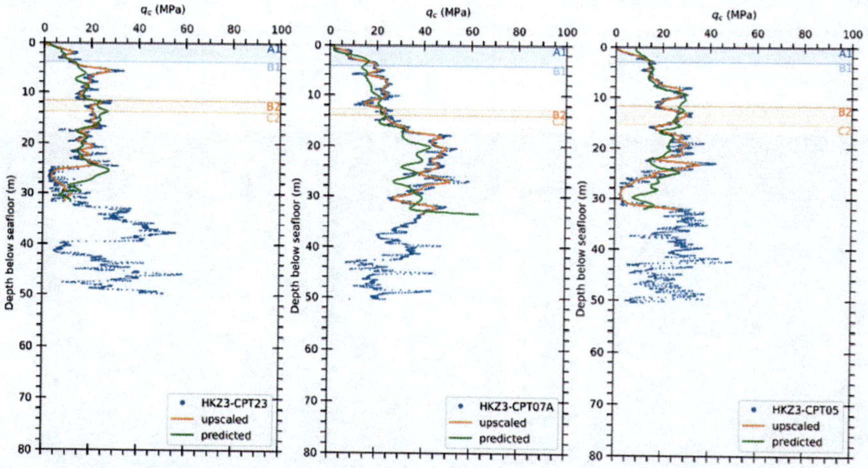

Fig. 7. Results from the 3D prediction of tip resistance, qc, using the Random Forest approach, on three selected CPT locations. The blue dots/lines are the measured data, the orange curves are the upscaled data to approximate the resolution of the seismic data, and the green curves are the predictions, after the given location was removed from the training set.

4 Room for Improvements and Novel Developments

4.1 Data Quality, Density, and Quality

Establishing reliable ground models for engineering applications requires a proper planning, execution and integration of the different data sets. One of the key observations is the fact that there are often inconsistencies and shortcomings with the off-the-shelf data that are typically collected for these purposes. Unfortunately, these introduce larger uncertainties in the final model. Some examples are highlighted underneath.

- It is not uncommon that geotechnical sampling and in situ testing locations do not fall along (cross-points of) seismic reflection profiles. Often, the distance between the geotechnical test location and the seismic data can be too large, such that the features observed cannot unambiguously be tied on the different data sets. Ideally, the locations should coincide, and be smaller than the geological features observed in the sub-surface. A recommendation would be say 10 m offset.
- The spacing between parallel seismic lines may be too large such that important geological features, like channels, can be missed. The smaller the line spacing, the better the understanding of the site. Over the years, the line spacing has decreased from several hundreds of m to about 70 m (e.g., Ter Noorden van de Waddeneilanden development off Holland), allowing the construction of improved ground models. Ideally, a full 3D survey would be recommended (see underneath).
- Improved time-to-depth conversions are also important to better link the geophysical and geotechnical data. Knowing that the geological complexity of offshore wind sites can be significant, dedicated seismic data processing, including removal of multiples and high-density velocity analysis, should be performed.
- The geophysical (comes first) and geotechnical (comes second) site investigations are often done separately. As a result, also the interpretation of the data is typically performed independently, with the geophysical data informing where the geotechnical data should be planned. There is critical value in re-assessing and updating the seismic data interpretation in a joint effort with geotechnical engineers, once the in situ geotechnical data become available, such that the units identified are consistent.
- More effort is required to obtain reliable site-specific and 2D to 3D time-depth relationships to optimate the integration between the seismic and geotechnical data. Often, the velocity models are too simplistic, adding uncertainty in matching the different data sets, and converting the seismic data from time to depth.

4.2 Added Value – Multi-sensor Core Logging

Multi-sensor core logging (MSCL) is a well-known data analysis technique used in research activities. Despite its added value, it is not often used for offshore engineering applications. The MSCL collects simultaneous, co-registered data from multiple sensor systems in an automated way on lined sediments or unlined rock samples. The tests can be performed on either full sample barrels or split cores. The sensors and methods described herein are developed and adapted by Geotek Ltd. The sensors currently used at NGI are (1) attenuated gamma density, (2) ultra-sonic *P*-wave velocity (and amplitude), (3) magnetic susceptibility, (4) non-contact electrical resistivity, and (5) total natural gamma. Additional sensors, including *S*-wave velocity scanning, can be performed as well. The results of these scans are continuous ultra-high-resolution (<1 cm) point measurements of various physical properties that can be used to understand the lithology

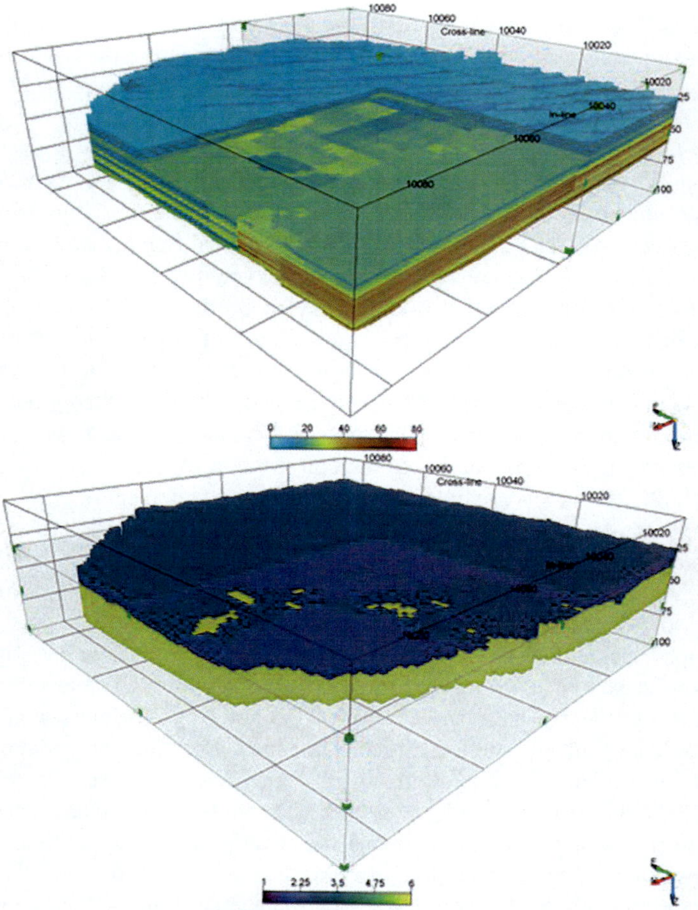

Fig. 8. Perspective view of the 3D prediction of tip resistance (q_c) (top) and its uncertainty (bottom) obtained from the Random Forest ground model approach.

and provenance of the sediments. They can also serve as calibration points for the seismic and geotechnical data typically collected in close proximity.

Figure 3 shows an example of the bulk density and magnetic susceptibility obtained with a MSCL system from a site in the immediate vicinity of a landslide complex, the 1996 Finneidfjord slide. The vertical resolution is superior to that of the conventional seabed CPTU data. In this case, the slip plane on which the landslide developed stands out more significantly on the MSCL data compared to the geotechnical data, and the synthetic seismic traces derived from the MSCL data, using a convolutional earth model, match well with the nearby trace from an UHR-3D chirp data set (Vardy et al. 2012; Vanneste et al. 2015).

Data acquired from the MSCL will contribute to a better understanding of the sediment properties in the different soil units and layers, and therefore enhance the ground model. The MSCL data will provide physical and geochemical quantitative data to characterize geological core material in more detail without destroying the sample. The data will also be helpful in identifying intervals for doing advanced geotechnical laboratory testing.

4.3 Added Value – UHR 3D Seismic Surveying

One of the main uncertainties in constructing the ground model is the spatial interpolation across the development area, due to the spacing between the seismic lines, which often is too large such that key features are missed. The collection of ultra-high-resolution 3D seismic data across development areas would be a significant step forward. Similar techniques are the standard for oil and gas exploration and their ultra-high-frequency counterparts should become the standard for offshore renewables in the near future.

There are several examples of such UHR-3D data collected for small subsets, often close to the sub-stations in offshore wind developments, for example on Dogger Bank, UK, and Ter Noorden van de Waddeneilanden (TNW, the Netherlands). The seismic profile shown in Fig. 3 is actually an in-line extracted from an ultra-ultra-high-resolution 3D chirp system with a bin length of 12.5 cm (Vardy et al. 2012; Vanneste et al. 2015). Despite them being more expensive to collect and process, these data allow much improved integration of the various data sets, facilitate the seismic data interpretation as well as understanding of the geological processes at different scales across the development sites.

4.4 Blending Seismic Inversion with Machine Learning

Seismic inversion is the process of deriving the best model of physical properties that can explain the measured data best, taking into account the uncertainty of the fit. Seismic inversion is commonly performed for reservoir purposes, but it is a computationally expensive intensive effort, particularly when applied to data that focus on to highlight the subtle detail in the shallow sub-surface, hence, having frequencies typically in the order of several hundreds of Hz.

The vast volume of data typically available for offshore developments also allow for another data-driven approach, i.e., using machine learning to invert the seismic data, and as such obtain information about the relative or absolute acoustic impedance (i.e., bulk density times *P*-wave velocity) which largely controls the wave propagation. To assess the possibilities and limitations of machine learning for this type of inversion, we performed a synthetic study on a complex sub-surface model, Marmousi2, i.e., an extension of the well-known Marmousi model with larger dimensions and variations in the properties (Martin et al. 2006). The model includes bulk density, *P*-wave and *S*-wave velocities, with seismic data generated using 2D finite-difference modelling program and an Ormsby wavelet with corner frequencies of 5, 10, 60 and 80 Hz. In this exercise, we only used the time-migrated sections.

The Marmousi2 model is an excellent example, as it contains a number of features that are characteristic of the challenging sub-surface conditions that one can encounter in offshore environment, including strongly dipping layers/stratigraphy, distinct velocity contrasts and variations in lateral and vertical directions, shallow gas features, faults, folds, etc. (see Fig. 9).

In this exercise, we adopted the approach by Mustafa et al. (2019), which is based on a Temporal Convolutional Network (TCN) architecture (Bai et al. 2018). This approach is a sequence-to-sequence type of Machine Learning architecture that allows linking a temporal sequence (here, the seismic trace) to another sequence (here, the acoustic impedance at that location). The method is based on convolutional layers, having two distinct characteristics: (1) the convolutions are causal, meaning that to predict a value at time t, only elements from time t and earlier are selected from the previous layer; and (2) it can benefit from long memory, meaning that it can look very far back to make a prediction using dilated convolutions. There are a number of advantages of TCNs. The use of those dilated convolutions allows the network to look at large parts of the input without having to use many layers. As such, TCNs capture long-term trends better than Recurrent Neural Networks (RNN). TCNs also avoid problems of exploding of vanishing gradients, which can be problematic for RNNs. Another advantage is the low memory requirement for training, particularly when long input sequences (like seismic data) are used. Keys parameters are the use of Rectified Linear Unit (ReLU) activation functions after convolutional layers and an Adam optimizer to backpropagate the error into the network during the training process.

For the inversion, we selected a limited number of reference traces and corresponding acoustic impedance from the synthetic data, in order to train the model and subsequently predict the acoustic impedance across the entire profile. While doing so, we wish to estimate the fit or uncertainty during the process. This is accomplished by selecting 25 traces uniformly along the profile and creating a series of 200 models, each model using a subset of 20 traces out of the 25 for the training, using 2000 epochs. Figure 10 shows the results of the machine learning inversion of three subsets of the Marmousi2 model, each with distinct features, and as the mean of the 200 models. As there are 200 results available, one can apply statistical methods to assess the quality of the results. For each segment, the left panel displays the true Acoustic Impedance whereas the right panel displays the mean of the 200 inverted model. This step is time consuming due the large number of models but is a preliminary approach in working toward uncertainties. Figure 11 illustrates that the mean values fairly well match the blue curve, corresponding to the models. The shading represents 1, 2 and 3 standard deviations from the mean value.

Despite it being a low-frequency model, the machine learning approach is applicable to higher-frequency data, as the results can be scaled.

4.5 Shear Wave Data: Tying the Knot

Whereas *P*-wave seismic data are essential to define the subseafloor architecture, these data may not tell too much about the geotechnical soil behavior at the site. *P*-waves sense both the soil skeleton and the pore fill, and are sensitive to saturation (e.g., shallow gas presence). *S*-waves, on the other hand, propagate essentially through the rock or soil skeleton, and are largely insensitive to the pore fill, unless the pore fluid experiences excess pore pressure. It is well-known that the joint analysis of *P*- and *S*-waves is beneficial for sub-surface characterization, fault mapping, lithology discrimination, pore pressure estimation, and shallow gas identification, amongst others.

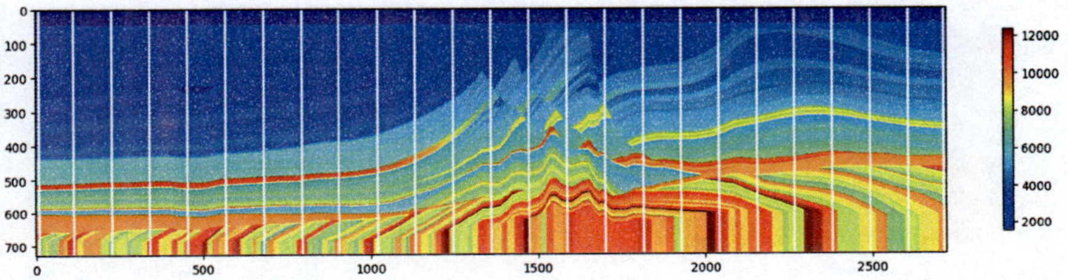

Fig. 9. The Marmousi2 model, acoustic impedance (bulk density times P-wave velocity). The horizontal white lines are key locations using in the analysis.

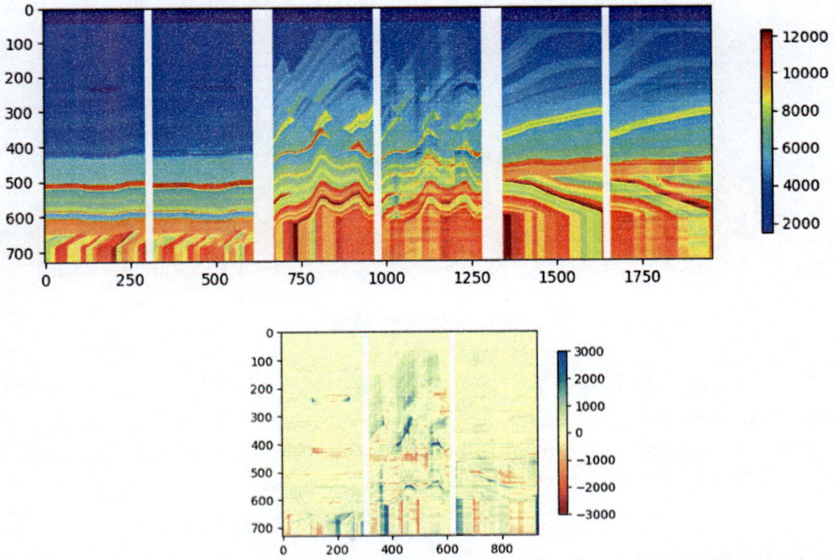

Fig. 10. (Top) Results obtained from the TCN for the Marmousi2 model, focusing on three different segments having distinctly different features, from flat-layered stratigraphy with a shallow gas accumulation (left), a faulted system with strong topographic changes and acoustic impedance changes both laterally and vertically (middle) and a segment with more gently dipping stratigraphy with strong acoustic impedance changes in mainly vertical direction (right). (Bottom) Differences between the machine learning solutions and the synthetic data for the subsets.

The main advantage of *S*-wave data is that they provide a more direct link with geotechnical and geomechanical properties compared to *P*-waves (Yilmaz 2001; Vanneste et al. 2007). Shear wave velocity is directly related to the rigidity or shear modulus of the soil or rock, which is a key parameter in geotechnical engineering, particularly for earthquake engineering, vibrations, fatigue assessment and design. In Fig. 12, we illustrate the relationship between undrained shear strength and the slowness, or the inverse of the propagation velocity, of *P*- and *S*-waves. The undrained shear strengths were derived from conventional CPTU data, whereas the *P*- and *S*-wave slowness were measured with a sonic logging tool. The figure clearly illustrates that there is virtually no relationship between *P*-wave slowness and undrained shear strength; however, the story is different for *S*-wave slowness. This figure provides clear evidence of the usefulness of collected shear wave data from the marine environment for geotechnical and engineering purposes.

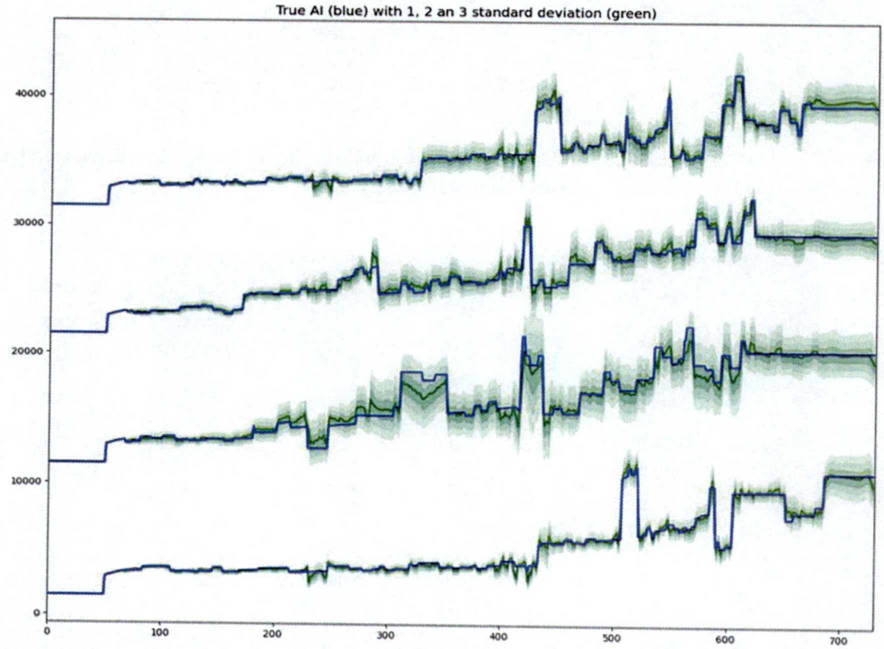

Fig. 11. Trace comparisons of the synthetic acoustic impedance profiles (blue curve) and the mean from the predictions (green curve) at four locations across the profile vs. depth (*x*-axis). The green shading corresponds to deviations of 1, 2 and 3 standard deviations from the mean.

S-waves do not travel through fluids, which implies that collecting these data is more challenging than collecting *P*-wave data. There are, however, various methods to obtain *S*-wave properties from geotechnical boreholes, including the use of Seismic CPT (SCPT) tools or PS logging tools, as mentioned above. For SCPT, the data often degrade beyond 25 m below the seafloor, and better source configurations and receiver systems are needed in the future. Other techniques can be the collection of marine surface waves or marine shear waves (see Vanneste et al. 2011; Socco et al. 2011). However, these experiments are inefficient and therefore expensive. New technology to capture *S*-wave data simultaneously with conventional multi-channel surface towed systems would provide great added value.

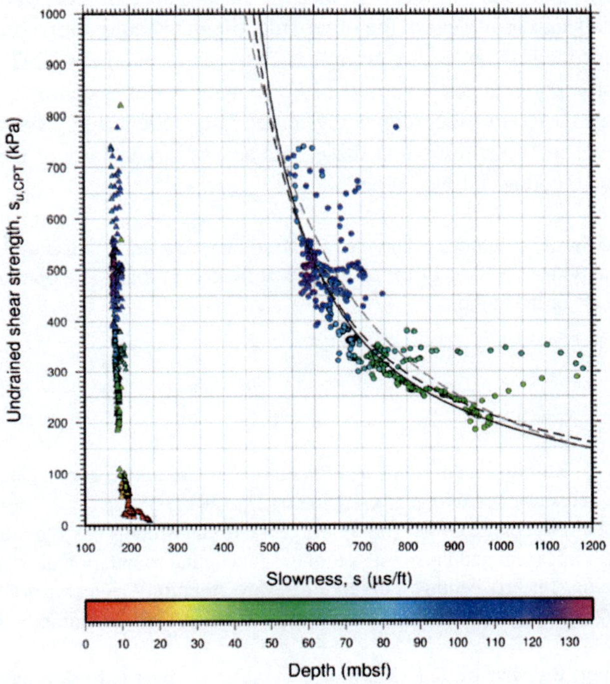

Fig. 12. Cross-plots showing undrained shear strength derived from cone penetration testing results (*y*-axis) and slowness for *P*-waves (triangles, low slowness values) and *S*-waves (circles, higher slowness values) from sonic logging data. The data illustrate that there is virtually no relationship between undrained shear strength and *P*-wave slowness. On the other side, there is a far more distinct relationship between undrained shear strength and *S*-wave slowness at the given site.

5 Summary

In this paper, we present the current practice as well as new developments to create data-driven ground models, with particular focus on offshore wind farm developments. The essence is the proper integration of geology, geophysics and geotechnical engineering into a consistent and robust framework in order to predict – with uncertainty – geotechnical soil properties in 3D. The vision is to use these ground models for geohazard and risk assessment, planning of follow-up geotechnical and geophysical site investigations, but also as input into engineering design, to assessing the implications for infrastructure costs (see also Sauvin et al. 2019).

These approaches come with different levels of complexity but require good control on the various input data, and streamlined communication and integration between the disciplines, for example, creating a consistent interpretation for the seismic stratigraphy that closely matches the soil units that can be identified from geotechnical data.

Whereas these methods are promising, more cases need to be performed to strengthen the developments and routines, and convince the various stakeholders of the added value that these models can provide. On the other side, the data acquisition and analysis tools can also be improved by collecting complementary data (like MSCL, shear-wave data) or optimizing the conventional methods, like moving towards denser 2D seismic line spacing or using 3D seismic techniques. Systematically collecting Seismic CPT data – or developing novel methods to collect these data across larger parts of the development sites – would also add additional benefits.

Acknowledgments. The authors acknowledge the reviews by Dr. Eric Cauquil (Total) and Dr. Knud Ho Karkov (Ørsted). Some of the research was funded by NGI. The opinions expressed in this paper are those of the authors.

References

Bai, S., Kolter, J.Z., Koltun, V.: An Empirical Evaluation of Generic Convolutional and Recurrent Networks for Sequence Modeling. arXiv:1803.01271 (2018)

Forsberg, C.F., et al.: Synthetic CPTs from Intelligent Ground Models based on the integration of geology, geotechnics and geophysics as a tool for conceptual foundation design and soil investigation planning. In: Proceedings Offshore Site Investigation Geotechnics 8th International Conference Proceeding, London, UK, Society for Underwater Technology, pp. 1254–1259 (2017)

Martin, G.S., Wiley, R., Marfurt, K.J.: Marmousi2: an elastic upgrade for Marmousi. Lead. Edge **25**(2), 156–166 (2006)

Mustafa, A., Alfarraj, M., AlRegib, G.: Estimation of acoustic impedance from seismic data using temporal convolutional network. SEG Technical Program Expanded Abstracts, pp. 2554–2558 (2019)

Sauvin, G., Vanneste, M., Vardy, M.E., Klinkvort, R., Forsberg, C.F.: Machine Learning and Quantitative ground models for Improving Offshore Wind Site Characterization, Offshore Technology Conference (OTC), Houston, TX, p. 18 (2019)

Socco, V.L., et al.: On the use of NGI's prototype seabed-coupled shear wave vibrator for shallow soil characterization - part II: joint inversion of multi-modal Love and Scholte Surface Waves. Geophys. J. Int. **185**(1), 237–252 (2011)

Vanneste, M., et al.: Integration of very-high-resolution seismic and CPTU data from a coastal area affected by shallow landsliding – the Finneidfjord natural laboratory. In: Meyer, V. (ed.) Frontiers in Offshore Geotechnics III (ISFOG): Oslo, pp. 1017–1022. Taylor and Francis, Norway (2015)

Vanneste, M., et al.: Finneidfjord, a field laboratory for integrated submarine slope stability assessment and characterization of landslide-prone sediments: a review. In: Proceedings Offshore Technology Conference (OTC 2013), Houston, TX, OTC, p. 17 (2013)

Vanneste, M., et al.: On the use of NGI's prototype seabed-coupled shear wave vibrator for shallow soil characterization - part I: acquisition and processing of surface waves. Geophys. J. Int. **185**(1), 221–236 (2011)

Vanneste, M., Westerdahl, H., Sparrevik, P., Madshus, C., Lecomte, I., Zühlsdorff, L.: Shear-wave source for offshore geohazard studies: a pilot project to improve seismic resolution and better constrain the shear strength of marine sediments. In: Proceedings Offshore Technology Conference, Houston, Volume 2007, OTC, p. OTC-18756 (2007)

Vardy, M.E.: Deriving shallow-water sediment properties using post-stack acoustic impedance inversion. Near Surf. Geophys. **13**, 143–154 (2015)

Vardy, M.E., Clare, M.A., Vanneste, M., Forsberg, C.F., Dix, J.K.: Seismic inversion for site characterization: when, where and why should we use it?. In: Offshore Technology Conference, Volume 2018, Houston, TX, OTC, p. 8 (2018)

Vardy, M.E., et al.: Multidisciplinary investigation of a shallow near-shore landslide, Finneidfjord, Norway. Near Surf. Geophys. (Spec. Issue: Appl. Mar. Geophys.) **10**(4), 267–277 (2012)

Vardy, M.E., Vanneste, M., Henstock, T.J., Clare, M.A., Forsberg, C.F., Provenzano, G.: State-of-the-art remote characterization of shallow marine sediments: the road to a fully integrated solution. Near Surf. Geophys. **15**, 387–402 (2017)

Vardy, M.E., Vanneste, M., Henstock, T.J., Morgan, E., Pinson, L.J.W.: Can high-resolution marine geophysical data be inverted for soil properties?. In: Seabed and Sediment Acoustics, Volume 37, pp. 24–29. Institute of Acoustics, Bath (2015)

Yilmaz, O.: Seismic data analysis: Processing, inversion and interpretation of seismic data. Society of Exploration Geophysicists, Investigation in Geophysics, Tulsa, 2027 p. (2001)

Experimental Observations on a Suction Bucket Foundation Under Vertical Pullout Loading

Sung-Ryul Kim(✉), Seongho Hong, and Ssenyondo Vicent

Department of Civil and Environmental Engineering,
Seoul National University, Seoul, South Korea
sungryul@snu.ac.kr

Abstract. The government of South Korea is currently implementing the national plan of developing offshore wind energy in the southwestern sea of the Korean peninsula. The aim is to decrease the nation's consumption of carbon emission technologies and increase renewable energy consumption. A tripod bucket foundation has been identified as a feasible foundation to support offshore wind turbine structures. This type of foundation is subject to failure by moment, horizontal, and pullout loadings due to the impact of wind and wave loads. The pullout of one bucket determines the stability of the entire tripod system, and should be investigated to fully satisfy the applicability of the tripod bucket foundation in the ongoing campaign. Therefore, a series of 1-g experimental tests were conducted on the model suction bucket to evaluate the pullout capacity under drained and undrained conditions. Soil deformations outside the bucket, soil heave, and suction force inside the bucket are discussed. A simple approach is proposed to guide the pullout capacity evaluation under drained and undrained conditions.

Keywords: Suction bucket · 1-g experimental test · Drainage condition · Pullout capacity · Sand

1 Introduction

In response to the United Nation's requirement for countries to mitigate greenhouse gas emissions, the South Korean government plans to increase renewable energy production from 4.8% to 20% by 2030 [1]. The latest 8th basic plan of South Korea for power supply and demand shows that offshore wind power is expected to contribute 23% of the total renewable energy production by 2030 [2]. The South Korean government, through the Ministry of Trade, Industry, and Energy, is currently progressively promoting the construction of offshore wind farms in the southwestern sea of the Korean peninsula. The development plan (Fig. 1) comprises the following three phases: phase 1-verification (2011–2018), phase 2-demonstration (2018–2020), and phase 3-expansion (2020–). This plan will increase the offshore wind capacity of the nation from the current 124 MW to 12 GW by 2030 [3]. The construction of 400 MW and 2 GW plants is scheduled to commence in 2022 and 2023, respectively. Nearly 2.46 GW will be generated by 2028, which is sufficient to supply electricity to 2.24 million households. Site investigations

D. V. K. Huynh et al. (Eds.): VSOE 2021, LNCE 208, pp. 22–34, 2022.
https://doi.org/10.1007/978-981-16-7735-9_2

revealed that the southwestern sea is an ideal site for wind power generation due to favorable wind speeds (>6.9 mm/s), shallow to deep waters (10–60 m), and short distances from the on-land sub stations [4].

Fig. 1. Development plan for offshore wind power projects in Korea [5].

A tripod bucket foundation has been identified as a feasible foundation to support offshore wind turbine structures [5]. The tripod bucket foundation is subject to failure by moment, horizontal, and pullout loadings due to the impact of wind and wave loads. The moment and horizontal capacities of the tripod bucket in this offshore environment have been addressed by previous studies. The pullout of a bucket in the tripod bucket foundation governs the stability of the entire tripod bucket system [6]. The pullout of one bucket can destabilize the entire tripod group, ultimately causing the breakdown of the entire foundation system. Figure 2 shows an illustration of the tripod bucket failure. The uplifting bucket is subjected to tension, while the downward buckets are subjected to compression. Therefore, evaluating the pullout capacity of the suction bucket is necessary. Moreover, the bucket has the least resistance to pullout when the pullout load acts in the vertical direction [7]. Tran et al. [8] discussed the relationship between the pullout and moment capacities of the bucket. The moment capacity can be calculated by considering the spacing of buckets, the pullout capacity of the upward bucket, and the vertical weight of the wind turbine structure.

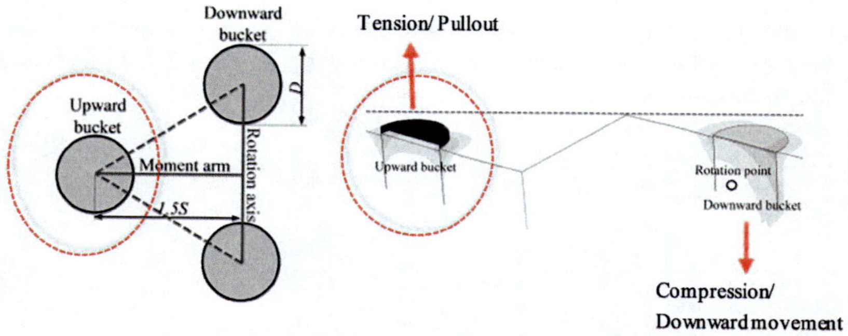

Fig. 2. Failure mechanism of a tripod bucket system [8].

2 Previous Studies on Pullout Capacity

The pullout capacity is considerably affected by the drainage condition of soils surrounding the bucket. Drainage conditions include drained, partially drained, and undrained conditions. Figure 3 shows failure modes of the bucket under the aforementioned drainage conditions. The drainage condition is influenced by suction developed inside the bucket during bucket pullout. This suction depends on factors, such as loading rate, drainage length, hydraulic conductivity, and shear strength properties of the soil [9].

A drained condition is induced by slow loading rates [10]. The suction pressure generated inside the bucket is substantially small under this drainage condition such that no soil is uplifted inside the bucket. The bucket resists uplift by friction resistance of the inner and outer faces of the bucket skirt. The skin resistance increases with the increase in friction angle of soil. Analysis of data from a centrifuge test study on model suction bucket foundations revealed that drained pullout capacity increases as a power function of bucket embedment depth at constant diameter [11]. The drained pullout capacity equals to the summation of the total skin resistance and the submerged weight of the bucket and its components because the bucket is pulled out by the sliding of the inner and outer faces of the bucket skirt [9, 12]. An accurate prediction of drained pullout capacity can be obtained by applying the equations proposed by Houlsby et al. [13], which account for the reduction of the vertical effective stress close to the bucket. The reduction of effective stress is due to friction forces along the bucket.

A partially drained condition, which exists between drained and undrained conditions, is characterized by partial uplift of the soil plug. A gap may form under the lid during the uplift of the bucket. Seepage into this gap leads to loss of suction. The generated suction increases with the loading rate, thus raising the volume of entrapped soil [14]. The pullout capacity under partially drained conditions can be predicted by summing up the submerged weight of the bucket, bottom resistance, and the outer skirt resistance [12]. Previous studies proposed that the pullout capacity in sandy ground can be predicted by equations suggested by Houlsby et al. [13]. The equations apply a factor to suction measured below the lid of the bucket. The factor accounts for the skin friction coefficient ($Ktan\delta$) and the change in embedment depth.

The undrained condition occurs at rapidly fast loading rates, which hinder the dissipation of suction. Gap formation under the lid is absent under undrained condition, and all soil trapped inside the bucket uplifts with the bucket. The pullout capacity increases as the uplifted soil plug volume increases. The generated suction force also increases and may contribute over 60% of the pullout capacity [15]. Previous studies suggested several models to evaluate the undrained pullout capacity. These models are summarised by the studies of Thorel et al. [16] and Patel and Singh [17]. The majority of these models were developed from studies on clay grounds (e.g., [18, 19]); therefore, their application to buckets in sandy ground might be limited. The general expression for undrained pullout capacity of a cylindrical caisson can guide the prediction of undrained bearing capacity in sandy ground [9]. The undrained pullout capacity is equal to the sum of submerged weights for the bucket and soil plug, the bottom resistance, and the outer skin resistance of the bucket. The outer skin resistance is calculated using the effective stress approach considering the increase in effective stress due to seepage along the outer skirt. The undrained pullout, which is a reversed end-bearing failure, and can be theoretically evaluated by applying the reversed bearing capacity theory [12]. Previous studies predicted that the undrained pullout capacity is approximately three times larger than the drained pullout capacity [9, 17]. Sandy ground occupies a considerable portion of the current project site. Until now, information to support the evaluation of pullout capacities in sandy ground is limited. [9, 20, 21]. Further studies are necessary to develop simplified solutions that can guide the prediction of the pullout capacity.

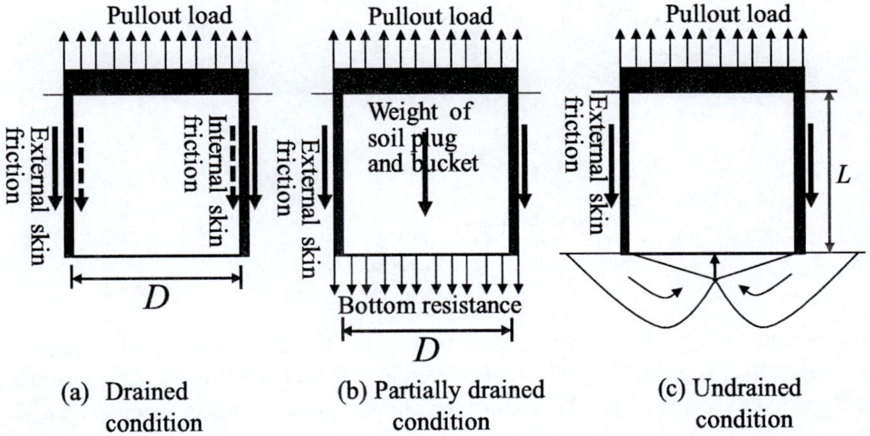

Fig. 3. Failure modes of the bucket during pullout loading (Modified after [12]).

3 Pullout Capacity According to Drainage Condition

A series of 1-g experimental tests were conducted on a single model bucket to evaluate the pullout capacity. The pullout load of the bucket was varied by applying different loading rates to induce bucket pullout under drained partially drained and undrained conditions. Soil heave and suction force inside the bucket are comprehensively discussed, and a simple approach is proposed to guide the pullout capacity evaluation under drained and undrained conditions.

3.1 Model Preparation and Testing

Figure 4 shows bucket models applied in the current study. Experimental data were obtained by applying the aluminum bucket. The polycarbonate bucket was purposely applied to enable visualization and monitoring of soil heave phenomena inside the bucket during the pullout process.

The aluminum model bucket had a Young's modulus (E = 70 Gpa), lid and skirt thicknesses of 5 and 1 mm, respectively. The model dimensions were obtained by scaling down geometric parameters of a commonly used steel prototype bucket (E = 210 Gpa) with L/D = 1.0 (L = 15 m). The scaling laws proposed by Wood [22] were adopted by applying the geometric scale ratio of 1:100 for length. The soil box had a diameter of 600 mm and height of 450 mm. Preliminary finite element analyses proved that, the boundary extents of the soil box had no effect on the pullout capacity. The model ground was saturated loose silica sand with the following properties: mean grain size (D_{mean} = 0.11 mm), permeability ($k = 2.51 \times 10^{-4}$ m/s), relative density (D_r = 28%) and specific gravity (G_s = 2.65).

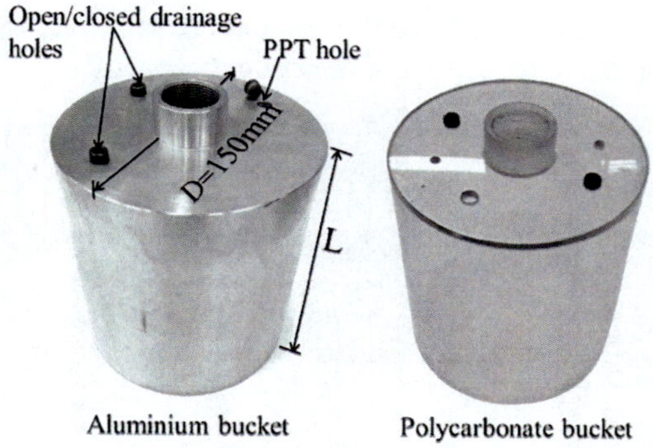

Fig. 4. Model suction bucket (L/D = 1.0).

Dry sand was initially submerged in water and allowed to saturate completely for a period of at least 48 h [23]. The model ground was prepared by pluviating saturated sand particles in water [24, 25] using the following procedure. A sieve (ASTM No. 40) was stacked on top of the soil box shown in Fig. 5. The soil box was filled with water, ensuring that the mesh is fully submerged. Saturated sand was dispersed over the submerged mesh. The sand particles dropped in water and settled with a uniform velocity inside the soil box. The model ground was loose with an average relative density (D_r) and saturated unit weight (γ_{sat}) equal to 28% and 17.3 kN/m^3, respectively). The same procedures were adopted to make the model ground for each experimental test. Drainage holes on the lid of the bucket were opened, and the bucket was slowly pushed into the ground at an average velocity of 0.0625 mm/s, which was sufficiently slow to limit excess pore water pressure generation in the soil. Bucket penetration was stopped when the lid touched the ground surface.

Figure 5 shows the test set-up used to conduct the experimental tests. The test equipment applied the load-controlled method. This test set-up was adopted because it allows the implementation of extremely slow loading rates e.g. 0.004 N/s. Extremely slow loading rates are difficult to implement using the existing test methods. The instrumentation included two linear variable differential transformers to measure bucket displacement, a load cell to measure the pullout load, and pore pressure transducers to measure suction pressure developed inside and below the bucket during uplift. A loading wire was set to pass over the pulley system connecting the bucket on one end while suspending the loading container on the other end. The control valve was opened to induce bucket pullout load and regulate a uniform flow of water into the loading container. As water accumulated in the loading container, the tension in the loading wire increased, thus applying a pullout force on the bucket. The above procedure was repeated for all loading rates in the test program. The test program was designed to apply loading rates of 0.004, 0.008, 0.08, 0.66, and 2.05 N/s.

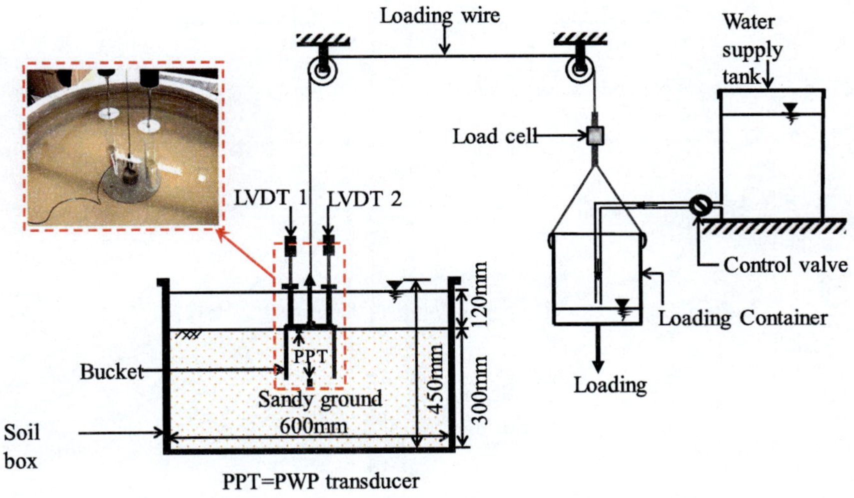

Fig. 5. Set-up of the test equipment for bucket pullout loading.

3.2 Results and Discussions

Pullout Capacity Considering Drainage Condition. Figure 6 shows the pullout load versus pullout displacement (w/D), where w is the upward displacement and D is the bucket diameter. The skin resistance is considerably mobilized at the onset of load application. When the pullout load exceeds some critical value, large displacements occur at small increments in pullout load. Figure 7 shows the suction pressure generated below the lid and tip of the bucket. The generated suction depends on the magnitude of applied loading rate. The larger the loading rate, the larger the generated suction force. The trend for the increase of suction force below the lid was similar to that of the load–displacement curve, indicating that the generated suction force governs the pullout resistance. The bottom suction resistance increased with the suction inside the bucket.

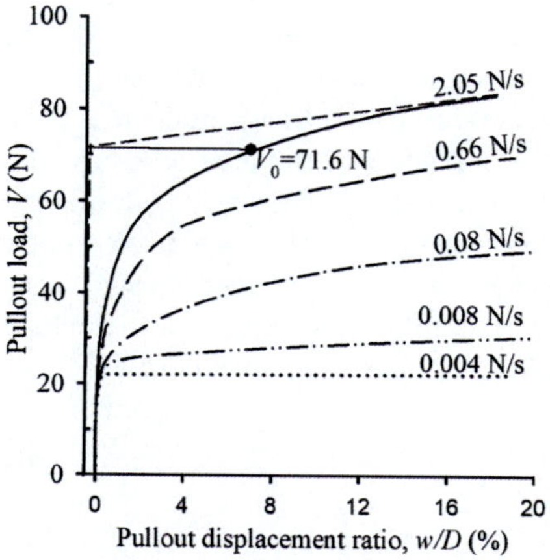

Fig. 6. Pullout load-displacement curves of the bucket.

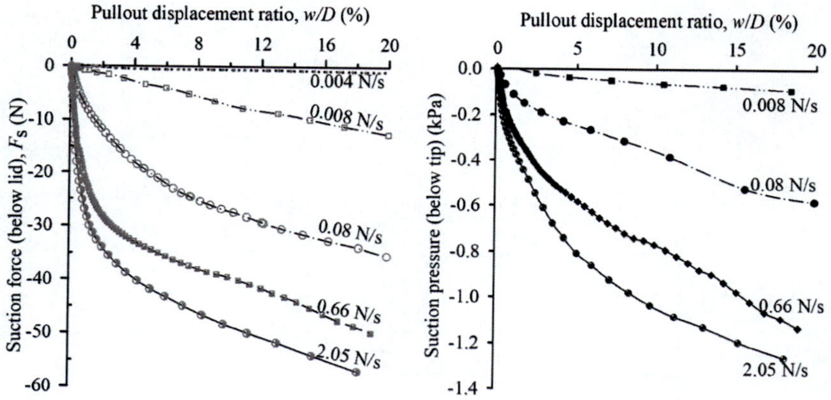

Fig. 7. Suction force below the lid and tip of the bucket.

The tangent intersection method was applied to estimate the pullout capacity (V_0) as illustrated in Fig. 6 [26, 27]. The displacement at V_0 was recorded and applied to estimate the suction force (F_{s0}) mobilized at pullout capacity. Figure 8 shows the increase of suction force during uplift. The suction force inside the bucket was not constant during bucket pullout for each loading rate, but increased with bucket uplift as shown at displacement ratios (*w/D*) of 3.3%, 10%, and 16%.

Figure 9 shows the variation of pullout capacities (V_0) and suction force (F_{s0}) with loading rate. The pullout capacity rapidly increased with loading rate from 0.004 N/s to 0.66 N/s. However, over 2.05 N/s indicates the absence of significant changes in the pullout capacity. The rapid increase in pullout capacity was due to the increase in suction force generated inside the bucket. The experimental data are plotted as shown in Fig. 9. A curve defined by a hyperbolic function was found to best fit experimental data. The drained pullout capacity (V_d) was 22.1 N and occurred at a loading rate of 0.004 N/s with negligible suction force. The undrained pullout capacity (V_u), the lid suction ($F_{su(lid)}$), and the tip suction ($F_{su(bottom)}$) were estimated to be 75, 48.3, and 18.1 N, respectively. The undrained pullout capacity occurred at loading rates larger than 2.05 N/s.

Table 1 summarizes the pullout capacities, top and bottom suction forces, and the percentage contribution of suction on pullout capacity. The suction force inside the bucket mainly contributed to the increase in pullout capacity at loading rates larger than 0.66 N/s.

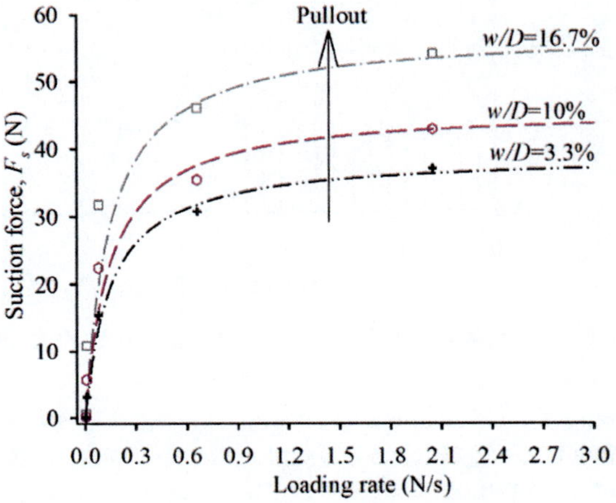

Fig. 8. Example of the increase in suction force (under lid) during bucket uplift.

Table 1. Pullout capacity and suction force inside the buckets.

Loading rate (N/s)	V0 (N)	Fs(lid) (N)	Fs(tip) (N)	$F_{s(lid)}/V_0$	$F_{s(tip)}/V_0$ (%)
0.004	22.1	≈0	≈0	*≈0*	*≈0*
0.008	26.66	3.34	1.32	*12.53*	*10.54*
0.08	41.97	18.51	5.11	*44.10*	*11.59*
0.66	57.07	32.53	10.06	*57*	*17.65*
2.05	71.6	44.48	15.92	*62.12*	*25.63*

As previously discussed, the suction pressure increased with the loading rate. The suction pressure below the bucket tends to resist uplift, but its percentage contribution to pullout capacity is small compared with the suction effects under the lid. Figure 9 shows a simple evaluation in which the sum of $F_{su(lid)}$, $F_{su(tip)}$, and $R_{s(outer\ skirt)}$ is approximately equal to V_u. $R_{s(outer\ skirt)}$ is estimated to be half of the difference between V_d and submerged bucket weight ($W_b' = 7.61$ N).

Drained (V_d) and Undrained (V_u) Pullout Capacities. Figure 10 shows the heaving of soil inside the bucket during pullout. The drained condition is observed in Fig. 10a,

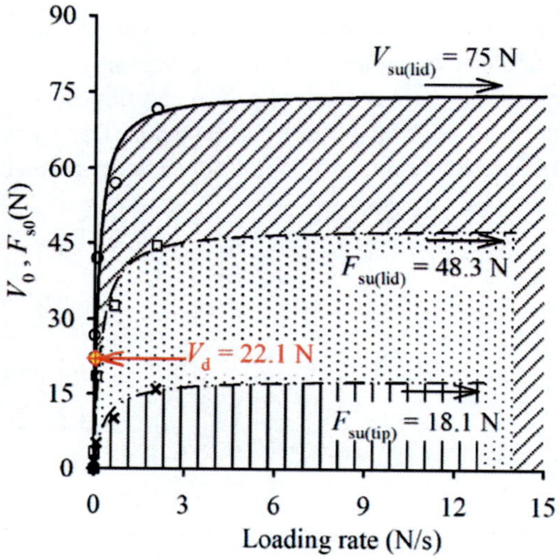

Fig. 9. Drained and undrained pullout capacities from the experimental test data.

which shows an empty bucket pullout. When the skin friction is exceeded, the bucket rapidly uplifts without the soil plug, and no settlement takes place in the soil surrounding the bucket. The drained pullout capacity can be evaluated by applying Eq. (1). The undrained condition is demonstrated in Fig. 10b, which shows that all the entrapped soil uplifts with the bucket. The mobilized suction force is larger than the weight of the soil plug. When a full soil plug was uplifted, the measured tip suction was approximately 50% of lid suction.

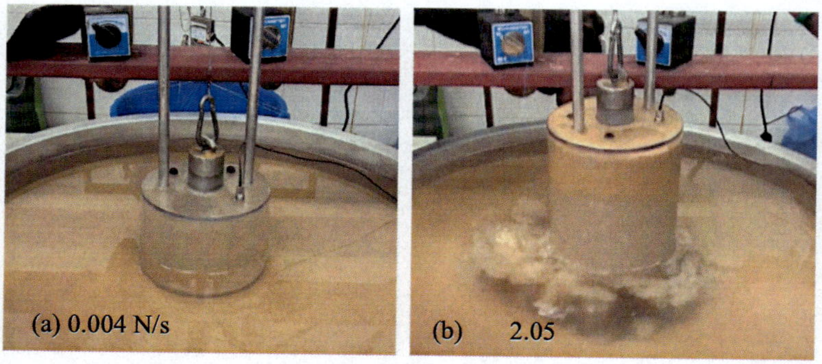

Fig. 10. Bucket pullout at slow and fast loading rates.

$$V_0 = R_{s(\text{outer skirt})} + R_{s(\text{inner skirt})} + W_b{}' \tag{1}$$

where $R_{s(\text{outer skirt})} = R_{s(\text{inner skirt})} = \frac{\gamma' L}{2} \times K \tan\delta \times \pi \times D \times L$, γ' is the effective unit weight of soil, K = lateral earth pressure coefficient, and δ is the soil–bucket skirt interface angle of friction. The submerged weight of the bucket (W_b') is 7.61 N, and $K \tan \delta$ is estimated as 0.182.

Figure 11 shows the balance of forces acting on the bucket during undrained pullout loading. The bucket with soil act as one body during the undrained pullout. The undrained pullout capacity can be estimated by applying Eq. (2). The maximum tensile force ($F_{s(bottom)}$) is assumed to be limited to the buoyancy at the bottom of the bucket ($\gamma_w LA$). It is also assumed that the increase in $R_{s(outer\ skirt)}$ due to downward seepage force equals to the decrease in $R_{s(inner\ skirt)}$.

$$V_u = W_b + W_{soil} + R_s + (\gamma_w \times H \times A) - (\gamma_w \times H \times A + \gamma_w \times L \times A) + F_{s(bottom)}$$

$$V_u = W_b + W_{soil} + R_s - (\gamma_w \times L \times A) + F_{s(bottom)} \quad (2)$$

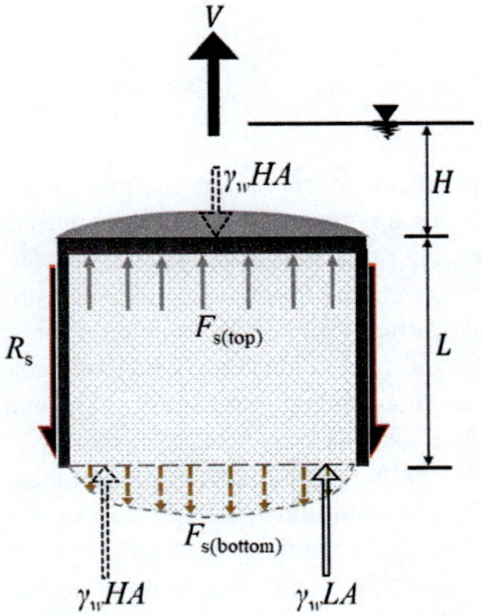

Fig. 11. Forces acting on the bucket during undrained pullout loading.

where W_b and W_s are the respective weights for the bucket and soil, $F_{s(bottom)}$ is the suction force at the bottom of the bucket, H is the water level from the top of the bucket, and L is the height of the bucket. The water pressure at the bottom of the bucket is equal to $\gamma_w \times (H + L) \times A$. At the ultimate pullout limit state, $F_{s(bottom)}$ is equal to $\gamma_w \times L \times A$, and the ultimate pullout capacity can be estimated by applying Eq. (3).

$$V_u = W_b + W_{soil} + R_s \quad (3)$$

R_s considers the $R_{s(outer\ skirt)}$ and $F_{seepage}$. $F_{seepage}$ is assumed equal to $R_{s(inner\ skirt)}$, and V_u can be predicted by applying Eq. (4).

$$V_u = W_b + W_{soil} + R_{s(inner\ skirt)} + R_{s(outer\ skirt)} \quad (4)$$

where W_b and W_{soil} are the respective weights for the bucket and trapped soil, and R_S can be evaluated by applying $K \tan \delta = 0.182$.

Equation 4 predicted $V_u = 68.9$ N, while Fig. 11 (Hyperbolic fit) predicted $V_u = 75$ N. The percentage difference between the two values is 8.5%, which is less than 10%. Therefore, Eq. (4) can be used to obtain a reasonably estimate for V_u.

The current study has a number of limitations. The bucket models of the current study are small and besides, experimental tests were conducted under 1-g conditions. As expected, the pullout loads are smaller compared to those usually encountered in field practice. Therefore, the effects of cavitation on the pullout capacity of model buckets are insignificant. In typical offshore bucket foundations, large pullout loads impact the suction bucket. The cavitation of the pore fluid beneath the bucket significantly limits the increase of pullout capacities at larger loading rates. Cavitation of the pore fluid is estimated to occur at a pressure of −100 kPa. Furthermore, bucket foundations are subjected to cyclic loads of wind and waves of various magnitudes. Factors that affect the cyclic response e.g. load inclination and number of cycles are not addressed by the current study.

4 Conclusion

This study discussed experimental observations and findings on the pullout capacity of a single bucket by conducting a series of 1-g experimental tests. The pullout load of the bucket was varied by applying different loading rates to induce bucket pullout under drained to undrained conditions. The following conclusions are drawn.

The tripod bucket foundation is a feasible foundation for offshore wind turbines in the southwestern sea of the Korean Peninsula. The tripod bucket foundation is subject to failure by moment, horizontal, and pullout loadings. The pullout capacity of a single bucket governs the design of a stable tripod bucket system.

The pullout capacity is significantly affected by the drainage condition of the soil surrounding the bucket. A drained condition is induced by a slow loading rate, and the suction force generated in the soil is substantially small and negligible. The undrained condition is induced by a fast loading rate. The suction force generated is large, which provides remarkable contributions to the pullout capacity. The partially drained condition, which exists between the drained and undrained conditions, is induced by intermediate loading rates. As the loading rate increases, the pullout capacity increases proportionally to the induced suction force.

The drained pullout capacity can be predicted by combining skin resistances along the inner and outer bucket skirts and the submerged weight of the bucket. Meanwhile, the undrained pullout capacity can be estimated by combining the soil plug weight, bucket weight, and the inner and outer skirt resistances of the bucket.

Acknowledgement. This work was supported by Research Resettlement Fund for the new faculty of Seoul National University and the Development of Life-cycle Engineering Technique and Construction method for Global competitiveness upgrade of Cable bridges (16SCIP-B119960) from the Smart Civil Infrastructure Research Program funded by the Ministry of Land, Infrastructure and Transport (MOLIT) of the Korean Government and the Korea Agency for Infrastructure. Technology Advancement (KAIA).

References

1. Park, J., Kim, B.: An analysis of South Korea's energy transition policy with regards to offshore wind power development. Renew. Sustain. Energy Rev. **109**, 71–84 (2019)
2. Park, J., Kim, B.: A comparative study on government's policy for offshore wind power development between major European country and Korea. New Renew. Energy **15**(3), 11–26 (2019)
3. MOTIE Homepage. http://www.motie.go.kr/motie/ne/presse/press2/bbs/bbsView.do?bbs_cd_n=81&bbs_seq_n=160584. Accessed 27 Nov 2020
4. Kim, J.Y., Kang, K.S., Oh, K.Y., Lee, J.S., Ryu, M.S.: A study on the site selection of offshore wind farm around Korean Peninsula. In: Proceedings of the 3rd International Conference on Ocean Energy, ICOE 2010, Bilbao, Spain (6), pp. 1–4 (2010)
5. Ryu, M.S., Lee, J.S., Kwag, D.J., Bang, S.: Comparison of two meteorological tower foundations for off-shore wind turbines. In: Proceedings of the 26th International Ocean and Polar Engineering Conference, pp. 723–730. International Society of Offshore and Polar Engineers, Rhodes (2016)
6. Kim, S.R., Oh, M.: Group effect on bearing capacities of tripod bucket foundations in undrained clay. Ocean Eng. **79**, 1–9 (2014)
7. Rao, S.N., Latha, K.H., Pallavi, B., Surendran, S.: Studies on pullout capacity of anchors in marine clays for mooring systems. Appl. Ocean Res. **28**(2), 103–111 (2006)
8. Tran, N.X., Hung, L.C., Kim, S.R.: Evaluation of horizontal and moment bearing capacities of tripod bucket foundations in sand. Ocean Eng. **140**, 209–221 (2017)
9. Iskander, M., El-Gharbawy, S., Olson, R.: Performance of suction caissons in sand and clay. Can. Geotech. J. **39**(3), 576–584 (2002)
10. Vicent, S., Kim, S.R., Tung, D.V., Bong, T.: Effect of loading rate on the pullout capacity of offshore bucket foundations in sand. Ocean Eng. **210**, 107427 (2020)
11. Wang, X., Zeng, X., Li, J.: Vertical performance of suction bucket foundation for offshore wind turbines in sand. Ocean Eng. **180**, 40–48 (2019)
12. Deng, W., Carter, J.P.: A theoretical study of the vertical uplift capacity of suction caissons. Int. J. Offshore Polar Eng. **12**(02), 89–97 (2002)
13. Houlsby, G.T., Kelly, R.B., Byrne, B.W.: The tensile capacity of suction caissons in sand under rapid loading. In: Proceedings of the International Symposium on Frontiers in Offshore Geomechanics, pp. 405–410. Taylor & Francis Group, Perth (2005)
14. Thieken, K., Achmus, M., Schröder, C.: On the behavior of suction buckets in sand under tensile loads. Comput. Geotech. **60**, 88–100 (2014)
15. Clukey, E.C., Morrison, M.J.: A centrifuge and analytical study to evaluate suction caissons for TLP applications in the Gulf of Mexico. In: Design and Performance of Deep Foundations: Piles and Piers in Soil and Soft rock, ASCE, pp. 141–156 (1993)
16. Thorel, L., Garnier, J., Rault, G., Bisson, A.: Vertical uplift capacity of suction caisson in clay. In: Proceedings of the International Symposium on Frontiers in Offshore Geotechnics (IS-FOG 2005), pp. 273–279 (2005)
17. Patel, S.K., Singh, B.: A parametric study on the vertical pullout capacity of suction caisson foundation in cohesive soil. Innov. Infrastruct. Solut. **4**(1), 1–11 (2018). https://doi.org/10.1007/s41062-018-0188-6
18. Rahman, M.S., Wang, J., Deng, W., Carter, J.P.: A neural network model for the uplift capacity of suction caissons. Comput. Geotech. **28**(4), 269–287 (2001)
19. Sgardeli, C.G.: A finite element analysis of the pullout capacity of suction caissons in clay (Doctoral dissertation, Massachusetts Institute of Technology) (2009)
20. Sawicki, A., Wachowski, Ł., Kulczykowski, M.: The pull-out capacity of suction caissons in model investigations. Arch. Hydro-Eng. Environ. Mech. **63**(2–3), 157–171 (2016)

21. Hung, L.C., Lee, S., Tran, N.X., Kim, S.R.: Experimental investigation of the vertical pullout cyclic response of bucket foundations in sand. Appl. Ocean Res. **68**, 325–335 (2017)
22. Wood, D.M.: Geotechnical Modelling, pp. 246–268. CRC Press, Boca Raton (2004). Reference to (Chapter 5)
23. Vicent, S., Hung, L.C., Kim, S.R.: Effects of long-term cyclic horizontal loading on bucket foundations in saturated loose sand. Appl. Ocean Res. **91**, 101910 (2019)
24. Kim, S.R., Jang, I.S., Chung, C.K., Kim, M.M.: Evaluation of seismic displacements of quay walls. Soil Dyn. Earthq. Eng. **25**(6), 451–459 (2005)
25. Vaid, Y.P., Negussey, D.: Preparation of reconstituted sand specimens. In: Advanced Triaxial Testing of Soil and Rock. ASTM International, pp. 405–417 (1988)
26. Vicent, S., Hung, L.C., Kim, S.R.: Evaluation of horizontal and vertical bearing capacities of offshore bucket work platforms in sand. Appl. Ocean Res. **101**, 102198 (2020)
27. Graham, J., Pinkney, R.B., Lew, K.V., Trainor, P.G.S.: Curve-fitting and laboratory data. Can. Geotech. J. **19**, 201–205 (1982)

Phu Cuong Soc Trang Offshore Wind Farm Development Some Lessons Learnt

Bernard Casey(✉)

Mainstream Renewable Power Ltd., Ho Chi Minh City, Vietnam
Bernard.Casey@mainstreamrp.com

Abstract. The first nearshore wind farm (99 MW) developed in Vietnam was commissioned in 2016 in Bac Lieu province, with the turbines connected to shore by walkways within which the array cables with onshore design protection were integrated. Since that time a number of other nearshore projects have commenced construction in Tra Vinh, Ben Tre, Soc Trang and Bac Lieu provinces. A common feature of all these projects is the deep soft marine deposits that necessitate long piles, and the difficult inter-tidal conditions that increase the time for and cost of construction.

This paper examines the design concepts being used at these projects and draws on learnings that Mainstream Renewable Power has made in developing the 1400 MW offshore wind farm in Soc Trang that is in preparation for a construction start in Q4 2021.

Keywords: Vietnam · Nearshore wind · Monsoon conditions · Renewable energy policy

1 Offshore Wind Resource in Vietnam

Vietnam has enormous potential for development of offshore wind energy, having a coastline of 3,400 km and high wind speeds focused along the south west coastal areas As published by the World Bank [1] in May 2020, the technical potential for fixed and floating offshore wind energy in Vietnam is 475 GW within 200 km off the coast (Fig. 1). The offshore wind sector is very nascent. As of May 2021, the only operating nearshore wind farm in Vietnam is the 99 MW wind power project in Bac Lieu province of the Mekong Delta region developed by a domestic investor. Several other nearshore wind farms are in construction at present.

The National Power Development Plan No. 8 (PDP 8), currently under finalization in government, promises to unlock the offshore wind potential of Vietnam with initial plans indicating 3–5 GW before 2030 and in excess of 30 GW before 2045.

D. V. K. Huynh et al. (Eds.): VSOE 2021, LNCE 208, pp. 35–47, 2022.
https://doi.org/10.1007/978-981-16-7735-9_3

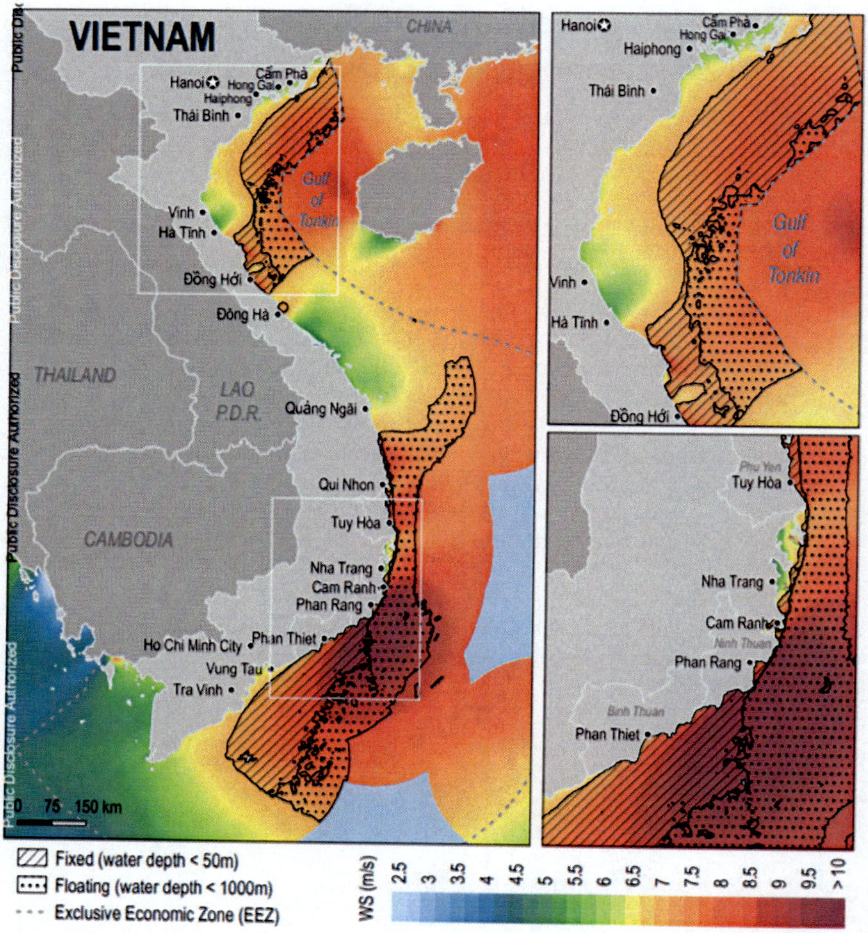

Fig. 1. World bank assessment of offshore wind potential in Vietnam

2 Features of Near Shore Windfarms Built to Date in the Mekong Delta Provinces

2.1 Ground Conditions

The Mekong River is approximately 4,900 km long rising in the Tibetan Plateau and then flowing through China, Myanmar, Laos, Thailand, Cambodia, before it reaches the sea in the low-lying southern provinces of Vietnam in an area known as the Mekong Delta.

The soil of the lower Delta consists mainly of sediment from the Mekong and its tributaries, deposited over thousands of years as the river changed its course due to the flatness of the low-lying terrain.

Geotechnical investigations for the Bac Lieu wind farm and the other wind farms under construction in the Mekong Delta provinces have shown similar ground profiles broadly as follows:

- Very soft to soft Silt/Clay between sea-bed and 15–25 m depth
- Inter-layered medium dense to dense SAND, and firm to hard CLAY to the investigated depths of up to 80 m below sea-bed

Based on the geological mapping of the Mekong Delta area in Vietnam [2], bedrock (conglomerate Sandstone and Shale) is expected more than 200 m below the surface, so was not encountered within the depths investigated for any of the wind farms.

2.2 Weather Conditions for Installation

Experience has shown that there are two main monsoon seasons in the Mekong Delta: the SW wind season (June to September), and the NE wind season (November to February) and transitional periods in between the two.

The most favorable time for offshore survey or installation activities is between March and September (SW wind Season) when wind speeds are generally <7 m/s and significant wave heights <1 m. The NW wind Season is characterized by higher wind speeds (>10 m/s) and higher significant wave heights (>1 m).

As a result, nearshore projects to date in this region have generally targeted the months between March and September for offshore foundations and wind turbine installation works, and some have borne substantial downtime for works that were conducted between the months of October and February.

2.3 Wind Turbine Foundation Solutions

On the basis of the soil profiles described above, all wind turbine foundation concepts on Mekong Delta projects are likely to require deep piles to reach the competent soils below 15–25 m depth. Pile driveability, however, is expected to be relatively straightforward with no drilling required and using readily available piling hammer sizes.

The Bac Lieu Phases 1 and 2 project (99 MW), and many of the near shore projects under construction at present have adopted raised concrete foundations on multi-piles (typically pretensioned spun high strength concrete (PHC) piles installed by driving) up to 60 m deep. Both raked and vertical piles have been adopted (Fig. 2).

This foundation solution has some perceived advantages and disadvantages:

Advantages:

1) Can be constructed by local Vietnamese contractors with equipment and materials sourced entirely in Vietnam
2) Are cost effective compared to other foundation solutions

Fig. 2. Raised concrete pile foundations at the Bac Lieu wind farm

Disadvantages:

1) Installation involves many steps including cofferdam construction to create a dry working area. Is therefore time consuming compared to, for example, monopiles. Experience to date has shown that each foundation can take up to one month to install.
2) Taking longer to install the installation period is therefore subject to more potential weather down-time, and impact on third party sea users.
3) Uncertainties on long term degradation of concrete piles due to damage to micro-bonds and fatigue due to hammering impact.
4) Higher visual impact of the wind farm from the coast.

A limited number of projects under construction are now using monopiles as the foundation choice. This foundation solution has some perceived advantages and disadvantages (It remains to be seen what construction lessons are learnt by these projects).

Advantages:

1) Can be installed very quickly (days) compared to the concrete multi-pile solution.
2) Taking shorter time to install, the installation period is therefore subject to less potential weather downtime.

Disadvantages:

1) Currently cannot be fabricated in Vietnam due to limitations on the rolling diameters available in steel fabrication facilities in Vietnam.
2) Installation equipment (hammers/cranes) for monopile installation currently not available within Vietnam, so requires international contractors.
3) Are typically more costly than concrete multi-piled foundations that can be installed by local contractors.

2.4 Operation and Maintenance Solutions

The Bac Lieu Phases 1 and 2 projects, and many of the near shore projects under construction at present have adopted concrete walkways to the turbines, supported on piles (Fig. 3). The walkways incorporate a recessed deck level that houses the array cable loops that bring the power to an onshore substation.

Fig. 3. Typical walkway arrangement

These walkways avoid the need for vessel-based access and permit a 'walk to work' facility for wind turbine technicians to do routine maintenance on the turbines. However the walkways are not sufficient to permit access for major maintenance that still requires vessel access to the project site.

The walkway solution has a number of perceived advantages and disadvantages

Advantages

1) Potentially lower costs (compared to vessel-based)

2) Convenient for array cable carriage
3) Can increase wind farm availability but allowing routine O&M to be completed in unfavourable weather conditions

Disadvantages

1) Blockage of maritime traffic, fishing and other activities
2) Higher visual impact of the wind farm from the coast
3) Potential negative impact on coastal processes (erosion and deposition)

3 Development of the Phu Cuong Soc Trang Offshore Wind Farm

3.1 Project Background

The Phu Cuong Soc Trang offshore wind farm is being developed by Mainstream Renewable Power in joint venture with local Vietnamese company, the Phu Cuong Group. The project is located in an offshore area near Vinh Chau in the Soc Trang province in the Mekong delta as shown in the blue box on the map (Fig. 4).

In 2015, the Phu Cuong Group secured a USTDA grant [3] that funded a feasibility study by DNV-GL to assess the feasibility of nearshore wind project in this area. Mainstream Renewable Power joined the project as a majority shareholder in 2018. The first 200 MW phase of the 1,400 MW project was approved in National Power Development Plan No.7. This first phase is about 4 km offshore with a minimum water depth of 2 m and a max water depth about 12 m at low tide.

3.2 Geotechnical Investigation

The geotechnical investigation at the site was conducted in 2 Stages:

Phase 1 (2019): By Fugro included 5no. boreholes to a maximum of 50 m depth below sea-bed with continuous sampling. The investigation works deployed a floating barge with the borehole drilling rig positioned over an edge platform (Fig. 5). The works were heavily hampered by weather, though conducted in the SW wind season.

Phase 2 (2020): By Technical World included 5no. boreholes to a maximum of 70 m depth below sea-bed with continuous sampling, and 17no. CPTs to a maximum of 40 m depth below sea-bed. The Phase 1 works deployed a Jack-up platform (Fig. 6) with the borehole and CPT rigs positioned centrally on the platform over a moon pool. The jack-up platform was less vulnerable to weather downtime in operation compared to the floating barge, although movement of the jack-up platform between investigation locations was subject to weather tighter restrictions.

Both investigations confirmed a relatively consistent soil profile as shown in Fig. 7 below.

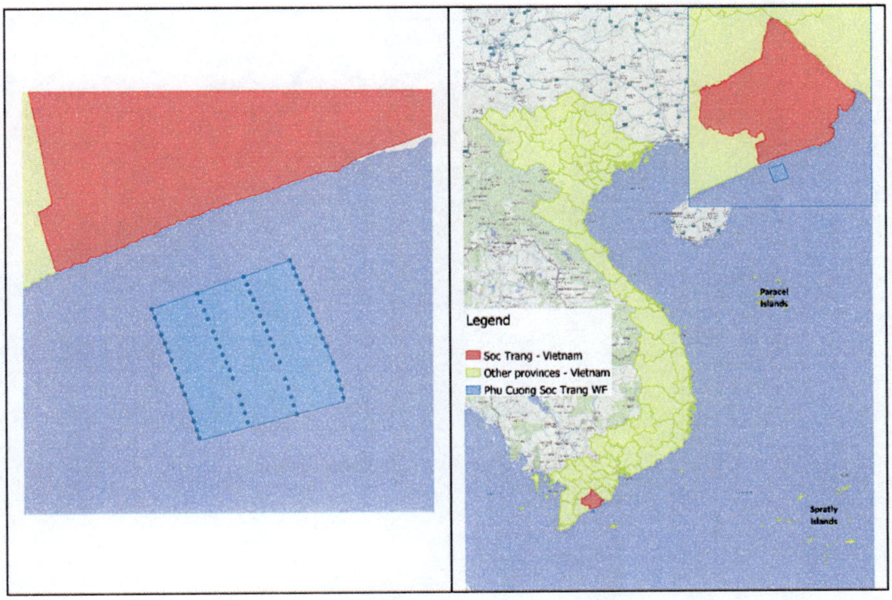

Fig. 4. Phu Cuong Soc Trang offshore windfarm location

Fig. 5. Phase 1 geotechnical investigation

Fig. 6. Phase 2 geotechnical investigation

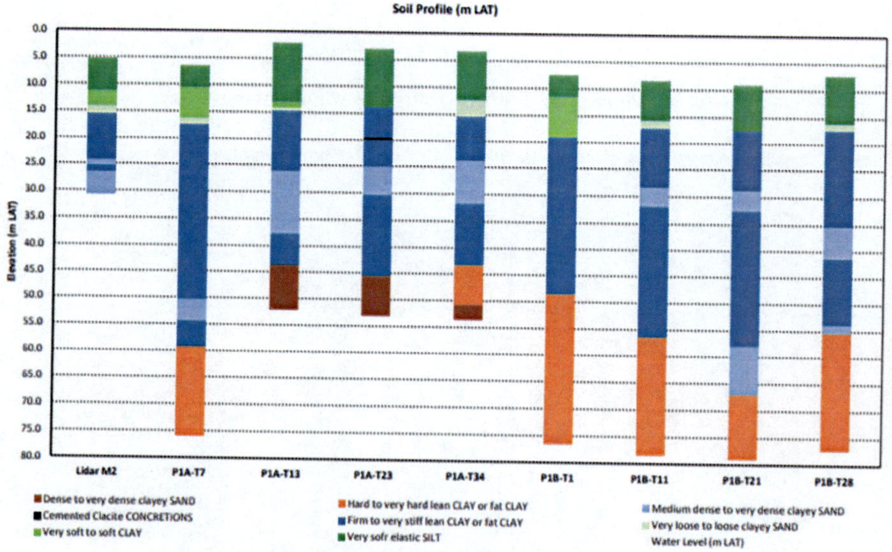

Fig. 7. Summary of soil profile

3.3 Metocean Surveys

Metocean surveys were completed by eCoast from New Zealand. A significant challenge was the collection of metocean data in this remote area with little infrastructure. Challenges included gathering year-round oceanographic data in persistently energetic conditions, sourcing equipment and vessels to undertake the work and navigating many kilometers of intertidal mud flats for instrument deployments and bathymetric surveys. Nonetheless, bathymetric data was collected covering an area of more than 360 km^2 and wave, current and sea level data were successfully recorded throughout the most energetic months. Surveys also quantified expected marine growth and mapped all permanent fishing structures in the area.

The field data were analyzed in a desktop study that provided information on wind, waves, currents, tsunamis and cyclones. Tsunami modelling considered worst case tsunamigenic currents and sea levels from all relevant sources. Non-cyclone conditions were studied through long term modelling of coupled waves and hydrodynamics. Cyclone activity in the area is sporadic leading to a sparse historical record. A numerical modeling approach was used to simulate all the historical cyclones to form the basis of subsequent extreme value analysis to arrive at final design criteria for wind farm design (Table 1).

Table 1. Metocean design parameters for the project area

Return period (yrs)	Parameter (unit)	Design value
50	Extreme wave height, Hs (m)	3.5
50	Extreme wave height, Hmax (m)	6.6
50	Storm surge (m)	0.94
50	Extreme surface current speed (m/s)	2.44
N/A	Marine growth thickness (mm)	100

3.4 Wind Measurement

An 80 m high mast was installed at an onshore location near the project site in August 2014 under the USTDA grant.

In 2018 studies were completed to assess the most cost-effective means to acquire offshore wind data at the project site. Construction of conventional met masts on offshore foundations were considered, as well as floating Lidar equipment. Both were ruled out on cost grounds. In the end a decision was made to build pile supported platforms at two locations on the project site, and install Lidar equipment on these:

- 1x LiDAR (M1) on offshore platform was installed in December 2018, after completing validation next to the onshore mast.
- 1x LiDAR (M2) was in September 2019, after completing validation next to the onshore mast.

The construction of the offshore platforms M1 and M2 provided an opportunity for assessing construction techniques in this nearshore environment.

LiDAR platform M1 was installed by Vietnamese contractor HTL in water depth of 3 m at high tide using a crane mounted barge, that could rest on the sea-bed at low tide. The platform is supported on 4 × 400 mm diameter PHC piles installed by driving to 35 m depth below sea-bed. A challenge for driving was the very soft ground conditions that meant the piles drop under their own self weight to a depth of about 10 m and maintenance of verticality with a supporting frame was crucial in this uncontrolled drop condition. The use of a 10-h window when the barge was resting on the sea-bed was used to install the piles and the platform.

LiDAR platform M2 was installed by Vietnamese contractor, OEC in water depth of 10 m at high tide using a crane mounted barge in a floating condition. The platform is supported on 4 × 500 mm diameter PHC piles installed by driving to 36 m depth below sea-bed (Fig. 8). Operating in 10 m water depth required construction of a supporting frame fixed to the sea-bed with shallow penetration piles. This enabled the operations to continue with decoupling of the barge movement from the piling operation.

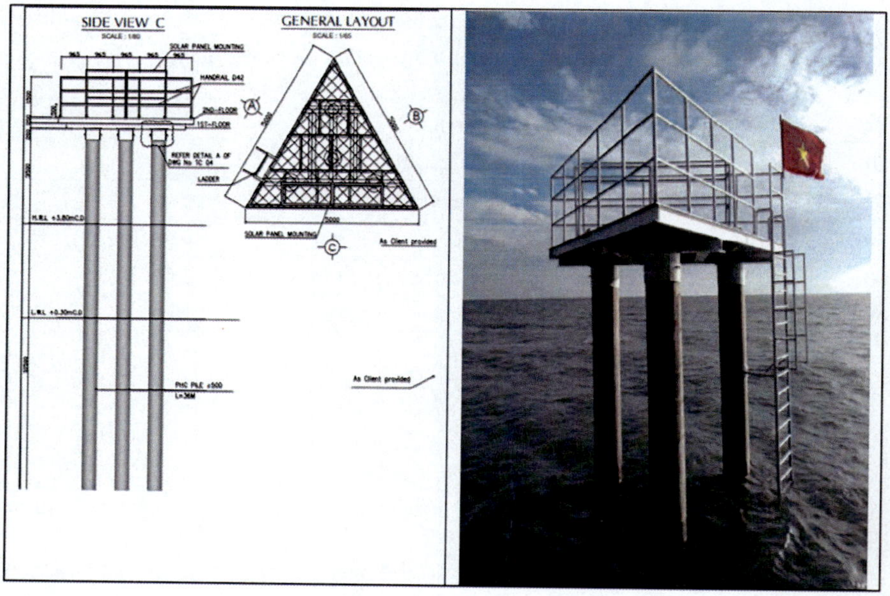

Fig. 8. LiDAR platform M2

3.5 Wind Turbine Foundation Design

Tenders for balance of plant, including wind turbine foundations for first 200 MW phase of the project were received in early 2021, with the selection of foundation concept open for offer by the bidding Contractors. Turbines between 4 MW and 6 MW in rated capacity are under consideration by the project.

Both monopile foundations and multi pile raised concrete foundations have been proposed and deemed suited to the ground conditions at the site. At the time of writing, a final selection has not yet been made.

3.6 Engagement with Fishing Industry

Site survey and Lidar platform construction activities in 2019 and 2020 highlighted the severe access constraints for fishing vessels using the harbour inlet from where the Stage 1 geotechnical investigation access vessels departed. The access limitations are at low tide/falling tide when the water depth is too shallow to permit boat movement. The jetty end will be designed as a co-use facility that can be used by both the project O&M vessels and the fishing vessels (Fig. 9).

Fig. 9. Fishing boats in use at the project site area

With submarine cables being deployed at the project site area, it will be important for safety considerations that fishing vessels remain outside the wind turbine area. In compliance with IFC Standards for sustainable development, the project team has held detailed consultations with the fishing community and is under agreement on a number of potential mitigating measures that will be implemented to ensure that the livelihood of the fishing community is not adversely affected:

- Upgrading of vessels to allow them to be deployed further from shore
- Re-training of fishing industry workers with skills that can be applied to offshore wind
- Supporting the development of craft skills

3.7 Operation and Maintenance

After an analysis of the available options, the project has elected to use crew transfer vessels rather than walkways for access to the turbines. This selection was made primarily on an economic basis when considering the high cost of the walkways for this project that is further from shore than other nearshore projects under construction. A further consideration was environmental impact of the walkways that has been questioned by a number of international lending institutions.

On this basis, a design decision has been made that buried submarine cables will be used to transmit the power to shore rather than onshore type cables housed in a walkway.

The project will build a 1.5 km long jetty from shore, the furthermost point of which will have a minimum water depth of 1.5 m at low tide. This will ensure 24-h berthing access will be available at the end of the jetty.

3.8 Policy Support for Offshore FIT in Vietnam

Vietnamese policymakers have recognised the potential of Offshore wind (OWF) as a vital component of Vietnam's future energy mix. This is evidenced through Resolution 55 [4], and translated into the draft planning document of PDP 8, which set an initial target in excess of 30,000 GW for 2045.

The current offshore feed in tariff (FIT) of 98 US cent per kwhr has provided an incentive to nearshore developers to build close to shore in shallow waters. This tariff comes to an end in November 2021, and the route to market after that time is as yet unclear for developers.

'True' OWF has not yet been built in Vietnam. Experience in Europe has shown that OFW technology cost curve will shift after an initial 3–4 GW of installation has been achieved in a number of countries, which is the volume required for local expertise to be gained among developers and a local supply chain to be formed (Fig. 10). In Europe the building of the first OWF projects was stimulated by government FIT policy, that transitioned to auctions after the first 3–4 GW had been completed successfully. This provides a valuable lesson for the future success of offshore wind in Vietnam.

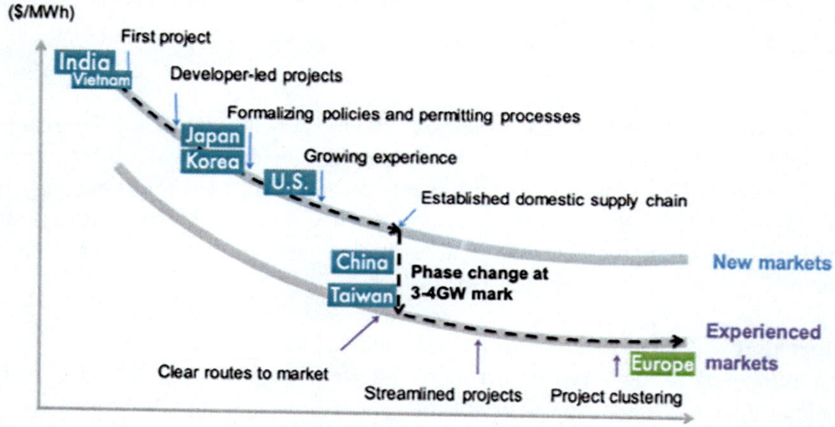

Fig. 10. Cost trajectory of offshore wind

3.9 Conclusions

Vietnam has enormous potential for development of offshore wind energy, and a strong policy direction from Resolution 55 by the Politburo. To date, however only limited nearshore wind projects have been built or are under construction.

Several lessons have been learnt from these nearshore projects including:

- Best time of year for offshore installation in this region.
- Advantages and disadvantages associated with monopile and raised concrete multi-pile foundation solutions for turbines in this nearshore shallow water environment.

– Visual and environmental impact of walkways used for the turbines.

The Phu Cuong Soc Trang project will incorporate these learnings, and promises to be the first Vietnam project to showcase best practice offshore wind solutions, including reaching a win-win agreement with the local fishing industry, ahead of the start of a true offshore wind industry in the country envisaged by PDP8.

The Vietnamese government policy makers should consider the lessons learnt in Europe, where the building of the first OWF projects was stimulated by government FIT policy, that transitioned to auctions after the first 3–4 GW had been completed successfully. This provides a valuable lesson for the future success of offshore wind in Vietnam.

References

1. The World Bank: Offshore Wind Technical Potential in Vietnam (2020)
2. Geological and Mineral Resources Map of Vietnam: Ca Mau - Bac Lieu area, sheets C-48-XXVII & C-48-XXVIII (2016)
3. USTDA Grant (2015). https://ustda.gov/wp-content/uploads/7.09.15-20th-Anniversary-of-US-Vietnamese-Relations.pdf
4. Politburo of Vietnam: Resolution 55-NQ/TW on the Orientation of the Viet Nam's National Energy Development Strategy to 2030 and outlook to 2045 (2020)

Planning, Policy and Integration for Sustainable Development of Offshore Wind Energy in Vietnam 2022–2030

Minh Ha-Duong[1](✉), Sven Teske[2], Dimitri Pescia[3], and Mentari Pujantoro[3]

[1] Centre International de Recherche sur l'Environnement et le Développement, CNRS, Paris, France
[2] Institute for Sustainable Futures, University of Technology of Sydney, Ultimo, Australia
[3] Agora Energiewende, Berlin, Germany

Abstract. The wind power sector took off in Vietnam after the feed-in tariff was raised to 8.5 US cents/kWh for onshore projects in 2018. As of March 2021, 113 wind projects with total capacity 6,038 MW have signed a power purchase agreement. Most are expected to enter commercial operation before December 2021. We explore here three scenarios for wind power development in Vietnam through 2030. It argues that by 2030 the wind power installed capacity in the New Normal could be around 17 GW onshore and 10 GW offshore. In a Factor Three scenario, offshore wind reaches 21 GW by 2030. This has three policy implications. First, Vietnam's next power development plan provides an important opportunity to increase at low costs the level of ambition of wind power development. Second, flexibility should be the guiding principle of the plan. Third, to realize the large potential of offshore wind power, infrastructure planning has to start soon.

Keywords: Wind energy · Vietnam · Scenarios

1 Introduction

Early 2021, Vietnam's 2025 targets for installed wind power capacity in the Power Development Plan VIII (PDP 8) draft were: 11.3 GW for onshore and nearshore, zero offshore. For 2030, these numbers increased respectively to 16.0 GW onshore/nearshore, and 2 GW offshore [1]. These numbers are not final, as the administration is still revising the draft. This manuscript perspective argues that Vietnam rely more on its excellent wind resources, which potential has been overlooked in the PDP 8 draft. Considering the pipeline of projects in the country, the upcoming Power Development Plan 8 (PDP8) still has the opportunity to increase the target to have several GW of installed offshore wind capacity by 2030.

Our argument follows the scenario analysis method. The next section exposes the natural and technical potential of wind energy in Vietnam. Section 3 presents an original project-level inventory of the wind power sector in Vietnam. The pipeline is large enough to meet the government targets by far. Still using our inventory dataset, Sect. 4 reviews the stated costs of wind power projects in the country, and why they are expected to

D. V. K. Huynh et al. (Eds.): VSOE 2021, LNCE 208, pp. 48–66, 2022.
https://doi.org/10.1007/978-981-16-7735-9_4

decline. Section 5 defines three scenario narratives to explore the range of possible futures. We quantified our scenarios with a top-down energy system modeling study led by the Institute for Sustainable Future, hereafter 'ISF' [2]. Sections 6 and 7 discuss, respectively, two implications of the expected surge in wind power: the need for flexibility in the whole system, and the urgency to plan the offshore transmission infrastructure. Section 8 summarizes our results and concludes by reviewing the specific Vietnamese policy changes called for by the wind power opportunities.

2 Vietnam Wind Resource is Excellent

Vietnam's wind resources mostly lie along its coastline of more than 3000 km, and in the hills and highlands of the northern and central regions. The Global Wind Atlas estimates that over 39% of Vietnam's area has annual average wind speed over 6 m/s at a height of 65 m, and over 8% of Vietnam's land area has annual average wind speed over 7 m/s (see Fig. 1 from [3]). This corresponds to wind resources physical potential of 512 GW and 110 GW.

In 2006 already, [4] found 3.5 GW economic potential at less than 6 US cents/kWh for onshore wind power. Figure 1 shows that Vietnam's the offshore wind potential is much larger than the onshore wind potential. Winds over sea are stronger, more stable. Land use constraints to sit turbines and transmission infrastructure are also lower.

Sean and Whittaker [5] presented estimates of the wind offshore potential for Vietnam, accounting only for wind speed greater than 7 m/s and less than 200 km from the coast. they found a potential of 261 GW in waters less than 50 m deep, and an additional potential of 214 GW in waters 50 to 1000 m deep.

Using the Weather Research and Forecasting (WRF) model (10-km resolution for 10 years, from 2006 to 2015), Doan et al., [6, 7] estimated the energy potential in the offshore area around the Phú Quý island (Bình Thuận province). This area alone can provide 38.2 GW of offshore wind power generation capacity. [8] precised the analysis, estimating that the five-year capacity factor over 2007–2011 of a LW 8 MW wind turbine over Phu Quy was over 54%.

The Institute of Energy based its study for the official Power Development Plan VIII on technical potentials 47 GW for onshore wind, 162 GW for offshore ([9], ch. 8, tables 8.14 and 8.15 p 332, see annexes 8.4 and 8.5 pp 32–35 for the distribution by province). This is at 80 m hub height, for wind speed over 5.5 m/s onshore, 6 m/s offshore, filtering out unsuitable land use.

The present manuscript is based on our own resource analysis [2]. Taking into account land-use restriction – excluding mountain areas with slopes of more than 30%, fractured spaces with a size of less than 1 km^2 and areas more than 10 km away from existing power lines – it found 42 GW of technical onshore wind energy potential for utility scale wind farms. It estimates the offshore wind resource to 609 GW, spreading over a total of 3000 km coastline and area of 150 000 km^2. This takes into account only coastal areas with a maximum water depth of 50 m and a maximum distance to the shore of 70 km, and uses 2015 meteorological data. The ISF study further estimated the regional wind energy potential and capacity factors (for details, see [10] annexes). While being a variable energy source, the estimated capacity factor of wind energy in Vietnam is up to 36% for onshore and 54% for offshore wind.

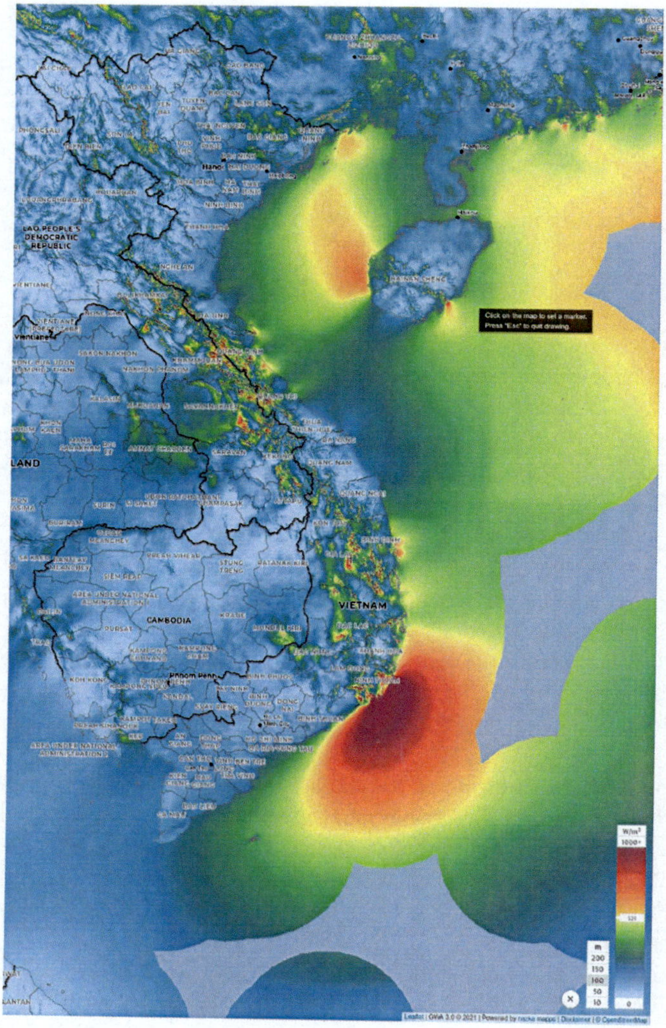

Fig. 1. Mean wind power density for Vietnam at 100 m, color scale from 0 to 1000+ W/m^2. Source: "Global Wind Atlas 3.1.

3 Towards 5 GW of Wind Power Operational at the End of 2021

As we exposed in [10], the history of the sector shows fifteen years of growing pains. The Bạch Long Vĩ island hybrid diesel + wind project inaugurated on October 30th, 2004 was a false start, not appropriate to the local capacities at the time. Vietnam's first high-capacity wind farm, 30 MW, was inaugurated by REVN, in the central province of Bình Thuận on April 18th, 2012. This is where the first turbine fire event in Vietnam happened on January 5th, 2020. The next two projects – the Phú Quý island hybrid grid with 6 MW and the near-shore Bạc Liêu phase 1 with 16 MW – both completed in 2013. These three pioneer projects had their lot of technical trouble: one turbine of the Phú

Quý system has been broken for years; one turbine in the REVN wind farm caught fired on January 5th, 2020; and the construction of Bac Liêu phase 2 took was much slower than the norm today.

No new capacity was added in 2014 or 2015. Provincial wind power development plans for eight provinces – Bạc Liêu, Bến Tre, Bình Thuận, Cà Mau, Ninh Thuận, Quảng Trị, Sóc Trăng and Trà Vinh – were published in 2016. That year the 24 MW Phú Lạc project in Bình Thuậnh province and the 83 MW Bạc Liêu phase 2 were completed. The following year, the 30 MW Hướng Linh 2 project in the Quảng Trị province was connected. By the end of 2018, the total installed wind power capacity in Vietnam had reached about 228 MW.

In 2019, Vietnam's wind power industry began to take off, although not as spectacularly as the solar power industry. By 31 May 2019, seven wind power plants were in operation, for a total capacity of 331 MW. By February 2020, we count eleven projects in operation, for a total capacity of 425 MW. To take one example: the Trung Nam project in Ninh Thuận province inaugurated in April 2019 a hybrid facility, co-locating a 40 MW wind farm with a 204 MW solar PV plant. This plant was being extended by 112 MW in phase two, by installing 28 turbines of 4MW, and after phase 3 added another 48 MW by mid April 2021, it became the largest in Vietnam. Installing this class of machines demonstrates that the local market has now reached the technological frontier.

We maintain a database of wind projects in Vietnam. To the best of our knowledge, this is the most complete and up to date list, even including commercial offers. The full database is available online [11]. Table 1 summarizes the pipeline of projects in Vietnam by mid 2021. Projects are classified by phase. The Operation phase is after the commercial operation date. The Implementation phase is from the groundbreaking ceremony to the commercial operation date. The Development phase is before building but after the administrative Decision of Investment, which usually given by provincial authorities soon after the project is added to the national Power Development Plan. The preliminary phase is from exclusivity MoU to administrative Decision of Investment.

Our database includes has all projects authorized by the Government. They are in Operation/Implementation/Development phases. Between these three rows, our numbers are only indicative. They are based solely on publicly available sources, we are only able to record projects advances when they are reported.

Table 1. The wind power sector in Vietnam, July 2021. Capacities by project type and stage. Source: Author, public sources

	Onshore	Intertidal	Offshore	Total
Operation	399 MW	99 MW	0	498 MW
Implementation	2 516 MW	1 139 MW	350 MW	4 004 MW
Development	4 265 MW	2 015 MW	700 MW	7 880 MW
Preliminary	20 525 MW	2 366 MW	31 812 MW	54 703 MW
Total	**27 704 MW**	**6 519 MW**	**32 862 MW**	**67 085 MW**

Row four adds up projects at the Preliminary phase. This is a wide category. It includes projects already proposed for inclusion in the Power Development Plan, but also projects that have not yet started the pre-feasibility study, or projects seeking an active investor. The 20 525 MW of Onshore wind projects at the Preliminary phase include 14 422 MW of capacity signed in the Gia Lai province investment projects [12]. The 31 812 MW of offshore wind projects at the Preliminary phase is less than the 59 720 MW capacity of potential projects listed by Institute of Energy in Table PL9.3 in [9, pp. 52–53].

According to [13], in March 2021, Electricity of Vietnam (EVN) informed the Ministry of Industry and Trade that, up to now, they signed power purchase agreements (PPAs) with 113 wind power projects, with a total capacity of 6,038 MW. Based on reports of the investors, progress of construction investment and commercial operation of the wind power projects as follows:

1. The number of projects that have already put into commercial operation as 12 projects with a total capacity of 582 MW.
2. The number of projects that expect to be put into commercial operation before October 31, 2021 as 87 projects with a total capacity of 4,432 MW.
3. The number of projects that cannot be put into commercial operation before December 31, 2021 as 14 projects with a total capacity of 1,024 MW.

The numbers from EVN confirms the main finding of Table 1, based only on public sources: over 4 GW of wind power projects are actively building by mid 2021.

The power development master plan PDP 7 revised, published in 2016, has set targets of 0.8 GW of wind power capacity by 2020, 2 GW by 2025 and 6 GW by 2030. Capacity in operation reached 538 MW at the end of 2020, so the 2020 target was missed [14]. However, EVN has signed PPA for more than 6 GW already, out of the 11 262 MW authorized by the government. It is certain that the 2025 objectives will be exceeded by the end of 2021.

The table confirms the potential estimates from the previous section: there is about 67.7 GW of wind power capacity in identified sites in Vietnam. Most of this potential comes from projects at the Preliminary stage. May will never be realized, but these numbers provide the best frame of reference to discuss 2030 scenarios for the sector.

The early 2021 draft of the Power Development Plan VIII proposed the revise the targets for wind power in 2030 as follows: In the low demand scenario, 16 GW onshore and 2 GW offshore. In the high demand scenario, increase the offshore target to 3 GW. ([9], ch 9, tables 1 and 2 pp 353–353). The analysis (Fig. 9.10, p 389) further examines several scenarios in which new wind power generation capacity substitutes for new imported coal generation capacities at the 2030 horizon.

Overall, the 16 to 19 GW target is realistic considering the amount of projects already accepted or proposed at the end of 2020. However, this draft of the plan does not include most large offshore projects under discussion at that moment.

In the offshore sector, the Ca Mau 1 (350 MW) project is under construction, while HBRE Vung Tau (500 MW) and Phu Cuong 1A, 1B (200 MW) are approved. But there are more than 31 GW of offshore projects at the preliminary stage. Offshore projects Phu Cuong 1C, 1D (200 MW) and Binh Dai Ben Tre (400 MW) are to be considered for PDP VIII. AIT and Mainstream's project in Bến Tre (500MW) is also proposed to the Plan. Phase 1 (600 MW) of the Thang Long project by Enterprize Energy (UK) has obtained a full year of wind data and is continuing its site survey. The La Gàn project by Copenhaguen Infrastructure Partners (Denmark) started strong, as is the Binh Dinh project by PNE AG (Germany). History will judge the 600MW offshore project in Bình Thuận between PV Power and IMPSA, signed in April 2011 with Argentina, was only a diplomatic gesture.

4 Cost of Wind Power is Rapidly Falling, but FIT is Uncertain

Offshore wind becomes more and more economically viable due to its higher yields, larger turbines and clusters. The auctions conducted to lease the rights to build wind farms offshore Massachusetts illustrate the cost decline in the industry. In 2014, they did not receive any bids. When the auction ran again in December 2018, the winning developers paid 405 million USD for the right to build wind farms [15, 16].

Costs will also improve because the average rated capacity size of offshore turbine units is increasing. Though it is around 7 MW per turbine today, the first 12 MW turbines are already being field tested, and the capacity is expected to reach 15 MW per turbine in 2025. The average capacity factor of current projects also ranges from 50–57% [17]. From 2020 to 2022, the cost of electricity from newly commissioned offshore wind power projects could range from 60 USD/MWh to 100 USD/MWh based on current trends and the winning bids from 2018 auctions – a significant decline compared to 140 USD/MWh in 2017.

For example, France aims to have 10 GW of offshore wind power by 2028, and recently conducted an auction for a 600 MW offshore wind farm near Dunkerque, scheduled to open in 2025–2026. The eight candidates proposed to sell electricity for 20 years between 44 €/MWh to 60,9 €/MWh (equivalent to 48 to 67 USD/MWh), with an average at 51 €/MWh (equivalent to 56 USD/MWh). Germany has held just two offshore wind auctions, one in 2017 and one in 2018. In both tenders, capacity was won by investors who offered to build parks without subsidies.

There is no auction for offshore wind electricity delivered in Vietnam around 2025. But if the Government were to hold one, there would be important differences with France and Germany. The bidders would factor in the risks of being the first to develop the technology in the country. And the bidders would also have to factor in whatever costs of the transmission infrastructure is required from them. According to experts surveyed in GWEC [39], "*With regards to support mechanisms for offshore wind, the current FiT level was seen as appropriate.*" which means that the industry expects to be able to deliver under 98 USD/MWh.

Since 2011, the Feed in tariff (FiT) for wind power project in Vietnam was 78 USD/MWh and that was not commercially viable for developers. In November 2019, the FiT was raised to 85 USD/MWh for onshore wind power projects and 98 USD/MWh for offshore wind power projects [18]. The new electricity tariffs are applied to a part or whole of the grid connected wind power projects with commercial operation date before 1 November 2021 for 20 years from the date of commercial operation. Already operating projects will benefit from the tariffs retroactively from 1 November 2018 for the remaining period of the signed PPA.

Future tariffs for wind power projects completed after 1/11/2021 remain uncertain. In 2020, the Ministry of Industry and Trade (MOIT) proposed to extend the current FIT scheme to the end of 2021, effectively giving a two months extra time for developers to finish their projects [19]. Later that year, MOIT proposed to extend the FIT scheme for already registered projects by two more years (Letter 8159/2020/BCT-DL cited in [20]), reducing the rates as described in Table 2 below.

From our perspective, extending by two years seems preferable to rushing all projects to the end of 2021. This can be justified considering the administrative delays in 2019 and the COVID crisis. More time allows a better execution of projects. Time is also needed to alleviate grid congestion issues. Renewable energy projects do not contribute to the supply when they are curtailed. And the auction mechanism, which is prepared to follow the FIT scheme, has not started to be deployed by June 2021.

5 Scenarios for Wind Power in Vietnam

The previous sections have shown that in mid 2021, the pipeline of wind power projects in Vietnam is much larger than the government targets. In [2, 10] we examined three visions for the development of the power sector in Vietnam. We estimated future demand trajectories and looked at how various energy mixes could satisfy it. In these visions, the role of wind power is defined as follows (see Table 3).

Table 2. Current and proposed FIT for wind projects in Vietnam (updated 2021–06-212).

Project type	Onshore	At sea
COD until Oct 2021	8.5 US cent/kWh	9.8 US cent/kWh
COD from Nov 2021 to Dec 2022	7.02 US cent/kWh	8.47 US cent/kWh
COD in 2023	6.81 US cent/kWh	8.21 US cent/kWh

The first scenario corresponds to the PDP7 revised, which planned for 6 GW of installed wind capacity in 2030, accounting for 2,1% of the power production. The other two scenarios respectively achieve that year a total installed wind capacity of 26 GW and 42 GW. In the most ambitious scenario, annual markets increase about 2 GW a year for both offshore and onshore wind during the next decade. While such annual market figures may seem ambitious, the country is expecting a high energy demand growth apparent from the PDP7 revised which planned to install 70 GW of additional generation capacity between 2020 and 2030 – mostly baseload plants. Indeed, high growth rates in the power supply infrastructure are required to lift a 100 million people from lower to higher middle income in ten years. Technological progress is rapidly reducing costs, leading to considerable shifts in investment patterns. For example the PDP7 solar PV objectives were 4 GW for 2025, but the installed capacity achieved more than 5 GW at the end of 2019. As explained in Sect. 3 above, a similar-sized wave of wind energy projects will arrive in 2021.

Table 3. Visions for Vietnam onshore and offshore wind energy installed capacity in 2030. Source: Authors.

Scenario	Onshore wind capacity in 2030	Annual market		Offshore wind capacity in 2030	Annual market
		2020–2025	2026–2030		
Old Plan	6.1 GW	300 MW/yr	1000 MW/yr	0.15 GW	negligible
New Normal	16.6 GW	730 MW/yr	3200 MW/yr	9.5 GW	1000 MW/yr in 2023–2025, then ramp up to 1400 MW/yr in 2030
Factor Three	21.6 GW	1530 MW/yr	3625 MW/yr	20.9 GW	1000 MW in 2023 1900 MW in 2024, then 3000 MW/yr from 2025 on

The following narratives flesh out the three scenarios, explaining various development rhythms:

In the *Old Plan* scenario, a wave of new wind farms connects to the grid in time to get the FIT before November 2021. After that, the government does not renew the FIT, and legal issues delay the first pilot auction to 2022. By then a global economic crisis affects Vietnam, reducing economic growth and, by extension, domestic electricity demand. International fossil fuels prices hit historical lows as global demand dips, and producing countries try to sell their remaining reserves before they are made useless. Vietnam policy priority is on GDP growth by international trade rather than climate protection. Natural gas plays the first role in the 2020–2030 decade. A rapid improvement in energy efficiency and a strong development of solar, both utility scale and behind the meter rooftops, leave little room for wind energy to expand above the PDP7 targets revised in 2016.

In the *New Normal* scenario, after the initial wave of wind projects in 2021 there is no second FIT period due to concerns on the part of policymakers that the market will overheat. The market is driven by direct power purchase agreements –multinational companies in Vietnam procuring green electricity directly from wind project developers – and by government auctions – the 2021 DPPA pilot is a success. The government credibly commits to an auction program for 1 GW of offshore wind per year. During the 2021–2025 period, the technology develops as a progressive extension of the near-shore wind farms, while the "true offshore" projects scale up. The expected market size leads many industries to choose Vietnam as their Southeast Asia base for manufacturing equipment and operating wind projects. The offshore wind industry is organized around two hubs: the port of Vũng Tàu serves farms in the zone facing Bình Thuận and Cà Mau coasts; and the port of Hải Phòng serves zones facing Quảng Ninh coasts.

In the *Factor Three* scenario, the Party decides to steer Vietnam into green development. The national oil and gas company PVN is redefined as a sustainable energy provider that mobilizes its offshore work capacities and the complementarity between gas and variable renewables. All projects meet their initially announced targets, which are extremely unlikely considering the situation today. The Thang Long Wind power project by Enterprize Energy, which intends to develop large-scale offshore near the Kê Gà area in Bình Thuận province, starts operating its first 600 MW phase at the end of 2022, and extends by 600MW every year. The three other projects – in Bac Lieu by KOSY group, in Vung Tau by HBRE group and in Phu Cuong by Mainstream – add 1.7 GW of offshore wind power capacity by 2024. By end of that year, the total offshore capacity is 2.9 GW, and the sector is competitive. This convinces the government to adopt a regional leadership strategy in the wind energy sector. A national scale offshore transmission infrastructure – power centers hubs and under sea lines – starts by connecting the existing projects. Starting in 2025, the backbone strategy to meet the power demand expansion in Vietnam is to auction 3 GW of offshore wind every year.

The key early signpost indicating which scenario Vietnam is heading towards will be the government's next decision about for wind power incentives. Feed In Tariff (FIT) is by far the proven policy mechanism, it worked in Germany between 1991 and 2018, and it is working in Vietnam now. Recent experience in Germany is also an example for the *Old Plan* scenario: auction design problems can cool down the wind market. Alternatively, a decision to offer FIT over 90 USD/MWh for offshore wind after 2022 would steer towards the third scenario. Costs would be high but tendering the rights to use offshore wind development zones could recover a portion of these costs.

Most previously published Vietnam power system development scenarios are close to the *Old Plan* scenario when it comes to wind power. The Renewable Energy scenario [21, p. 39] sees 8.5 GW of wind installed capacity in 2030. The Danish Energy Agency [22, p. 45] scenarios project wind power capacity to grow to 6.2 – 6.4 GW in 2030, which is close to the PDP7 revised. The 2019 update to this Vietnam Energy Outlook [23] sees more wind power generation than solar in 2030 in all cases; offshore wind enters at 2.5 GW in 2030. The ECA Made in Vietnam Energy Plan 2.0 [24, p. 11] assumes that a doubling of the PDP7A target is possible, for a 12 GW of installed wind power by 2030, not mentioning offshore wind. The low-carbon development pathway report for Vietnam by the Asian Development Bank [25, p. 35] targets 16 GW of wind capacity by 2050, giving a preference to solar, nuclear and biomass energy. To sum up, up to 2018 even the most progressive previously published scenario for Vietnam targeted a maximum of 12 GW in 2030, way below the new trajectories assessed by ISF, while still overlooking the offshore wind potential. Only the Renewables-Led Pathway [26] foresees 39 GW of wind capacity installed in 2030.

World Bank published an Offshore Wind Roadmap for Vietnam [27] to argue that building 10 GW in 2030 is much cheaper in the long run than building only 5GW. This provocative result arise because the cost of electricity in the high growth scenario reach grid parity earlier. It drops to 7 US cent/kWh in 2030 as opposed to 8 US cent/kWh in the low growth scenario. Two reasons explain the economies of scale. One is capital expenditure reduction with learning by doing and increased competition. The other is the cost of capital reduction with increased market confidence and less risk.

GWEC [28] reports that "2020 saw global new wind power installations surpass 90 GW, a 53% growth compared to 2019, bringing total installed capacity to 743 GW, a growth of 14% compared to last year. The world's top five markets in 2020 for new installations were China, the US, Brazil, Netherlands and Germany." The share of offshore installations is growing [28, p. 51]. Offshore wind installation reached its first GW in 2007. According to REN21 [29] globally about 6 GW of offshore wind capacity was connected in 2020, half of it in China. In this growing market, there are opportunities to develop domestic manufacturing and engineering capacities. Although the result is surprising, the World Bank arguments that the economic policy objectives are as important as energy policy objectives has some value. The question is not just how to produce the cheapest possible electricity, but also how to integrate Vietnam in the global value chain of a growing sector.

Now that Taiwan and Japan have both announced that they will install more than 5 GW of offshore wind power, manufacturers are already redesigning their machines and platforms to resist the typhoons and earthquakes prevalent in the Pacific area.

The *Old Plan* scenario (and previously published studies) neglects the role that offshore wind could play in 2030 in Vietnam, while the Factor Three scenario might seem extremely ambitious today: the commission of 42 new offshore wind farms in ten years. Both scenarios are extreme and unlikely. The most plausible future scenario is something in between. Connecting the realm of possible with extreme cases is useful for analysis.

Party resolution 55 orienting the national energy development strategy of Vietnam [30] prioritize wind and solar energy for electricity generation, and mentions building 8 billion m3 of LNG import capacity by 2030. This suggests that the middle scenario is the most plausible of the three. Given the worldwide trends, the abundance of wind in Vietnam, the full pipeline of projects and the urgent need to increase electricity supply, we believe that the government will revise upwards the wind targets in the next PDP and commission several 500 MW offshore wind farms over the next ten years.

6 Variability and Flexibility Needs

Wind power is a variable energy source: wind speed varies by day, region and season. The high wind speed season is during the north-eastern monsoon months (December, January, February) and low wind speed seasons is during the inter-monsoon months (March, April, and May). [6] observed that the seasonal variability[1] in Vietnam associated with monsoon onsets and the daily variability[2] associated with the wind diurnal cycles is in the 30–50% range, and the inter-annual variability[3] can reach up to 10%.

The wind regime in the North differs from the wind regimes in the South and Center. The central and southern regions have high correlation (<80%), while the northern region shows weak correlation with the south (<5%). These fairly low correlations suggest that the wind generation would not peak at the same time within the country,

1 Normalized standard deviation of the monthly mean values of wind power density at 105 m height.

2 Normalized standard deviation of hourly data.

3 Normalized standard deviation of yearly mean from 10-year mean.

as Fig. 2 illustrates. Distributing wind generation deployment across the region and across intra-regional exchanges creates balancing effects that considerably reduces the variation of wind power output. The maximum change of wind generation within the central region from one hour to the next is 30%, but the Vietnam-wide maximum change is 2%.

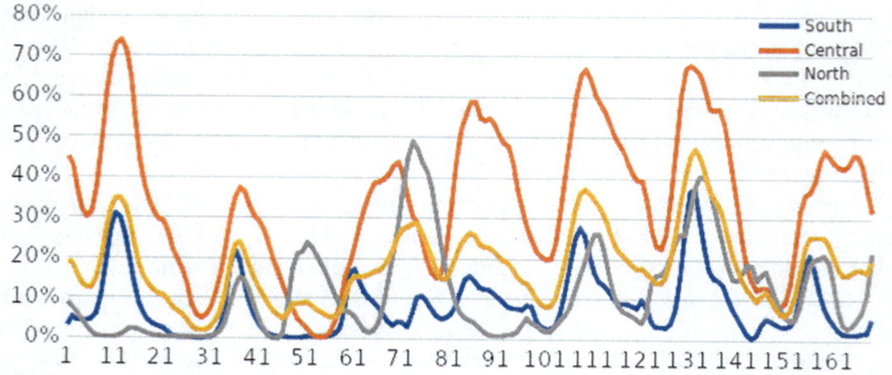

Fig. 2. Wind generation profile in 3 regions, first week of January. Generation profile from Renewable Ninja.

To illustrate how the power system copes with variability, we compared simulations of the power supply under the Factor Three scenario in 2030 for the Red River Delta region during a windy and a windless week. By choosing this area, we want to stress that while the southern and central provinces attracted most solar and wind energy projects to date, the power grid must be balanced, and increasing power generation from renewable sources in the northern regions is an option.

In a windy week in May 2030 (Fig. 3), wind and solar provide a major share of electricity generation, while fossil fuels make up only a minor share. In this case, wind power gets shunted from the windy Northeast and/or North Central Coast to the Red River Delta and displaced coal generation in three evenings.

Figure 4 shows the windless week simulation, in August 2030. Most of the power supply during this week comes from coal and gas, while solar photovoltaic and wind have minor contributions. The Red River Delta region is self-sufficient for almost the entire week, importing only minor amounts of electricity from other regions during evening hours.

The variability of the two situations described in these figures is already being experienced by countries that have a much larger share of wind power in their energy mix than Vietnam does. In 2019, wind power represented 55% of the electricity produced in Denmark; and 20% in Germany according to IEA statistics. Because of intra-annual variability, the maximum share of wind at monthly and weekly time scale is higher in some months. For example, wind satisfied 30% of Germany's electricity demand in January 2019 and only 18% in June 2019.

These simulations only aim to showcase the potential magnitude of the variability issue, in the hope that solutions will be found. The major flexibility solutions include pumped storage, battery storage, more flexible thermal and interconnection. Experience from Europe shows that anti-correlation of the seasonal weather patterns of monthly onshore wind and PV generation yields a more stable total variable renewable energy output. There is also a seasonal-scale balancing effect between variable renewables in Vietnam. Statistically, from November to February wind speed is high, from February to June solar radiation reaches a peak, and during the wet season from July to November the water inflows are high [31].

Researchers have observed that offshore wind has more stable output throughout the year than onshore. This alleviates the variability issue and contributes to reliable capacity in the medium and long term. Another lesson from these countries is that variable renewable energy capacity is faster to install than flexibility solutions, in particular those involving additional grid infrastructure. That can lead to system congestion, high level of curtailment and sub-optimal period of negative electricity prices.

Between 2008 and 2018, China increased its renewable power generation capacity from 15 GW to 370 GW [32], but failed to keep the same pace in transmission capacities. As a result, wind farms across China had an average curtailment rate of 17% in 2011 and 20% in 2012 [33]. Increased investment in transmission in recent years succeeded to bring down curtailment rates significantly, to 17.1% in 2016 and further in 2018 to only 7% [32].

This challenge also occurs in the Vietnam energy system. Over 30% curtailment rates are frequent in provinces where installed solar capacity exceed grid capacity [34]. Looking at the ability to release power from wind and solar sources up to 2022, [35] found that in the Central Highlands area, only 40% of the total available capacity of wind power plants could be used, and there are also acute curtailment issues in the South Central area. The increasing penetration of solar and wind power requires the adoption of technical and economic flexibility solutions. The flexibility solutions are not a distant hypothetical; they are a clear and near-term need. It is time to adapt the system now. Flexibility should be the guiding principle for Vietnam's 2020–2030 power development strategy.

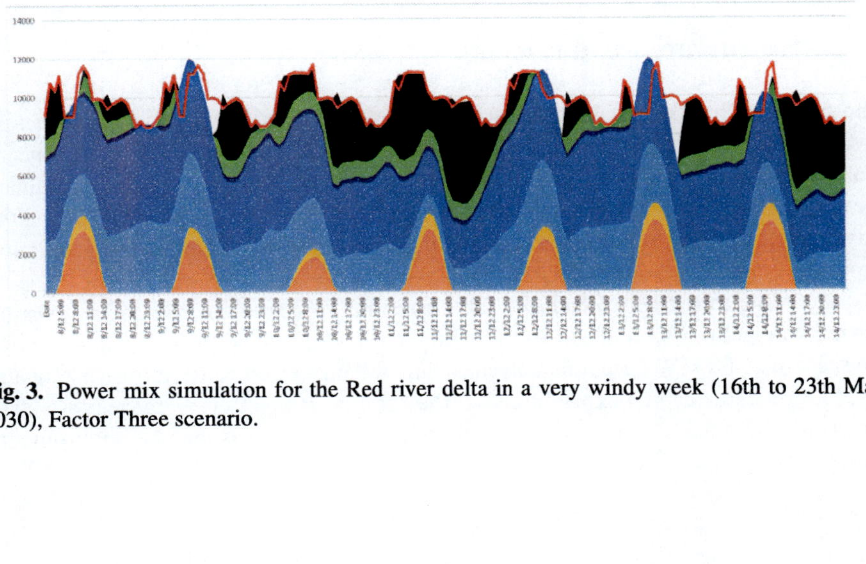

Fig. 3. Power mix simulation for the Red river delta in a very windy week (16th to 23th May 2030), Factor Three scenario.

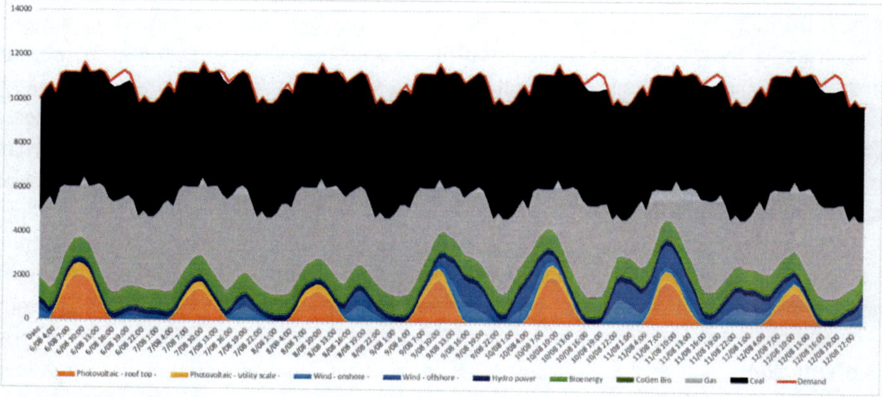

Fig. 4. Power mix simulation for the Red river delta in a very low wind week (6th to 13th August 2030), Factor Three scenario. Imported electricity can be seen as white field between the red load curve and the black marked coal generation.

Economic flexibility solutions aim to level the playing field for all technologies, replacing must-run and priority dispatch rules with competitive electricity markets. Mature markets include liquid intra-day trading, the balancing and ancillary services market, and the future and derivative markets. It is paramount that curtailment decisions remain transparent.

Technological flexibility solutions include: the flexible operation of thermal power plants; smart demand management; stronger interconnections with neighboring provinces and countries; hydro or chemical storage; and switching from coal to gas.

All the solutions cited above are or should be considered the PDP8. In preparation for 2050, Vietnam must also consider more innovative technical solutions such as power-to-gas; power-to-hydrogen; or vehicle to grid storage. Floating gas power plants, although not a part of the Masterplan, offer high flexibility at multiple timescales and complement offshore wind well at low investment cost.

7 Infrastructure Requirements

The rapid development of Vietnam electricity grid needs national-scale infrastructure investment. The next power development plan is an opportunity to optimize the conceptual design and prepare the financing of this infrastructure. The plan serves to coordinate the anticipations of the government, the State owned enterprises EVN and PVN and the private sector, towards the prospect of having many GW of offshore wind generation capacity installed in the next decade.

According to the IEA Wind, the 2017 baseline infrastructure cost for an existing offshore project was around 638 €/kW, with total capital expenditure of 3500 €/kW. At the scale of a 500 MW farm, this translates into 320 million euros for infrastructure and 1.75 billion euros capital expenditure. At the national scale, 10 GW capacity by 2030 means to invest 6.4 billion euros in offshore infrastructure. It is thus highly profitable to find ways to optimize these costs. Besides link capacity, distance from shore and technological progress, network topology is a key driver of the cost.

The naive approach is to connect each offshore wind farm to the inland grid independently. The ISF study [2] argues that the new infrastructure should not only be seen as East-West transmission lines from wind farms offshore to the land. As an alternative topology, consider the offshore busbar approach. An undersea cable runs parallel to the coast line, and wind farms connect to it. In any case, Vietnam needs to reinforce its electricity transmission network integrating the North, Center and South electric systems to accommodate its expansion. Part of the new network North-South capacity could be submarine. This National Offshore Power Link would transport the power to the two load centers in the north and south and – with lower capacity interconnects – to the other coastal provinces in between.

Figure 5 shows a possible position of this offshore wind cable which has interconnections with six currently existing grid knots to each of the six coastal regions. The total length of this cable would be 1775 km, including land connections.

Existing undersea projects confirm that such infrastructure is technically possible. To date, Ultra High Voltage cables have the largest transmission capacity of around 1 GW with voltage levels of typically 1000 kV per cable. The NordLink between Norway and Germany is over 620 km long, has a 1400 MW capacity, and is estimated to cost 1.5 to 2 billion euros (Skopljak 2019). The North Sea Link between Norway and UK at 730 km long and a 1400 MW capacity is estimated to cost 2 billion euros (North Sea Link 2019). The EuroAsia Interconnector cable will have 2000 MW capacity and a

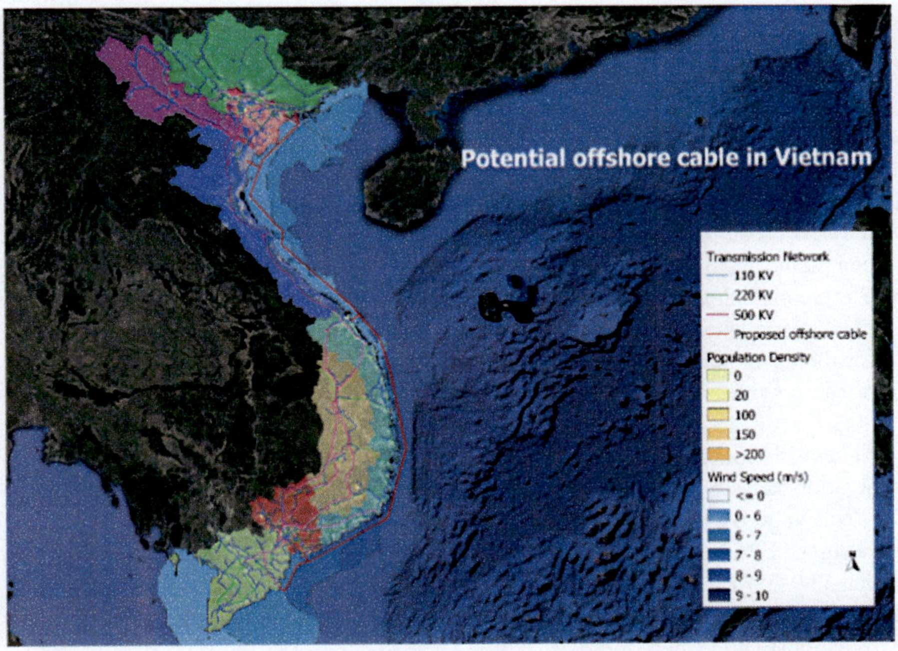

Fig. 5. How an offshore busbar could fit in the national power grid topology.

submarine segment of 1208 km. The construction cost of stage 1, with half capacity, is expected to cost 2.5 billion euros (EuroAsia interconnector 2019).

Having a strategic infrastructure, not only may decrease the offshore projects cost efficiency, it would also send signal that Vietnam is committed to generate electricity from offshore and near shore wind farms, and increase the' confidence of investors to build up a local industry. Diminishing returns due to the scarcity of the good onshore site will be felt in the long term, which increases the attractiveness of offshore wind. An undersea cable to reinforce the North-Center-South power system interconnection does not require sensitive discussions about land-use. The undersea interconnectors may feed discussions about the regional ASEAN super grid, which stands to be an important source of system flexibility and reliability at the 2050 planning horizon.

8 Conclusion and Policy Implications

Vietnam is in the early stage of forming a marine economic development strategy in general and an offshore wind development plan in particular. Near-shore and offshore wind development is one of the key levers to boost the marine economy and protect the sovereignty of Vietnamese seas and islands. The following national plans define the role and orientation of wind offshore development strategies:

- The National Marine Spatial Plan for 2021–2030 with a view to 2050. This plan provides the legal basis for the management and exploitation of offshore wind potential, for many aspects (i) the development and selection of marine spatial exploitation scenarios, to co-construct the near-shore and offshore power generation policies; (ii) maps for resource exploitation and maps for infrastructure development within marine space, to allocate zones for near-shore and offshore wind farms, for undersea cables and for seaports that serve the energy sector.
- The National Power Development Plan for 2030 with vision to 2045 should specify annual targets for offshore wind development, consider the needs and opportunities of having submarine segments in the power transmission network, and coordinate the new offshore infrastructure with the new LNG power centers to maximize their complementarity.
- The National Seaports Plan for 2030 with a vision to 2045 should improve the capacity of the seaport systems to support the energy sector. The plan should have specific goals for the development of seaports that provide services for the energy sector such as offshore wind combined with logistics and business services for other energy types.

There is economic and technical potential to build dozens of GW of wind power generation capacity in Vietnam. After fifteen years of growing pains, the sector is taking off decisively. By mid 2021, only 538 MWp of wind power are operating, but 6 GW have signed PPA. And the total potential capacity based on projects is over 67.7 GW, of which 32.9 GW offshore. EVN expect to have about 5 GW of wind power connected by the end of 2021, before November to benefit from the preferential FIT of 85 USD/MWh.

Previously published studies, the Old plan scenario and the third draft of the Power Development Plan VIII underestimate the role of offshore wind power. Offshore wind energy can contribute significantly to the mid- to long-term Vietnamese energy mix. We believe that the government will auction several 500 MW offshore wind farms during the next decade. The most plausible of the three scenarios discussed in this paper is the New Normal: an installed capacity in 2030 of 15 GW for wind onshore and 10 GW for wind offshore. Wind offshore in the scenario includes two actions: a phase of progressive seaward extension of existing near-shore projects, and the construction of new large-scale "true offshore" projects delivering from 2025.

While offshore has more potential in the long run, in the next few years the onshore and near-shore will dominate installations because the technology is more mature and smaller scale projects are easier to conduct and integrate in the grid.

These numbers are orders of magnitude, more detailed analysis is needed to better understand how they will fit into Vietnam's fast moving energy supply and demand landscape. Moreover, one must identify the policy mechanisms for implementing the scenario at a reasonable cost to consumers [36] The two actions in the New normal scenarios may require two different mechanisms. For "true offshore projects" the industry [39] proposes a FIT with a 2026 deadline and a 4–5 GW capacity cap, before moving into an auction system. This negotiated tariff would compare to the negotiated PPAs terms given to thermal power projects. The land-use constraints for onshore wind require innovative solutions that respect agricultural needs. Finally, the reliability of offshore wind turbines in East Asian conditions must be demonstrated.

When it comes to policy discussion for 2030, while a scenario can be a useful anchor, the indication of a plausible range instead of a point estimate allows for more robust planning. To propose such ranges, we have to keep in mind a few points. First, today's pipeline recognizes 67.7 GW of potential wind power projects. Second, offshore has more potential to scale up than onshore. Third, 2030 is in ten years, while it takes 2 to 5 years to realize a projects. Considering these, we propose that the target for wind power in Vietnam disregards the old objectives in the PDP7, and focuses on 2030 installed capacity ranges of: 12–15 GW for onshore, 10–12 GW for offshore.

If the government sets an ambitious goal for 2030, we propose a few elements for the roadmap to get there, see [36–38] for more detailed studies. To help the pioneer true offshore wind projects in Vietnam, they should be located in priority in the most windy area shown on Fig. 1 [7], the South Central coast. However the Northern region also has attractive conditions for offshore wind power generation, they should be considered as soon as the technology is demonstrated.

To conclude, we offer three recommendations for the PDP8. First, wind energy targets for 2030 should be increased considerably over those in previous plans. Second, considering the amount of variable power generation sources already under construction and the number already in planning, increased flexibility of the electric system must be paramount in PDP8. Third, an infrastructure development plan for deploying several additional GW of offshore wind energy per year should start immediately.

Acknowledgements. Lead author HDM is responsible for the framing, data collection and the scenario narratives. The ISF study team is responsible for modeling simulations. All authors contributed to the reviewing. Authors declare no conflict of interest. We acknowledge financial support from the European Climate Foundation grants.

References

1. Ha-Duong, M.: Quy hoạch điện 8: Ưu tiên gì để không phải chịu hệ quả đắt giá? Tia Sáng, no. 6, pp. 8–12, 25 March 2021 (2021)
2. Teske, S., Morris, T., Nagrath, K.: Renewable Energy for Viet Nam - A proposal for an economically and environmentally sustainable 8th Power Development Plan for the Viet Nam Government. Institute for Sustainable Futures (ISF). University of Technology Sydney (UTS), Australia, October 2019. https://doi.org/10.5281/zenodo.3515786. Accessed 22 Oct 2019
3. Technical University of Denmark (DTU), "Global Wind Atlas 3.1. The Global Wind Atlas 3.1 is released in partnership with the World Bank Group, utilizing data provided by Vortex, using funding provided by the Energy Sector Management Assistance Program (ESMAP), 21 June 2021. https://globalwindatlas.info
4. Nguyen, K.Q.: Wind energy in Vietnam: resource assessment, development status and future implications. Energy Policy **35**(2), 1405–1413 (2007). https://doi.org/10.1016/j.enpol.2006.04.011
5. Whittaker, S.: Global picture of offshore wind. Presented at the Consultation workshop on Developing Offshore Wind in Vietnam, Ha Noi, Viet Nam, 17 October 2019
6. Doan, V.Q., et al.: Usability and challenges of offshore wind energy in Vietnam revealed by the regional climate model simulation. SOLA **15**, 113–118 (2019). https://doi.org/10.2151/sola.2019-021
7. Van Toan, D., Van, D.Q., Pham, D.A.L., Van Nguyen, D.: The zoning of offshore wind energy resources in the Vietnam sea. In: Proceedings of the 1st Vietnam Symposium on Advances in Offshore Engineering, vol. LNCE 18, pp. 1–7 (2019). https://doi.org/10.1007/978-981-13-2306-5_34
8. Quang, V.D., et al.: Evaluation of resource spatial-temporal variation, dataset validity, infrastructures and zones for Vietnam offshore wind energy. VJSTE **62**(1), 3–16 (2020). https://doi.org/10.31276/VJSTE.62(1).03-16
9. Institute of Energy: Quy hoạch phát triển điện lực quốc gia thời kỳ 2021–2030 tầm nhìn đến năm 2045 - Báo cáo dự thảo Lần 3 - TẬP 1: THUYẾT MINH CHUNG. Ha Noi, Viet Nam: Ministry of Industry and Trade, February 2021
10. Minh, H.D., Teske, S., Dimitri, P., Mentari, P.: Options for wind power in Vietnam by 2030. VIET, October 2019. http://www2.centre-cired.fr/IMG/pdf/options_for_wind_power_vn_2030.pdf
11. Ha-Duong, M.: List of wind power projects in Vietnam. Zenodo, 23 January 2021. https://doi.org/10.5281/ZENODO.3698080
12. Thành, V.N.: Decision 68/QD-UBND on approving the list of investment of the province in the period 2020–2021. People's Committee of the Province, Gia Lai, Decision 68/QD-UBND, February 2020
13. VietnamEnergy.vn: What situation of the wind power projects in 2021? http://nangluongvietnam.vn/, 27 March 2021. http://nangluongvietnam.vn/news/en/nuclear-renewable/what-situation-of-the-wind-power-projects-in-2021.html. Accessed 21 June 2021
14. Dũng, T.D.: Letter 196/2021/TTg-CN. Prime Minister, The Government of Vietnam, Letter 196/TTg-CN, February 2021
15. Gerdes, J.: Record-Breaking Massachusetts Offshore Wind Auction Reaps $405 Million in Winning Bids, 17 December 2018. https://www.greentechmedia.com/articles/read/record-breaking-massachusetts-offshore-wind-auction. Accessed 19 Oct 2019
16. Asimov, N.: In Race for Offshore Wind, Three New Bids. The Vineyard Gazette - Martha's Vineyard News, 16 September 2019. https://vineyardgazette.com/news/2019/09/16/race-offshore-wind-farms-around-marthas-vineyard-continues. Accessed 19 Oct 2019

17. Noonan, M., et al.: IEA Wind TCP Task 26: Offshore Wind Energy International Comparative Analysis. IEA Wind, NREL/TP-6A20-71558, October 2018. www.nrel.gov/publications
18. Phúc, N.X.: Decision 39/2018/QD-TTg - Amending several articles of decision No.37/2011/QD-TTG dated June 29, 2011 of the prime minister on provison of assistance in development of wind power projects in Vietnam. The Government of Vietnam, Hanoi, Vietnam, Decision 39/2018/QD-TTg, September 2018. http://vanban.chinhphu.vn/portal/page/portal/chinhphu/hethongvanban?class_id=1&mode=detail&document_id=194743
19. Nangluongvietnam.vn: Bộ Công Thương trả lời kiến nghị VEA về 'giá FIT cho điện gió, mặt trời'. Nangluongvietnam.vn, 06 January 2021. http://nangluongvietnam.vn/news/vn/dien-hat-nhan-nang-luong-tai-tao/nang-luong-tai-tao/bo-cong-thuong-tra-loi-kien-nghi-vea-ve-gia-fit-cho-dien-gio-mat-troi.html. Accessed 21 June 2021
20. Định Công: Đề xuất kéo dài cơ chế giá FIT điện gió đến hết năm 2023. Tạp chí Công Thương, 13 November 2020. http://tapchicongthuong.vn/bai-viet/de-xuat-keo-dai-co-che-gia-fit-dien-gio-den-het-nam-2023-76485.htm. Accessed 21 June 2021
21. GreenID: Analysis of future generation capacity scenarios for Vietnam. GreenID, Vietnam, June 2018. http://en.greenidvietnam.org.vn/view-document/5b13fb0a5cd7e89403c94bda
22. Danish Energy Agency: Vietnam Energy Outlook Report 2017, MOIT (2017). https://ens.dk/sites/ens.dk/files/Globalcooperation/Official_docs/Vietnam/vietnam-energy-outlook-report-2017-eng.pdf. Accessed 5 Oct 2019
23. Lundsager, J.S., Venturini, G., Linh, N.H., Sørensen, S.S.: Vietnam Energy Outlook Report 2019, Hanoi, November 2019. https://ens.dk/en/our-responsibilities/global-cooperation/country-cooperation/vietnam
24. Economic Consulting Associates (ECA): Made in Vietnam Energy Plan 2.0. Report prepared for the Vietnam Business Forum, Power and energy working group, Ha Noi, Viet Nam, December 2019. http://auschamvn.org/wp-content/uploads/2016/10/Made-in-Vietnam-Energy-Plan-MVEP-v12.pdf
25. ADB: Pathways to low-carbon development for Vietnam (2017). https://www.adb.org/sites/default/files/publication/389826/pathways-low-carbon-devt-viet-nam.pdf
26. Breu, M., Castellano, A., Frankel, D., Rogers, M.: Exploring an Alternative Pathway for Vietnam's Energy Future. McKinsey & Company (2019)
27. Tran, K.H., Kitchlu, R., Chu, T.B.: Offshore wind roadmap for Vietnam. The World Bank, Washington DC, June 2021. https://documents1.worldbank.org/curated/en/261981623120856300/pdf/Offshore-Wind-Development-Program-Offshore-Wind-Roadmap-for-Vietnam.pdf. Accessed 10 June 2021
28. Lee, J., Zhao, F.: Global Offshore Wind Report 2021. Global Wind Energy Council (GWEC), Brussels, Belgium, March 2021
29. REN21: Renewables 2021 Global Status Report. REN21 Secretariat, Paris, ISBN 978-3-948393-03-8, June 2021. https://www.ren21.net/reports/global-status-report/. Accessed 15 June 2021
30. Vietnam's Politburo: Resolution 55-NQ/TW on the orientation of the National Energy Development Strategy of Vietnam to 2030, vision to 2045. Central Executive Committee, Communist Party of Vietnam, Resolution of the Political Bureau, February 2020
31. Intelligent Energy Systems Pty Ltd and Mekong Economics. Socialist Republic of Viet Nam Power Sector Scenarios, vol. 6. WWF (2016)
32. Richard, C.: Curtailment reduction 'could double Chinese wind fleet'. Wind Power Monthly, 23 August 2019. https://www.windpowermonthly.com/article/1594722?utm_source=website&utm_medium=social. Accessed 30 Sept 2019
33. Liu, Z., Zhang, W., Zhao, C., Yuan, J.: The economics of wind power in china and policy implications. Energies **8**(2), 1529–1546 (2015). https://doi.org/10.3390/en8021529

34. Kenning, T.: Vietnam leapfrogs Australia in solar deadline boom. PV Tech, 05 July 2019. https://www.pv-tech.org/news/vietnam-leapfrogs-australia-in-solar-deadline-boom. Accessed 19 Oct 2019
35. Le, Q.A., Duong, V.D.: Ability to release capacity for wind power and solar power up to 2022.VIET, Hanoi, Research report RR/04-VIET06.2020/ENG, June 2020
36. Nguyễn, Đ.V., Nguyễn, X.H., Nguyễn, T.T.H.: Cơ hội và thách thức trong phát triển điện gió ở Việt Nam - Opportunities and challenges to wind energy development in Vietnam. Tap Chi Khoa Hoc va Cong Nghe Viet Nam, May 2019. http://congnghiepcongnghecao.com.vn/tin-tuc/chinh-sach/t21907/co-hoi-va-thach-thuc-trong-phat-trien-dien-gio-o-viet-nam.html. Accessed 11 Oct 2019
37. Nguyen, H.P., Ngo, T.T.N., Nguyen, T.H.A.: Offshore Wind Integration Study – Review International Experience on Offshore Wind Integration. Vietnam Initiative for Energy Transition (VIET), April 2020. https://vietse.vn/en/publication/research-on-offshore-wind-power-integration-review-international-experience-on-offshore-wind-power-integration/
38. Tran, K.H., Kitchlu, R., Ba Chu, T.: Offshore Wind Roadmap for Vietnam. The World Bank, Washington DC, June 2021. https://documents1.worldbank.org/curated/en/261981623120856300/pdf/Offshore-Wind-Development-Program-Offshore-Wind-Roadmap-for-Vietnam.pdf
39. Li, C.-H., McCausland, A., Taylor, J., Stephenson, M.: Vietnam's Future Transition to Offshore Wind Auctions. International Best Practices and Lessons Learned. Global Wind Energy Council (GWEC), 22 July 2021. https://gwec.net/vietnams-future-transition-to-offshore-wind-auctions-international-best-practices-and-lessons-learned/

Site Characterization and Ground Modelling

A Calculation Model for Borehole Collapse Volume in Deviated Openhole in Formation with Weak Plane

Nguyen Van Hung[(✉)] and Bui Thi Thuy Linh

PetroVietnam University, 762 Cach Mang Thang 8, Long Toan Ward, Ba Ria, Vietnam
hungnv@pvu.edu.vn

Abstract. Borehole breakout is well known from drilling operations. The borehole often assumes an elongated or elliptical shape mainly because the normal borehole stresses are anisotropic. This very same mechanism applies to sand production which is actually a wellbore collapse. In order to quantify the risk associated to wellbore collapse a reliable estimate of the collapse volume is necessary. In this study, we investigate the effects of in-situ stresses and rock properties combined with the optimal well geometry. The well is initially circular, but the collapsed state should be modelled with elliptical geometry. The study shown that borehole collapse is a shear-type failure that occurs at low wellbore pressure when the tangential stress become large, ultimately resulting in failure. The volume of produced sand is then computed from the equilibrium state of the borehole as the volumetric difference between the ellipse and a circular hole. The collapse volume decreases with the increase of borehole fluid density and is more sensitive to the change of fluid density when fluid density is near the equivalent density of pore pressure.

Keywords: Borehole breakout · Sanding · Weak plane

1 Introduction

During the drilling wells in the world, the wellbore collapse is the main instability problems. The borehole collapse problems include tight hole while tripping, stuck logging tools, and subsequent fishing, stuck pipe, and sidetracking are time consuming and expensive. Many times, because of severe stuck pipe problem and unsuccessful fishing operations, the well has to be drilled with more than one sidetrack or, in the worst case, the wellbore has to be abandoned (Abbas et al. 2018). Borehole collapse is a shear type wellbore failure that occurs at low wellbore pressure when the tangential stress becomes large, ultimately resulting in failure. Rock fragments fall off the wellbore wall, often leaving an oval shaped hole due to anisotropic normal stresses acting on the wellbore. In fact, wellbore collapse and sand production are the same type of failures; they just take place at different operational phases of well drilling and operation. While the sand production is initiated during drilling phase or during well production time, the wellbore collapse is responsible with the decrease of wellbore pressure. As the result, the models

D. V. K. Huynh et al. (Eds.): VSOE 2021, LNCE 208, pp. 69–79, 2022.
https://doi.org/10.1007/978-981-16-7735-9_5

are used to describe the sand production, but they are equally applicable for wellbore collapse studies (Aadnoy 2010).

The effects of in situ stresses and rock properties combined with the optimal well geometry (horizontal, vertical or inclined wells) are usually used to investigate wellbore collapse features. The well is initially circular using the Kirsch equation, but the collapsed state should be modelled with elliptical geometry as developed and discussed in several papers (Liu et al. 2014; Abbas et al. 2019; Aadnoy and Looyeh 2011; Gholami et.al. 2016; Wu et al. 2005; Aadnoy and Kaarstad 2010; Alkamil et al. 2017). After a wellbore fails, borehole stresses are anisotropy. Recently the elliptical solution, the volume of sand production produced in a depletion phase is then computed from equilibrium state of the borehole. The models are explicitly analytical and simple to use.

In solid mechanics, the effect of biaxial loading on circular and elliptical holes including internal pressure has been studied for many years. Studying the stress concentration is important when, for example, designing the optimal shape of an airplane window in the biaxially loaded cabin. Similarly, the stress concentration around a borehole is crucial for determining the optimal shape. In reality, the borehole is always drilled as a circular hole, and the stress concentration around the hole is affected by the in-situ stresses, formation pore pressure, and irregularities on the borehole wall. If the resulting stress concentration around the borehole wall varies sufficiently, the borehole will try to change its geometry such that the hole becomes stable with a minimum of stress concentration variation. This will happen either as a deformation of the borehole, or as wellbore collapse (Aadnoy and Looyeh 2011).

The drilling mud is normally chosen in such a way to resist the formation pore pressure, hence preventing the formation fluid from flowing into the wellbore. Not always the drilling mud can resist the compressive stresses of the wellbore. In this case, shear failures of rock or borehole collapse will occur due to the imbalance between stress and rock strength, where stresses around the wellbore exceed the compressive strength of the rock due to lack of pressure support from the drilling fluid. The breakout pressure is defined as the least mud pressure that can avoid this enlargement and elongation of the wellbore. According to this definition, the enlargement of the wellbore cross section does not occur if the mud pressure is larger than the breakout pressure. The wellbore enlargement issue, if not treated in time, can lead to catastrophic wellbore enlargement cases such as borehole washout. The wellbore enlargement can be identified using 4- or 6-arm caliper tools, resistive based borehole image log, acoustic image log, optical imaging tool and downhole video (Abbas et al. 2019).

A better understanding of the share failure mechanism should allow for prediction of the volume of borehole collapse more realistically. However, this is a complex mechanism as shear failure is impacted by various parameters, including geological, geomechanical and fluid characteristics of the formations as well as the initial state of stresses, pressure conditions, wellbore completions and boundary conditions (Van de Hoek et al. 2000). In recent years, attention has now focused on establishing methodologies for predicting the volume of borehole collapse or the extent of rock failure around the borehole during the drilling and depletion stages of oil and gas reservoirs. In 2011, Jin et al. concluded that when pore pressure is very large, the effective radial stress may be tensile stress. However, this study did not take into account the circumferential stress analysis and collapse volume. As shown by Escobar et al. in (2014), the uses of the caliper reading as an indicator of the breakout radius illustrated that the area of a triangle as the caving area of wellbore. However, there was a significant discrepancy of 27% between their analytical and numerical solution.

The redistributed stresses are normally referred to as the hoop stress or circumferential stress, σθ (tangential stress), the radial stress, σr, and the axial stress σa. The radial stress acts in all directions perpendicular to the wellbore wall, the tangential stress circles the borehole, and the axial stress acts parallel to the wellbore axis (Fig. 2). In directional wells an additional shear component, τθz is generated.

The aim of this paper is to present a methodology based on changes in the geometry of boreholes to calculate wellbore collapse volume in openhole during drilling and depletion phases of horizontal well with the strength anisotropic formation.

2 Borehole Stresses

Before a well is drilled the rock is in a state of equilibrium. The stresses in the Earth under these conditions are known as the far field stresses. One excavated, the static stress state becomes disturbed as the support originally offered by the drilled out rock is replaced by the hydraulic pressure of the drilling mud, so causing instability in the rock formation, and the disturbed in-situ stress state therefore imposes a different set of stresses in the area of excavation. Figure 1 illustrates a schematic of in-situ stresses existing in the formation around a wellbore (Aadnoy and Looyeh 2011). The stresses can be resolved into a vertical or overburden stress, σv, the maximum horizontal in-situ stress, σH, and the minimum horizontal in-situ stress, σh, which are generally unequal. The direction well is modeled as shown in the Fig. 3. The stresses at the wellbore all can be calculated in the following steps: (1) Identify the principal in-situ stress state (σv, σH, σh); (2) Transform the stress state (σv, σH, σh,) to the stress state (σx, σy, σz). defined with respect to the Cartesian coordinate system attached to the wellbore (Eq. 1); (3) Find the local stress state (σr, σθ, σa) with respect to the cylindrical coordinated system attached to the wellbore at the distance of a from the center of well, in terms of the stress state (σx, σy, σz) (Eq. 2); (4) Find the stress state at the wellbore (σr, σθ, σa) (a = r). For a vertical well, the local stresses can be calculated from the Eq. 3, as below Eq. 4.

$$\begin{aligned}
\sigma_x &= \left(\sigma_H cos^2 a_w + \sigma_h sin^2 a_w\right) cos^2 i_w + \sigma_v sin^2 i_w \\
\sigma_y &= \sigma_H sin^2 a_w + \sigma_h cos^2 a_w \\
\sigma_z &= \left(\sigma_H cos^2 a_w + \sigma_h sin^2 a_w\right) sin^2 i_w + \sigma_v cos^2 i_w \\
\tau_{xy} &= \frac{1}{2}(\sigma_H - \sigma_h) sin2a_w cosi_w \\
\tau_{xz} &= \frac{1}{2}\left(\sigma_H cos^2 a_w + \sigma_h sin^2 a_w - \sigma_v\right) sin2i_w \\
\tau_{yz} &= \frac{1}{2}(\sigma_H - \sigma_h) sin2a_w sini_w
\end{aligned} \tag{1}$$

$$\sigma_r = \frac{1}{2}(\sigma_x + \sigma_y)\left(1 - \frac{a^2}{r^2}\right) + \frac{1}{2}(\sigma_x - \sigma_y)\left(1 + 3\frac{a^4}{r^4} - 4\frac{a^2}{r^2}\right) cos2\theta + \tau_{xy}\left(1 + 3\frac{a^4}{r^4} - 4\frac{a^2}{r^2}\right) sin2\theta + \frac{a^2}{r^2} P_w$$

$$\sigma_t = \frac{1}{2}(\sigma_x + \sigma_y)\left(1 + \frac{a^2}{r^2}\right) - \frac{1}{2}(\sigma_x - \sigma_y)\left(1 + 3\frac{a^2}{r^2}\right)cos2\theta - \tau_{xy}\left(1 + 3\frac{a^2}{r^2}\right)sin2\theta - \frac{a^2}{r^2}P_w$$
$$\sigma_a = \sigma_z - 2\nu(\sigma_x - \sigma_y)\frac{a^2}{r^2}cos2\theta - 4\nu\tau_{xy}\frac{a^2}{r^2}sin2\theta$$
$$\tau_{\theta z} = (\tau_{yz}cos\theta - \tau_{xz}sin\theta)(1 + \frac{a^2}{r^2})$$
$$\tau_{r\theta} = [\frac{1}{2}(\sigma_x - \sigma_y)sin2\theta + \tau_{xy}cos2\theta](1 - 3\frac{a^4}{r^4} + 2\frac{a^2}{r^2})$$
$$\tau_{rz} = (\tau_{xy}cos\theta + \tau_{yz}sin\theta)(1 - \frac{a^2}{r^2}) \quad (2)$$

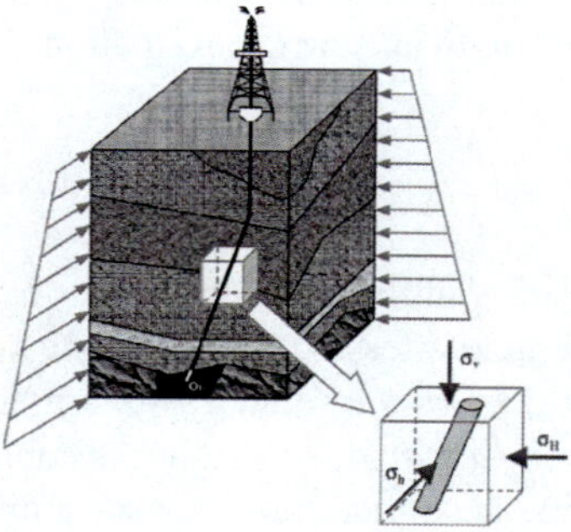

Fig. 1. A schematic showing in-situ stresses around a wellbore (Aadnoy and Looyeh 2011)

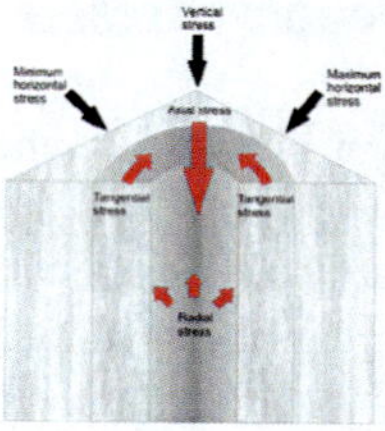

Fig. 2. Position of stresses around a wellbore in the rock formation (Borivoje et al. 2007)

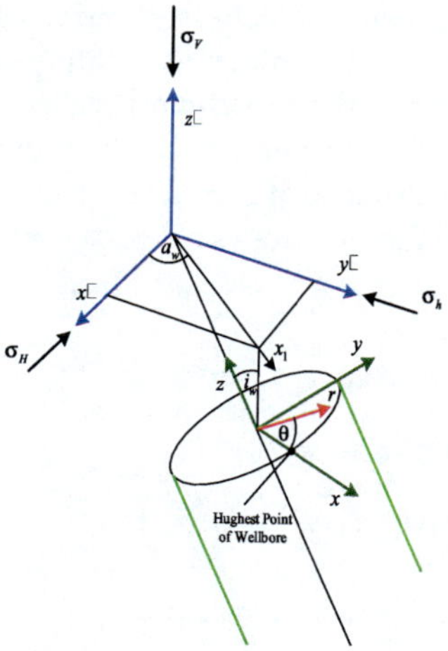

Fig. 3. The coordinate system for the in-situ stress display (Borivoje et al. 2007)

$$
\begin{aligned}
&\sigma_r = P_w \\
&\sigma_t = (\sigma_x + \sigma_y) - 2(\sigma_x - \sigma_y)cos2\theta - 4\tau_{xy}sin2\theta - P_w \\
&\sigma_a = \sigma_z - 2v(\sigma_x - \sigma_y)cos2\theta - 4v\tau_{xy}sin2\theta \\
&\tau_{\theta z} = 2(\tau_{yz}cos\theta - \tau_{xz}sin\theta) \\
&\tau_{r\theta} = 0; \tau_{rz} = 0)
\end{aligned}
\tag{3}
$$

$$
\begin{aligned}
&\sigma_r = P_w \\
&\sigma_t = (\sigma_x + \sigma_y) - 2(\sigma_x - \sigma_y)cos2\theta - P_w \\
&\sigma_a = \sigma_z - 2v(\sigma_x - \sigma_y)cos2\theta \\
&\tau_{\theta z} = 0; \tau_{r\theta} = 0; \tau_{rz} = 0)
\end{aligned}
\tag{4}
$$

3 Elliptical Wellbore for Collapse Volume Calculation

Borehole are expected to have a circular shape, but in practice, due to stress concentrations around the borehole, the borehole tends to change its geometry to reach a new state of stability. To maintain the rock mass in a state of equilibrium, the borehole is filled with drilling fluid to build a new stress pattern around the borehole. Generally, the optimal shape is circular during the life of a well and may become elliptical due to shear failure. This concept is depicted in Fig. 4.

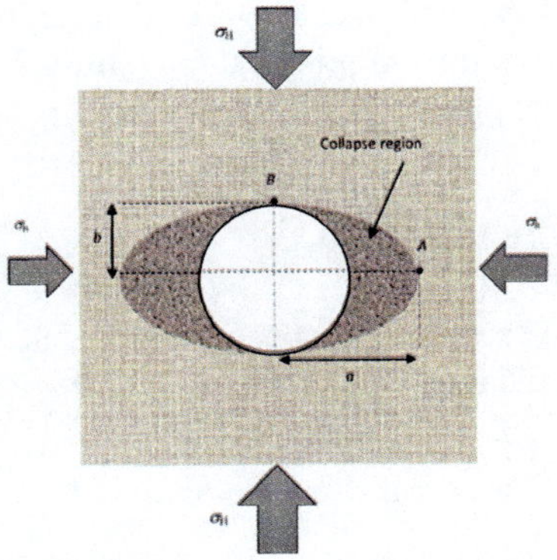

Fig. 4. Initial circular hole and final elliptical hole (Abbas et al. 2019)

In case of normal fault stress state (σH = σh), the equilibrium is obtained for a circular hole. However, the in-situ stress is anisotropic (i.e. the principal stresses normal to the borehole axis not be equal). The result is that the hole will be stable when it obtains an elliptical shape as shown in Fig. 4. In addition, when the fluid pressure in the borehole is equal to the pore pressure, there is no external load on the formation. Therefore, the external load exerted by the wellbore fluid is equal to the pressure difference between the wellbore pressure and the pore pressure. One of the approaches conventionally used for determination of the maximum and minimum tangential stresses in the elliptical borehole is developed by Lekhnitskii (1968) and formulated as Eq. 5:

$$\sigma_A = (1 + 2c)\sigma_H - \sigma_h - (\frac{2}{c} - 1)P_w$$
$$\sigma_B = \left(1 + \frac{2}{c}\right)\sigma_h - \sigma_H - (2c - 1)P_w \quad (5)$$

where σ_A and σ_B are the tangential stresses at points A and B around the wellbore, respectively and c is a function of borehole diameters which is defined as:

$$c = \frac{b}{a} \quad (6)$$

The borehole is considered to be stable when the tangential stress is uniform around the ellipse. Setting $\sigma_A = \sigma_B$, the parameter c now defines as below:

$$c = \frac{\sigma_h + P_w}{\sigma_H + P_w} \quad (7)$$

Borehole collapse is induced when the applied mud pressure is less than the breakout pressure, the compression condition is dominant, and borehole breakout failure may occur because the mud pressure supporting the borehole wall is not high enough. Shear failure during drilling occurs once the maximum induced tangential stress point A (σ_A), as shown in Fig. 4, is higher than the radial stress (Eq. 3). The worst scenario for shear failure is applied in order to determine the parameter c during drilling. This means that the tangential ($\sigma_A\ or\ \sigma_\theta$), axial ($\sigma_z$) and radial ($\sigma_r$) stresses are, respectively, the maximum (σ_1), intermediate (σ_2) and minimum principal (σ_3) stresses around the borehole. This is because collapse will take place at point A.

4 Volume of Borehole Collapse

From the Fig. 4, the collapse volume produced can be calculated from the difference between the shape before and after collapse:

$$V = \pi\,\text{ab} - \pi b^2 = \pi\frac{1 - c}{c}b^2 \quad (8)$$

where V is the rate of sand production per unit length of the wellbore (m^3/m).

Having the parameter c (Table 1), the volume of borehole collapse can be calculated using the Eq. 8.

When the shape of borehole is circular (c = 1), the most dangerous point for wellbore collapse is the point A: $\sigma_A = 3\sigma_H - \sigma_h - P_w$). At the initial stage of production, the pore pressure at the borehole wall becomes equal to the wellbore pressure and, as a result, the effective radial stress becomes zero ($\sigma_r = P_w - P_p$, where P_p is the pore pressure).

Table 1. Parameter c for difference failure criteria

No.	Failure criteria	Note
1	*Mohr-Coulomb*	Angle of internal friction is equivalent to the angle of inclination of a surface sufficient to cause sliding of a superincumbent block of similar material down the surface
	$\tau' = \tau_0 + \sigma' tan\phi$ $\tau = \frac{1}{2}(\sigma_1' - \sigma_3{}')cos\phi$ $\sigma' = \frac{1}{2}(\sigma_1' + \sigma_3') - \frac{1}{2}(\sigma_1' - \sigma_3')sin\phi$	
	$\sigma'_3 = \sigma'_r = P_w - P_p = 0$ $\sigma'_1 = 2\tau_0 \frac{cos\phi}{1-sin\phi}$ $\sigma_A = (1+2c)\sigma_H - \sigma_h - (\frac{2}{c} - 1)P_w$	
	$\sigma'_1 = \sigma_\theta = \sigma_A$	
	$c = \frac{-Y+\sqrt{Y^2-4XZ}}{2X}$	
	$X = 2\sigma_H$ $Y = \sigma_H - \sigma_h - P_w - P_p - 2\tau_0 \frac{cos\phi}{1-sin\phi}$ $Z = -2P_w$	
2	*Hoek-Brown*	Reasonable match brittle failure, Naturally fractured formation; Poor results in ductile failure
	$\sigma'_1 = \sigma'_3 + \sqrt{I_f \sigma_c \sigma'_3 + I_i \sigma_c^2}$ I_f: Frictional index; σ_c: crack stress parameter; I_i: intact index	
	$\sigma'_3 = \sigma'_r = P_w$	
	$\sigma_A = (1+2c)\sigma_H - \sigma_h - (\frac{2}{c} - 1)P_w = \sigma'_1$	
	$c = \frac{-Y+\sqrt{Y^2-4XZ}}{2X}$	
	$X = 2\sigma_H$ $Y = \sigma_H - \sigma_h - P_w - P_p - \sqrt{I_f \sigma_c \sigma'_3 + I_i \sigma_c^2}$ $Z = -2P_w$	
3	*Mogi-Coulomb*	Best fits the triaxial and polyaxial test data
	$\tau_{oct} = k + m\sigma_{oct}$ $\tau_{oct} =$ $\frac{1}{3}\sqrt{(\sigma_1 - \sigma_2)^2 + (\sigma_1 - \sigma_3)^2 + (\sigma_2 - \sigma_3)^2}$ $= \sqrt{J_2}$ $\sigma_{oct} = \frac{1}{3}(\sigma_1 + \sigma_2 + \sigma_3)$ k, m: rock material constants	
	$c = \frac{-Y+\sqrt{Y^2-4XZ}}{2X}$	

(continued)

Table 1. (*continued*)

No.	Failure criteria	Note
	$X = 4\sigma_H$ $Y = \frac{(ab-A)+\sqrt{A^2-CF}}{C} - \sigma_H - \sigma_h - P_w$ $A = \frac{m^2}{2}\sigma_z - 2P_w - 2\sigma_z - ab$ $C = \frac{8}{9} - 2m^2$ $F =$ $\left(\frac{2}{9} - \frac{m^2}{4}\right)\sigma_z + \frac{2}{9}P_w - 2\sigma_z P_w - k^2 + mkP_w$	

5 Results and Discussion

In this analysis, we study the influence of weak planes in the formation. For more definitions and the effects of weak bedding planes can be seen in He et al. (2015).

Based on the model presented previously, the effect of fracture trend and dip angle of weak plane in the formation on the volume of wellbore collapse were analyzed. The results are presented below (see Table 3).

Table 2. Defines the properties of a horizontal well properties

Property	Value
Depth (m)	6000
Overburden stress (MPa/100 m)	2.4
Major horizontal stress (MPa/100 m)	2.08
Minor horizontal stress (MPa/100 m)	1.9
Pore pressure (MPa)	70
Cohesion of the weak planes (MPa)	3.5
Internal friction angle of weak planes (°)	14
Cohesion strength of rock (MPa)	18
Internal friction angle (°)	32
Poisson's ratio	0.22
Biot's coefficient	0.9
Wellbore diameter (mm)	140
Direction of the maximum horizontal principle in-situ stress (°)	115
Fluid density (g/cm^3)	1.22

The ultimate ellipse defined by Ea.7 yields c = 0.95 while a well having the properties of Table 2 yield c* = 0.97. Because the equations of c in the Table 2 are the function of the cohesion strength, the degree of consolidation will play an important effect on how much the ellipse is allowed to develop. Wellbore stability or sand production depends on the dip of weak plane and the azimuth of wellbore. If the fracture trend angle changes from 00 to 900 the collapse volume varies between 0.004 to 0.028 m^3/m, 0 to 0.033 m^3/m, 0 to 0.04 m^3/m for the dip angle 300, 450, 600 respectively. Thus, collapse around the wellbore may occur so that the hole becomes larger in all directions. The change in wellbore collapse volume will increase as the increase of weak plane dip. The volume of wellbore collapse is smaller when the dip of weak plane is less than 300.

It is the stress anisotropy that is the major critical factor for the elliptic shape. If the well was drilled vertical, the stress anisotropy between the major horizontal stress and minor horizontal stress would result in ellipses of c = 0.86 and c* = 0.97.

The effect on borehole collapse is that a hole first will be broken until a stable elliptic hole is obtained, then as depletion or decrease of wellbore pressure take place a further change in shape of the hole will result in some more borehole collapse. Depletion is a slow process, so the second phase of borehole collapse will also be very slow. In addition, the formation is not completely homogeneous, the borehole collapse as a step function even if the wellbore pressure or borehole fluid density decreases. It is interesting to note that during depletion the horizontal stresses change while the overburden stress remains constants. This results in a higher degree of anisotropy for deviated and horizontal boreholes. As a result, the change in shape of the elliptic borehole will be larger for deviated well profile.

When the azimuth of wellbore in the range of 1100–2300, the dip angle of weak plane is equal 450, the wellbore stability is good. And when the dip of weak plane is of 600,

Table 3. Influence of wellbore azimuth and occurrence of weak plane on the volume of wellbore collapse (Dip = 30°)

Wellbore azimuth (radian)	Ω (rad)	Beta				V (m^3/m)			
		Fracture trend							
		0°	30°	60°	90°	Fracture trend			
		0	0.523599	1.047198	1.570796327	0°	30°	60°	90°
0	−2.094395	0.261799	0.406378	0.661374	0.911738291	0.006	0.025	0.027	0.027
0.174533	−1.919862	0.281807	0.334411	0.574191	0.831502391	0.016	0.022	0.026	0.027
0.349066	−1.745329	0.334411	0.281807	0.488133	0.747641	0.021	0.015	0.024	0.025
0.523599	−1.570796	0.406378	0.261799	0.406378	0.66137389	0.022	0.004	0.022	0.024
0.698132	−1.396263	0.488133	0.281807	0.334411	0.57419066	0.021	0.012	0.018	0.022
0.872665	−1.22173	0.574191	0.334411	0.281807	0.488132857	0.02	0.016	0.01	0.019
1.047198	−1.047198	0.661374	0.406378	0.261799	0.406377781	0.018	0.016	0.001	0.016
5.759587	3.665191	0.406378	0.661374	0.911738	1.120096327	0.025	0.027	0.027	0.027
5.934119	3.839724	0.334411	0.574191	0.831502	1.057042359	0.023	0.027	0.027	0.027
6.108652	4.014257	0.281807	0.488133	0.747641	0.987259971	0.017	0.027	0.028	0.028
6.283185	4.18879	0.261799	0.406378	0.661374	0.911738291	0.006	0.025	0.027	0.027

the azimuth of wellbore is in the range of 1100–1800, the wellbore is also comparatively stable.

We further studied the effect of fluid density on the volume of wellbore collapse under the condition of any weak planes and a given wellbore azimuth. The results illustrated that the volume of wellbore collapse increases with the decrease of fluid density, and collapse volume is more sensitive to the change of fluid density (Table 4).

Table 4. Influence of wellbore azimuth and occurrence of weak plane on the volume of wellbore collapse (Dip = 45°)

Wellbore azimuth (radian)	Ω (rad)	Beta				V (m^3/m)			
		Fracture trend							
		0°	30°	60°	90°	Fracture trend			
		0	0.523599	1.047198	1.570796	0°	30°	60°	90°
0	−2.0944	1.49E−08	0.3681	0.722734	1.047198	0.031	0.025	0.027	0.027
0.174533	−1.91986	0.123335	0.246197	0.606946	0.943665	0.032	−0.014	0.026	0.027
0.349066	−1.74533	0.246197	0.123335	0.488533	0.835226	−0.014	0.031	0.024	0.025
0.523599	−1.5708	0.3681	1.49E−08	0.3681	0.722734	0.021	0.027	0.021	0.024
0.698132	−1.39626	0.488533	0.123335	0.246197	0.606946	0.021	0.027	−0.014	0.022
4.886922	2.792527	1.092924	1.426183	1.686606	1.822976	0.019	0.017	0.024	0.022
5.061455	2.96706	0.972789	1.32099	1.610879	1.794547	0.021	0.02	0.034	0.024
5.235988	3.141593	0.850324	1.209429	1.523411	1.748507	0.023	0.023	0.015	0.027
5.410521	3.316126	0.726989	1.092924	1.426183	1.686606	0.025	0.025	0.022	0.03
5.585054	3.490659	0.604726	0.972789	1.32099	1.610879	0.026	0.026	0.025	0.04
5.759587	3.665191	0.486695	0.850324	1.209429	1.523411	0.026	0.027	0.027	0.018
5.934119	3.839724	0.379108	0.726989	1.092924	1.426183	0.025	0.027	0.027	0.025
6.108652	4.014257	0.295617	0.604726	0.972789	1.32099	0.02	0.027	0.028	0.027
6.283185	4.18879	0.261799	0.486695	0.850324	1.209429	0.006	0.026	0.027	0.027

Acknowledgment. The authors are grateful to the Petrovietnam University for financial support of the work.

References

Abbas, A.K., Flori, R.E., Anssari, A.A.L., Alsab, M.: Laboratory analysis to assess shale stability for the Zubair formation, Southern Iraq. J. Nat. Gas Sci. Eng. **56**, 315–323 (2018). https://doi.org/10.1016/j.jngse.2018.05.041

Aadnoy, B.S.: Modern Well Design. 2nd edn., 304 p. Francis and Taylor, Leiden (2010)

Aadnoy, B.S., Kaarstad, E.: Elliptical geometry model for sand production during depletion. In: IADC/SPE Asia Pacific drilling technology conference and exhibition, Ho Chi Minh City, Vietnam, 1–3 November. IADC/SPE 132689 (2010)

Aadnoy, B., Looyeh, R.: Petroleum rock mechanics: drilling operations and well design. 1st edn. 376 p. Gulf Professional Publishing (2011)

Abba, A.K., Manhalawi, A., Alameedy, U., Flori, R.E.: Using an analytical model to predict collapse volume during drilling: a case study from Southern Iraq. In: Proceedings of 53rd US Rock Mechanics/Geomechanics Symposium, New York, NY, USA, 23–26 June 2019. ARMA 19-1525 (2019)

Alkamil, E.H.K., Flori, R.E., Eckert, A.: A novel approach to predict collapse volume using image processing. In: Proceedings of 51st US rock mechanics/Geomechanics symposium, San Francisco, California, USA, 25–28 June 2017. ARMA 17-976 (2017)

Borivoje, P., Nediljka, G.M., Davorin, M.: Wellbore instability: causes and consequences. Rudarsko-GeološKo-Naftni Zbornik **19**, 87–98 (2007)

Escobar, R.G., Carrillo, H.C., Pena, Y.A.Q.: New approach for estimating cavings volume to avoid wellbore instabilities. In: International Society for Rock Mechanics, Proceedings of ISRM Conference on Rock Mechanics for Natural Resources and Infrastructure, Goiania, Brazil, 9–13 September 2014 (2014)

Papamichos, E., Stenebraten, J., Cerasi, P., et al.: Rock type and hole failure pattern effects on sand production. In: Proceedings of 42nd US rock mechanics symposium, Alexandria, Virginia (2008)

He, S., Wang, W., Zhou, J., Huang, Z., Tang, M.: A model for analysis of wellbore stability considering the effects of weak bedding planes. J. Nat. Gas Sci. Eng. **27**, 1050–1062 (2015)

Jiu, Y., Chen, K.P., Chen, M.: Development of tensile stress near a wellbore in high pressure gas flows. Int. J. Rock Mech. Min. Sci. **48**(8), 1313–1319 (2011)

Lekhnitskii, S.G.: Anisotropic Plates. Gordon and Breach, New York (1968)

Liu, Z., Chen, M., Yang, X., Lu, Y., Xiong, Q.: Calculation model for borehole collapse volume in horizontal openhole in formation with multiple weak planes. Pet. Explor. Dev. **41**(1), 113–119 (2014)

Gholami, R., Aadnoy, B., Rasouli, V., Fakhari, N.: An analytical model to predict the volume of sand during drilling and production. J. Rock Mech. Geotech. Eng. **8**(4), 521–532 (2016)

Van de Hoek, P., Kooijman, A., de Bree, P., Kenter, C., Sellmeyer, H., Willson, S.: Mechanisms of downhole sand cavity re-stabilisation in weakly consolidated sandstones. In: Proceedings of SPE European Petroleum conference, Paris, France, 24–25 October 2000 (2000)

Wu, B., Choi, S.K., Tan, C.P.: Effect of stress anisotropy on cavity strength and implications in wellbore stability and sand production analyses. In: Proceedings of 40th US Symposium on rock mechanics, Anchorage, Alaska, 25–29 June 2005. ARMA/USRMS 05-773 (2005)

A Practical Application Using Industrial Waste for Enhancing the Mechanical Properties of Expansive Soil

Liet Chi Dang[1,2](✉) and Hadi Khabbaz[2]

[1] Department of Civil Engineering, Gwangju University, 277 Hyodeok-ro, Nam-gu, Gwangju 61743, Korea
danglieet@gwangju.ac.kr

[2] School of Civil and Environmental Engineering, University of Technology Sydney (UTS), 15 Broadway, Ultimo, NSW 2007, Australia

Abstract. In this study, a series of laboratory tests was conducted to investigate the possibility of enhancing the mechanical properties of expansive soil using bagasse fibre (BF, a waste by-product of sugar industry) integrated without or with lime stabilisation as a novel, practical application of reuse of industrial waste materials for sustainability. Soil samples reinforced with three different contents of bagasse fibre ranging from 0% to 2% without or with lime combination in a range of 0–6%, were systematically prepared to assess their effect on improved engineering mechanism of expansive soil. The results revealed that BF reinforcement produced the shear strength development of reinforced soils. Moreover, a lime-BF combination provided better improvement in the shrink-swell behaviour and the compressibility of reinforced soils as compared to soils treated with lime or bagasse fibre alone. The findings also indicated that adding BF into lime-soil mixtures reduced the compressible properties of lime-treated soils. Meanwhile, excessively increasing bagasse fibre content greater than 1% caused a minor decrease in the compressibility improvement of reinforced soils. Hence, an appropriate combination of lime and BF should be determined and used as an environmental-friendly, cost-effective and green solution for stabilisation of expansive soil to facilitate sustainable civil infrastructure development.

Keywords: Expansive soil · Soil stabilisation · Bagasse fibre · Industrial waste recycling · Fibre reinforcement · Sustainable civil infrastructure

1 Introduction

Bagasse fibre is an abundant fibrous waste of the sugar-cane industry, left after the crushing of sugar-cane for juice extraction. Bagasse fibre has been used for many purposes as a combustible material for energy supply in sugar-cane factories, a pulp raw material in paper industries and building materials. Besides, BF waste is used as fuel in the cogeneration boiler to generate steam for the production of sugar as well as electricity, while producing bagasse ash. Nonetheless, due to increasing sugar production

D. V. K. Huynh et al. (Eds.): VSOE 2021, LNCE 208, pp. 80–88, 2022.
https://doi.org/10.1007/978-981-16-7735-9_6

worldwide, a huge amount of bagasse fibre and its ash are simply disposed in a landfill and now becoming an environmental burden. It is noted that bagasse fibre is a renewable, environmentally friendly, low density and low-cost material that has physical and mechanical properties similar to other natural fibres such as jute, sisal, pineapple and coir fibres. Hence, BF has a highly potential application in the area of construction as reinforced building materials and reinforcing components for soil reinforcement in support of subgrade and subbase beneath pavement and road systems.

Expansive soil is classified as a type of problematic soil, showing significant volume change when exposed to the fluctuations of moisture content. Frequent soil movements due to the moisture change can generate cracks and the damage of superstructures, freeways, seaports, airports and other civil constructions directly placed on this type of problematic soil. The average annual cost of structural destruction associated with light structures and footings constructed on expansive soils was estimated about £400 million in the UK, $15 billion in the USA, and many billions of dollars worldwide [1]. There are many ground modification methods that have been applied in construction practice to minimise the adverse effects of expansive soils. Among them, special attention has been paid to stabilisation of expansive soil utilising agricultural and industrial wastes because it has been identified to be effective in curtailing the negative impacts of shrinkage and expansion behaviour of expansive soils, while minimising the environmental impacts of industrial wastes [2–9]. Moreover, it is worth mentioning that fibre reinforcement of cemented clays could be useful for various construction practices such as fibre-reinforced load-transfer platform (FRLTP) to support highway embankments over soft grounds [10–19]. This has increasingly become a focused research interest in recent years.

This research aimed to evaluate the potential use of recycled BF waste for expansive soil reinforcement. An array of comprehensive laboratory experiments was conducted on non-reinforced expansive soil, and three different contents (e.g. 0.5%, 1% and 2%) of BF reinforced expansive soils together with various contents of lime stabilisation ranging between 0% and 6% after various curing times. The experimental results were presented and analysed to comprehend the mechanical characteristics of BF reinforcement of expansive soil without or with lime combination for sustainability.

2 Materials and Experimental Programs

2.1 Materials

The soil samples used in this investigation were collected from a road construction site in Queensland, Australia, which was thoroughly described in previous studies [6, 8, 20]. According to the high swell potential, linear shrinkage, and plasticity index, the collected soil can be categorised as highly expansive soil (CH) in accordance with the Unified Soil Classification System (USCS), AS 1726 [21]. It should be noted that for preparation of soil samples, particles larger than 2.36 mm were removed resulting in more consistent samples. Hydrated lime has high quality and quantity of calcium hydroxide (>85%) that was used in this investigation. The hydrated lime is locally purchased in Sydney, Australia.

Bagasse fibre adopted in this research was obtained from ISIS Central Sugar Mill Co., Ltd, Queensland in Australia [22–24]. The BF, as depicted in Fig. 1, had a diameter ranging from 0.3–3.1 mm and a length ranging from 0.3–13.8 mm. The specific gravity of BF (G_f) was about 1.25–1.55, and their average tensile strength was 96.24 ± 29.95 MPa. The obtained fibre was air dried at a controlled room environment with a temperature of 25 °C and a relative humidity of 60% until its mass remained constant. Then, the dried fibre was carefully sieved and passed through 9.5 mm aperture sieve and retained on 300 μm aperture sieve, which was selected for this research. It is known that natural fibres are biodegradable with time. Hence, the durability of natural fibres such as coir fibre and BF, can be improved by chemical treatment using sodium sulphite, sodium hydroxide, or silane coating [25] to prevent the decay of natural fibres from water absorption. Yet, these experimental results were identified as future studies, which are beyond the scope of this paper.

Fig. 1. Photograph of BF used in this investigation

2.2 Experimental Programs

Soil samples were prepared by thoroughly mixing the pulverised natural soil with individual hydrated lime, BF or their combination in the examined ranges to become homogeneous mixtures before tap water was added at the target water content by the dry mass of each mixture. In case of untreated soil specimens, air-dried soils were thoroughly mixed with distilled water to obtain the optimum moisture content of 36.5%. A mechanical blender was used for the mixing of the expansive soil with BF, lime and water. After mixing of the materials, fibre-lime-soil specimens were prepared for both the conventional and comprehensive geotechnical experiments. To investigate the shear strength properties of BF reinforced soils, isotropically consolidated-undrained (CU) tests were conducted on soil samples after 7 days of curing using triaxial compression apparatus following the standard procedure of Australian Standard, AS 1289.6.4.2 [26]. Meanwhile, swelling and consolidation tests were carried out on untreated and treated soil specimens after 3 days of curing using conventional oedometer apparatus following the testing procedure, AS 1289.6.6.1 [27]. The details of testing can be found in previous research by Dang et al. [5, 22]. It should be noted that soil specimens were compacted at the optimum moisture content and maximum dry unit weight [5, 22–24, 28]. For each type of mixtures, the reported test value was obtained as the average of three experimental tests and should not deviate by more than 10% from the mean strength.

3 Test Results and Discussion

3.1 Ultimate Shear Strength of Bagasse Fibre Reinforced Soil

Figure 2 displays variations of ultimate (peak) deviatoric strength versus BF content at different confining pressures. As illustrated in Fig. 2, the ultimate deviatoric strength of reinforced soils positively and significantly depended on the BF content. An increase in the BF content resulted in a corresponding increase in the ultimate deviatoric strength of reinforced soils as observed in Fig. 2 for any given confining pressures. This finding confirms the effectiveness of BF for expansive soil reinforcement in terms of the ultimate shear strength. The shear strength improvement of BF reinforced soils compared to non-reinforced soil might result from the interaction effect between fibre surface and clay particles due to the compaction energy. Such interaction effect improved with increasing BF content, which plays a key factor in increasing the ultimate shear strength of reinforced soils.

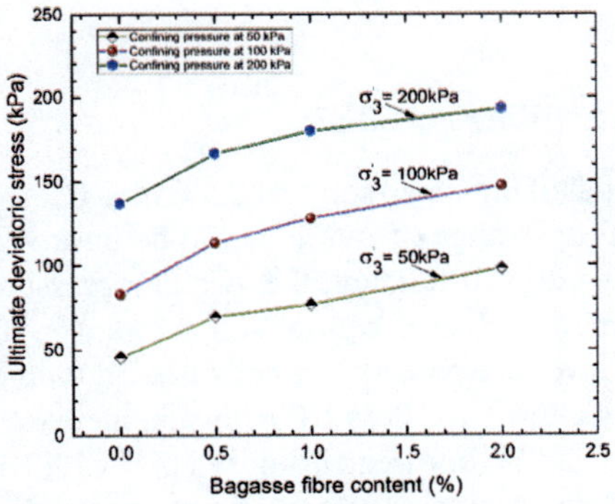

Fig. 2. Ultimate deviatoric strength of soil reinforced with different BF contents

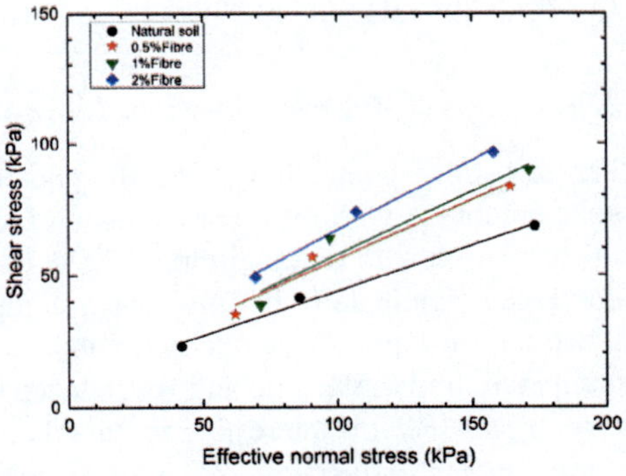

Fig. 3. Peak strength envelopes of soil reinforced with different BF contents

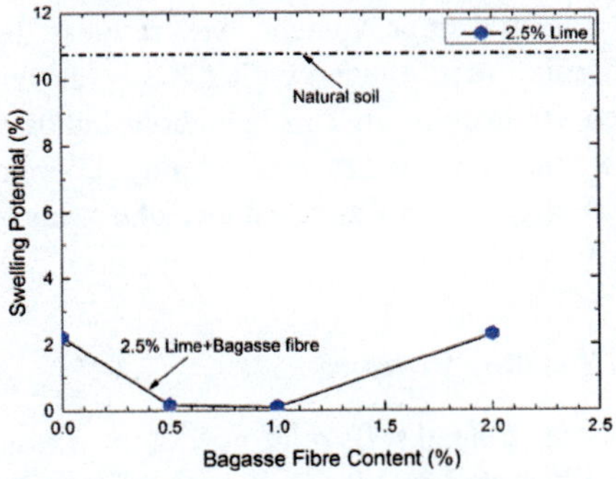

Fig. 4. Swelling potential of BF reinforced expansive soil with 2.5% lime

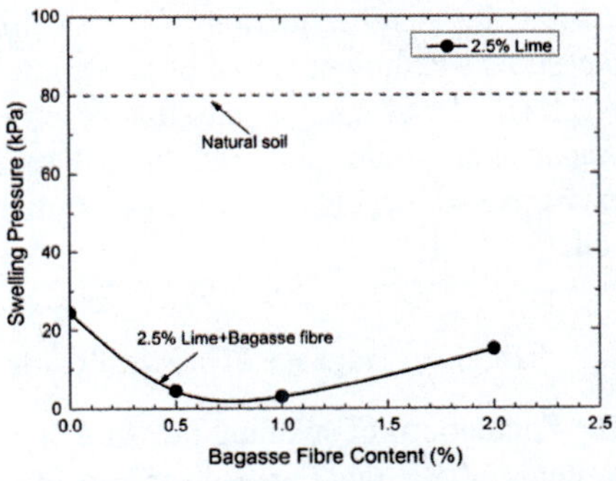

Fig. 5. Swelling pressure of BF reinforced soil with 2.5% lime

It should be noted that a nonlinear relationship between the ultimate deviatoric strength and the BF content was clearly depicted in Fig. 2 as the predominant mechanism for soils reinforced with BF. Moreover, the ultimate deviatoric strength was found to be significantly influenced by the applied confining pressures varying from a relatively low pressure level of 50 kPa to a high pressure level of 200 kPa. As expected, an additional increase in the confining pressure ranging from 50 kPa to 200 kPa improved the ultimate deviatoric strength regardless of adopted BF content. The improved ultimate shear strength was observed for both non-reinforced and reinforced soil specimens, which could be attributable to the effective stress increase corresponding to the confining pressure increase. From the test results, the peak shear strength envelops for both non-reinforced and BF reinforced soils are presented in Fig. 3. Overall, the slopes of the peak shear strength envelopes are steeper and their locations are higher for BF reinforced soils than those of non-reinforced soil. This pattern reveals the beneficial effect of BF reinforcement on the improvement in the peak shear strength of reinforced soils. From observation of Fig. 3, the peak shear strength parameters consisting of effective internal friction angle (φ') and intercept cohesion (c') were derived using linear regression analysis and noted that both φ' and c' values were found to increase with increasing BF inclusions. For instance, approximately 45% increase in the peak friction angle from 19.5° to 28° was observed when BF content increased from 0% for non-reinforced soil to 2% for soils reinforced with BF.

3.2 Effects of Bagasse Fibre and Lime on Swelling Potential

Figure 4 shows fluctuations in swelling potential of untreated soil and 2.5% lime treated soils combined with different contents of BF in a range of 0–2%. It can be observed in Fig. 4 that with the addition of 2.5% lime into soil mixture, the swelling pressure decreased significantly by 79% when compared to that of natural soil. Moreover, as illustrated in Fig. 4, it can be noted that the swelling potential of soils treated with a combination of 2.5% lime and BF reduced by almost 100% as BF inclusion increased from 0% to 1%. It is interesting to state that the 2.5% lime treated soil samples with BF reinforcement in the range of 0.5%–1% could be considered as non-expansive soil due to the negligible (almost zero) swelling potential obtained. However, when an additional increase of BF exceeded 1%, the soils stabilised with 2.5% lime and BF reinforcement exhibited a small reduction of the swelling potential improvement. Nevertheless, the measured swelling potential of 2.5% lime-soil mixture reinforced with 2% BF was very much smaller as compared to that of untreated expansive soil. This behaviour indicates that an appropriate combination of BF reinforcement and lime inclusion for treatment of expansive soil would be very effective in improving the swelling resistance of expansive soil.

3.3 Effects of Bagasse Fibre and Lime on Swelling Pressure

The fluctuations of swelling pressure of 2.5% lime treated soils reinforced with various contents of BF ranging from 0% to 2% are illustrated in Fig. 5. It is noted that the swelling pressure of about 80 kPa for expansive soil without chemical treatment and fibre reinforcement is also presented in Fig. 5 for better comparison purpose. As Fig. 5 shows, the addition of a relatively small content of 2.5% lime into soil was found to cause an approximately 70% reduction of the swelling pressure compared with untreated soil. Moreover, as observed in Fig. 5, the reduction of the swelling pressure was noted to be more significant as BF was inserted into 2.5% lime-soil mixtures. For example, the swelling pressure decreased to a nominal value of less than 5 kPa for 2.5% lime treated soil sample with BF reinforcement as the BF increased to 1%. It is possible to conclude that the soil sample treated with 2.5% lime in combination of BF reinforcement in the range of 0.5–1% could be considered as non-problematic soil because of its negligible swelling pressure as evident in Fig. 5. However, when BF insertion increased beyond 1%, an increase in the swelling pressure to a certain extent (e.g. 15 kPa) was observed for 2.5% lime treated soils reinforced with (e.g. 2%) BF as depicted in Fig. 5. Obviously, the swelling pressure change illustrates a similar trend of the swelling potential of 2.5% lime treated soils reinforced with increasing contents of BF.

3.4 Effects of Bagasse Fibre and Lime on the Compression Characteristics

Figure 6 shows changes in the compression curves of lime treated soils with BF inclusion after 3 days of curing and 4 days soaking (so-called 7 days of curing). As Fig. 6a displays, to examine the effect of BF reinforcement on the compressible property of lime-soil mixtures, a fixed lime content of 2.5% was added to BF-soil mixtures, while the BF content was varied from 0% to 2%. The compression curve of untreated soil is also plotted in Fig. 6a in order to compare with lime treated soils with BF reinforcement. Observation of the compression results reveals that adding BF into lime treated soil mixtures was found to reduce the compression characteristics of reinforced soils as BF content increased from 0% to 1%. Subsequently, the compressibility of lime treated soils reinforced with BF indicated a small increase when the addition of BF exceeded 1%. By comparing with untreated soil and 2.5% lime treated soil, the lower slope reduction of the virgin compression curves was found for soil samples treated with 2.5% lime in combination with BF reinforcement. Referring to Fig. 6a, the combination 2.5% lime and 1% BF was observed to cause the most notable improvement in the virgin curve of reinforced soils.

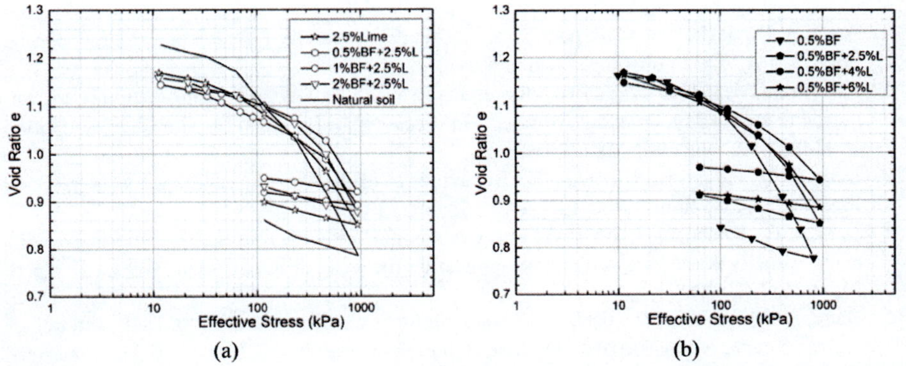

Fig. 6. Effective stress-void ratio curves of expansive soil reinforced with (a) various contents of BF and 2.5% lime (L); (b) various contents of lime and 0.5% BF

Furthermore, to comprehend the effect of lime stabilisation on the compressibility of expansive soils reinforced with BF, a fixed content of 0.5% BF was inserted to soils mixed with various contents of lime from 0% to 6% as plotted in Fig. 6b. The results indicate that a combination of 0.5% BF and various lime contents was found to decrease the compression characteristics of treated soils when increasing lime content from 0% to 4%. This trend was followed by a slight increase in the compressibility with a higher lime inclusion (e.g. 6%). As Fig. 6b illustrates, the most reduction of compression curve was observed for 4% lime treated soil sample reinforced with 0.5% BF.

4 Conclusion

This investigation has explored the potential reuse of a recycled and low-cost waste material (bagasse fibre) for sustainable development of civil infrastructure construction projects. Based on the test results, it was found that the ultimate deviatoric strength of reinforced soils was considerably depended on the BF content. The interfacial friction between fibre surface and clay particles due to compaction energy plays an important role in improving the ultimate shear strength of reinforced soils compared to the control sample. Adding lime and BF into soil mixtures resulted in a remarkable improvement in the swelling and compressible properties of reinforced soils. More interestingly, adding a relatively small content of 2.5% lime with 0.5% BF could result in non-swelling of treated soil. The compressibility of lime treated soils with fibre reinforcement were found to decrease with increasing lime content from 0% to 4% followed by a slight increase as lime content increased further to 6%. This finding reveals that adding a certain amount of BF into lime-soil mixtures is very effective in improving the compressibility characteristics of treated soil.

Acknowledgement. The results presented in this paper are partially funded by the Australian Technology Network (ATN) and Arup Pty Ltd. The authors gratefully acknowledge their support.

References

1. Jones, L.D., Jefferson, I.: Expansive soils. In: ICE Manual of Geotechnical Engineering, pp. 413–441. ICE Publishing, London (2012)
2. Bhadriraju, V., Puppala, A.J., Enayatpour, S., Pathivada, S.: Digital imaging technique to evaluate shrinkage strain potentials of fiber reinforced expansive soils. In: Site Characterization and Modeling, vol. GSP 138 (2005)
3. Dang, L.C., Fatahi, B., Khabbaz, H.: Behaviour of expansive soils stabilized with hydrated lime and bagasse fibres. Proc. Eng. **143**, 658–665 (2016)
4. Dang, L.C., Khabbaz, H., Ni, B.-J.: Improving engineering characteristics of expansive soils using industry waste as a sustainable application for reuse of bagasse ash. Transp. Geotech. **31**, 100637 (2021)
5. Dang, L.C., Khabbaz, H., Fatahi, B.: Evaluation of swelling behaviour and soil water characteristic curve of bagasse fibre and lime stabilised expansive soil. In: Second Pan-American Conference on Unsaturated Soils (PanAm-UNSAT 2017). ASCE, Texas, USA, vol. GSP 303, pp. 58–70 (2017)
6. Dang, L.C., Khabbaz, H.: Assessment of the geotechnical and microstructural characteristics of lime stabilised expansive soil with bagasse ash. In: The 71st Canadian Geotechnical Conference and the 13th Joint CGS/IAH-CNC Groundwater Conference (GeoEdmonton 2018), Alberta, Canada. Canadian Geotechnical Society, Alberta (2018)

7. Dang, L.C., Hasan, H., Fatahi, B., Khabbaz, H.: Influence of bagasse ash and hydrated lime on strength and mechanical behaviour of stabilised expansive soil. In: The 68th Canadian Geotechnical Conference and the 7th Canadian Permafrost Conference (GEOQuébec 2015), Québec City, Canada. Canadian Geotechnical Society, Québec City (2015)
8. Dang, L.C., Hasan, H., Fatahi, B., Jones, R., Khabbaz, H.: Enhancing the engineering properties of expansive soil using bagasse ash and hydrated lime. Int. J. GEOMATE **11**(25), 2447–2454 (2016)
9. Dang, L.C., Hasan, H., Fatahi, B., Jones, R., Khabbaz, H.: Effects of bagasse ash and hydrated lime addition on engineering properties of expansive soil. In: Fifth International Conference on Geotechnique, Construction Materials and Environment (GEOMATE 2015), Osaka, Japan, pp. 90–95 (2015)
10. Dang, L.C., Dang, C.C., Khabbaz, H.: Behaviour of columns and fibre reinforced load transfer platform supported embankments built on soft soil. In: The 15th International Conference of the International Association for Computer Methods and Advances in Geomechanics, Wuhan, China, pp. 1233–1246 (2017)
11. Dang, L.C., Dang, C.C., Khabbaz, H.: Numerical analysis on the performance of fibre reinforced load transfer platform and deep mixing columns supported embankments. In: Bouassida, M., Meguid, M.A. (eds.) GeoMEast 2017. SCI, pp. 157–169. Springer, Cham (2018). https://doi.org/10.1007/978-3-319-63889-8_13
12. Dang, L.C., Dang, C.C., Khabbaz, H.: Modelling of columns and fibre-reinforced load-transfer platform-supported embankments. Ground Improv. **173**(4), 197–215 (2020)
13. Dang, L.C., Dang, C.C., Khabbaz, H.: A parametric study of deep mixing columns and fibre reinforced load transfer platform supported embankments. In: Khabbaz, H., Youn, H., Bouassida, M. (eds.) GeoChina 2018. SCI, pp. 179–194. Springer, Cham (2019). https://doi.org/10.1007/978-3-319-95771-5_15
14. Dang, L., Dang, C., Fatahi, B., Khabbaz, H.: Numerical assessment of fibre inclusion in a load transfer platform for pile-supported embankments over soft soil. In: Chen, D., Lee, J., Steyn, W.J. (eds.) Geo-China 2016: Material, Design, Construction, Maintenance, and Testing of Pavement. ASCE, pp. 148–155 (2016)
15. Dang, C., Dang, L., Khabbaz, H., Sheng, D.: Numerical study on deformation characteristics of fibre reinforced load transfer platform and columns supported embankments. Can. Geotech. J. **58**, 328–350 (2021)
16. Dang, C.C., Dang, L.C.: Influence of fibre-reinforced load transfer platform supported embankment on floating columns improved soft soils. In: Tutumluer, E., Chen, X., Xiao, Y. (eds.) Advances in Environmental Vibration and Transportation Geodynamics. LNCE, vol. 66, pp. 215–227. Springer, Singapore (2020). https://doi.org/10.1007/978-981-15-2349-6_14
17. Dang, C.C., Dang, L.C.: Evaluation of the at-rest lateral earth pressure coefficient of fibre reinforced load transfer platform and columns supported embankments. In: Ha-Minh, C., Dao, D.V., Benboudjema, F., Derrible, S., Huynh, D.V.K., Tang, A.M. (eds.) CIGOS 2019: Innovation for Sustainable Infrastructure. LNCE, vol. 54, pp. 647–652. Springer, Singapore (2020). https://doi.org/10.1007/978-981-15-0802-8_102
18. Dang, L.C., Dang, C.C., Khabbaz, H.: Numerical modelling of embankment supported by fibre reinforced load transfer platform and cement mixed columns reinforced soft soil. In: The XVII European Conference on Soil Mechanics and Geotechnical Engineering (ECSMGE-2019). ISSMGE, Reykjavik (2019)
19. Dang, C.C., Dang, L.C.: Numerical investigation on the stability of soil-cement columns reinforced riverbank. In: Correia, A.G., Tinoco, J., Cortez, P., Lamas, L. (eds.) ICITG 2019. SSGG, pp. 879–888. Springer, Cham (2020). https://doi.org/10.1007/978-3-030-32029-4_74
20. Dang, L.C.: Enhancing the engineering properties of expansive soil using bagasse ash, bagasse fibre and hydrated lime. Ph.D., University of Technology Sydney, NSW, Australia (2019)

21. AS, Geotechnical Site Investigations. Standards Australia: Sydney, Australia, vol. AS 1726-1993 (1993)
22. Dang, L.C., Khabbaz, H., Fatahi, B.: An experimental study on engineering behaviour of lime and bagasse fibre reinforced expansive soils. In: 19th International Conference on Soil Mechanics and Geotechnical Engineering (19th ICSMGE), pp. 2497–2500. ISSMGE, Seoul (2017)
23. Dang, L.C., Khabbaz, H.: Experimental investigation on the compaction and compressible properties of expansive soil reinforced with bagasse fibre and lime. In: McCartney, J.S., Hoyos, L.R. (eds.) GeoMEast 2018. SCI, pp. 64–78. Springer, Cham (2019). https://doi.org/10.1007/978-3-030-01914-3_6
24. Dang, L.C., Khabbaz, H.: Enhancing the strength characteristics of expansive soil using bagasse fibre. In: Wu, W., Yu, H.-S. (eds.) Proceedings of China-Europe Conference on Geotechnical Engineering. SSGG, pp. 792–796. Springer, Cham (2018). https://doi.org/10.1007/978-3-319-97112-4_177
25. Tian, H., Zhang, Y.X.: The influence of bagasse fibre and fly ash on the long-term properties of green cementitious composites. Constr. Build. Mater. **111**, 237–250 (2016)
26. AS, Determination of compressive strength of a saturated specimen tested in undrained triaxial compression with measurement of pore water pressure. Standards Australia: Sydney, Australia (1998)
27. AS, Determination of the one-dimensional consolidation properties of a soil - Standard method. Standards Australia: Sydney, Australia (1998)
28. Dang, L.C., Khabbaz, H.: Shear strength behaviour of bagasse fibre reinforced expansive soil. In: International Conference on Geotechnical and Earthquake Engineering 2018 (IACGE 2018). ASCE, vol. GSP 304, pp 393–402 (2018)

Dynamic Properties of a Calcareous Sand in UAE

Man T. Bui[1](✉), Emad Sharif[2], and Hossam E. Ali[3]

[1] GTC Laboratory, GTC Soil Analysis Services, Dubai, UAE
man.buitan@gtc-lab.com
[2] GTC Soil Analysis Services, Dubai, UAE
[3] Artelia Group, Saint-Ouen, France

Abstract. Calcareous sand is sedimentary soil, consisting of abundant calcium carbonate mineral or lime/chalky carbonate particles, such as shell fragments, or remains of marine organisms. The sand exhibits some special behaviours that are different from those of silica sand, e.g. relatively high specific gravity, special surface texture, irregular particle shape, local voids, and highly crushable susceptibility.

Due to the high risks, hence high cost implication, involved to structures built over land reclaimed by mean of hydraulic fills of calcareous sand e.g. for the Palms of the Emirates and other artificial islands, it is crucial to study the dynamic properties of calcareous sand for e.g. liquefaction potential analysis to achieve safe and economical foundation solutions. In this paper we present and discuss on cyclic resistance and dynamic properties of a local calcareous sand obtained from various tests, including cyclic triaxial, resonant column, and advanced triaxial tests with bender elements.

Keywords: Calcareous sand · Dynamic soil properties · Liquefaction potential

1 Introduction

Calcareous sand is prevailing in UAE. It has been a main source of material for land reclamation in UAE, such as Palm Islands, the World Islands, the Dubai Marina, the Burj Al Arab, Yas Island, etc. There are high risks thanks to the liquefaction potential and compressibility of calcareous sand for the land reclamation.

Dynamic properties of calcareous sand have been studied around the world during the last 30 years (see e.g. [1–4]). This paper presents some key results of a preliminary testing program (Table 1) undertaken to determine the cyclic resistant ratio (CRR) together with their small strain stiffness, soil shear strength, and compressibility of a local calcareous sand collected at a land reclamation site in Abu Dhabi, UAE. The testing techniques included triaxial tests with bender elements (BE), resonant column tests (RCT), and cyclic triaxial tests. The relative density of the test specimens was varied in a relatively narrow range from 60% to 65%, which was expected at the site. In addition, oedometer tests and creep measurement in triaxial cell were also conducted, however, they are out of the scope of this paper. Due to confidentiality, the site name cannot be disclosed.

D. V. K. Huynh et al. (Eds.): VSOE 2021, LNCE 208, pp. 89–96, 2022.
https://doi.org/10.1007/978-981-16-7735-9_7

Table 1. Preliminary laboratory testing program

Test techniques	Test abbreviation	Relative density	Effective stress (kPa)	No of tests
Multiple stage isotropic consolidation undrained triaxial tests	CIUm	60%	100, 200, 400	1
Drained triaxial tests with measurement of V_s using bender elements	CID3+BE	60%, 65%	50, 100, 200	4
Multiple stage resonant column tests	RCTm	60%, 65%	50, 100, 200	4
Undrained cyclic triaxial tests at various cyclic stress ratios	CTXL	60%, 65%	100	10

2 Laboratory Testing

Testing Material. Some properties of the calcareous sand were summarised in Table 2. The sand has 89% carbonate content. Based on the particle sieve analysis, the sand can be described (as per BS 5930:2015 [5]) as *light brown to brown, slightly silty/slightly clayey well graded fine to coarse calcareous SAND with occasional gravel.*

Table 2. Index properties of the calcareous sand

Min dry density	Max dry density	G	Particle sieve analysis				
(Mg/m^3)	(Mg/m^3)	(Mg/m^3)	Fine (%)	Sand (%)	Gravel (%)	D_{50} (mm)	D_{max} (mm)
1.15	1.60	2.77	6	92	2	0.4	6.3

Sample Preparation. Best efforts were made to prepare uniform cylindrical specimens to the target dry densities using moist tamping method. The material was dried in oven for at least 48 h at 60 °C, and only the right amount of dried material were used. The dried material was mixed with an appropriate amount of sea water and then carefully compacted into a cylindrical split mould (in five equal layers). The mould diameter is 83.0 mm and the mould height is approximately twice the diameter. The reconstituted specimen was then frozen at −16 °C before removing the mould. The sample height was measured at four positions around the perimeter, and the sample diameter was measured for six locations along the length. The average height and diameter were taken to estimate the sample volume.

Sample Saturation. After specimen setup, a suction up to −45 kPa was applied from the top of the specimen. Increments of cell pressures and back pressures were appropriately applied to saturate the sample until Skempton's B value not less than 95%. During applying back pressure, the vertical deformation of the sample was closely monitored to ensure the sample volume was not expanded when compared to its initial volume.

Isotropic Consolidation. After saturated, the specimen was consolidated overnight to the effective stress scheduled in Table 2. The completion of the consolidation was assessed by careful monitoring of the rates of the specimen vertical deformation and changes in volume of water flow.

Drained and Undrained Triaxial Test. A set of three drained triaxial test was conducted to access the shear strength of the material. For verification, a multiple stage triaxial test at 100 kPa, 200 kPa, and 400 kPa. as per BS1377-8:1990 [8] was performed. The advantages of multiple stage triaxial test was the less time consuming (when compare to the set of three CIU tests). In addition, as the test conducted on a single specimen, the possible variations in sample size and/or sample void ratio were also removed. During the early stages, the excess PWP and the effective stress ratio σ_1'/σ_3' were closely motioned, allowing the stages were stopped at the right time, when it just about to reach the strength envelop[1].

Measurement of Soil Mechanical Properties at Small Strain. Resonant column test (RCT) is one of the most popular and advanced techniques that can precisely estimate the shear modulus, G_0, Damping ratio, D, shear modulus degradation G/G_0, from very small to medium strain (up to 0.1%). In this paper, RCTs were conducted using a Stokoe's type resonant column apparatus (RCA, made by GDS Instrument, UK), which is a fixed-free resonant column, where one end of the cylindrical soil specimen is excited and the other is (assumingly) fixed. The test base was bolted to a heavy and sturdy bench to increase the system fixity and mass moment inertia of the system [6]. GDS-RCA V2.10 software was used to precisely determine the resonant frequency and damping ratio of the material using free decay vibration method. The software was well designed to control (hence minimise) the number of excitation cycles required to establish resonance. The shear modulus was derived with consideration of the system compliance [6, 7]. RCTs can be classified as a non-destructive technique, therefore, resonant vibration stage was allowed to be repeated at different effective stresses, i.e. 50 kPa, 100 kPa, and 200 kPa.

Measurement of Vs Using Bender Elements in Triaxial Cell. Bender elements are composite sheets of piezo-electric material bonded together. They were inserted into

[1] The axial strains induced in the first two stages of the test conducted in the research were relatively small, and much far from reaching the peak deviator stress. The effects of the previous stages were somewhat removed when the sand sample was consolidated at higher effective stresses.

both ends of a cylindrical soil specimen in a triaxial system. An applied voltage across a triggering bender element causes the sheets to flex. Conversely, flexure of the sheets causes a voltage to be induced in the receiving bender element.

Based on the waveforms recorded, we picked up the shear wave travel times correspondent to the duration measured from the first peak of the trigger wave to the first peak of the receiving wave (Fig. 1). The advantages of first peak-to-peak techniques over other techniques (e.g. first arrived, first crossing, first deflection, etc.) are consistency, objectivity, and less dependency on excitation frequencies and the signal noises.

Cyclic Triaxial Test. Cyclic tests were conducted using a GDS MiniDyn system to understand the cyclic behaviours of the sand and to evaluate the cyclic resistance ratio. On the completion of the consolidation, sinusoidal loading cycles were applied to the sample at frequency of 0.5 Hz in undrained conditions until liquefied.

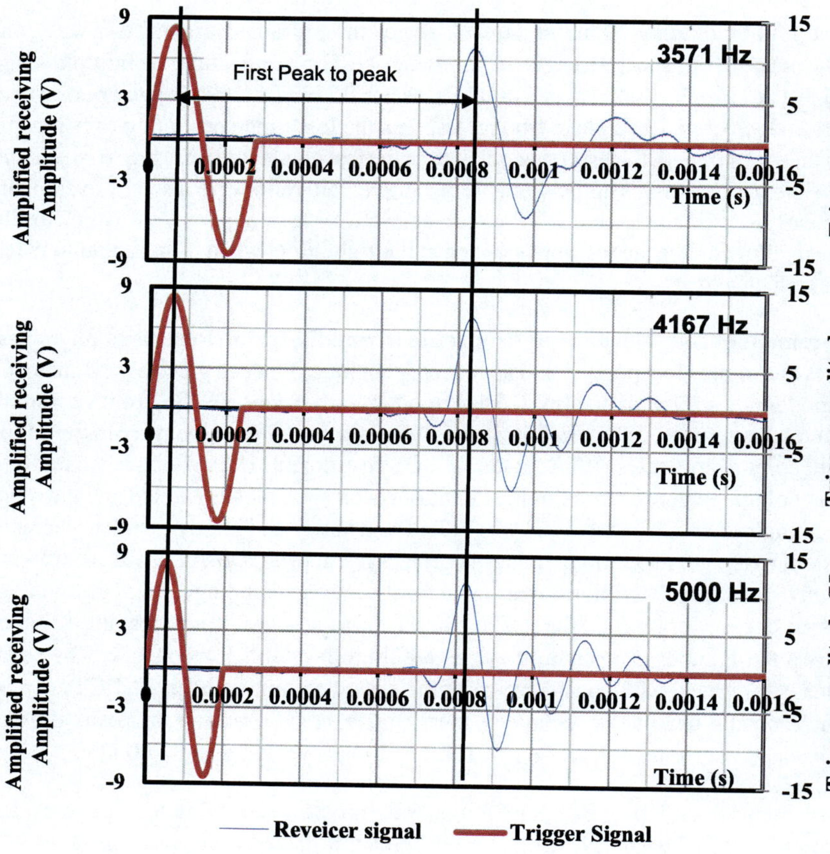

Fig. 1. First peak to peak technique.

3 Test Results and Discussions

3.1 Shear Strength

It appears that both drained and undrained tests reached the same shear strength envelope (CSL) correspondent to the internal friction angle of 40° and the cohesion of 20 kPa, derived from the Mohr circles and the shear strength envelope (Fig. 2). The relatively high apparent cohesion (for sand) can be contributed by the existence of sea salt and fine particles that are rich in calcium. This may also be the reasons for the relatively high internal friction angle of the sand, despite of the 60% relative density.

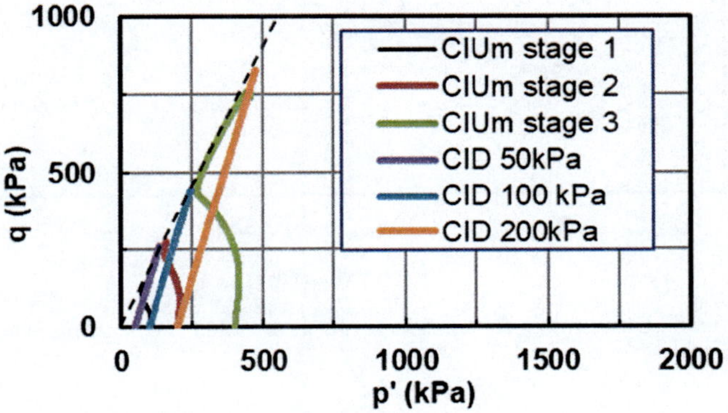

Fig. 2. Drained and undrained stress-path during shear

3.2 Small Strain Stiffness

The G_0 values measured for the calcareous sand can be curve-fitted using Eq. (1) below:

$$G_0 = A\sigma_0'^{n}(1+e)^{-3} \tag{1}$$

Where,

G_0 is small strain shear modulus, in MPa;
e is void ratio;
σ'_0 is the mean effective tress, in kPa;
n is exponent number; and

A is a coefficient, dependent on particle characteristics e.g. particle shape, surface roughness, particle gradation, cementation, etc.

Based on curve fitting technique applied for the data obtained from both RCT and bender element tests (Fig. 3), we derived Eq. 2 below:

$$G_0 = 49\sigma_0'^{0.60}(1+e)^{-3} \tag{2}$$

The relatively high exponent number (n = 0.6) suggested that calcareous sand is quite susceptible to the change in effective stress. This was attributed by particle crushing and particle shape, which is relatively angular and also rough at the surfaces. It should be noted that Eq. 2 is valid for the local sand tested with range of void ration and effective stress applied in this project only.

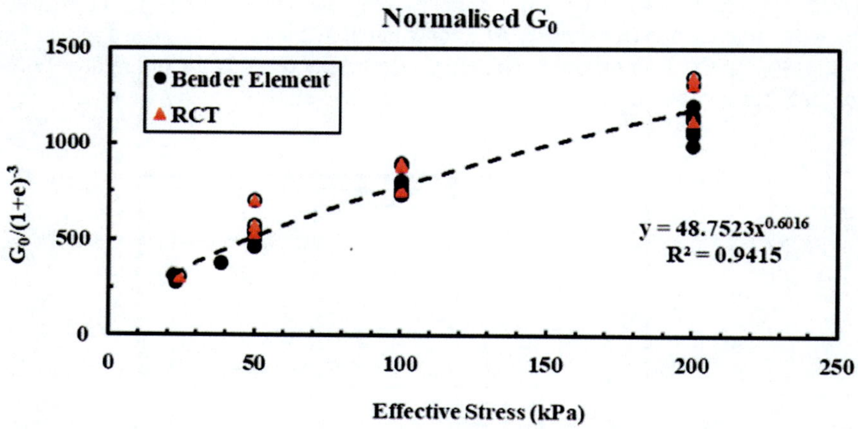

Fig. 3. Curve fitting for small strain shear modulus of calcareous sand

3.3 Damping Ratio

Damping ratios of the sands were measured using decay vibration method. The system damping was inconsiderable and the measured material damping included the system damping. It can be seen in Fig. 4 that the damping ratios follow the common trends i.e. decreasing with an increase in effective stress or dry density. At 0.001% shear strain, the damping ratio varied from 0.5% to 1.0%[2]. Generally, the material damping can be fitted to a hyperbolic family curves, see e.g. [1]. The authors have also developed a hyperbolic family curve with four curve fitting coefficients, however, due to the limitation of paper length, this will be presented in another publication.

3.4 Cyclic Resistance Ratio

Figure 5 presents typical stress-strain, stress-path, excess PWP ratio, and cyclic stiffness degradation of a calcareous sand sample. The sample was failed after 12 cycles. The liquefaction criteria applied were (i) excess PWP ratio greater than 95% (See Fig. 5), or the double peak (axial) strain amplitude greater than 5%.

The number of loading cycles to fail a sample, *Nf*, versus the cyclic stress ratios applied for all tests were plotted in Fig. 6. From the best fit curve, the cyclic resistant ratio (CRR of the calcareous sand correspondent to 10 cycles was estimated about 0.25.

[2] The damping ratios measured for a 60%-relative-density specimen at 50 kPa was somewhat erratic and may not representative for the sand.

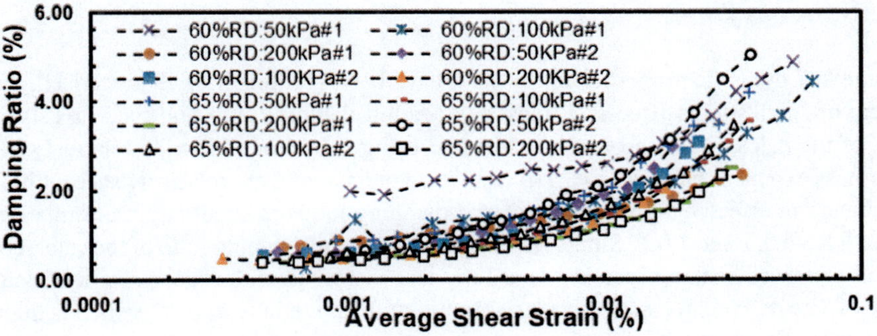

Fig. 4. Damping ratios vs shear strain of the calcareous sand

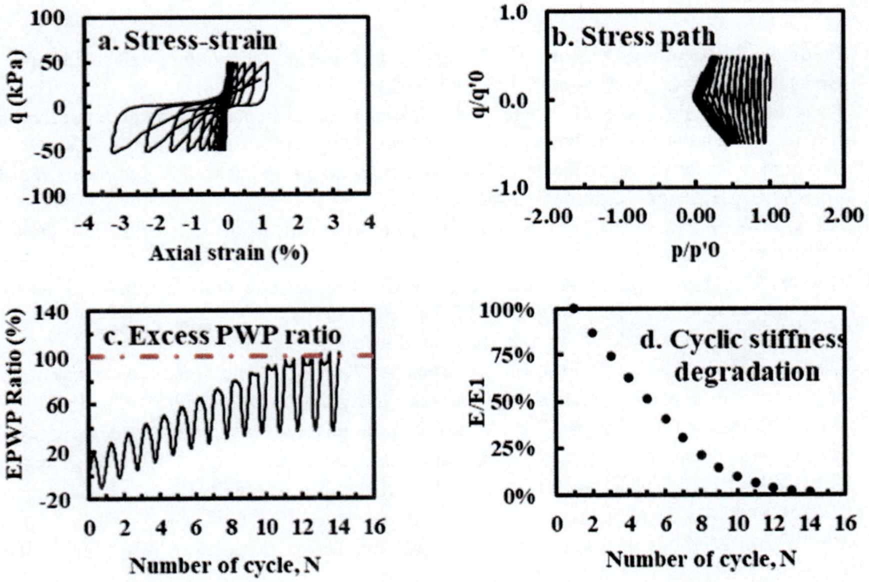

Fig. 5. Liquefaction of a calcareous sand sample at 65% relative density

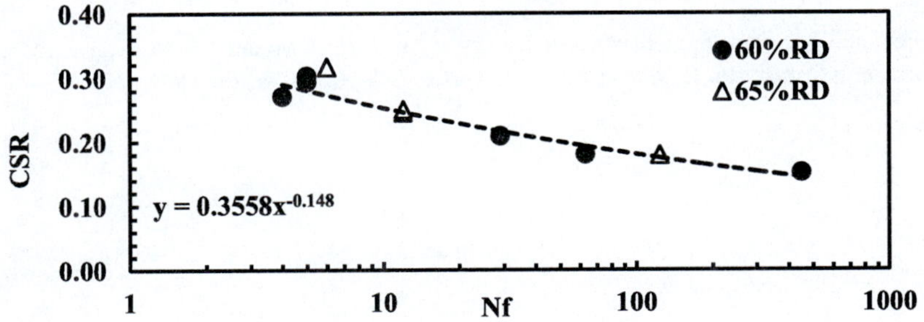

Fig. 6. Cyclic stress ratio (CSR) applied vs. N_f.

4 Summary

Some key dynamic properties of the sand have been preliminarily characterised, with focus on small strain stiffness, stiffness degradation, damping ratio, and cyclic resistance. G_0 of the calcareous sand can be estimated using Eq. 2, in which the relatively high stiffness exponent number ($n = 0.6$) confirmed the material are relatively susceptible to a change in effective stress. The material damping ratios at small strain are relatively small, from 0.5% to 1.0%. Similar to other materials, the damping ratio of the calcareous sand increased with an increase void ratio and increased with a decrease in effective stress. The derived cyclic stress resistant ratio of the material was 0.25, which is higher than that of a silica sand with the similar particle size distribution.

References

1. Frydman, S., Hendron, D., Horn, H., Steinbach, J.: Liquefaction study of cemented sand. J. Geotech. Eng. Div. ASCE **106**(3), 275–297 (1980)
2. Kaggwa, W.S., Poulos, H.G., Carter, J.P.: Response of carbonate sediments under cyclic triaxial test conditions. In: Jewell, R.J., Andrews, D.C. (eds.) Engineering for Calcareous Sediments. Proceedings of the International Conference on Calcareous Sediments, Perth, pp. 97–107. Balkema, Rotterdam (1988)
3. Airey, D.W., Fahey, M.: Cyclic response of calcareous soil from the North West shelf of Australia. Geotechnique **41**(1), 101–121 (1991)
4. Hyodo, M., Hyde, A.F.L., Aramaki, N.: Cyclic strength and deformation of crushable carbonate sand. Soil Dyn. Earthq. Eng. **15**, 331–336 (1996)
5. BS 5930:2015+A1: Code of practice for ground investigations (2020)
6. Clayton, C.R.I., Priest, J.A., Bui, M.T., Zervos, A., Kim, S.G: The Stokoe resonant column apparatus: effects of stiffness, mass and specimen fixity. Géotechnique **59**(5), 429–437 (2009)
7. Bui, M.T., Priest, J.A., Clayton, C.R.I: A new calibration technique to improve data reduction for Stokoe resonant column test. In: Randolph, M., Doan, D., Tang, A., Bui, M., Dinh, V. (eds.) VSOE 2018. LNCE, vol. 18, pp. 43–48. Springer, Singapore (2019). https://doi.org/10.1007/978-981-13-2306-5_3
8. BS 1377-8, Methods of test for soils for civil engineering purposes. Shear strength tests (effective stress), Clause 7 (1990)
9. Bui, M.T., Clayton, C.R.I., Priest, J.A.: The universal void ratio function for small strain shear modulus. In: International Conferences on Recent Advances in Geotechnical Earthquake Engineering and Soil Dynamics and Symposium in Honor of Professor I.M. Idrisss, San Diego, California, 24–29 May 2010 (2010)
10. Darendeli, M.: Development of a New Family of Normalized Modulus Reduction and Material Damping Curves (Ph.D. dissertation). University of Texas at Austin, USA (2001)

Estimation of Damping Factors for a Better Assessment of Offshore Driven Piles Capacity

M. Maron[1(✉)] and Z. L. Delimi[2]

[1] FUGRO France S.A.S, Le Bourget-du-Lac, France
m.maron@fugro.com
[2] FUGRO Survey Middle East, Abu Dhabi, UAE

Abstract. Dynamic Monitoring System (DMS) and Signal matching analysis is widely used in the offshore industry to validate the axial capacity of foundation piles. While the uniqueness of the signal matching solution is still being discussed (Buckley et. al. 2017), field experience highlights that an accurate choice of the damping parameter is key to evaluate correctly the static capacity of the foundation. This paper, proposes to review two independent set of signal matching analyses performed using the same recorded data at Platform A, field X, in the Arabian Gulf, to appreciate fully the importance of the damping factor in the analysis and confront the results to literature and past experience. Findings presented in this paper aims to guide the DMS user towards a more accurate selection of the damping and hence a better assessment of the pile's axial capacity.

Keywords: Pile driving · DMS · Pile monitoring · Capacity · Damping · Driveability · Setup · Signal matching · CAPWAP

1 Presentation of the Studied Cases

The jacket of Platform A is six-legged jackets that was installed in January 2020 in the Arabian Gulf. 72″ piles were driven using a hydraulic hammer until the target penetration of 75 m. The soil consisted mainly of firm to very stiff calcareous Clays with layers of weak to moderately strong calcarenite. A layer of very weak to moderately weak Calcisiltite was encountered at target penetration. Due to weather standby, the two re-strike tests were performed after an unusually long delay ranging from 60 to 125 h following the end of driving, respectively. All pile driving sequences were recorded using a set of strain gauges and accelerometers and signal matching analysis was performed at target penetration for all piles as well as for each re-strike test (Fig. 1).

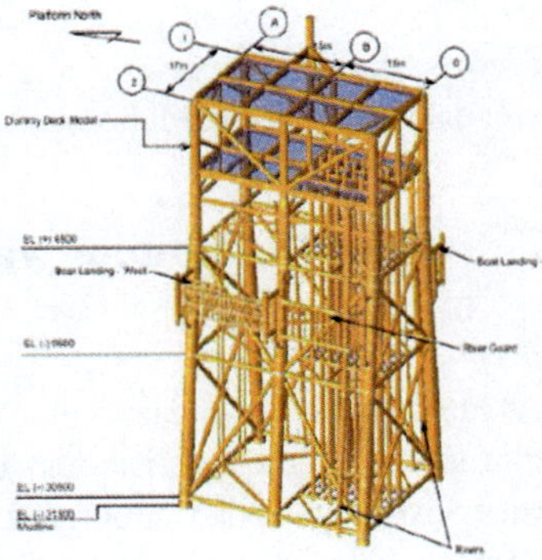

Fig. 1. Illustration of jacket A

D. V. K. Huynh et al. (Eds.): VSOE 2021, LNCE 208, pp. 97–104, 2022.
https://doi.org/10.1007/978-981-16-7735-9_8

2 Problem Statement

The signal matching software CAPWAP, used for the capacity analyses, models the soil resistance as a succession of individual soil elements i. During driving, the hammer induced energy develops a total resistance R_{ti} for each soil element. This total resistance is composed of a velocity dependent dynamic resistance component R_{di} and a static resistance component R_i such as $R_{ti} = R_{di} + R_i$. For each element of soil i along the pile, E.A.L. Smith (1960) defines a soil damping parameter J as per the equation below in order characterize the dynamic resistance of the soil as a function of the static resistance and the pile velocity v.

$$R_{di} = R_i.J_i.v_i \tag{1}$$

$$R_{ti} = R_i(1 + J_i.v_i) \tag{2}$$

In the signal matching software CAPWAP, the damping factor is divided into a unique shaft damping J_{shaft} linked to the static shaft resistance R_{shaft} and a toe damping J_{toe} linked to the static resistance at toe R_{toe}.

From the Smith definition of the damping factor, it is clear that the evaluation of the static resistance is directly linked to the evaluation of the damping factor. In other words, for a given total resistance R_t recorded by the DMS, several combinations of static resistance and damping can be found. Care must be taken not to over or under evaluate the damping factors as it will result in an under or over estimation of the static resistance, respectively.

The difficulty to perform a representative and accurate evaluation of the capacity based on signal matching analysis is then increased by the fact that recommendations on damping factors are empirical (Roussel H.J. 1979; GRL WEAP Manual 2005) with a scale effect that is not well known for large diameter piles, the Smith damping factor itself is dependent of the pile velocity (Rausche 2018). Additionally, Svinkin (2010) reports that despite several attempts, damping parameters cannot be determined from standard geotechnical laboratory or in-situ test.

3 Impact of Damping Variations on Capacity Assessment at End of Drive

At Platform A, the capacity assessments were performed by the Contractor's appointed pile monitoring specialists and independently re-assessed for review purposes by the supervision specialist appointed by the main Client. Care was taken to use the same hammer blow, deemed of good quality, for each of the analyses in order to compare results and understand the sources of the differences.

The Table 1 below presents the comparative results of the capacity analyses performed and associated damping values as reported by the Contractor's specialist and by the Reviewer at End Of Drive (EOD) and Beginning of Restrike (BOR). Restrike test was performed for pile P1 only. Both set of analyses report a good match quality parameter below 3.09 and a reasonable difference between the observed and modelled

Table 1. Platform A capacity assessments results (From CAPWAP)

Pile	Section and test status	Capacity			Smith Damping	
		Shaft [MN]	Toe [MN]	Total [MN]	Skin [s/m]	Toe [s/m]
P1	EOD	8.00 *(11.12)*	9.50 *(8.44)*	17.50 *(19.56)*	1.23 *(0.88)*	0.61 *(0.47)*
	BOR[a]	43.20 *(37.66)*	9.30 *(8.66)*	52.50 *(46.32)*	0.34 *(0.52)*	0.54 *(0.40)*
	BOR[b]	43.50 *(38.38)*	9.50 *(8.08)*	53.00 *(46.45)*	0.48 *(0.66)*	0.79 *(0.30)*
P2	EOD	10.00 *(11.39)*	13.00 *(15.16)*	23.00 *(26.55)*	1.17 *(0.85)*	0.81 *(0.32)*
P3	EOD	9.10 *(11.15)*	10.00 *(8.50)*	19.10 *(19.65)*	1.10 *(0.70)*	0.60 *(0.64)*
P4	EOD	12.50 *(13.95)*	14.50 *(16.44)*	27.00 *(30.39)*	1.21 *(0.86)*	0.18 *(0.23)*
P5	EOD	10.80 *(12.33)*	16.00 *(16.02)*	26.80 *(28.35)*	1.30 *(0.83)*	0.08 *(0.25)*
P6	EOD	7.70 *(9.93)*	8.50 *(6.38)*	16.20 *(16.31)*	1.12 *(0.75)*	0.76 *(0.70)*

[a] First restrike on platform A after 60 h interruption period.
[b] Second restrike on platform A after 125 h interruption period.

set per blow below 1.04 mm at all times. These are CAPWAP indication of a signal matching of good quality.

Despite software's indication of a good match for all analyses, a variation of the damping factors is observed between Contractor's and Reviewer's assessments. The extent of the variation is presented in the Fig. 2, showing the shaft damping factors J_{shaft} versus the shaft resistance R_{shaft} and the toe damping factor J_{toe} versus the toe resistance R_{toe}. No justifications were provided on the Contractor's choice of damping whilst Reviewer focused on obtaining a good signal matching match with consistent damping between piles of the same jacket at EOD.

At the end of driving, the shaft damping selected by Reviewer are between 38% and 57% lower than the one selected by the Contractor. Due to the nature of the damping according to Smith, presented in Eq. (1) above, these differences result in a corresponding higher shaft capacity R_{shaft} and therefore a total capacity R assessed by Reviewer up to 12% higher than the capacity assessed by Contractor at the end of driving.

At the beginning of restrikes, the shaft damping selected by Reviewer were however 27% to 35% higher than the ones selected by Contractor, therefore resulting in a total capacity R assessed by Reviewer 12% lower than the capacity assessed by Contractor at the beginning of restrike. It shall be noted that whilst the contractor's assessment of the shaft resistance is generally above the reviewer assessment, the differences of toe resistances assessed by contractor and reviewer also impact the differences in the total capacity R.

The section below demonstrates how the differences of total capacities R are maximizing the uncertainty for the evaluation of long-term capacity.

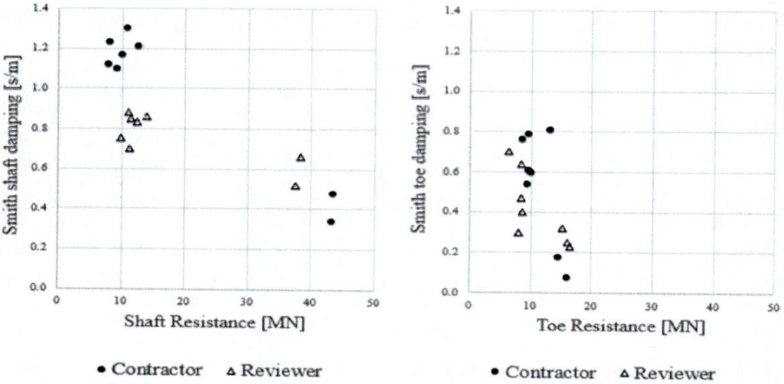

Fig. 2. Platform A, shaft and toe damping versus shaft and toe resistances

4 Impact of Damping Variations on Long-Term Capacity Estimation

It is now proposed to present the impact of the observed variations of damping factors on the setup effect and on the estimation of the long-term capacity. Offshore practice defines the setup effect as the gain of shaft resistance with time. The setup effect observed between the end of driving $R_{shaft,EOD}$ and the restrike tests $R_{shaft,BOR}$ is a portion of the long-term setup effect and can be quantified by a setup factor $S_{restrike}$ calculated as follow:

$$S_{retrike} = \frac{R_{shaft,\mathrm{BOR}}}{R_{shaft,EOD}} \tag{3}$$

In offshore practice, restrike test is currently not performed on all piles due to time constraints. With the setup up effect calculated for the pile where the restrike test has been performed, it is accepted that this setup effect is applied to the shaft resistance of other piles of the jacket in order to estimate the lower bound of the pile resistance after setup R_{setup} by using the following equation:

$$R_{setup} = S_{retrike}.R_{shaft,EOD} + R_{toe,EOD} \tag{4}$$

the setup effect $S_{retrike}(P1)$ on the tested pile P1 is calculated at 544% by Contractor and 345% by Reviewer. The Fig. 3 below presents the comparative estimation of lower bound pile resistance after setup R_{setup} estimated for all piles of the jacket based on the setup factor calculated on pile P1.

The resulting uncertainty on the pile resistance after restrike is ranging from 13% (pile P1) to a significant 28% (piles P4 and P5). We therefore demonstrated that an unknown accuracy on the shaft damping J_{shaft} can lead to significant differences in the calculated setup effect. Statement of Svinkin (2010) "multiplication of set-up factor on pile capacities with unknown accuracy cannot enhance the accuracy of the pile capacity after restrike" is therefore maximized in offshore practice where the unknown error on damping is reported to all piles of the jacket in the calculation of pile capacity after setup.

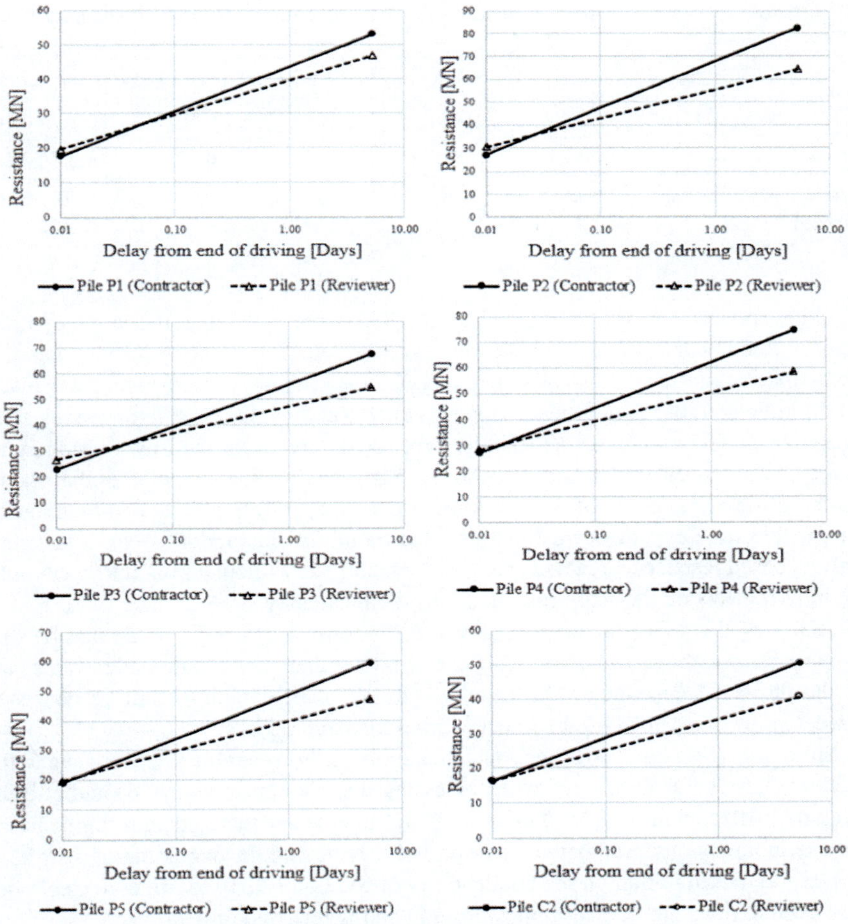

Fig. 3. Comparison of capacity extrapolated after setup

The observed variations demonstrate the importance of an accurate estimation of the shaft damping for signal matching analyses. Deviation from expected values shall be thoroughly assessed and well documented. The rest of the paper concentrates on various means available to the DMS user to guide him toward the selection of the best damping factor to be used during his signal matching assessment.

5 Expected Damping Factor from the Literature

Recommendations from literature can help identifying that a signal matching analysis has suspicious shaft damping and shall be the starting point for a re-assessment. However, limitations apply as most of the recommendations available are empirical and were established based on records from the Gulf of Mexico. Recommendation from Smith (1960), Roussel (1979) and extracted from the GRLWEAP manual (2005) are however commonly used worldwide (Table 2).

Table 2. Reference recommendations for shaft damping in the offshore industry

	Soft Clay	Firm Clay	Stiff Clay	V. Stiff Clay	Hard Clay	Silt	Sand
Smith (1960)	0.16	0.16	0.16	0.16	0.16	0.16	0.16
Roussel (1979)	0.36	0.23	0.20	0.16	0.10	0.23	0.26
GRLWEAP Manual (2005)	0.65	0.65	0.65	0.65	0.65	0.40	0.16

Other studies such as Coyle and Gibson (1970) have shown that the Smith damping factor is velocity dependent and is expected to be greater for smaller pile velocities (that is why they suggested the use of an exponential damping law). More recently, Rausche (2018) indicated that the influence of the velocity on the damping factor is greater for highly plastic Clays. This effect shall however be moderate for the Low to Moderately plastic Clays present at Platform A locations. Rausche (2018) also indicates that "signal matching overestimations of static soil resistance due to rate effects can be expected to be within the range of overestimations that traditional factors of safety or statistical analysis based resistance factors cover", however, this is without taking into account the maximized effect on the setup ratio and long-term capacity.

Applying the above recommendations to Platform A, it can be noted that (i) shaft damping factors above 1 s/m are significantly higher than the recommended values and (ii) despite some dependency to the pile velocity, shaft damping shall be consistent between piles of a same jacket showing similar driving records.

Study from Svinkin (2010) reports that damping is however time dependent and that different damping shall be used when assessing the pile capacity at EOD and at BOR. Svinkin (2010) concluded from the results of all 5 piles tested that a general augmentation of the damping is observed between the end of driving and the beginning of restrike.

Despite the limitation on the applicability of Svinkin (2010) to offshore steel piles, these observations are in clear opposition to what is reported on platform A where the shaft damping is reduced by a factor of between 2.5 and 3.6 times (for contractor) for restrikes BOR2 and BOR respectively. Such contradiction shall lead to a thorough evaluation of the signal matching and damping assessment.

6 Expected Damping Factors from Local Experience

Platform A was installed as part of the first development campaign at Field X. At the time of installation, the field X was therefore considered as a green field with no historic pile installation data. The nearest available data has been extracted from three jackets installed at field Y located approximately 85 km away in fairly similar interbedding of firm to stiff Clay and calcareous Sand. Local experience at Field Y consisted in three independent installation campaigns with 48″ to 60″ diameter piles driven using MHU 500 and MHU 600 hammers. Analyses for each campaign were performed by a different DMS user.

Smith shaft damping extracted from the signal matching analyses performed at Field Y are presented alongside analyses at Platform A and B, Field X in the Fig. 4 below.

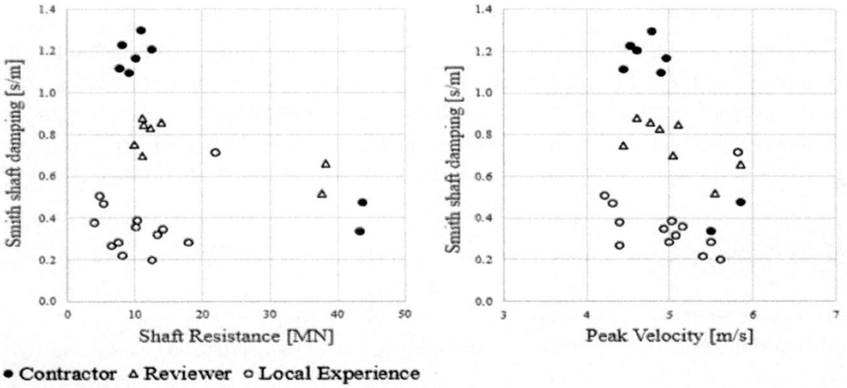

Fig. 4. Comparison of Shaft damping values obtained at Field Y (Local Experience) and at Platform A, Field X (Contractor and Reviewer) versus shaft resistance (left) and peak velocity (right)

Review of local experience from Field Y indicates that shaft damping factors are found to be generally lower than for Platform A, Field X. While some of the difference may be attributed to the different pile size geometry and different hammer used for driving, the main cause of variation is attributed to the DMS user.

The amount of experience available at Field Y (13 analyses over 3 jackets performed by 3 independent DMS users) as well as the consistency of the selected damping parameters between each other and with the recommendations from literature are indicators of a representative selection of the damping and resistance parameters. However, a possible over-estimation of the side damping of the Field Y data cannot be ruled out and limits the use of local experience for recommendations. Additional analyses from the local experience are needed to better support the guidance concluded from local experience.

7 Conclusions and Recommendations

Uncertainty remains in the selection of damping and resistance parameters when using the signal matching software CAPWAP despite software's indication of a good match. Based on the case study presented in this paper, it is recommended that the following guidance points are followed when performing a capacity analysis by signal matching.

- Based on literature and past experience, shaft damping factors shall remain below 1 s/m for open ended offshore steel piles;
- Despite some dependency to the pile velocity, shaft damping shall be consistent between piles of a same jacket showing similar driving records;
- The shaft damping shall generally follow an increasing trend with time.

Any deviation from the above guidance points shall trigger a detailed evaluation of the selected parameters. This evaluation should be performed by an experienced geotechnical engineer having a sound knowledge of the regional geology and soil conditions.

Moreover, the current practice of using a unique restrike test and setup effect to estimate the long-term capacity on all piles of a jacket can only magnify the uncertainty on the final capacity and shall be only valid where pile driving behavior and signal matching analyses are consistent. Additional restrike tests shall be recommended accordingly.

Re-evaluation of signal matching analyses for past installation following the guidance highlighted in this paper falls in the context of life extension of offshore facilities.

References

Buckley et al.: Common pitfalls of pile driving resistance analysis – a case study of the Wikinger offshore windfarm. In: Proceedings of the Offshore Site Investigation and Geotechnics (OSIG) Conference, pp. 1246–1253 (2017)

Coyle, H., Gibson, G.: empirical damping constants for sand and clays. J. Soil Mech. Found. Div. ASCE **96**(SM3), 949–965 (1970)

GRL and Associates, Inc.: GRLWEAP –Wave Equation Analysis of Pile Driving, Manual, Cleveland, Ohio (2005)

Roussel, H.J., Jr.: Pile Driving Analysis of Large Diameter high Capacity Offshore Piles, Ph. D. Dissertation, Department of Civil Engineering, Tulane University (1979)

Smith, E.A.L.: Pile Driving Analysis by the wave equation. J. Soil Mech. Foundation Div. ASCE **86**, 35–61 (1960)

Svinkin, M.R.: The Variable Damping Concept in Pile Capacity Prediction by Wave Equation Analysis (2010)

Finite Element Model of Wave Loading on a Soil Seabed Part I: Multi-layered Anisotropic Gassy Soil Conditions

Stephen D. Thomas(✉)

OGI Groundwater Specialists, Durham DH7 8TP, UK
admin@ogi.co.uk

Abstract. Ocean wave loading on a gassy soil seabed, i.e. a soft soil sediment containing biogenic methane gas bubbles, results in harmonic changes in stresses, displacements and pore pressures over depth.

For a multi-layered seabed soil with homogeneous properties within each layer, the behaviour of the dependent variables of displacement and pore pressure can be represented by a harmonic function in both horizontal space (x) and time (t).

The amplitude of the horizontal & vertical displacement and pore-water pressure will vary over depth, dependent on the particular distribution of the multi-layered properties, together with the wave loading amplitude, wave number and angular velocity, together with the boundary conditions at the seabed surface and the soil rock interface.

To simulate the behaviour of the seabed in dynamic conditions, this first paper of a two-part paper series, presents a series of governing equations based on a modified Biot theory for poro-elastic gassy soil. These equations have been further modified to take into account the harmonic behaviour using real & imaginary complex number theory.

The Finite Element Galerkin formulation combined with Green's Theorem is applied to these harmonic equations which results in six degrees of freedom per node, i.e. x-displacement, z-displacement and pore-water pressure, in both real (in-phase) & imaginary (out-of-phase) conditions.

Finally, the finite element model developed is used to present a case study of the displacement, stresses & pore pressures in a two-layer soil subject to wave loading.

Keywords: Wave loading · Marine sediment · Gassy soil · Multi-layer anisotropy

1 Introduction

This is the first of a two-part series of papers that describes a Finite Element formulation of the equations that govern the stresses, displacements and pore-water pressures induced in a multi-layered poro-elastic soft gassy soil seabed when subject to a surface wave loading.

D. V. K. Huynh et al. (Eds.): VSOE 2021, LNCE 208, pp. 105–113, 2022.
https://doi.org/10.1007/978-981-16-7735-9_9

This first paper simulates the soil behavior under the condition that the soil properties are uniform & homogeneous in the horizontal x and y directions. However, the soil structure can be anisotropic, i.e. variable in the downward z-direction as in a multi-layered soil.

Under such conditions, the behaviour of the dependent variables is harmonic in both time and x-direction, as governed by the wave period and wave length. As a consequence, only the downward z-direction requires finite element discretization, resulting in the need only for one-dimensional elements in the vertical domain, with each node having six degrees of freedom.

2 Governing Differential Equations for Multi-layer and Anisotropic Poro-Elastic Soil Conditions

Based on the Biot (1941) equations that govern the displacement, stresses and pore water pressures in a poro-elastic soil, as modified by Thomas (1987) for a Double Compressibility gassy soil, the equations of equilibrium in the x-direction, equilibrium in the z-direction and from continuity principles incorporating Darcy's Law of pore-fluid flow through a deformable poro-elastic medium, the following can be written for small strain conditions as:

$$(\lambda + 2G)\frac{\partial^2 w_x}{\partial x^2} + G\frac{\partial^2 w_x}{\partial z^2} + \lambda\frac{\partial^2 w_z}{\partial x \partial z} + G\frac{\partial^2 w_z}{\partial z \partial x} - B\frac{\partial u_w}{\partial x} = 0 \tag{1}$$

$$(\lambda + 2G)\frac{\partial^2 w_z}{\partial z^2} + G\frac{\partial^2 w_z}{\partial x^2} + \lambda\frac{\partial^2 w_x}{\partial z \partial x} + G\frac{\partial^2 w_x}{\partial x \partial z} - B\frac{\partial u_w}{\partial z} = 0 \tag{2}$$

$$\begin{aligned} \frac{\partial}{\partial x}\left[\frac{k_{xx}}{\gamma_w}\frac{\partial u_w}{\partial x} + \frac{k_{xz}}{\gamma_w}\frac{\partial u_w}{\partial z}\right] + \frac{\partial}{\partial z}\left[\frac{k_{zx}}{\gamma_w}\frac{\partial u_w}{\partial x} + \frac{k_{zz}}{\gamma_w}\frac{\partial u_w}{\partial z}\right] \\ = C\frac{\partial u_w}{\partial t} + B\frac{\partial}{\partial t}\left[\frac{\partial w_x}{\partial x} + \frac{\partial w_z}{\partial z}\right] \end{aligned} \tag{3}$$

In the above equations, w_x and w_z are the soil displacements in the x-horizontal and the z-vertical directions respectively, t is time, u_w the pore-water pressure within the soil matrix, k_{ij} the coefficient of permeability tensor, λ the Lame's constant and G the shear modulus. The gassy soil parameters B and C are derived from gassy soil theory (Thomas 1987) as follows:

$$B = \left(1 + K''/K^o\right)^{-1} \tag{4}$$

$$C = \left(K^o + K''\right)^{-1} + (1 - H)c_w n_w \tag{5}$$

The parameter K^o is the component of undrained bulk modulus governed by compression of the discrete gas bubbles, caused by an increase in total stress. Contributing to undrained compressibility is also water compressibility, c_w, water porosity, n_w, and Henry's constant H.

The parameter K'' is the component of drained bulk modulus governed by consolidation of the saturated soil matrix. This results in two independent strain components,

(i) caused by an increase in the total stress σ^o, and (ii) caused by an increase in the operative stress $(\sigma^o - u_w)$.

The modified equilibrium equations for a poro-elastic gassy soil are based on a Double Compressibility Model (Thomas 1987) where the compression of the discrete gas bubbles is governed by the increase in the pressure of the gas bubbles, which in turn is governed by the average total stress within the soil. The compression of the surrounding saturated soil matrix, however, is governed by the difference between the total stress and the pore water pressure.

This difference between total stress and pore water pressure $(\sigma^o - u_w)$ is normally known as *effective stress* (Terzaghi 1944). However, for a gassy soil, effective stress does not, on its own, govern the total volume change. For a gassy soil, this difference between the mean total stress on a soil sample and the water pressure has been coined the *operative stress* (Sills et al. 1991). Such a gassy soil structure is depicted in Fig. 1. according to Wheeler et al. (1990).

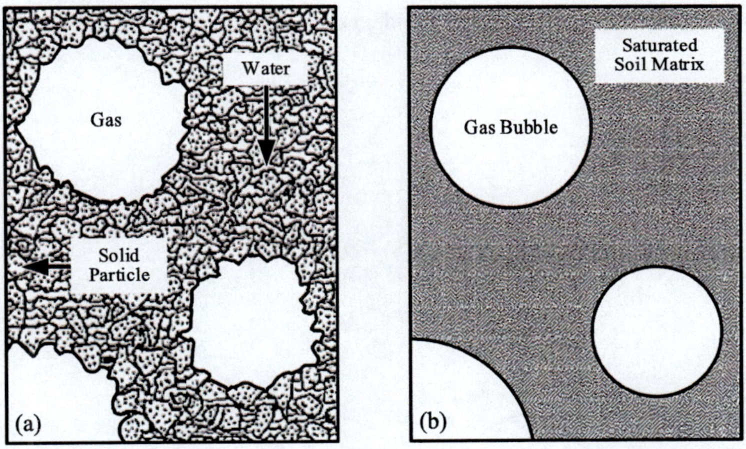

Fig. 1. Soils containing large gas bubbles: (a) Soil structure; (b) Continuum model. Wheeler et al. (1990)

This results in two independent strain components caused by (i) an increase in the total stress σ^o, and (ii) an increase in the operative stress $(\sigma^o - u_w)$,

$$\epsilon^o = \Delta V^o = \Delta\sigma^o / K^o \tag{6}$$

$$\epsilon'' = \Delta V'' = \Delta\sigma'' / K'' = (\Delta\sigma^o - \Delta u_w) / K'' \tag{7}$$

where ϵ^o is the volumetric strain due to the compression of the discrete gas bubbles caused by an increase in total mean stress $\Delta\sigma^o$, with ϵ'' as the volumetric strain due to the compression of the soil matrix caused by an increase in operative stress $\Delta\sigma''$. The main principle of the Double Compressibility Model (Thomas, 1987) for a gassy soil is that these two components of volumetric strain for a gassy soil, ϵ^o & ϵ'', can then be added directly together to provide the total volumetric strain, as follows:

$$\epsilon = \epsilon^o + \epsilon'' = \Delta\sigma^o/K^o + \Delta\sigma''/K'' = \Delta\sigma^o/K^o + (\Delta\sigma^o - \Delta u_w)/K'' \quad (8)$$

from which the change in total stress can be defined in terms of strain and pore pressure as:

$$\Delta\sigma^o = \left(\epsilon + \Delta u_w/K''\right)/\left(1/K^o + 1/K''\right) \quad (9)$$

3 Harmonic Approximation of Governing Equations

On the basis that the wave loading on the seabed is travelling in the x-direction along the surface of the seabed as depicted in Fig. 2 below, under poro-elastic conditions, based on the equations by Yamamoto et al. (1978) and Madsen (1978), any particular variable ψ which is dependent on x, z and t can be defined by a variable dependent in z only, multiplied by a harmonic function dependent on x & t.

$$\psi(x, z, t) = \psi_c + i\psi_s \quad (10)$$

where

$$\psi_c = \overline{\psi}(z)\cos(ax - \omega t) \quad (11)$$

$$\psi_s = \overline{\psi}(z)\sin(ax - \omega t) \quad (12)$$

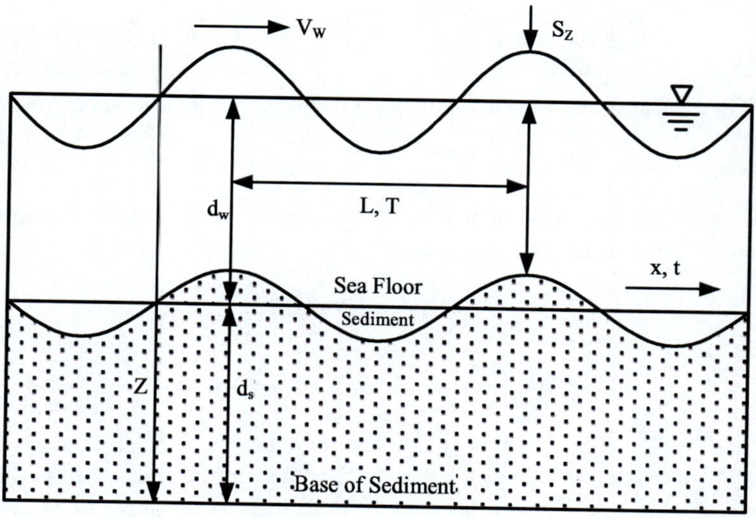

Fig. 2. Harmonic wave loading on a seabed sediment

and where $\overline{\psi}(z)$ is the peak variable function amplitude at depth z, from which:

$$\overline{\psi}(z) = \sqrt{\psi_c^2 + \psi_s^2} \tag{13}$$

$$\frac{\partial \psi}{\partial x} = +ia\psi = +ia\psi_c - a\psi_s \tag{14}$$

$$\frac{\partial \psi}{\partial t} = -i\omega\psi = -i\omega\psi_c + \omega\psi_s \tag{15}$$

In the above equations $\overline{\psi}$ is the amplitude of a particular harmonic variable at depth z, a is the wave number $2\pi/L$, and ω is the angular velocity $2\pi/T$, in which L is the wavelength and T is the angular velocity.

This results in the harmonic behaviour of displacements and pore-water pressure in a homogeneous, anisotropic soil sediment being represented as follows:

$$w_x(x, z, t) = w_{xc} + iw_{xs} = \overline{w}_x(z)[\cos(ax - \omega t) + i\sin(ax - \omega t)] \tag{16}$$

$$w_z(x, z, t) = w_{zc} + iw_{zs} = \overline{w}_z(z)[\cos(ax - \omega t) + i\sin(ax - \omega t)] \tag{17}$$

$$u_w(x, z, t) = u_{wc} + iu_{ws} = \overline{u}_w(z)[\cos(ax - \omega t) + i\sin(ax - \omega t)] \tag{18}$$

Substituting the above three harmonic functions of displacements and pore pressure into the governing equilibrium and continuity Eqs. (1), (2) and (3) produces three differential equations which can be written as a mixture of real & imaginary harmonic variables:

$$-a^2(\lambda + 2G)w_x + G\frac{\partial^2 w_x}{\partial z^2} + ia(\lambda + G)\frac{\partial w_z}{\partial z} - iaBu_w = 0 \tag{19}$$

$$+(\lambda + 2G)\frac{\partial^2 w_z}{\partial z^2} - a^2 G w_z + ia(\lambda + G)\frac{\partial w_x}{\partial z} - B\frac{\partial u_w}{\partial z} = 0 \tag{20}$$

$$-a^2\frac{k_{xx}}{\gamma_w}u_w + \frac{k_{zz}}{\gamma_w}\frac{\partial^2 u_w}{\partial z^2} + ia\frac{k_{xz}}{\gamma_w}\frac{\partial u_w}{\partial z} + ia\frac{k_{zx}}{\gamma_w}\frac{\partial u_w}{\partial z} = -i\omega C u_w + a\omega B w_x - i\omega B\frac{\partial w_z}{\partial z} \tag{21}$$

4 Galerkin Finite Element Formulation

Applying the Galerkin Finite Element formulation to the above equations in the vertical dimension, combined with Green's Theorem results in:

$$\int\left[a^2(\lambda + 2G)N_I N_J w_{xJ} + G\frac{\partial N_I}{\partial z}\frac{\partial N_J}{\partial z}w_{xJ} + iaG\frac{\partial N_I}{\partial z}N_J w_{zJ} - ia\lambda N_I\frac{\partial N_J}{\partial z}w_{zJ}\right]dz + \int iaBN_I N_J u_{wJ}\,dz = S_{xI} \tag{22}$$

from equilibrium in the x-direction

$$\int\left[(\lambda+2G)\frac{\partial N_I}{\partial z}\frac{\partial N_J}{\partial z}w_{zJ}+a^2GN_IN_Jw_{zJ}+ia\lambda\frac{\partial N_I}{\partial z}N_Jw_{xJ}-iaGN_I\frac{\partial N_J}{\partial z}w_{xJ}\right]dz$$
$$-\int B\frac{\partial N_I}{\partial z}N_Ju_{wJ}dz=S_{zI} \tag{23}$$

from equilibrium in the z-direction, and

$$\int\left[\frac{k_{zz}}{\gamma_w}\frac{\partial N_I}{\partial z}\frac{\partial N_J}{\partial z}-ia\frac{k_{xz}}{\gamma_w}N_I\frac{\partial N_J}{\partial z}+ia\frac{k_{zx}}{\gamma_w}\frac{\partial N_I}{\partial z}N_J+a^2\frac{k_{xx}}{\gamma_w}N_IN_J-i\omega CN_IN_J\right]u_{wJ}dz$$
$$+\int\left[a\omega BN_IN_Jw_{xJ}-i\omega BN_I\frac{\partial N_J}{\partial z}w_{zJ}\right]dz=Q_{wI} \tag{24}$$

from continuity principles incorporating Darcy's Law.

Substituting Eqs. 16, 17, 18 into the above three mixed real & imaginary differential equations, then by separating the real & imaginary terms, results in six differential equations, each with nodal harmonic dependent variables as follows:

$$\left(a^2(\lambda+2G)\bar{\Gamma}_{\text{IJ}}+G\bar{A}_{\text{IJ}}^{\text{zz}}\right)\bar{w}_{xcJ}-\left(aG\bar{B}_{\text{IJ}}^{\text{zN}}-a\lambda\bar{B}_{\text{IJ}}^{\text{Nx}}\right)\bar{w}_{zsJ}-aB\bar{\Gamma}_{\text{IJ}}\bar{u}_{wsJ}=\bar{S}_{xcI} \tag{25}$$

$$\left(a^2(\lambda+2G)\bar{\Gamma}_{\text{IJ}}+G\bar{A}_{\text{IJ}}^{\text{zz}}\right)\bar{w}_{xsJ}+\left(aG\bar{B}_{\text{IJ}}^{\text{zN}}-a\lambda\bar{B}_{\text{IJ}}^{\text{Nx}}\right)\bar{w}_{zcJ}+aB\bar{\Gamma}_{\text{IJ}}\bar{u}_{wcJ}=\bar{S}_{xsI} \tag{26}$$

$$\left((\lambda+2G)\bar{A}_{\text{IJ}}^{\text{zz}}+a^2G\bar{\Gamma}_{\text{IJ}}\right)\bar{w}_{zsJ}+\left(a\lambda\bar{B}_{\text{IJ}}^{\text{zN}}-aG\bar{B}_{\text{IJ}}^{\text{Nx}}\right)\bar{w}_{xcJ}+B\bar{B}_{\text{IJ}}^{\text{zN}}\bar{u}_{wsJ}=\bar{S}_{zsI} \tag{27}$$

$$\left((\lambda+2G)\bar{A}_{\text{IJ}}^{\text{zz}}+a^2G\bar{\Gamma}_{\text{IJ}}\right)\bar{w}_{zsJ}+\left(a\lambda\bar{B}_{\text{IJ}}^{\text{zN}}-aG\bar{B}_{\text{IJ}}^{\text{Nx}}\right)\bar{w}_{xcJ}+B\bar{B}_{\text{IJ}}^{\text{zN}}\bar{u}_{wsJ}=\bar{S}_{zsI} \tag{28}$$

$$\left(a^2\frac{k_{xx}}{\gamma_w}\bar{\Gamma}_{\text{IJ}}+\frac{k_{zz}}{\gamma_w}\bar{A}_{\text{IJ}}^{\text{zz}}\right)\bar{u}_{wcJ}-\left(a\frac{k_{zx}}{\gamma_w}\bar{B}_{\text{IJ}}^{\text{zN}}-a\frac{k_{xz}}{\gamma_w}\bar{B}_{\text{IJ}}^{\text{Nz}}-\omega C\bar{\Gamma}_{\text{IJ}}\right)\bar{u}_{wsJ}$$
$$+a\omega B\bar{\Gamma}_{\text{IJ}}\bar{w}_{xcJ}+\omega B\bar{B}_{\text{IJ}}^{\text{Nz}}\bar{w}_{zsJ}=\bar{Q}_{wcI} \tag{29}$$

$$\left(a^2\frac{k_{xx}}{\gamma_w}\bar{\Gamma}_{\text{IJ}}+\frac{k_{zz}}{\gamma_w}\bar{A}_{\text{IJ}}^{\text{zz}}\right)\bar{u}_{wsJ}+\left(a\frac{k_{zx}}{\gamma_w}\bar{B}_{\text{IJ}}^{\text{zN}}-a\frac{k_{xz}}{\gamma_w}\bar{B}_{\text{IJ}}^{\text{Nz}}-\omega C\bar{\Gamma}_{\text{IJ}}\right)\bar{u}_{wcJ}$$
$$+a\omega B\bar{\Gamma}_{\text{IJ}}\bar{w}_{xsJ}-\omega B\bar{B}_{\text{IJ}}^{\text{Nz}}\bar{w}_{zcJ}=\bar{Q}_{wsI} \tag{30}$$

Where

$$\bar{A}_{\text{IJ}}^{\text{zz}}=\int\frac{\partial N_I}{\partial z}\frac{\partial N_J}{\partial z}dz \tag{31}$$

$$\bar{\Gamma}_{\text{IJ}}=\int N_IN_Jdz \tag{32}$$

$$\bar{B}_{\text{IJ}}^{\text{xN}}=\int\frac{\partial N_I}{\partial x}N_Jdz \tag{33}$$

$$\bar{B}_{\text{IJ}}^{\text{Nx}}=\int N_J\frac{\partial N_I}{\partial x}dz \tag{34}$$

5 Incorporation into a Finite Element Model

Using a simple two-noded linear finite element shape function is sufficient to approximate the above matrix equations (Thomas 1989). This provides the ability to simulate accurately the displacement, stress and pore water pressure, both in-phase and out-of-phase of the harmonic loading wave (Thomas 1995).

The above Finite Element formulation has been programmed in to the software WISPP (Wave Induced Stresses & Pore Pressures) and is freely available at http://ogi.co.uk/software.

For selected input of soil conditions, together with applied boundary conditions of surface loading, wavelength and angular velocity, this software produces a graphical output of the in-phase and out-of-phase x & z-displacements and pore-pressure at each of the nodes over the depth of the seabed, and the x-x, z-z & x-z operative stresses at the centroid of each element.

6 Example of Wave Loading on a Two-Layered Soil Seabed

As an example, Fig. 3 below represents the finite element modelling of a multi-layered saturated soil with a 5m upper layer with shear modulus $G_1 = 750$ kPa & drained bulk modulus $K_1 = 1$ MPa, overlying a second 10m layer with higher stiffness of $G_2 =$ 3 MPa & $K_2 = 4$ MPa.

A harmonic wave load of 20 kPa amplitude is applied to the soil surface with a wavelength of 20 m and wave period of 20 s. The vertical domain has been discretized into 300 finite elements, with 100 elements in the upper layer and 200 elements in the lower layer.

The coefficient of permeability is 1×10^{-5} m/s for both layers. The output demonstrates that the heterogeneity resulting from two layers, with the stiffness of the underlying soil only four times that of the upper soil, produces a variation in behaviour as depicted below.

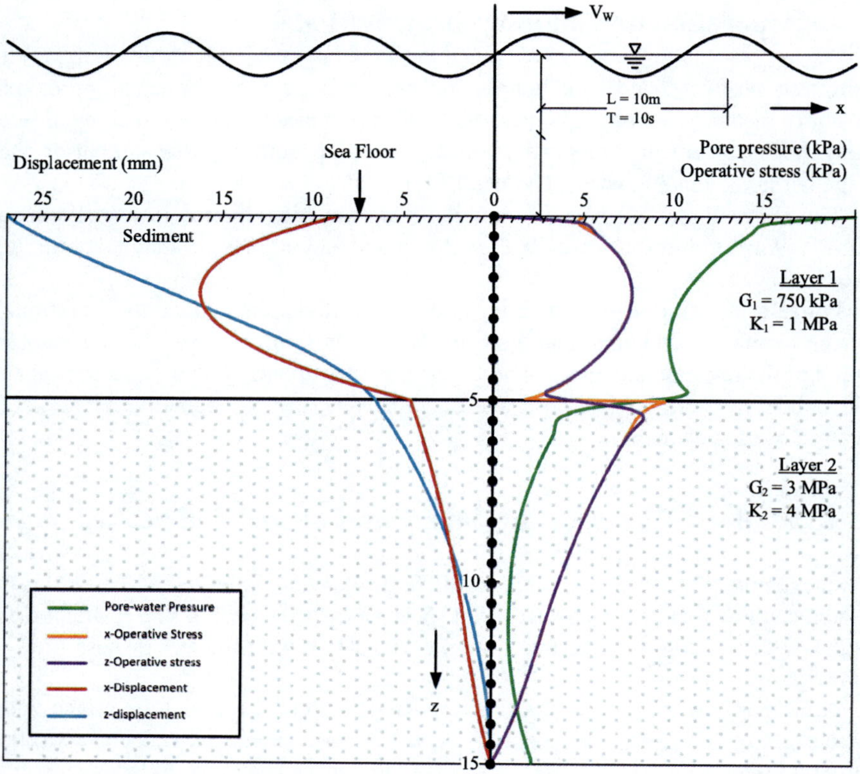

Fig. 3. Wave induced x & z displacements, x & z operative stresses and pore-water pressures in a two-layer soil, simulated using the one-dimensional harmonic finite element WISSP model (Thomas 1988)

7 Conclusions

Ocean wave loading on waves on a multi-layered seabed of varying properties can result in unexpected displacements, stresses and pore-pressure behaviour. This is especially the case for a sediment containing gassy soil where significant undrained soil compression and shear strain can occur.

As a consequence, such deformation, stress and pore-water pressure variations can result in unforeseen behaviour. For this reason, a Finite Element Model has been developed to simulate such behaviour under harmonic wave loading conditions. This model is freely available for public use.

Acknowledgements. My gratitude to colleagues George French, who regularly incorporates my original FORTRAN algorithms into more modern software, and Megan Thomas who converts my text and drawings into a more professional presentation. Finally, recognition always to then my supervisor, now colleague, Professor Gilliane Sills, who initiated at the University of Oxford her research into soft soil behaviour, including gassy soils and wave loading on the seabed.

References

Biot, M.A.: General theory of three-dimensional consolidation. J. Appl. Phys. **12**, 155–164 (1941)

Madsen, O.S.: Wave-induced pore pressures and effective stresses in a porous bed. Geotechnique **28**(4), 377–393 (1978)

Sills, G.C., Wheeler, S.J., Thomas, S.D., Gardner, T.N.: Behaviour of offshore soils containing gas bubbles. Geotechnique **41**(2), 227–241 (1991)

Terzaghi, K.: Theoretical Soil Mechanics. Wiley, Hoboken (1944)

Thomas, S.D.: The Consolidation Behaviour of Gassy Soil. D. Phil. Thesis, University of Oxford (1987)

Thomas, S.D.: WISPP: A finite element model for the analysis of wave induced stresses, and pore pressures in a variably saturated seabed. University of Oxford, Department of Engineering Science, Report OUEL 1750/88 (1988)

Thomas, S.D.: A finite element model for the analysis of wave induced stresses, displacements and pore pressures in an unsaturated seabed. I: theory. Comput. Geotech. **8**, 1–38 (1989)

Thomas, S.D.: A finite element model for the analysis of wave induced stresses, displacements and pore pressures in an unsaturated seabed. II: model verification. Comput. Geotech. **17**, 107–132 (1995)

Wheeler, S.J., Sham, W.K., Thomas, S.D.: Gas pressure in unsaturated offshore soils. Can. Geotech. J. **27**, 79–89 (1990)

Yamamoto, T., Koning, H.L., Sellmeiher, H., Hijum, E.V.: On the response of a poro-elastic bed to water waves. J. Fluid Mech. **87**(1), 193–206 (1978)

Zienkiewicz, O.C.: The Finite Element Method, 3rd edn. McGraw-Hill, New York (1977)

Finite Element Model of Wave Loading on a Soil Seabed Part II: Heterogeneous Gassy Soil Conditions

Stephen D. Thomas(✉)

OGI Groundwater Specialists, Durham DH7 8TP, UK
admin@ogi.co.uk

Abstract. Ocean wave loading on a soil seabed which contains a buried pipeline results in harmonic changes in stresses, displacements and pore pressures in the soil. When the soil is gassy, i.e. a soft soil containing biogenic gas bubbles, larger displacements are produced.

This is the second in a two-paper series which presents the application of the modified Biot equations to simulate the changes in stress, displacement and pore-water pressure in a gassy soil containing a buried pipeline when loaded by wave pressures on the seabed.

In a similar approach to the first paper in this series, the behaviour of the dependent variables of displacement & pore pressure can be represented by a harmonic function, but only in time (t) as the variables in horizontal space (x) cannot be represented by a harmonic function.

The governing modified Biot equations have been further modified to take into account the harmonic behaviour using real & imaginary complex number theory.

A Finite Element Galerkin formulation, combined with Green's Theorem, is applied to these harmonic equations, resulting in the calculation of six degrees of freedom per node, these being x-displacement, z-displacement and pore-water pressure, in both real (in-phase) & imaginary (out-of-phase) time conditions.

Finally, a two-dimensional finite element model is used to present an example simulation of the modelling of displacement, stress & pore-pressure in a seabed containing a buried pipeline.

Keywords: Wave loading · Marine sediment · Gassy soil · Heterogeneous · Buried pipeline

1 Introduction

This is the second of a two-part series of papers that describes a finite element formulation of the equations that govern the stresses, displacements and pore-water pressures induced in a multi-layered poro-elastic soft gassy soil seabed when subject to a surface wave loading.

D. V. K. Huynh et al. (Eds.): VSOE 2021, LNCE 208, pp. 114–122, 2022.
https://doi.org/10.1007/978-981-16-7735-9_10

This second paper simulates the soil behavior under the condition that the soil properties can be heterogeneous in the horizontal x-direction and in the vertical z-direction, but homogeneous in the third y-direction as required by plane-strain conditions. These conditions also apply to a soil that contains a buried structure, such as a pipeline, running orthogonally to the direction of wave travel as depicted in Fig. 1.

Under such conditions, the behaviour of the dependent variables is harmonic only in time, governed by the wave period. Subsequently, the variation in space both in the vertical z-direction and in the horizontal x-direction requires finite element discretization in both directions, resulting in the need for two-dimensional elements, with each node having six degrees of freedom.

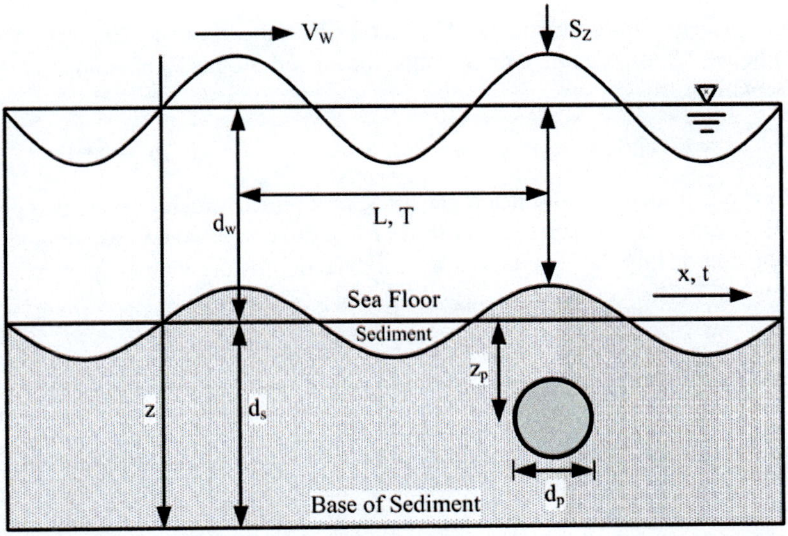

Fig. 1. Plane strain harmonic wave loading on a seabed containing a buried pipeline.

2 Governing Differential Equations for Heterogeneous and Anisotropic Poro-Elastic Soil Conditions

Based on the Biot (1941) equations that govern the displacement, stresses and pore water pressures in a poro-elastic soil, as modified by Thomas (1987) for a Double Compressibility gassy soil, the equations of equilibrium in the x-direction, equilibrium in the z-direction and from continuity principles incorporating Darcy's Law of pore-fluid flow through a deformable poro-elastic medium, the following can be written for small strain conditions as:

$$(\lambda + 2G)\frac{\partial^2 w_x}{\partial x^2} + G\frac{\partial^2 w_x}{\partial z^2} + \lambda\frac{\partial^2 w_z}{\partial x \partial z} + G\frac{\partial^2 w_z}{\partial z \partial x} - B\frac{\partial u_w}{\partial x} = 0 \tag{1}$$

$$(\lambda + 2G)\frac{\partial^2 w_z}{\partial z^2} + G\frac{\partial^2 w_z}{\partial x^2} + \lambda\frac{\partial^2 w_x}{\partial z \partial x} + G\frac{\partial^2 w_x}{\partial x \partial z} - B\frac{\partial u_w}{\partial z} = 0 \tag{2}$$

$$\frac{\partial}{\partial x}\left[\frac{k_{xx}}{\gamma_w}\frac{\partial u_w}{\partial x}+\frac{k_{xz}}{\gamma_w}\frac{\partial u_w}{\partial z}\right]+\frac{\partial}{\partial z}\left[\frac{k_{zx}}{\gamma_w}\frac{\partial u_w}{\partial x}+\frac{k_{zz}}{\gamma_w}\frac{\partial u_w}{\partial z}\right] = C\frac{\partial u_w}{\partial t}+B\frac{\partial}{\partial t}\left[\frac{\partial w_x}{\partial x}+\frac{\partial w_z}{\partial z}\right] \quad (3)$$

In the above equations, w_x and w_z are the soil displacements in the horizontal (x) and vertical (z) directions respectively, t is time, u_w is the pore-water pressure within the saturated soil matrix. The remaining parameters of the above equations are described in paper 1 of this series.

3 Harmonic Approximation of Governing Equations

Paper 1 presents the development of a harmonic finite element formulation of the governing equilibrium and continuity equations based on the Yamamoto et al. (1978) representation:

$$\psi(x,z,t)=\psi_c+i\psi_s=\overline{\psi}(z)[\cos(ax-\omega t)+i\sin(ax-\omega t)] \quad (4)$$

However, if there are soil heterogeneities, or a buried structure such as a pipeline (Fig. 1) within a poro-elastic soil, the above harmonic representation can only be made as a function of time. This results in the modification of the Yamamoto et al. (1978) equation to:

$$\psi(x,z,t)=\overline{\psi}(x,z)[\cos(-\omega t)+i\sin(-\omega t)] \quad (5)$$

$$=\psi_c+i\psi_s \quad (6)$$

where

$$\psi_c=\overline{\psi}(x,z)\cos(-\omega t) \quad (7)$$

$$\psi_s=\overline{\psi}(x,z)\sin(-\omega t) \quad (8)$$

and where $\overline{\psi}(x,z)$ is the peak variable function amplitude at coordinate (x, z).

$$\overline{\psi}(x,z)=\sqrt{\psi_c^2+\psi_s^2} \quad (9)$$

and where

$$\frac{\partial\psi}{\partial t}=-i\omega\psi=-i\omega\psi_c+\omega\psi_s \quad (10)$$

Using the harmonic functions representing the displacements and pore-pressures,

$$w_x(x,z,t)=\overline{w}_x(x,z)[\cos(-\omega t)+i\sin(-\omega t)]=\overline{w}_{xc}+i\overline{w}_{xs} \quad (11)$$

$$w_z(x,z,t)=\overline{w}_z(x,z)[\cos(-\omega t)+i\sin(-\omega t)]=\overline{w}_{zc}+i\overline{w}_{zs} \quad (12)$$

$$u_w(x,z,t)=\overline{u}_w(x,z)[\cos(-\omega t)+i\sin(-\omega t)]=\overline{u}_{wc}+i\overline{u}_{ws} \quad (13)$$

these functions can be used to convert the time derivative into a harmonic function. However, as only the continuity Eq. (3) has time derivatives, only the continuity equation results in a mixture of real and imaginary components when substituting the functions in Eqs. 11, 12, 13.

$$(\lambda + 2G)\frac{\partial^2 w_x}{\partial x^2} + G\frac{\partial^2 w_x}{\partial z^2} + \lambda\frac{\partial^2 w_z}{\partial x \partial z} + G\frac{\partial^2 w_z}{\partial z \partial x} - B\frac{\partial u_w}{\partial x} = 0 \tag{14}$$

$$(\lambda + 2G)\frac{\partial^2 w_z}{\partial z^2} + G\frac{\partial^2 w_z}{\partial x^2} + \lambda\frac{\partial^2 w_x}{\partial z \partial x} + G\frac{\partial^2 w_x}{\partial x \partial z} - B\frac{\partial u_w}{\partial z} = 0 \tag{15}$$

$$\begin{aligned}\frac{\partial}{\partial x}\left[\frac{k_{xx}}{\gamma_w}\frac{\partial u_w}{\partial x} + \frac{k_{xz}}{\gamma_w}\frac{\partial u_w}{\partial z}\right] + \frac{\partial}{\partial z}\left[\frac{k_{zx}}{\gamma_w}\frac{\partial u_w}{\partial x} + \frac{k_{zz}}{\gamma_w}\frac{\partial u_w}{\partial z}\right] \\ = -i\omega C u_w - i\omega B\left[\frac{\partial w_x}{\partial x} + \frac{\partial w_z}{\partial z}\right]\end{aligned} \tag{16}$$

4 Galerkin Finite Element Formulation

Applying the Galerkin Finite Element formulation to the above equations in both the x & z dimensions, combined with Green's Theorem (Zienkiewicz 1977) results in:

$$\begin{aligned}\int\left[(\lambda + 2G)\frac{\partial N_I}{\partial x}\frac{\partial N_J}{\partial x} + G\frac{\partial N_I}{\partial z}\frac{\partial N_J}{\partial z}\right]w_{xJ} + \left[\lambda\frac{\partial N_I}{\partial x}\frac{\partial N_J}{\partial z} + G\frac{\partial N_I}{\partial z}\frac{\partial N_J}{\partial x}\right]w_{zJ}dR \\ - \int\left[B\frac{\partial N_I}{\partial x}N_J\right]u_{wJ}dR = \int N_I S_x d + \int \mathrm{N_I F_x dR}\end{aligned} \tag{17}$$

from equilibrium in the x-direction,

$$\begin{aligned}\int\left[G\frac{\partial N_I}{\partial x}\frac{\partial N_J}{\partial z} + \lambda\frac{\partial N_I}{\partial z}\frac{\partial N_J}{\partial x}\right]w_{xJ} + \left[G\frac{\partial N_I}{\partial x}\frac{\partial N_J}{\partial x} + (\lambda + 2G)\frac{\partial N_I}{\partial z}\frac{\partial N_J}{\partial z}\right]w_{zJ}dR \\ - \int\left[B\frac{\partial N_I}{\partial z}N_J\right]u_{wJ}dR = \int N_I S_z d + \int \mathrm{N_I F_z dR}\end{aligned} \tag{18}$$

from equilibrium in the z-direction, and

$$\begin{aligned}\int\left[\left(\frac{k_{xx}}{\gamma_w}\frac{\partial N_I}{\partial x}\frac{\partial N_J}{\partial x} + \frac{k_{xz}}{\gamma_w}\frac{\partial N_I}{\partial x}\frac{\partial N_J}{\partial z} + \frac{k_{zx}}{\gamma_w}\frac{\partial N_I}{\partial z}\frac{\partial N_J}{\partial x} + \frac{k_{zz}}{\gamma_w}\frac{\partial N_I}{\partial z}\frac{\partial N_J}{\partial z} - i\omega C N_I N_J\right)u_{wJ}\right]dR \\ - \int\left[i\omega B N_I\frac{\partial N_J}{\partial x}w_{xJ} + i\omega B N_I\frac{\partial N_J}{\partial z}w_{zJ}\right]dR = \int q_w d\end{aligned} \tag{19}$$

from continuity principles incorporating Darcy's Law

Substituting the harmonic representation of the x-displacement, z-displacement and pore-pressure (Eqs. 11, 12, 13) into the above three differential equations, then separating real & imaginary term, results in six differential equations, with six harmonic dependent variables as follows:

$$\left((\lambda + 2G)\overline{A}_{\text{IJ}}^{\text{xx}} + G\overline{A}_{\text{IJ}}^{\text{zz}}\right)\overline{w}_{xcJ} + \left(\lambda\overline{A}_{\text{IJ}}^{\text{xz}} + G\overline{A}_{\text{IJ}}^{\text{zx}}\right)\overline{w}_{zcJ} - \left(B\overline{B}_{\text{IJ}}^{\text{xN}}\right)u_{wcJ} = S_{xcI} + F_{xcI} \tag{20}$$

$$\left((\lambda + 2G)\overline{A}_{\text{IJ}}^{\text{xx}} + G\overline{A}_{\text{IJ}}^{\text{zz}}\right)\overline{w}_{xsJ} + \left(\lambda\overline{A}_{\text{IJ}}^{\text{xz}} + G\overline{A}_{\text{IJ}}^{\text{zx}}\right)\overline{w}_{zsJ} - \left(B\overline{B}_{\text{IJ}}^{\text{xN}}\right)u_{wsJ} = S_{xsI} + F_{xsI} \tag{21}$$

$$\left((\lambda + 2G)\overline{A}_{\text{IJ}}^{\text{zz}} + G\overline{A}_{\text{IJ}}^{\text{xx}}\right)\overline{w}_{zcJ} + \left(\lambda\overline{A}_{\text{IJ}}^{\text{zx}} + G\overline{A}_{\text{IJ}}^{\text{xz}}\right)\overline{w}_{xcJ} - \left(B\overline{B}_{\text{IJ}}^{\text{zN}}\right)u_{wcJ} = S_{zcI} + F_{zcI} \tag{22}$$

$$\left((\lambda + 2G)\overline{A}_{\text{IJ}}^{\text{zz}} + G\overline{A}_{\text{IJ}}^{\text{xx}}\right)\overline{w}_{zsJ} + \left(\lambda\overline{A}_{\text{IJ}}^{\text{zx}} + G\overline{A}_{\text{IJ}}^{\text{xz}}\right)\overline{w}_{xsJ} - \left(B\overline{B}_{\text{IJ}}^{\text{zN}}\right)u_{wsJ} = S_{zsI} + F_{zsI} \tag{23}$$

$$\begin{aligned}&\left(\frac{k_{xx}}{\gamma_w}\overline{A}_{\text{IJ}}^{\text{xx}} + \frac{k_{xz}}{\gamma_w}\overline{A}_{\text{IJ}}^{\text{xz}} + \frac{k_{zx}}{\gamma_w}\overline{A}_{\text{IJ}}^{\text{zx}} + \frac{k_{zz}}{\gamma_w}\overline{A}_{\text{IJ}}^{\text{zz}}\right)\overline{u}_{wcJ} + \omega C\overline{\Gamma}_{\text{IJ}}\overline{u}_{wsJ}\\ &\qquad + \omega B\overline{B}_{\text{IJ}}^{\text{Nx}}\overline{w}_{xsJ} + \omega B\overline{B}_{\text{IJ}}^{\text{Nz}}\overline{w}_{zsJ} = \overline{Q}_{wcI}\end{aligned} \tag{24}$$

$$\begin{aligned}&\left(\frac{k_{xx}}{\gamma_w}\overline{A}_{\text{IJ}}^{\text{xx}} + \frac{k_{xz}}{\gamma_w}\overline{A}_{\text{IJ}}^{\text{xz}} + \frac{k_{zx}}{\gamma_w}\overline{A}_{\text{IJ}}^{\text{zx}} + \frac{k_{zz}}{\gamma_w}\overline{A}_{\text{IJ}}^{\text{zz}}\right)\overline{u}_{wsJ} - \omega C\overline{\Gamma}_{\text{IJ}}\overline{u}_{wcJ}\\ &\qquad - \omega B\overline{B}_{\text{IJ}}^{\text{Nx}}\overline{w}_{xcJ} - \omega B\overline{B}_{\text{IJ}}^{\text{Nz}}\overline{w}_{zcJ} = \overline{Q}_{wsI}\end{aligned} \tag{25}$$

Where

$$\overline{A}_{\text{IJ}}^{\text{xx}} = \iint \frac{\partial N_I}{\partial x}\frac{\partial N_J}{\partial x}\,\text{dxdz} \qquad \overline{A}_{\text{IJ}}^{\text{xz}} = \iint \frac{\partial N_I}{\partial x}\frac{\partial N_J}{\partial z}\,\text{dxdz}$$

$$\overline{A}_{\text{IJ}}^{\text{zx}} = \iint \frac{\partial N_I}{\partial z}\frac{\partial N_J}{\partial x}\,\text{dxdz} \qquad \overline{A}_{\text{IJ}}^{\text{zz}} = \iint \frac{\partial N_I}{\partial z}\frac{\partial N_J}{\partial z}\,\text{dxdz}$$

$$\overline{B}_{\text{IJ}}^{\text{xN}} = \iint \frac{\partial N_I}{\partial x}N_J\,\text{dxdz} \qquad \overline{B}_{\text{IJ}}^{\text{Nx}} = \iint N_I\frac{\partial N_J}{\partial x}\,\text{dxdz}$$

$$\overline{B}_{\text{IJ}}^{\text{zN}} = \iint \frac{\partial N_I}{\partial z}N_J\,\text{dxdz} \qquad \overline{B}_{\text{IJ}}^{\text{Nz}} = \iint N_I\frac{\partial N_J}{\partial z}\,\text{dxdz}$$

$$\overline{\Gamma}_{\text{IJ}} = \iint N_I N_J\,\text{dxdz} \qquad \overline{F}_I = \iint N_I\,\text{dxdz}$$

5 Incorporation into a Finite Element Model

A simple first order two-dimensional finite element shape function is sufficient to approximate the above matrix equations. This provides the ability to simulate accurately the displacement, stress and pore water pressure, both in-phase and out-of-phase of the harmonic loading wave surrounding a buried structure or other heterogeneous soil conditions.

As the boundary conditions on the down-wave face and up-wave face of the two-dimensional soil model are critical to the simulation of the soil behaviour surrounding the pipeline, careful treatment of these boundaries is required. In the WISDOM model, the boundaries are set to be sufficiently distance from the buried pipeline to mitigate the boundary affects. Furthermore, at the specific location of the down-wave or up-wave boundaries, the one-dimensional harmonic model WISSP (Thomas 1988, 1989) is applied. This model calculates the nodal displacements and pore-pressure at each boundary node in the vertical dimension as though the buried pipeline was not present. These calculated boundary nodal values are then applied to calculate the in-phase and out-of-phase stresses and pore-pressures on the boundaries to the two-dimensional model that simulates the soil behaviour over the domain area specified.

These have been programmed into the software WISDOM (Wave Induced Stresses & Displacements around Offshore Marine pipelines, Thomas 1992) using a finite element mesh comprising a combination of triangular and rectangular elements as depicted in Fig. 2.

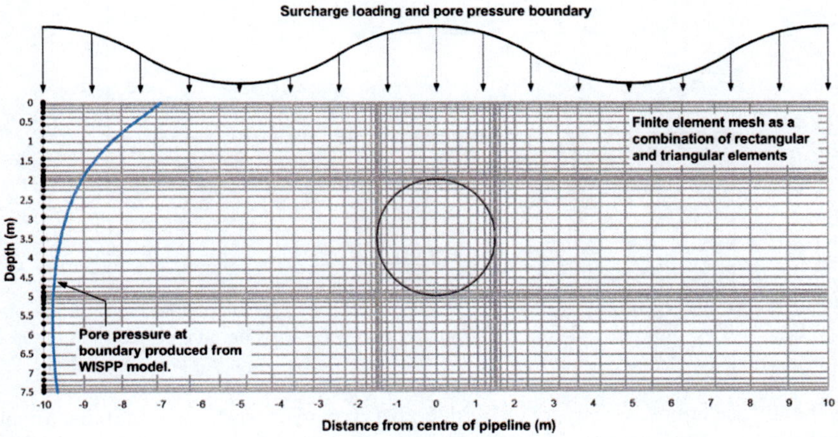

Fig. 2. Two-dimensional model produced by WISDOM (boundary conditions by WISPP)

For selected input of soil conditions, together with applied boundary conditions of surface loading, wavelength and angular velocity, this software produces a graphical output of the in-phase and out-of-phase x-displacement & z-displacement and pore-pressure at each of the nodes over the depth of the seabed, and the x-x, z-z & x-z stresses at the centroid of each element.

For selected input of soil conditions, together with applied boundary conditions of surface loading, wavelength and angular velocity, this software produces a graphical output of the in-phase and out-of-phase x-displacement, z-displacement and pore-pressure at each of the nodes over the depth of the seabed, together with the stresses at the centroid of each element.

6 Example Simulation of Wave Loading on a Buried Pipeline

Figure 3 below represents the WISDOM model applied to a 7.5 m thick saturated sediment layer (G = 750k Pa, K = 1 MPa, k = 1 × 10^{-5} m/s) subject to a 20 kPa harmonic wave amplitude loading. The loading has a wavelength of 10 m and wave period of 20 s. The modelled domain is discretized into a triangular and rectangular finite element mesh as depicted in Fig. 2.

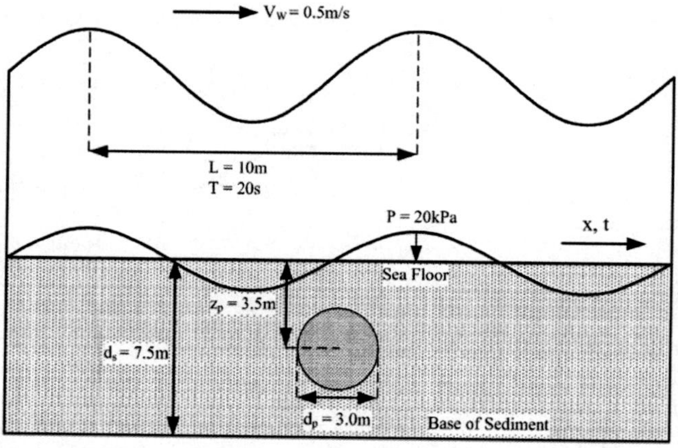

Fig. 3. Two-dimensional conceptual model of a wave loading on a buried pipeline

The surface boundary is specified with both a 20 m vertical stress amplitude, together with a fixed pore water pressure of 20 kPa amplitude. The interface between the sediment and the underlying rock is specified as a rough boundary condition, i.e. zero displacement both in the vertical and horizontal direction.

The application of the WISDOM Model to the conceptual model of Fig. 3 results in the calculation of nodal values of in-phase and out-of-phase displacements and pore-water pressures. Operative stresses are calculated at the element centroids. Figure 4 below depicts a contour plot of the increased pore-water pressure amplitude as the wave passes over the pipeline. This figure demonstrates that exponential decay with depth of the pore-water pressure amplitude is broken above and below the pipeline due to the high stiffness of the structure.

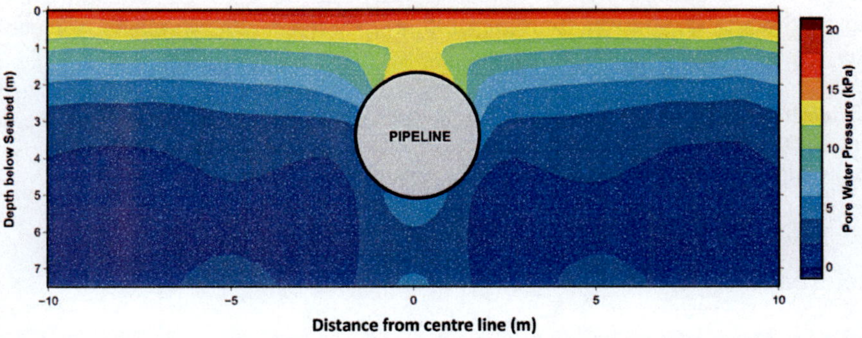

Fig. 4. Simulated wave induced pore-water pressure amplitude surrounding a buried pipeline

The graphical output depicted below in Fig. 5 demonstrates that the presence of a pipeline impacts on the pore pressure behaviour both below and above the pipeline, and that the pipeline has displacement, both in-phase and out-of-phase with the wave loading.

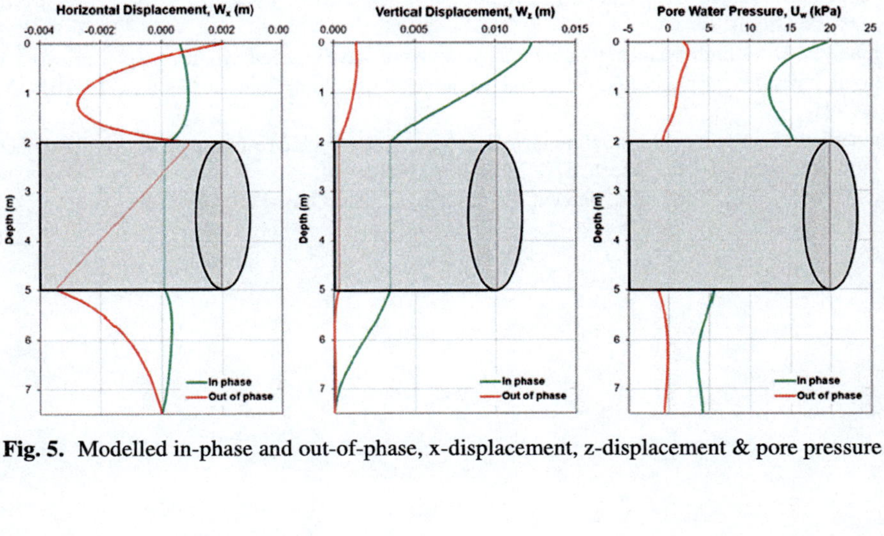

Fig. 5. Modelled in-phase and out-of-phase, x-displacement, z-displacement & pore pressure

7 Conclusions

Ocean wave loading on waves on a heterogeneous seabed of varying properties can result in unexpected displacements, stresses and pore-pressure behaviour. This is especially the case for a sediment containing a buried structure where significant concentrations of stress and pore-pressure can occur, potentially resulting in local liquefaction.

As a consequence, such deformation, stress and pore-water pressure variations can result in unforeseen behaviour. For this reason, a two-dimensional Finite Element model (WISDOM) has been developed to simulate such behaviour under harmonic wave loading conditions. This model is in preparation to become freely available for public use.

Acknowledgements. My gratitude to colleagues George French, who regularly incorporates my original FORTRAN algorithms into more modern software, and Megan Thomas who converts my text and drawings into a more professional presentation.

References

Biot, M.A.: General theory of three-dimensional consolidation. J. Appl. Phys. **12**, 155–164 (1941)

Thomas, S.D.: The Consolidation Behaviour of Gassy Soil. D. Phil. Thesis, Oxford University (1987)

Thomas, S.D.: WISPP: a finite element model for the analysis of wave induced stresses, and pore pressures in a variably saturated seabed. University of Oxford, Department of Engineering Science, Report OUEL 1750/88 (1988)

Thomas, S.D.: A finite element model for the analysis of wave induced stresses, displacements and pore pressures in an unsaturated seabed. I: theory. Comput. Geotech. **8**, 1–38 (1989)

Thomas, S.D.: WISDOM: Wave Induced Stresses and Displacements around Offshore Marine Pipelines. Oxford Geotechnica Ltd Report (ref: 018/wi015) prepared for the University of Oxford (1992)

Thomas, S.D.: A finite element model for the analysis of wave induced stresses, displacements and pore pressures in an unsaturated seabed. II: model verification. Comput. Geotech. **17**, 107–132 (1995)

Yamamoto, T., Koning, H.L., Sellmeiher, H., Hijum, E.V.: On the response of a poro-elastic bed to water waves. J. Fluid Mech. **87**(1), 193–206 (1978)

Zienkiewicz, O.C.: The Finite Element Method, 3rd edn. McGraw-Hill, New York (1977)

Geotechnical Characteristics of Seabed Deposits at Su Tu Trang Site, Cuu Long Basin, Offshore Vietnam

Huy Dong Ngo[1], Quy Ngoc Pham[2], Tien Dung Nguyen[3(✉)], and Huy Hien Doan[2]

[1] FECON Corporation, Hanoi, Vietnam
[2] Vietnam Petroleum Institute, Hanoi, Vietnam
[3] VNU Vietnam Japan University, Hanoi, Vietnam
nt.dung@vju.ac.vn

Abstract. The increasing demand for oil and gas exploration and use of wind energy in recent years has led to increase in construction of nearshore and offshore structures, which requires thorough understanding of soil characteristics in the areas for foundation design. This paper presents some geological features and soil characteristics at an offshore site in Cuu Long basin, southern shelf of Vietnam. Under the water depth of 56.3 m, the soil profile up to 140 m consists of mainly silty and clayey soils. The CPTu-based classification chart proposed by Robertson (2009) reflects soil types well compared with that obtained from laboratory samples. The s_u/σ'_{v0} profile of the silty clay layer (40.0–100.0 m) is relatively constant with depth and has an average value of $s_u/\sigma'_{v0} = 0.19$.

Keywords: Offshore site investigation · Soil classification · CPTu · Sampling

1 Introduction

Vietnam has huge oil and gas basins along the coast with 1.4 billion cubic meters of oil equivalent, in which the oil reverse is about 700 million cubic meters and gas reverse is about 700 million cubic meter of oil equivalent (Bojesen-Koefoed et al. 2009). The increasing demand for oil and gas exploration and use of wind energy in recent years has led to significant increase in construction of structures, such as drill rigs, pipe line system, and wind turbines, in nearshore and offshore areas of Vietnam. The construction of these structures requires rigorous design of foundation, in which the understanding of soil characteristics plays a vital role in the foundation design. However, to the best knowledge of the authors, there has been very little information of soil characteristics in the nearshore and offshore areas of Vietnam. Therefore, studies and reports on geological features and soil characteristics in such areas would be very valuable references for foundation design of nearshore and offshore structures.

This paper presents some geological features and soil characteristics at Su Tu Trang site, Cuu Long basin, the southern shelf of Vietnam. The geotechnical investigation at the site was for construction of drill rigs of Petro Vietnam (PVN). The outcome of this study would enrich the understanding in offshore geological characteristics and serve as a good reference for geotechnical engineers in designing foundations in the area.

D. V. K. Huynh et al. (Eds.): VSOE 2021, LNCE 208, pp. 123–130, 2022.
https://doi.org/10.1007/978-981-16-7735-9_11

2 Field Test Program

2.1 Site Location

The geotechnical investigation was carried out at Su Tu Trang site, Cuu Long Basin, which is sometimes referred to as the Mekong Basin in the southern shelf of Vietnam. It is well-known as the major hydrocarbon basin in Vietnam with a total area of approximately 25,000 km^2. The site location for the investigation program is shown in Fig. 1. The water depth in this area was measured at each borehole location upon commencement of the drilling activities using a combination of electronic sensor and echo sounder. It was shown that the average measured water depth at the site was approximately 56.30 m.

Fig. 1. Location of the investigated site (Fugro 2007)

2.2 In-Situ and Laboratory Tests

In this study, in-situ tests at the site included one sample borehole to a depth of 114.5 m (STT-BH), one Cone Penetrometer Test (CPTu)/sampling composite borehole to a maximum depth of 141.9 m (STT-CPT), one CPTu/sampling composite borehole to a maximum depth of 30.6 m (STT-Jackup). The three boreholes were located in a North-East line, in which the STT-BH (in between) was 11.18 m from the STT-BH and 25.23 m from the STT-Jackup. Table 1 shows some key information and functions of the boreholes at the site.

Drilling operations were carried out through a motion compensated Furgo Offshore Drilling Rig over a moon pool in the centre of the vessel. The heave compensated soil drilling method consisted of a straight flush rotary system with an open-center

Table 1. Information of the boreholes

Borehole	Depth (m)	Function
SST-BH	114.5	Sample collection and visual classification
SST-CPT	141.9	Sample collection and CPTu test (composite borehole)
SST-Jackup	30.6	Sample collection and CPTu test (composite borehole)

21.5 cm. On the other hand, sampling was conducted through the open-centre bit using Wison BHA and collars, which facilitated umbilical piston, push and hammer sampling operations, as well as downhole CPTu test. In cohesive materials, seawater was used to clean and stabilize the borehole. In cohesionless materials, drilling mud composed of guar gum and seawater ballast was used to stabilize the hole and to aid retention of cuttings. Samples were then obtained by piston push sampling techniques using Wison push thin-walled (76-mm in outer diameter, 72-mm in inner diameter).

The CPTu was conducted using Fugro's Downhole Wison system as schematically shown in Fig. 2. The test was performed using a 10 cm^2 base area, 60° cone with a 150 cm^2 cylindrical sleeve. The filter element for pore pressure measurement was located on the shoulder of the cone tip (u_2 position). The rate of penetration during testing was kept between 20 mm to ±5 mm per second throughout the 3.0 m continuous stroke.

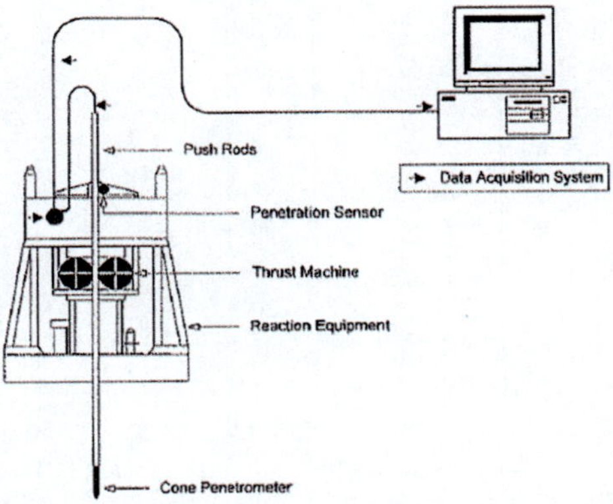

Fig. 2. The downhole CPT system used in the field test program (Fugro 2007)

In the laboratory, a number of tests were carried out, including classification tests (unit weight, water content, Atterberg limits, particle size analysis, specific gravity), undrained shear strength test (pocket penetrometer, torvane, miniature vane, unconfined compression test), triaxial tests (UU, CU, and CD), 1-D consolidation test, and direct shear tests. All the tests were carried out according to the specific ASTM or BS standards.

3 Soil Characterization

3.1 Soil Profile

Typical soil parameters obtained from the classification tests were used to graph the soil profile as shown in Fig. 3. Up to the depth of 140.0 m, the soil profile contains many alternate sub-layers of silty clay, clayey silt, sandy silt and silty sand. However, the soil profile might be broadly divided into four main layers: (1) fine sand, silty sand, silt (0–40.0 m); (2) silty clay, stiff clay, silt (40.0–100.0 m); (3) fine sand, sandy silt and silt (100.0–128.4 m); and (4) silty clay (128.4–140.0 m).

It is interesting to note that the unit weight (γ_t) value of the soil profile varies slightly with depth (19.7 to 20.9 kN/m^3), indicating that the soil layers are rather homogeneous in type. Similar to the unit weight, the water content (w_n) also varies in a narrow range (15.0 to 30.0%). The clay layers at the site are typically classified as low to medium plasticity ($PI = 10$ to 20%). The liquidity index (I_L) of the clay layers varies inconsistently with depth and most of the values are less than 1.0, indicating that that the liquid limit is higher than the water content at most depths.

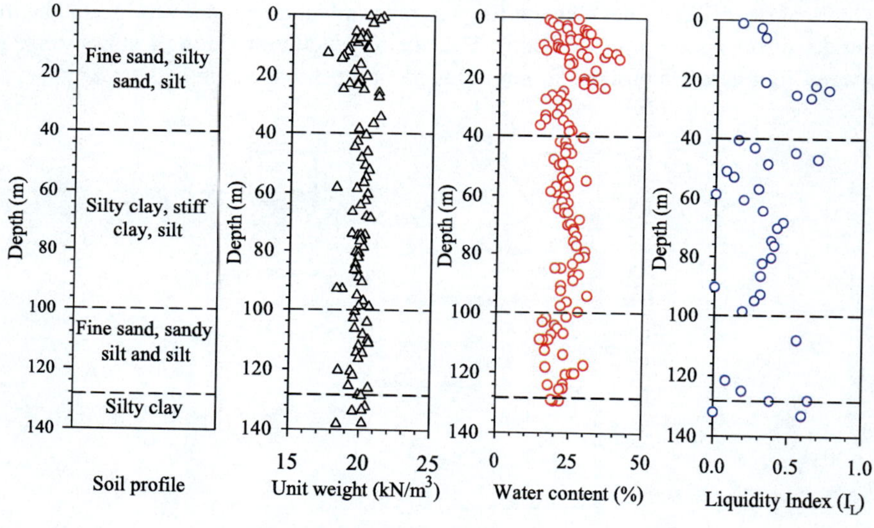

Fig. 3. Physical properties profile at the site

The cone resistance (q_c), sleeve friction (f_s), pore water pressure (u_2) diagrams, and soil behavior type index (I_c) from CPTu test are presented in Fig. 4. It is interesting to note from q_c and u_2 diagrams that the four main soil layers can easily be identified and are well matched with the classification from physical parameters profile shown in Fig. 3. The u_2 diagrams show an unusual characteristic, that is, the pore pressure is mostly negative in silty sand and sandy silt layers (typically 0–40.0 m, and 100.0–110.0 m). This unusual characteristic is typically found in sandy deposits in offshore environment due to the cavitation phenomenon (Lunne et al. 1997). On the other hand, the dilative behavior of the soil within this depth might cause the negative pore water pressure in the soil.

As shown in Fig. 4, the soil behavior type index (I_c) (Robertson 2009) also helps to classify the soil profile well. For the depths of 40.0 to 100.0 m, the I_c is pronouncedly larger than 2.6, indicating that the layer is clayey soil. The index value of other layers is mostly equal or less than this boundary value, indicating the layers are silty to sandy soils.

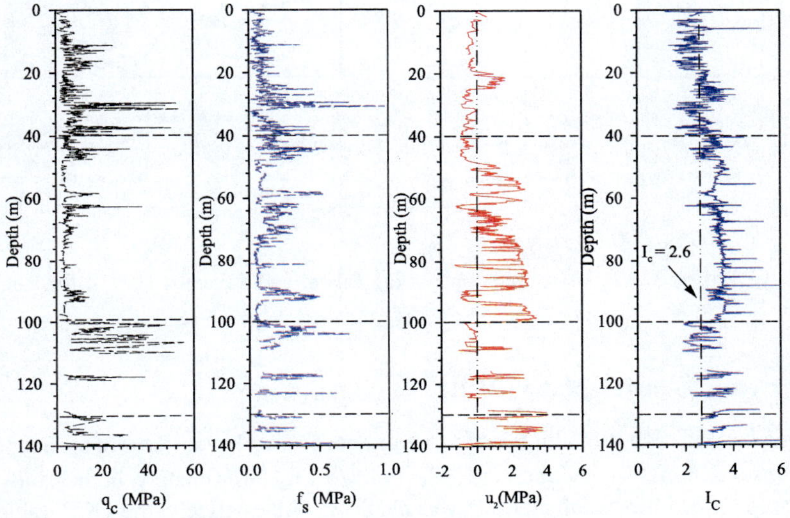

Fig. 4. CPTu test results at the test site

It is interesting to examine the applicability of CTPu-based soil classification charts to classify the soil profile in this study. Figure 5 shows the classification charts obtained from the ones proposed by Eslami and Fellenius (1997) and Robertson (2009). Note from Fig. 5(a) that $q_E = q_t - u_2$ and from Fig. 5(b) that the normalized cone resistance $Q_{tn} = [(q_t - \sigma_{v0})/p_a](p_a/\sigma'_{v0})^n$, the normalized friction ratio $F_r = [f_s/(q_t - \sigma_{v0})]100\%$, the stress exponent $n = 0.381(I_c) + 0.05(\sigma'_{v0}/p_a) - 0.15$. For better recognition, the data points from the four main layers were plotted using different symbols. As shown, the Eslami and Fellenius (1997) chart tends to indicate soil types that are coarser than ones obtained from the Robertson (2009) chart. For example, the first layer (0–40.0 m) is classified mostly as silt mixture, sand mixture and clean sand (zones 4,5, and 6) in Robertson (2009) chart whereas the layer is classified as sandy silt, silty sand and sand and/or sandy gravel (zones 4a, 4b and 5) in Eslami and Fellenius (1997) chart. Especially, for the second main layer (40–100.0 m), it is classified as clays (clay to silty clay) (Zone 3) in Robertson (2009) chart whereas the layer is classified as silty clay and/or clayey silt, sandy silty and silty sand (Zones 3, 4a and 4b) in Eslami-Fellenius chart. Taking classification from the soil samples in the lab as a reference, the Robertson (2009) chart results in better classification in this study.

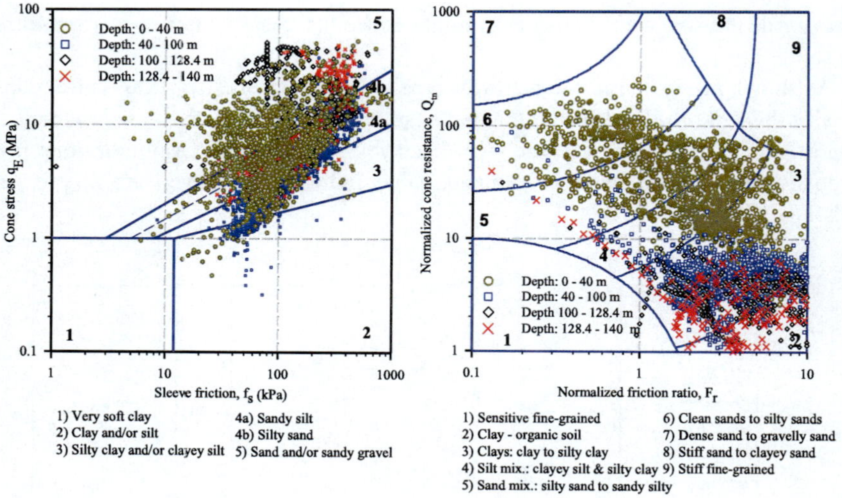

(a) (b)

Fig. 5. CPTu-based soil classification charts: (a) Eslami and Fellenius (1997); (b) Robertson (2009)

3.2 Overconsolidation Ratio (OCR)

The preconsolidation stress (σ'_p) is often determined from *e*-log σ'_v curve of the oedometer (1D consolidation) test, and then the OCR ($=\sigma'_p/\sigma'_{v0}$). Unfortunately, in this study only one sample from the oedometer test was available. Alternatively, the OCR value of a clayey soil may be indirectly determined from CPTu data by using the following two general equations (Mayne 2007).

$$OCR = \frac{\sigma'_p}{\sigma'_{v0}} = \frac{k_1(q_t - \sigma_{vo})}{\sigma'_{vo}} \quad (1)$$

$$OCR = \frac{\sigma'_p}{\sigma'_{v0}} = \frac{k_2(q_t - u_2)}{\sigma'_{vo}} \quad (2)$$

where σ'_p is preconsolidation stress of the soil; k_1 *and* k_2 are empirical factors. For intact clays, a first-order estimate can be made by using $k_1 = 0.33$ and $k_2 = 0.6$ (Mayne 2007).

Figure 6(a) shows the CPTu-based OCR profiles for the layer 2 (40.0–100.0 m) obtained from Eq. (1) and Eq. (2) with the use of suggested values of $k_1 = 0.33$ and $k_2 = 0.60$, respectively. The OCR value obtained from the oedometer test is also plotted in the figure. Note that the profiles are rather spiky with depth due to the existence of thin lenses of silty soil. As shown, Eq. (1) with typical $k_1 = 0.33$ slightly underestimates the OCR value (i.e., OCR < 1.0) from the depth of 75.0–100 m, whereas Eq. (2) with the typical $k_2 = 0.6$ results in OCR values of approximately 1.0 in these depths. Both method results in OCR value of larger than 1.0 in the upper depths (40.0 to 75.0 m). The oedometer-based OCR value (of around 2.0) represent an average of the two values from Eqs. (1) and (2). From the OCR profiles, it may be concluded that, for this layer, the soil is slightly overconsolidated to overconsolidated (OC) in the depths from 40.0 to 75.0 m and normally consolidated (NC) in the depths from 75.0 to 100.0 m.

3.3 Undrained Shear Strength (s_u)

The s_u values of the clayey soil at various depths were obtained from Torvane, pocket penetrometer, unconsolidated undrained triaxial (UU) and consolidated undrained triaxial (CU) tests. Figure 6(b) shows the normalized undrained shear strength values (s_u/σ'_{v0}) obtained from the tests for the layer 2. As shown, among the tests presented, the UU and pocket penetrometer tests produced rather scattered values with depths, whereas the CU and torvane tests produced rather similar results throughout the soil profile.

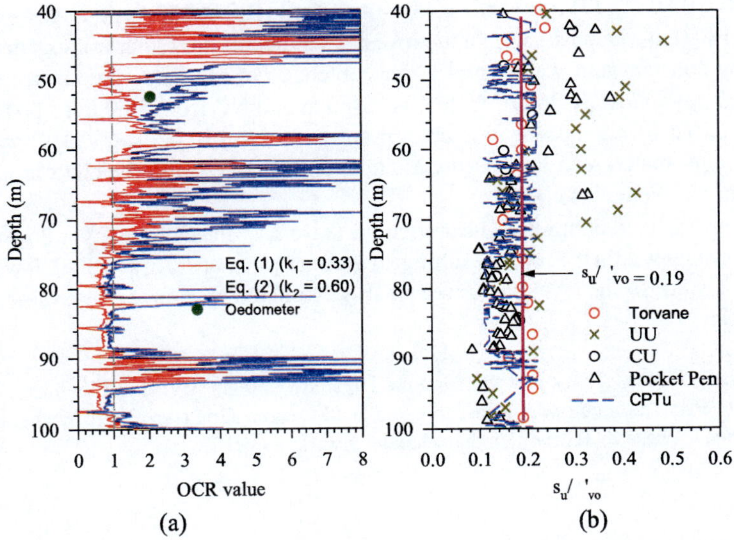

Fig. 6. Variation of (a) OCR values with depth and (b) s_u/σ'_{v0} ratio with depth

Putting aside the s_u values from the UU and pocket penetrometer tests, the s_u value of the silty clay layer obtained from the CU and torvane tests increases almost linearly with the increase in effective vertical stress (σ'_{v0}), which results in an average value of $s_u/\sigma'_{v0} = 0.19$ as shown in the figure.

The s_u value might be indirectly determined from the CPTu data as follows (Lunne et al. 1997):

$$s_u = \frac{q_t - \sigma_{vo}}{N_{kt}} \tag{3}$$

Where q_t is the corrected cone resistance [$q_t = q_c + (1 - a)u_2$], σ_{v0} is the total overburden stress; N_{kt} is a cone factor, an empirical value that depends very much on geologic conditions (Lunne et al. 1997). For a new site, the N_{kt} value must be calibrated from reference s_u values at the site obtained direct shear tests (e.g., field vane test).

Taking s_u values from the CU and torvane tests as references ones, an average $N_{kt} = 20$ was found reasonable for the layer 2 as shown in the figure. The N_{kt}-based s_u/σ'_{v0} profile fluctuates with depth but in general the ratio is rather constant in the depths of 40.0–100.0 m.

4 Conclusions

This paper presented a study on some geological features and soil characteristics at Su Tu Trang site, Cuu Long basin, the southern shelf of Vietnam. The following conclusions can be made from the study.

Under the water depth of 56.3 m, the soil profile up to 140.0 m at the site consists of four main soil layers: (1) fine sand, silty sand, silt (0–40.0 m); (2) silty clay, stiff clay, silt (40.0–100.0 m); (3) fine sand, sandy silt and silt (100.0–128.4 m); and (4) silty clay (128.4–140.0 m). Deposits within the investigated depths are rather homogeneous with relatively constant unit weight and water contents. For the silty clay layer, the soil is classified as slightly OC to OC from 40.0–75.0 m and NC from 75.0 m to 100.0 m.

The CPTu-based classification chart proposed by and Robertson (2009) was found to give better match with lab-based classification compared with the chart proposed by of Eslami and Fellenius (1997).

The normalized undrained shear strength (s_u/σ'_{v0}) values of the layer 2 were found rather constant within the layer, resulting an average value of $s_u/\sigma'_{v0} = 0.19$. It was found that the s_u value of the layer can be reasonably estimated with an average value value of $N_{kt} = 20$.

Acknowledgements. This study was supported by PetroVietnam (PVN) and Vietnam Petroleum Institute (VPI). The study was also supported by Research Grant provided by the Technical Cooperation – Phase 2 (TC2) of Vietnam Japan University (VJU).

References

Bojesen-Koefoed, J.A., Nytoft, H.P., Dau, N.T.: Petroleum composition in the Cuu long Basin (Mekong Basin) offshore southern Vietnam. Mar. Pet. Geol. **26**(6), 899–908 (2009)

Eslami, A., Fellenius, B.H.: Pile capacity by direct CPT and CPTu methods applied to 102 case histories. Can. Geotech. J. **34**(6), 886–904 (1997)

Fugro: Geotechnical survey – Su Tu Trang early production system, Block 15–1. Report No. 07046S-1 (2007)

Lunne, T., Robertson, P.K., Powell, J.J.M.: Cone penetration testing. Blackie Academic & Professional (1997)

Mayne, P.W.: Cone penetration testing. NCHRP Synthesis No. 368 (2007)

Robertson, P.K.: Interpretation of cone penetration tests—a unified approach. Can. Geotech. J. **46**(11), 1337–1355 (2009)

Investigation of Long-Term Settlement of Flexible Steel Oil Storage Tanks on Calcareous Sand in UAE

Emad Sharif[1] and Man Bui[2](✉)

[1] GTC Soil Analysis Services, Dubai, UAE
[2] GTC-Laboratory, GTC Soil Analysis Services, Dubai, UAE
man.buitan@gtc-lab.com

Abstract. Middle East has one of the largest numbers of oil storage worldwide. Most of the tank farms have been built along coastlines of Arabian Gulf and Gulf of Oman. Commonly, the tanks shallow foundations were seated on very dense gravelly gabbro fills, which was laid above a layer of dynamic-compacted calcareous (natural) sand. It has been observed for several cases that different settlements beneath tank bottom annulus after 10-year operation exceeded acceptable limited as per API standard [1].

In this paper, the long-term settlement of large oil tanks on shallow foundation was investigated using both in-situ tests and laboratory tests. In laboratory, creep of calcareous sand was measured using triaxial cell. Effects of oil tank filling and emptying on foundation settlement was also evaluated. 100 loading cycles under drained conditions were applied on the gravel extracted underneath one of the tank foundations. The loading frequency was relatively slow, 120 min per cycle, to simulate the long-term tank loading operation. The accumulative residual settlement and soil stiffness vs. loading cycles were used to explain the settlement phenomenon of the tank.

Keywords: Settlement · Calcareous sand · Cyclic test · Oil tank

1 Introduction

Middle East has one of the largest numbers of oil storage facilities in the world. Tank farms are mostly located at ports along coastlines of the Arabian Gulf, where calcareous sand is prevailing, and flexible steel storage tanks having diameters up to 100 m and heights up to 30 m has been constructed.

Differential settlement, measured at adjacent points along the perimeter of the tank, is of the greatest concerns, as high settlement and/or high differential settlement beyond the acceptable limits affect the walls verticality and tank performance. To evaluate the tank performance after constructed, the tank was undergone a hydrostatic test, in which the tank is incrementally filled with water up to the highest pressure. The water is then emptied to achieve a permanent settlement. During the operation lifetime, the subsequent foundation settlements are assumingly inconsiderable hence ignored.

D. V. K. Huynh et al. (Eds.): VSOE 2021, LNCE 208, pp. 131–139, 2022.
https://doi.org/10.1007/978-981-16-7735-9_12

In this paper, we present a geotechnical investigation and assessment to understand the reasons and mechanism to cause the relatively large settlements observed in four oil fixed cover, flexible steel oil – storage tanks. The tank diameters are from 33 m to 40 m, and the tank height is 19 m, constructed in 2008 (Fig. 1). After more than 10 years of operation, the tanks experienced relatively large settlement. Tank settlement inspection as per API-653 [2] reported that the accumulative settlement has exceeded 50 mm (Fig. 2). The excessive settlement can be observed at various locations along the annular plate seating area. The annular plate itself was also deformed, hence the tanks were emptied, and remediation was required.

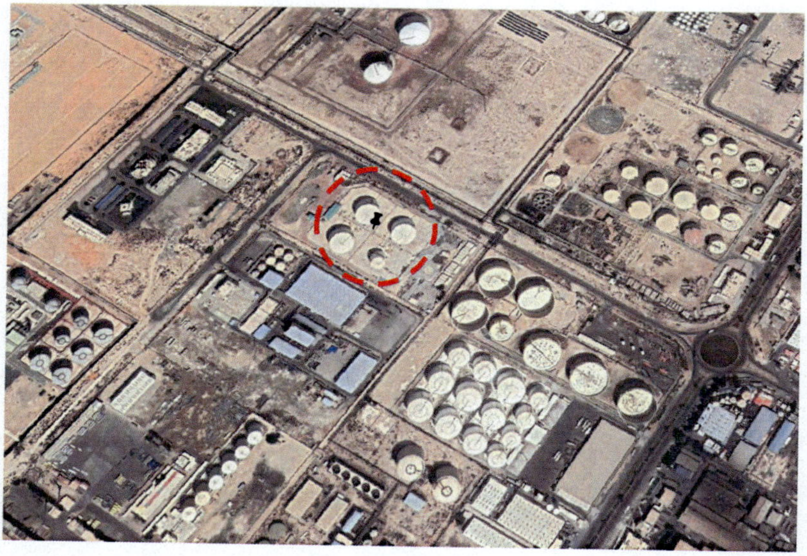

Fig. 1. The investigation site with 4 large oil tanks in a tank farm near a shoreline in UAE. Image from Google Earth Pro, accessed on 04/02/2020

The recorded "permanent" settlement at eight monitoring points, evenly distributed along the perimeter of each tank, due to hydrostatic test conducted in 2008, were from 20 mm to 25 mm. This relatively low settlement resulted from the hydrostatic test was well confirmed by the results of our finite element (FE) model of the tank subjected to an incremental uniform static load of 200 kPa, which is higher than the peak pressure applied in the hydrostatic test. This indicates that the soils beneath the foundation were sufficient stiff. The predicted settlement, that could occur further during operational lifetime of the tanks due to cycles of compression and uplift and a long-term creep of foundation on sand [5, 6] as estimated and given in the foundation design report was less than 15 mm.

The causes of the excessive settlement were not known. We believe that it could be linked to the factual time dependent creep, which can be higher than the estimated value, and/or the accumulated cyclic settlement caused by cycles of loading and unloading during tank operation. Therefore, we undertook a geotechnical investigation including

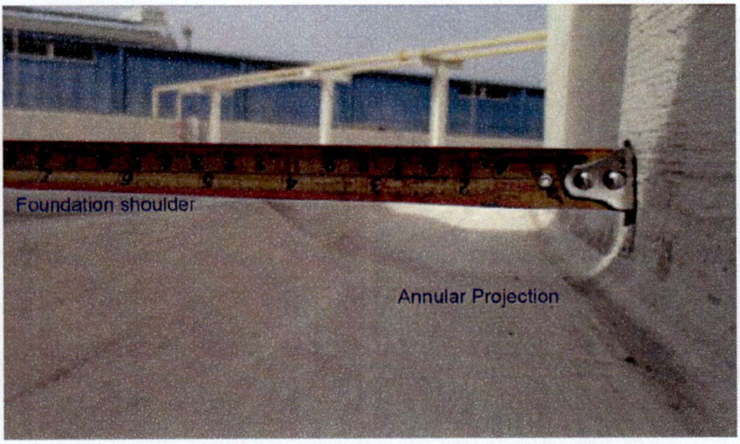

Fig. 2. Excessive oil tank annular projection settlement

in-situ tests, Non-destructive geophysical survey, and advanced laboratory tests to verify the above hypothesis, to understand the mechanism of the settlement, and also to support the design of remediation works. Due to confidential reasons, the name and location of the tanks investigated cannot be disclosed.

2 General Tank Foundation and Ground Condition

Figure 3 shows a tank foundation seated on dense crushed rock embankment over well compacted granular engineered fill with a total thickness up to 5.0 m. The tank shallow foundation was designed as per API Standard 650 [2]. Beneath the granular engineered fill is a layer of natural calcareous sand, with the thickness up to 10 m. The natural calcareous sand, underlain eventually by sedimentary rock, was compacted by mean of vibro-compaction or dynamic compaction[1].

We reassessed the sand density based on the available data obtained from CPT tests conducted previously and also the shear wave velocity measured using MASW technique in this study (presented below). The sand can be classified as medium dense to dense fine to coarse calcareous SAND, with loose/medium dense thin localities. Bedrock of extremely weak sandstone at the site is found at the depth of 15 m below the ground level (bgl). The ground water table at the site is at about 5 m (bgl) with an approximately seasonal fluctuation in range of ±0.5 m (Fig. 4).

3 Assessment in Situ Soil Density and Stiffness

Three trial pits with 2 m depth around the tank perimeter were excavated to assess the quality and compactness of engineered fill. At the bottom of trial pits, some in-situ tests i.e. in-situ density using sand replacement [7], manually operated dynamic

[1] Assessment of the sand relative density in situ is challenging and questionable, However, this was common requirement in UAE to densify the calcareous sand to the average relative density of 80% (see e.g. [3, 4]).

penetration test [8], and dynamic plate loading test [9] were conducted. The measured in-situ dry density varies from 2.01 to 2.19 T/m^3. The in-situ test results suggested that the engineered fill was very well compacted with the dry density from 2.01 to 2.19 T/m^3 and the degree of compaction more than 95%–100%. Multichannel analysis of surface waves (MASW) was conducted to evaluate the stiffness of the foundation soils up to 28 m (bgl) as shown in Fig. 5.

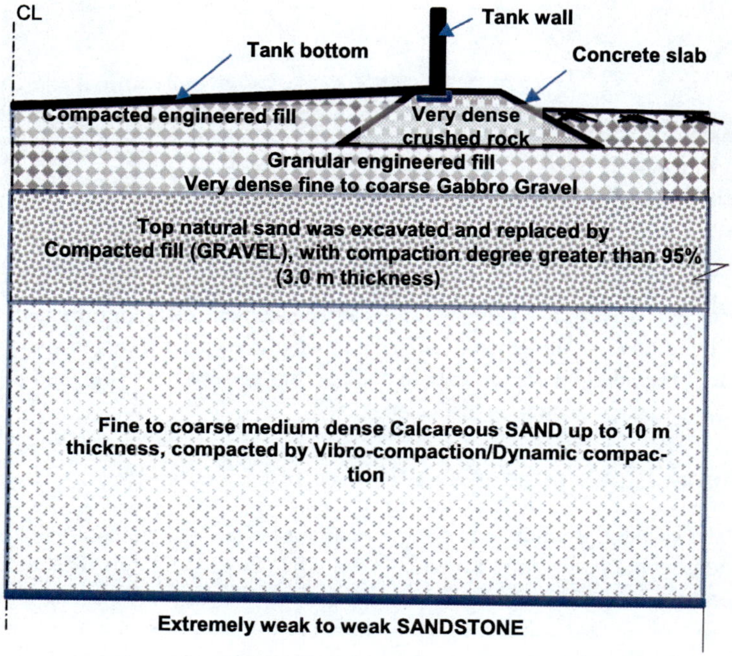

Fig. 3. Sketch of Tank foundation and soil stratigraphy at the site.

4 Settlement Evaluation Using Laboratory Tests

4.1 Measurement of Creep for Calcareous Sand Using Triaxial Test

A representative sample of calcareous sand was collected and tested in triaxial cell to investigate the time depended creep. Triaxial tests were preferred selected over Oedometer tests due to the relatively larger in size (e.g. triaxial sample height is more than 8 time thicker than oedometer sample thickness). The sand was moistly tamped in a split mould in five equal layers to achieve a uniform specimen at 80% relative density, which represents the average in-situ relative density of the sand. The sample were saturated until Skempton's B value was greater than 0.95.

Initially, the samples were isotopically consolidated at the overburden effective stress of 145 kPa, representing average depth of the sand layer. Then it was anisotropically

Fig. 4. Trial pit reveals the competent materials beneath the oil tank foundation.

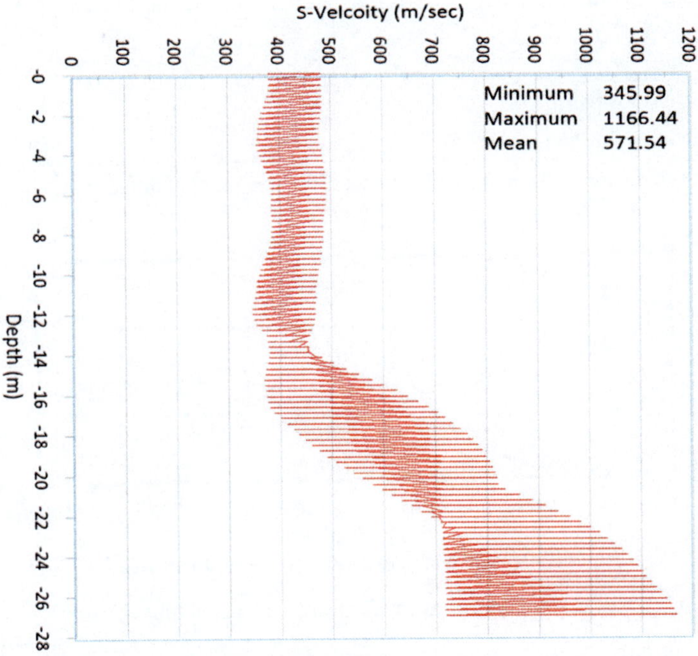

Fig. 5. 2D shear wave velocity images along a survey line obtained from MASW.

consolidated to the deviator stress of 190 kPa as detailed in Table 1, to mimic the pressure applied during the hydrostatic test. The effective stress conditions were maintained constantly for a sufficient period of time i.e. 18 days and the vertical deformation of the specimen was monitored using an external LVDT.

From the trendline for the most linear part of the cure (Fig. 6), the derived creep rate (α) for the calcareous sand is 0.095% per logarithmic cycle. The total creep strain measured in laboratory after 18 days was approximately 0.19% (Fig. 3). The settlement was almost ceased after that as shown by the plateau at the end of the statement curve. This level of vertical strain induced will correspond to 19 mm settlement for a 10 m layer of sand in the field. This measured creep settlement was higher than the value estimated in design of 15 mm. It should be noted that the stress level of 190 kPa applied in triaxial cell is higher than the possible maximum oil pressure during operation (about 170 kPa). The results obtained from the creep test suggest that (i) creep of calcareous sand could contribute to the settlement of the tank under a constant load, however, (ii) creep may not be the major factor causing the tank large settlement.

Table 1. Stresspath applied during anisotropic consolidation

Stage	For granular fill		For calcareous sand	
	σ'_r (kPa)	σ'_v (kPa)	σ'_r (kPa)	σ'_v (kPa)
Overburden pressure	45	45	145	145
Superstructure	45	75	145	175
Hydrostatic test loading	45	265*	145	345*
Hydrostatic test unloading	45	75	145	175

*Note: The summation of the hydrostatic test pressure, dead weight of the tank

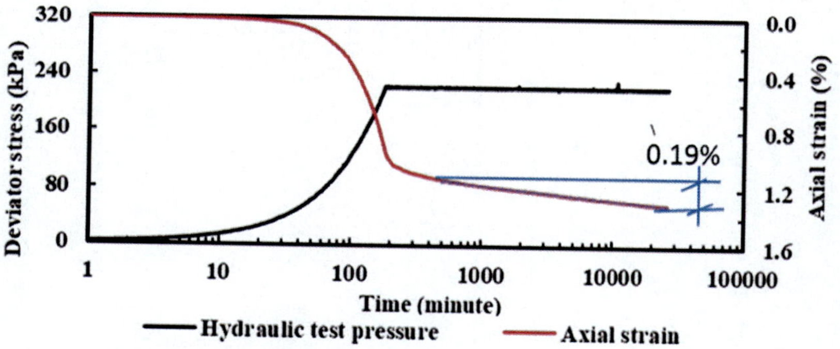

Fig. 6. Monitoring creep of calcareous Sand in triaxial cell

4.2 Assessment of Cyclic Foundation Settlement

Triaxial tests in drained conditions were conducted to simulate the cyclic loading of both granular engineered fill and calcareous sand to study the deformation behaviour of the materials. The associated residual deformation from such cyclic loading is typically ignored in common designing practice. Similar to test procedure applied for the creep test above, the samples were prepared and anisotropically consolidated in triaxial cell to the deviator stress of 190 kPa to simulate the pressure applied during the hydrostatic test. The specimens were then unloaded to the tank deadload (Table 1) prior to the cyclic loading stage. A number of 100 cycles of loading and unloading were applied to the specimen in drained condition to mimic the processes of filling and emptying of oil tank during the last 10 years, with an average of one cycle per month. The peak-to-peak deviator stress amplitude applied was 170 kPa (Fig. 7), which is as same as the possible highest oil filling pressure. The loading frequency was sufficiently slow i.e. 0.00014 Hz or 120 min/cycle, allowing water to be drained freely without any EPWP.

During the cyclic load application, the strain gradually accumulated as shown in Fig. 7 and Fig. 8, and the soils became stiffer thanks to the effects of draining conditions. The residual strain accumulated vs. number of cycles were plotted in Fig. 7. It is observed that when the number of loading cycles exceeded 75 (for granular fill) and 95 cycles (for calcareous sand), the accumulated strains were almost constant, meaning that the settlement could be no longer taking place in the future.

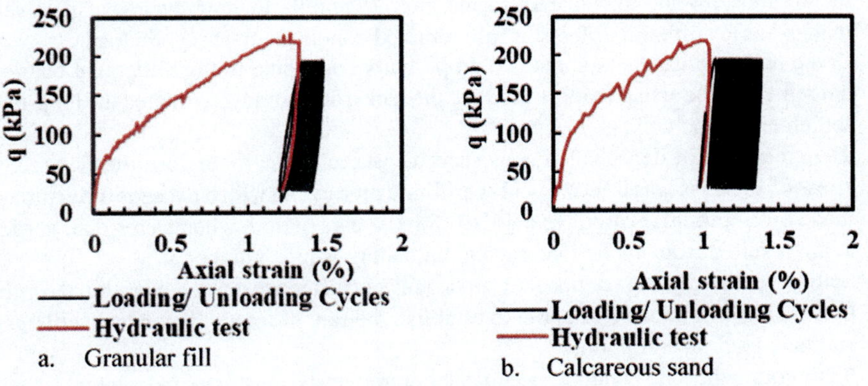

Fig. 7. Stress-strain curve resulted from process of oil 100 filling/emptying cycle

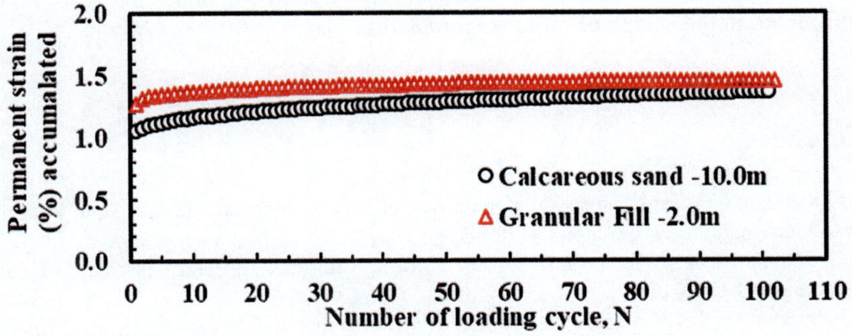

Fig. 8. Permanent strain (settlement) accumulated during cyclic loading

The total accumulative cyclic vertical strain induced was about 0.2% for granular fill and 0.3% for sand (Fig. 8). These cyclic strains are equivalent to 10 mm settlement for 5.0 m thickness of granular fill, and 30 mm settlement for 10 m thickness of calcareous sand. The total permanent cyclic settlement therefore was up to 40 mm, which was significantly higher than the creep settlement under static load measured above. It should also be noted that the level of accumulated vertical strains illustrated in this study by no meant can be deemed as representatives for all calcareous sands in UAE. The accumulated settlement appeared to be terminated after a larger number of cycles for sand compared to granular fill. This can be attributed to the lower relative density and particle crushability of the calcareous sand.

5 Concluding Remarks and Recommendations

- The process of filling and emptying tank during its operation caused the major excessive settlement with time. This was observed for both dense calcareous sand and very dense granular engineering fill.
- For the studied tanks, it is likely that no further settlement could be taken place in the granular fill and in calcareous sand after 75 and 95 loading cycles, respectively, thanks to the stiffening effects in fully drained condition under cyclic load.
- Creep under constant static load could partially contribute to the settlement but does not represent the actual loading mode of the tanks, and it underestimated the long-term settlement of the tank.
- The creep rate of dense calcareous sand measured in the laboratory under effective stress of 190 kPa, was 0.095% per logarithmic time cycle, which exceeds the estimates made by empirical relations (e.g. [5, 6]) for 10 years period, whereas residual vertical strain resulted from cyclic loading and unloading is much higher.
- Settlement monitoring results of a typical tank performance during operational service if available will be very valuable to establish the rate of time – dependent settlement and tank loading condition.
- It is recommended to conduct specialist laboratory tests, in which sand relative density, and the in-situ stress history shall be considered, i.e. cyclic tests to determine the residual settlements caused by various loading modes and creep monitoring under a constant pressure. These cyclic settlement and creep measured shall be then appropriately considered in the design of oil-tanks foundation.

References

1. API Standard 653: Tank Inspection, Repair, Alteration, and Reconstruction, 3rd edn. American Petroleum Institute Washington DC (2003)
2. API Standard 650: Welded Tanks for Oil Storage, 12th edn. American Petroleum Institute Washington DC (2013)
3. Al-Homoud, A.S., Wehr, W.: Experience of vibrocompaction in calcareous sand of UAE. Geotech. Geol. Eng. **24**, 757 (2006)
4. Kirsch, K., Kirsch, F.: Ground Improvement by Deep Vibratory Methods, 2nd edn. CRC Press (2019)
5. Burland, J.B., Burbidge, M.C.: Settlement of foundations on sand and gravel. Proc. Inst. Civil Eng. **1**(78), 1325–1381 (1985)

6. Schmertmann, J.H.: Static cone to compute static settlement over sand. J. Soil Mech. Found. Div. ASCE **96**(SM3), 1011–1043 (1970)
7. ASTM D1556/D1556M - 15e1: Standard Test Method for Density and Unit Weight of Soil in Place by Sand-Cone Method. ASTM, PA, USA (2018)
8. ASTM D6951/D6951M – 09: Standard Test Method for Use of the Dynamic Cone Penetrometer in Shallow Pavement Applications. ASTM, PA, USA (2015)
9. ASTM E2835 – 11: Standard Test Method for Measuring Deflections using a Portable Impulse Plate Load Test Device. ASTM, PA, USA (2015)

Investigation of the Long-Term Cyclic Behaviour of Monopile Foundation by Impact and Vibratory Installation

Viet Hung Le(✉) and Frank Rackwitz

Soil Mechanics and Geotechnical Engineering Division, Technische Universität Berlin, Gustav-Meyer-Allee 25, 13355 Berlin, Germany
hung.le@grundbau.tu-berlin.de
https://www.grundbau.tu-berlin.de

Abstract. This paper presents the results of the numerical investigation of the cyclic behaviour of the monopile which has been installed by impact and vibratory driving. By the numerical investigation, two different phases were considered: installation process (phase 1) and lateral cyclic loading (phase 2). In the first phase, a model with MMALE formulation for the simulation of large deformation of soil due to penetration was used. Hereby the impact and vibratory driving were modelled. In the second phase, a model with Lagrangian formulation was used. In this model, an explicit accumulation model (HCA-Model) for the simulation of the long-term cyclic loading was applied. The influences of the installation were considered in the second model in form of the initial condition of the void ratio and stress states after the penetration of the pile which were adapted from the first model. The investigation showed a significant effect of the installation on the behaviour of the pile. In the small scale models, the vibratory pile indicated a stiffer behaviour on the lateral cyclic loading in comparison to the impact driven pile.

Keywords: Installation effects · Vibratory driving · Impact driving · Long-term cyclic loading · Accumulation deformation · Explicit accumulation model · MMALE formulation

1 Introduction

In geotechnics, the penetration of a pile in the ground is one of the most complicated processes with different complex mechanical phenomenon. Furthermore, for different soil types and different boundary conditions, various installation methods can be used (e.g.: jacking, drilling, impact or vibratory driving etc.). Therefore, it is still a big challenge for engineers to predict the installation and to consider the influence of this process in the design phase.

During the penetration of a pile in sand, the initial stress condition and void ratio of the sand is changed significantly. This can have a crucial influence on the bearing capacity and deformation behaviour of the installed pile. However, in many analytical

D. V. K. Huynh et al. (Eds.): VSOE 2021, LNCE 208, pp. 140–147, 2022.
https://doi.org/10.1007/978-981-16-7735-9_13

and numerical design approaches this process is usually ignored (calculation with K_0-stress condition and constant void ratio) or strongly simplified. In the present study, the behaviour of the monopile which has been installed by impact and vibratory driving was investigated using different numerical models. With the aim to validate the simulation results, small scale model tests were conducted.

2 Numerical Calculation Concept

Due to the fundamental difference of the mechanical phenomenon and the behaviour of soil due to the installation and the loading of a monopile, different numerical formulations and calculation methods were required for the simulation of these processes. Therefore, in the present study, a concept with two separate simulations were carried out for the installation and loading process (Fig. 1).

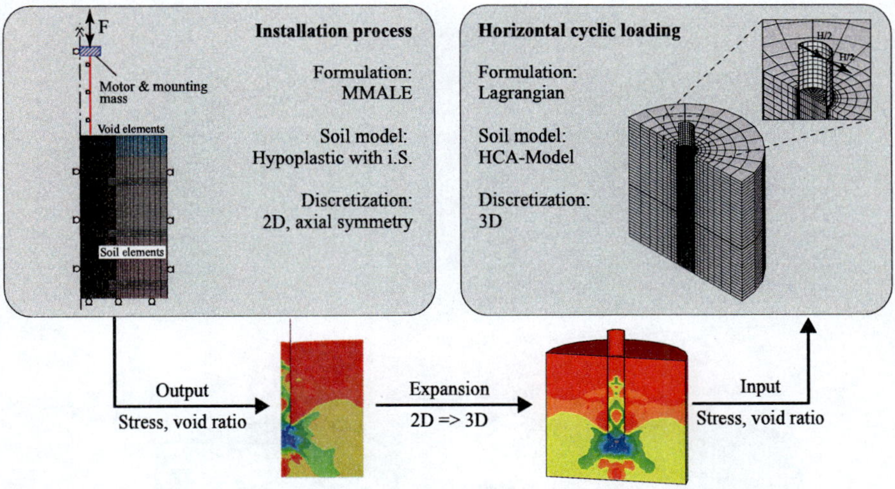

Fig. 1. Schematic view of the calculation concept with models for installation process (left) and for horizontal cyclic loading (right)

In the first simulation, a model with MMALE formulation [1] was used to simulate the large deformation by the penetration of the pile in sand. Hereby two installation methods (impact and vibratory driving) were considered. In the second phase, a model with Lagrangian formulation (small strain) was used. The influences of the installation process were considered in the second model in form of the initial condition (void ratio and stress states) which is based on the condition of soil after the penetration of the pile in the ground. The state variables of all integration points at the end of the first model (2D axis symmetric model) simulation were extrapolated in to a 3D model. Afterward, these variables were applied in the second model for the long-term cyclic loading. While in the first model, the Hypoplasticity with Intergranular Strain (Niemunis & Herle [2]) was used to describe the behaviour of sand, in the second model, the explicit calculation method with HCA-Model was applied [3]. This calculation method is a combination of different material models which enables to simulate the boundary condition with a large number of load cycles effectively. Details of the calculation steps with the explicit accumulation model were presented by Le & Rackwitz [4] and Le et al. [5]. All calculations in this study were conducted under fully drained condition.

3 Soil Conditions After the Installation Process

In the first part of this study, the modelling of the pile penetration process using vibratory and impact pile driving was conducted. The theoretical backgrounds and details regarding numerical algorithms and procedures by using the MMALE technique were published in previous works of the research group at TU Berlin. The results used in the following sections and several investigations were presented by Daryaei et al. [6] in detail. For further information about the calculations, the authors would like to refer to this paper.

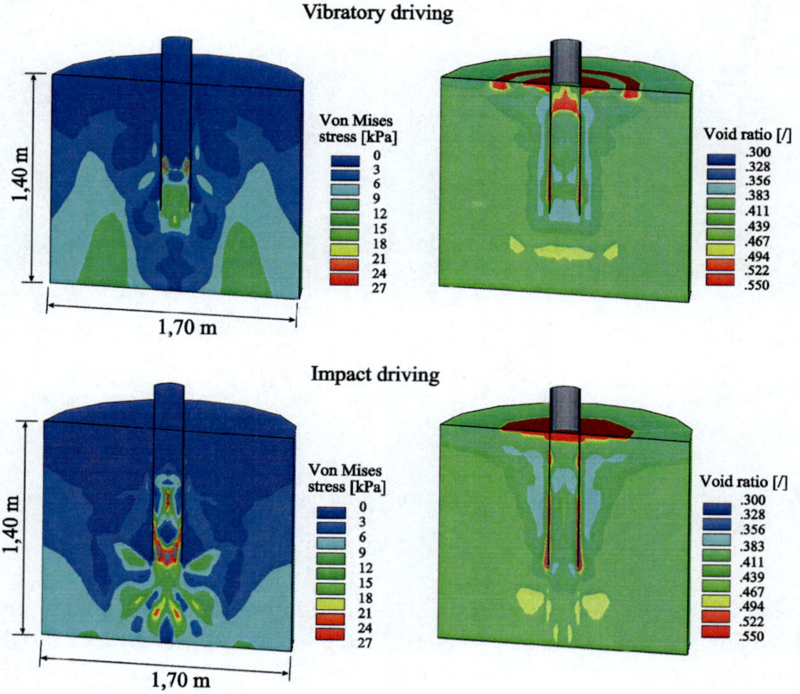

Fig. 2. Distribution of the Von Mises stress and void ratio before the horizontal loading (after the installation process)

As an output of these simulations, the soil conditions after the installation including the distribution of the void ratio and stress are delivered. Figure 2 shows the initial condition of the soil before the cyclic loading in the second model adapted from the results of the installation simulations. The presented stress condition is von Mises stress, which is defined with $q(\sigma) = \sqrt{3/2} \cdot \|\sigma_{dev}\|$. It can be observed that there are significant differences in stress as well as in void ratio between vibratory and impact driving. In comparison to the condition of the soil before the pile installation, stress condition and void ratio distribution are changed. By impact driving, it can be seen that the stress values under the pile tip were significantly higher compared to vibratory driving. This can be explained with the well-known plugging effect that can occur during impact penetration. This effects were also observed in the conducted small scale model tests. The disturbed zone around the pile by the void ratio which is developed during the installation seems to be larger for the pile with impact driving. In both simulations, a thin loosening zone arose directly at the pile shaft.

4 Simulation of the Cyclic Loading

4.1 Numerical Model and Load Program

In this study, for the modelling of the loading phase with a large number of load cycles, the explicit accumulation model (HCA-Model) was used. The discretisation of the model is shown in Fig. 1 (right) as well as presented by Le et al. [7] in detail. Figure 3 presents the schematic view of the load program, which is identical in both calculations for impact and vibratory driving. The horizontal force was applied in two outer points of the pile head and 30 cm over the sand surface. At the beginning of the test a static preloading of 450 N was applied. Afterwards two load packages with 1,000 cycles each followed. The first package had a force amplitude of $F^{ampl} = 225\,N$ with an average force $F^{avg} = 122{,}5\,N$. In the second package the force amplitude was $F^{ampl} = 450\,N$ and the average force $F^{avg} = 225\,N$.

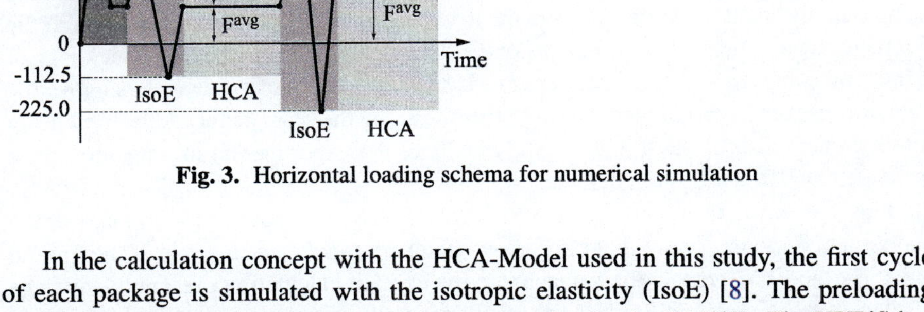

Fig. 3. Horizontal loading schema for numerical simulation

In the calculation concept with the HCA-Model used in this study, the first cycle of each package is simulated with the isotropic elasticity (IsoE) [8]. The preloading was calculated with Hypoplasticity with Intergranular Strain (HYPiS). The HYPiS has already demonstrated its ability to represent soil behaviour under monotonic loads down to a few loading cycles. Due to the "ratchetting" problem by un- and reloading, according to Wichtmann [9], the simulation with HYPiS in the second cycle can lead to unrealistic deformation predictions. Therefore, for the second cycle and the update cycles performed during explicit calculation, the Isotropic Elasticity (IsoE) was preferred.

4.2 Material and Parameters for the Constitutive Models

For this study, Berlin Sand was used for the investigations. This is a medium to coarse sand with a uniformity coefficient $C_u = 3.3$ and an average grain diameter $d_{50} = 0.55$ mm. The minimum and maximum density was determined according to the German standard DIN 18126 with $\rho_{d,max} = 1.906\,\text{g/cm}^3$ ($e_{min} = 0.391$) and $\rho_{d,min} = 1.570\,\text{g/cm}^3$ ($e_{max} = 0.688$). The parameter of this material for the HYPiS and HCA-Model are presented in Le et al. [7]. In comparison to [7], the record cycle was calculated in this study with the IsoE (the implicit cycle for the evaluation of ε^{ampl}). The parameters of this material model for Berlin Sand are listed in Table 1. These parameters were calibrated using results of the cyclic triaxial tests. The material used for the pile was steel. In the numerical simulation it was modelled with the linear elastic material model.

Table 1. Parameter of Berlin Sand for isotropic elasticity

C_{1c}	C_3	C_4	C_5
725	1.70	0.50	0.40

4.3 Validation of the Numerical Results

For validation of the numerical model the results of the small scale model tests were used. In a test setup at the TU Berlin, cyclic tests on the pile with impact and vibratory driving were carried out. The test set up consisted of a steel container with inner base dimensions of 1.70 m by 0.90 m and a height of 1.15 m plus an additional 0.25 m of drainage layer at the bottom of the container. This dimensions and boundary conditions were considered in the numerical simulations presented above. After preparation using dry pluviation technique, the sand placed in the container was saturated from the bottom through the drainage layer. The steel pile has an outer diameter of 200 mm, a wall thickness of 4 mm and achieved a penetration depth of 0.87 m. A system of guide rollers above the container ensured vertical penetration of the pile along the glass panel and restricted any tilt. Technical data of the test setup and results of the experimental investigation were presented in [10] and [11] in detail.

Figure 4 shows the results of the FEM simulations in comparison with those of the experimental model. These diagrams show the displacements of the pile at the ground level. By the FEM-results, the values from the explicit calculation present the average displacement of each cycle, while the results from the experimental model test include the complete movement of the pile (with oscillation due to un- and reloading). In both installation methods the FEM results were in good accordance with the conducted model tests. At the end of the second package the difference in displacement at the ground level was less than 5%. The results of the FEM calculation as well as of the experimental model tests indicate a stiffer behaviour of the vibratory driving pile in case of cyclic lateral loading.

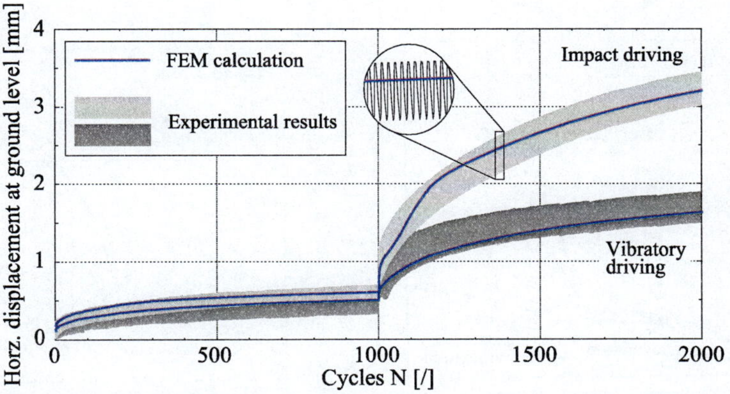

Fig. 4. Validation of the FEM calculation using results of the small scale model tests for impact and vibratory driving

5 Parameter Study

In this part of the study, the influence of the separate state parameters from the results of the installation simulation are investigated. Figure 5 presents results of a test series with different initial conditions. In the first simulation the installation effects were not taken into account (black curves). The initial condition of the soil was assumed in these investigations with a constant density and homogeneous stress distribution under K_0-conditions. At the beginning of the cyclic loading in this calculation, a condition of the soil with the relative density $I_{D0} = 75\%$ and $K_0 = 0.5$ was applied (the same density at the beginning of the pile penetration). In the second simulation the stress condition was transferred from the first model of pile penetration while the void ratio remained constant (magenta dashed curves). Here, only the influence of the change in stress was considered. In the next calculation, only the effect of the change in density was considered. The distribution of the stress condition here was homogeneous (according to K_0-condition) while the void ratio was adapted from the penetration model (orange pointed curves). For the last model, both variations of stress and density parameters were considered (blue curves, same simulation as in Fig. 4).

The development of the pile deformation in Fig. 5 showed the difference when installation effects were considered. In the literature, the installation effects are usually not considered, i.e. the so called "wished in place" technique is used. Obviously, the installation process had a significant effect on the pile deformation. In both cases of vibratory and impact driving, the penetration of the pile led to stiffer soil behaviour. The same trend was observed by the numerical study from Staubach et al. [12]. It can be seen for both installation methods that the change in density resulted in a larger reduction of horizontal pile head displacement compared to the change in stress. Especially in the first loading package the effects of the installation process were mainly dominated from the change in void ratio.

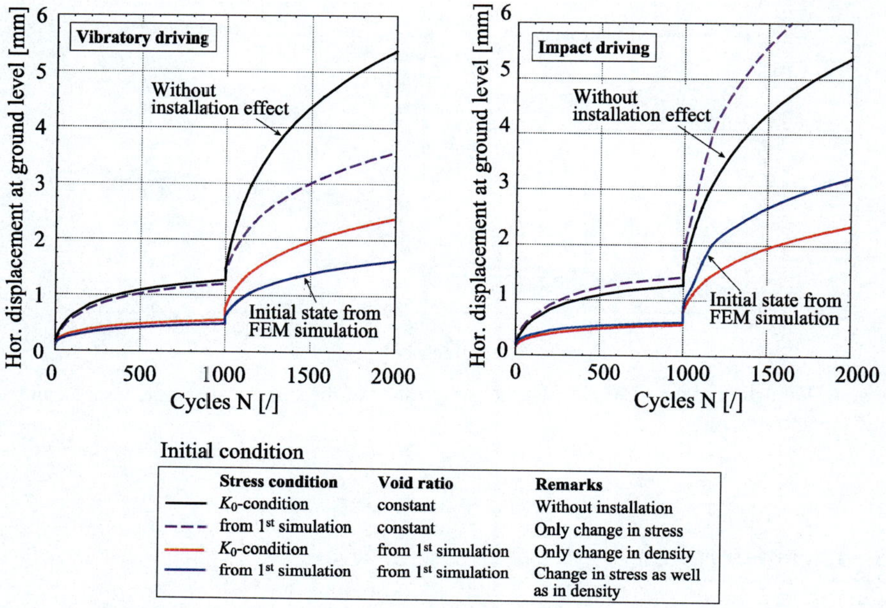

Fig. 5. Parameter study for the influence of the installation effects

6 Conclusions

In this study a numerical methodology with two separate models was carried out to consider the influence of the installation process on the cyclic behaviour of the monopile. The effects of the impact and vibratory driving were investigated. The comparison between numerical and experimental results showed good agreement for vibratory as well as for impact driving. In comparison with impact driving, vibratory driven piles had smaller deformations under the same cyclic load (stiffer behaviour). In total, the calculation without installation effect (using "wished in place" technique) resulted in larger deformations of the pile and was more conservative.

Acknowledgement. The investigation presented was partly supported by INNOGY SE as part of the research project "Cyclic and Near Field Effects of Vibratory Driving in Sand - CAFE", which is gratefully acknowledged here.

References

1. Savidis, S.A., Aubram, D.: Lagrange-Euler Formulierungen in der Bodenmechanik. In: Schanz, T., Hettler, A. (eds.) Aktuelle Forschung in der Bodenmechanik 2015, pp. 29–46. Springer, Heidelberg (2015). https://doi.org/10.1007/978-3-662-45991-1_2
2. Niemunis, A., Herle, I.: Hypoplastic model for cohesionless soils with elastic strain range. Mech. Cohesive-Frictional Mater. **2**, 279–299 (1997)
3. Niemunis, A., Wichtmann, T., Trianstafyllidis, T.: A high-cycle accumulation model for sand. Comput. Geotech. **32**(4), 245–263 (2005)
4. Le, V.H., Rackwitz, F.: Prediction of deformation for monopile foundation under horizontal cyclic multi directional loading. In: Geotechnics for Sustainable Infrastructure Development - Geotec Hanoi, Phung Duc Long (edit.) (2016)
5. Le, V.H., Rackwitz, F., Savidis, S.: Verformungsakkumulation bei Monopilegründungen in Sand mit Lastrichtungswechseln. Bauingenieur **92**, 263–270 (2017)
6. Daryaei, R., Bakroon, M., Aubram, D., Rackwitz, F.: Numerical evaluation of the soil behavior during pipe-pile installation using impact and vibratory driving in sand. Soil Dyn. Earthquake Eng. **134** (2020)
7. Le, V.H., Remspecher, F., Rackwitz, F.: Numerical investigation of installation effects on the cyclic behaviour of monopile foundation under horizontal loading. In: Vietnam Symposium on Advances in Offshore Engineering, Hanoi - Vietnam (2018)
8. Zachert, H.: Zur Gebrauchstauglichkeit von Gründungen für Offshore-Windenergieanlagen. Ph.D. thesis, Pubplication of the Institute for Soil and Rock Mechanics, Karlsruher Institute for Technology (KIT), Triantafyllidis, Th. (edit.), no. 180 (2015)
9. Wichtmann, T.: Explicit accumulation model for non-cohesive soil under cyclic loading. Ph.D. thesis, Ruhr Universität Bochum (2005)
10. Labenski, J., Remspecher, F., Le, V.H., Moormann, C., Rackwitz, F.: Lateral-bearing behaviour of vibratory driven monopiles in different model test set-ups. Int. J. Phys. Model. Geotech. 07 August 2019
11. Remspecher, Le, V.H., Rackwitz, F., Herwig, V., Matlock, B.: Vibratory driven installation of monopiles – an experimental investigation of the soil-pile interaction. In: Vietnam Symposium on Advances in Offshore Engineering, Hanoi - Vietnam (2018)
12. Staubach, P., Machacek, J., Moscoso, M.C., Wichtmann, T.: Impact of the installation on the long-term cyclic behaviour of piles in sand: a numerical study. Soil Dyn. Earthq. Eng. **138** (2020)

Jack up Spudcan Penetration Prediction for Murmanskaya Platform at X-Block, Nam Con Son Basin, Offshore Vietnam

Nguyen Van Hung(✉), Tran Viet Kha, and Nguyen Viet Khoi Nguyen

PetroVietnam University, 762 Cach Mang Thang 8, Long Toan Ward, Ba Ria City, Vietnam
hungnv@pvu.edu.vn

Abstract. Jack up rigs are mobile drilling units that used in offshore for hydrocarbon exploration and production activities in water depths up to about 150 m. The flexible and mobile characteristics of the platforms enable them to have other applications in the offshore oil and gas industry in the role of supporting facilities for marine operations in general. The operation procedure carries a high accident probability, in which soil failure accounts for the highest proportion of risks (about 25%) [1]. Therefore, the precise prediction of soil bearing capacity before installing the platform will help to propose an appropriate operating procedure to minimize the risks that may occur. It is a critical problem for an offshore Jack up drilling platform to estimate exactly the penetration depth of spudcans under the complex geologic conditions at sea. In practical operation, the penetration depth prediction is restricted with the limit value calculated from the empirical formula and the normal calculation methods cannot avoid the phenomena of the excessive penetrating depth. In view of this problem, the article presents some calculation methods, current analysis and typical results has actually implemented for practice operations in offshore Viet Nam for Murmanskaya Jack up at X-Block, Nam Con Son Basin.

Keywords: Jack up · Penetration prediction · Nam Con Son Basin · Offshore Vietnam

1 Introduction

Jack up is a structure composed of a hull, legs and a lifting system that allows it to be towed to a site, lower its legs into the seabed and elevate its hull to provide a stable work deck capable of withstanding the environmental loads. After an appropriate and rigorous site assessment the unit will be arranged for operation. Jack up locations have water depths up to 150 m and exploration drilling takes up to 3 months per well location. A typical modern drilling Jack up is capable of working in harsh environments (wave heights up to 80 m, wind speeds in excess of 100 knots) [2].

Jack up platforms are designed and constructed to satisfy the rules of classification societies and classification is a comprehensive verification procedure. In this respect they may be regarded as mobile units together with ships and other marine structures.

D. V. K. Huynh et al. (Eds.): VSOE 2021, LNCE 208, pp. 148–155, 2022.
https://doi.org/10.1007/978-981-16-7735-9_14

This paper reviews current practice on assessing Jack up foundation stability during installation. Particular emphasis is given to the background and capabilities of a computational model developed to optimize leg penetration analyses.

2 Theoretical Background

Structural Features

A Jack up is an offshore structure, a type of mobile platform and is regarded as a self-elevating unit. It consists of a buoyant hull fitted with a number of moveable legs (most usually 3, but early designs have 4 legs), which are capable of raising the hull above the surface of the sea. The three main structural features of a Jack up which are to be considered in a design are the hull, the legs and foundations. The buoyant hull enables the transportation by wet or dry tow of the unit and the variable deck load, including machinery, to the desired location. The legs of the unit may be designed to penetrate the seabed, may be fitted with large sections or footings for each leg or may be attached to a bottom mat. The majority of offshore Jack up have three lattice legs which typically have center to center spacings of 40 to 60 m. The gross spudcans cross-sectional area is in the order of 100 to 200 m^2 and maximum bearing pressure ranges from 200 to 550 kPa [3].

Jacking System

All Jack ups have mechanisms for lifting and lowering the hull. The most basic type of elevating system is the pin and hole system, which allows for hull positioning only at discrete leg positions. However, the majority of Jack up in use today are equipped with a rack and pinion system for continuous jacking operations.

In general, there are two systems for elevating the deck of a rig. The first uses hydraulic cylinders equipped with moving and stationary pins. The cylinders extend and retract to climb up and down the legs of the Jack up rig. The second system employs a rack and two pinion gears that are turned to move the legs up and down for continuous jacking. An optimal design requirement is to achieve an even distribution to increase overall reliability and to improve the fatigue life of specific pinion teeth. Smooth, constant speed control systems with variable torque are to be recommended in this respect.

Operation of Jack ups

By definition, Jack up platforms are mobile rigs and have three modes of operation: Transit from one location to another; Elevation on its legs and Jacking up and down between afloat and elevated modes. Each of the above has specific precautions and requirements to be followed to ensure a safe operation. The stages for transportation may be summarized as follows: Preparing for going off location; Transit from one location to another; Arriving on location; Soft pinning the legs; Final going on location and jacking; Preload operations.

To achieve the elevated operating condition, the unit is jacked up to the full air gap. Transit may be achieved using wet tow for which the hull floats and is towed directly by barges or dry tow when the hull is placed on a floating vessel which is towed. This is a critical stage in the operation of a Jack up and there have been many instances of rigs being lost due to instability.

Jack ups are preloaded when they first arrive at a site to ensure the soil is capable of withstanding the maximum expected footing reaction without experiencing additional leg penetration or soil failure. The amount of leg penetration is determined by soil properties, vertical reaction of the legs, and footing area. However, this value is generally not relevant for design loading conditions during drilling operations, since reduced bearing capacities due to horizontal environmental loads are applicable.

Stability of Jack up

For the features of Jack up structures, they are excluded from the classification of fixed rigs or construction entity, because the platform foundations are not designed to specific standards for each site. Therefore, they must be designed to be highly stable under any ground conditions, because offshore areas around the world have different engineering properties of soils. The flexible operating envelops of a Jack up implies that a platform can function in many locations.

The significant amount of foundation failures and resulting direct and consequential financial losses has led to an increased number of geotechnical investigations for Jack up rig siting. Also, the quality of the investigations is gradually improving: An increasing number of oil companies and rig operators are adopting the North Sea practice to perform soil investigation at a future Jack up rig site using a dedicated soil investigation vessel [4]. With this approach the most reliable assessment of soil conditions can be done using high quality drilling/sampling, in-situ and laboratory testing techniques. Moreover, this investigation approach allows a thorough review of soil conditions and more advanced analyses in the specialist geotechnical office.

The overall stability of Jack up rigs during preloading and during its operational life has generally not been considered in past routine practice. However, this is expected to change in the future as a result of increasing safety requirements in the offshore industry.

Methodology for Spudcans Penetration Analysis

The geotechnical analysis of the spudcan foundation is straight forward. The vertical bearing capacity of the spudcan foundation is evaluated considering a number of possible penetration depths, and the resulting curve of soil capacity versus leg penetration is plotted. The planned preload is compared to the predicted capacity to determine the expected penetration depth. Punch through risk is assessed by checking the resistance at eventual strong layers in the profile. The industry standard approach follows the SNAME recommendations, based on simple bearing capacity formula with minor modifications for offshore conditions [5]. The following sections outline the foundation failure classification and formulae used for vertical bearing capacity for each soil foundation types.

Types of Foundation Failure

After reviewing the accidents of the Jack ups, Young et al. [6] pointed out the causes of foundation failure as follows:

- The soil is punctured during preloading, the soil cross-section in this case consists of a fine layer on the soft soil layer;

- Excessive settlement during a storm, because the ground preloaded under the maximum load is less than the storm survival load;
- Seabed instability is caused by displacement of sediments on the seabed and causing horizontal forces to damage the rig;
- The legs is pulled up, when the pulling resistance is less than the lifting force.

Analysis of Load Capacity for Each Soil Foundation Structure Types
Ultimate unit bearing capacity at footing spudcan of soil foundation is calculated depending on the spudcan footing penetration character.

Table 1 illustrates the different models for calculating load capacity of Jack up.

Table 1. Different models for calculating load capacity [Modified Ref. 7]

Type of foundation	Model	Formula	Parameters
Clay soils	Skempton, 1951	$q_u = 6C_u(1 + 0.2D/B)$	$q_u, C_u/S_u$, D, B
Granular soils	McClelland et al. 1981	$q_u = 0.3\gamma' BN_\gamma + \gamma' DN_q$	$q_u, N_q, N_\gamma, \gamma'$, D, B
Two layered soils	(Brown and Meyerhof, 1969)	$q_u = Cu_t N_{mc}$	Cu_t, N_{mc}
	Hana and Meyerhof, 1980	$q_u = 6C_u + \frac{2\gamma'_1 H^2}{B}\left(1 + \frac{2D}{H}\right)K_s tg\phi + \frac{\gamma' V}{A} < q_u(t)$ $q_b = 6C_u(1+0.2D/B) + \gamma' V/A$	$q_u, q_b, q_u(t), K_s$, D, H

3 Results and Discussion

3.1 Geomorphology Seabed Hazard

a. Bathymetry
The bathymetric data were reduced to Mean Sea Level (MSL) in meters based on predicted tides at Lan Tay tidal station. The contours in 0.5 m interval are shown in the bathymetry chart to indicate the general trend of the seabed [7].

- Water depth: 75.7 m MSL;
- Minimum water depth: 71.8 m MSL;
- Maximum water depth: 81.7 m MSL;
- Seabed gradient: <1°.

Seabed topography within the site survey area is characterized by undulating due to present of large seabed depressions and sand wave. The overall gradient within the survey area is <1° while maximum seabed gradient observed is up to 5°. There are no other bathymetric anomalies observed within 200 m radius of the location that could have an adverse effect on Jacket Installation/Jack up rig operations.

b. Seabed Features

Mapping of the seabed using side scan sonar provided seabed information, including all discernible seabed features. The seabed within the survey area exhibits generally low to moderate acoustic reflectivity, indicating predominantly fine textured sediments. Based on the samples collected within the survey area, the seabed sediments within the site consist of loose fine sand with shell fragments.

The main seabed features noted within the site area are sonar contacts, seabed depression, sand wave area, seabed scars and high reflective seabed.

3.2 Soil Conditions at Jack up Rig Location

The soil strata disclosed at the Sampling Borehole are based on results of data processing laboratory analysis, which carried out by one service company (Table 2).

3.3 Footing Penetration Prediction

a. Footing Dimensions of Jack up Rig

- Equivalent spudcan diameter: 15.00 m;
- Spudcan area: 177.00 m^2;
- Spudcan tip below widest section: 0.75 m;
- Spudcan reaction at full preload: 62.24 MN.

b. Spudcan Tip Penetration

Depth penetration of the Murmanskaya Leg has been estimated for the maximum preload of 62.24 MN and the graph penetration refers to the base height of spudcan. The spudcan may be penetrate to the depths from 1.0 m to 15.0 m that mean the tip may be rest at any depth in the range of 1.8 m to 15.8 m (Fig. 1). The most expected depth that the Starboard and Port Legs spudcan rest is about 3.0 m (tip at 3.8 m). Because of the difference in soil condition encountered in borehole No. 1 near to the Starboard & Port Leg and CPT no 1 near to the Bow Leg of proposed Jack up location, the penetration of each Jack up leg could be differed from one another [7].

Experience has shown that the scatter in soil parameters may give a large uncertainty in leg penetration predictions. The causes of error in the calculation result of soil bearing capacity are:

- The capacity prediction model of foundation is based on the assumption;

- Total actual load placed on the rig may differ from calculated load;
- The surrounding ground soil is deformed, damaged the structure due to the impact of the legs penetrating the ground;
- Heterogeneity of soil layers according to distribution;

Table 2. Soil parameters selected for calculating Jack up rig footing penetration [7]

Calculation parameters						
Strata no	Soil type	Thickness of layer	Height of layer	γ	φ	c
		H	Z			
		m	m	KN/m^3	degree	KN/m^2
1	SAND	1,0	1,0	9,0	30,0	
2	SAND	1,0	2,0	9,0	30,0	
3	SAND	1,0	3,0	9,0	30,0	
4	SAND	1,0	4,0	9,0	30,0	
5	SAND	1,0	5,0	9,0	30,0	
6	SAND	1,8	6,8	9,7	23,0	
7	SAND/CLAY	1,0	7,8	7,7	14,0	17,2
8	SAND/CLAY	2,1	9,9	8,1	25,0	12,8
9	SAND/CLAY	0,9	10,8	8,1	16,0	38,0
10	CLAY	1,0	11,8	7,4		26,0
11	SAND	1,1	12,9	8,1	23,0	
12	SAND/CLAY	1,1	14,0	7,1	14,0	31,0
13	CLAY	0,8	14,8	7,2		30,0
14	CLAY	4,4	19,2	8,2		60,0
15	SAND/CLAY	3,0	22,2	9,3	28,0	8,4,0
16	CLAY	0,9	23,1	8,6		45,6
17	SAND	0,7	23,8	9,5	32,0	
18	CLAY	1,0	24,8	8,5		47,8
19	CLAY	5,9	30,7	8,4		58,0
20	SAND	4,0	34,7	9,2	32,0	

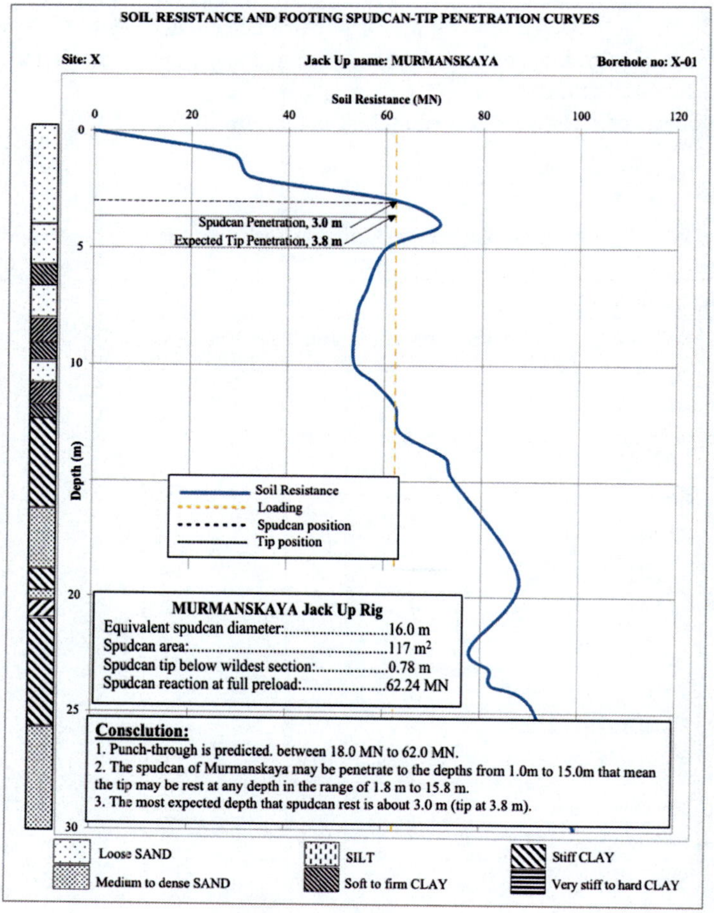

Fig. 1. Leg penetration analysis.

4 Conclusions

Prediction of leg penetration is based on conventional ultimate bearing capacity criteria for a variety of failure modes for shallow foundations. A significant number of potential failure modes need to be investigated in layered soils to find the lower bound resistance. Experience has shown that the scatter in soil parameters may give a large uncertainty in leg penetration predictions. The calculation of the bearing capacity allows only a very approximate estimation to the nearest ±20 to 25% of the observed penetration of the spudcans in the soil at the X-Block of Nam Con Son Basin, offshore Viet Nam.

A criterion procedure is recommended to devise an appropriate preloading operation taking into consideration a proper preloading sequence, carefully maintaining the hull level during preloading, adopting minimum air gap during preloading or preloading with draft, monitoring rack phase difference regularly, avoid preloading delay and prevent leaning instability during preloading and measures to be adopted should uncontrolled rapid leg penetration occur. Leg reserve length shall be checked and in-place stability during operation should be studied. Maintain a more loading time than usual practice is obligatory request for this situation.

References

1. Ullah, S.N.: Jackup Foundation Punch-Through in Clay with Interbedded Sand. School of Civil, Environmental and Mining Engineering Centre for Offshore Foundation Systems, The University of Western Australia, Crawley WA 6009, Australia, April 2016
2. Bennett & Associates, L.L.C.: Jack Up Units A Technical Primer for The Offshore Industry Professional. Offshore Technology Development Inc., 1140 St. Charles Ave. New Orleans, LA 70130, USA, pp. 1–12 (2005)
3. Fugro, McClelland Engineers B.V, A Computer Program for Jack up Rig Leg Penetration Analyses. International Society for Soil Mechanics And Geotechnical Engineering, Leidschendam, The Netherlands, pp. 1433–1436 (1994)
4. Tirant, P.L., Pérol, C.: Stability and Operation of Jack ups. Editions Technip, Paris (1993)
5. Society of Naval Architects and Marine Engineers (SNAME), Recommended Practice for Site Specific Assessment of Mobile Jack-Up Units, Technical and Research Bulletin 5-5A, Rev. 1, Jersey City, NJ (2007)
6. Dier, A.: Guidelines for Jack up rigs with particular reference to foundation integrity. MSL Engineering Limited for the Health and Safety Executive, Platinum Blue House, 18 The Avenue, England (2004)
7. Hi´êu, T.V.: Geotechnical Engineering Analysis for installing Mobile Offshore Drilling Unit Murmanskaya at site X. Research and Engineering for Oil and Gas, Vung Tau, pp. 1–10

Marine Operability Prediction Using Phase-Resolved Wave-Fields

Bernard Voon Ee How(✉), Yun Zhi Law, Hongchao Wang, Harrif Santo, and Eng Soon Chan

Technology Centre for Offshore and Marine, Singapore, 12 Prince George's Park, #04-01, Singapore 118411, Singapore
bernard_how@tcoms.sg

Abstract. Many offshore operations are carried out under restrictive operational limits due to uncertainties in the operating environment. Operators are looking for methods to enhance the operation envelopes and safety of their operations. With the advent of wave radar, optical and respective processing technology, it is now possible to measure incident phase-resolved wave-fields a distance away from a floating system. The incident wave-fields can be used to predict the response of the floating system ahead of time taken for the wave-fields to evolve into the floater location. In this paper, we demonstrate the concept via simulations of evolving wave-fields and the induced response of a generic floater. It is shown that the floating system response, and subsequent exceedance of the operational limits, can be predicted in advance. This methodology can help to reduce uncertainties related to the operating environment and the system response and forms a building block for operability decision support systems.

Keywords: Marine operations · Phase-resolved · Wave-fields

1 Introduction

Many offshore operations are carried out under restrictive operational limits due to uncertainties in the operating environment. Operators are looking for methods to enhance the operation envelopes and safety of their operations. Helicopter landing, human or cargo transfer operations with at least one floating system involved are examples of operations of which the critical phase (touch down or lift off) is conducted preferably during a time window with low waves. The ability to predict the approach of a freak wave, which seem to occur much more frequently than previously thought, can help to control the operation in a safer way. In higher sea-states, workable time windows may occur as well making their prediction very valuable to increase operability [1].

With the advent of spatio-temporal optical measurement techniques [2], wave radar and respective processing technology [3–5], it is now possible to measure and infer incident phase-resolved wave-fields a distance away from a floating system. We use the knowledge to derive the phase resolved wave field at the floating system location, and predict its response ahead of time taken for the wave-fields to evolve into the floater location.

D. V. K. Huynh et al. (Eds.): VSOE 2021, LNCE 208, pp. 156–163, 2022.
https://doi.org/10.1007/978-981-16-7735-9_15

The contributions of this paper are (1) the generalized procedure to achieve operation advisory from phase-resolved wave information, (2) evolving the waves from a measurement point a distance away to the floater location, (3) simulation of the induced floater response, (4) demonstration on the prediction of exceedance of the operation envelop during a low sea-state and (5) demonstration of the availability of response meeting the operation envelop criteria for short duration during periods of a high sea-state, which, depending on the operation itself, could enable the operation to proceed. This methodology can help to reduce uncertainties related to the operating environment and the system response and forms a building block for operability decision support systems.

2 Concept and Methodology

In this section we detail the concept and methodology to achieve the prediction for marine operations from phase-resolved wave information.

2.1 Concept

The generalized procedure to achieve operation advisory from phase-resolved information shown in Fig. 1 is as follows:

Step 1. Measure incident wave-fields a distance away from the floating system.
Step 2. Resolve the amplitude spectrum and phase information.
Step 3. Evolve the wave-fields to the location of the floating system ahead of real-time.
Step 4. Predict the motion response of the vessel using motion or force Response Amplitude Operators (RAOs).
Step 5. Feed the response prediction into an operability module for operation advisory.

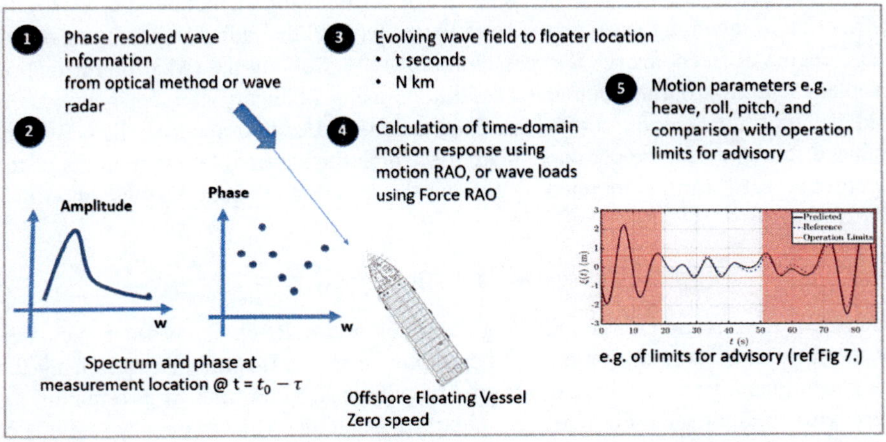

Fig. 1. Illustration of the methodology from phase-resolved wave-fields to operation advisory

2.2 Methodology

Evolving Waves

We start from the assumption that there is a remote sensor measuring phase-resolved wave-fields at an upstream distance d from the floating system. Let $\eta^m(t), t \in [t_0 - \tau, t_0]$ be the surface elevation measured up to current time t_0 for the last τ seconds. It can be written as a linear superposition of waves at different frequencies, i.e.

$$\eta^m(t) = \sum_{n=1}^{N} A_n \cos[\omega_n(t - t_0 + \tau) + \phi_n], \tag{1}$$

where A_n and ϕ_n are the magnitude and phase of the Fourier coefficient obtained by performing Fast Fourier Transform (FFT) for angular frequency ω_n. Equation 1 can be generalized to describe the entire wave-fields, $\eta(x, t)$, by invoking linear dispersion:

$$\eta(x, t) = \sum_{n=1}^{N} A_n \cos[\omega_n(t - t_0 + \tau) - k_n(x + d) + \phi_n], \tag{2}$$

with the floating system location taken as $x = 0$, and k_n is the wavenumber for wave component at angular frequency ω_n. Note that $\eta(-d, t) = \eta^m(t)$ for $t \in [t_0 - \tau, t_0]$.

The objective is to predict the surface elevation at the floating system location for the next τ_p seconds, i.e. $\eta^v(t), t \in [t_0, t_0 + \tau_p]$. Its expression can be obtained from Eq. 2 as follows:

$$\eta^v(t) = \eta(0, t) = \sum_{n=1}^{N} A_n \cos[\omega_n(t - t_0 + \tau) - k_n d + \phi_n]. \tag{3}$$

Motion Response

The RAOs of a floating vessel are usually calculated using linear potential flow theory, by treating the fluid as inviscid, incompressible and irrotational such that the velocity potential Φ satisfying the Laplace equation $\nabla^2\Phi = 0$ exists. The velocity potential is considered as a superposition of potentials due to the undisturbed incoming wave Φ_I, the diffraction of incoming wave on the fixed vessel Φ_D and the radiation potentials due to the 6 degree of freedoms (DOFs) vessel motions Φ_R. Solving the radiation-diffraction problem with appropriate boundary conditions using boundary element method, the added mass $\boldsymbol{A}(\omega)$, damping coefficients $\boldsymbol{B}(\omega)$ and first-order zero-mean oscillatory wave-induced forces $\boldsymbol{\tau}_1$ can be obtained. With these quantities available, the motions of the floating vessel $\boldsymbol{\xi}$ can be obtained by solving the differential equations of motion of a linear system:

$$[\boldsymbol{M}_{RB} + \boldsymbol{A}(\omega)]\ddot{\boldsymbol{\xi}} + \boldsymbol{B}(\omega)\dot{\boldsymbol{\xi}} + \boldsymbol{C}\boldsymbol{\xi} = \boldsymbol{\tau}_1$$

where $\boldsymbol{C}$ is the restoring matrix. The amplitudes of motion RAOs can be then calculated as the ratio of the vessel motions to the amplitude of the incoming wave. Using the aforementioned wave evolution model, the amplitude A_n and phase ϕ_n information for each wave component at the vessel are obtained. Then, the wave frequency motions of the vessel can be evaluated as

$$\xi^q(t) = \sum_{n=1}^{N} A_n \cdot R^q(\omega_n) \cdot \cos\left[\omega_n(t - t_0 + \tau) - k_n d + \phi_n + \epsilon^q(\omega_n)\right] \tag{4}$$

where $R^q(\omega_n)$ and $\epsilon^q(\omega_n)$ are the amplitude and phase of the motion RAOs for q^{th} degree of freedom at angular frequency ω_n. In this study, we focus on the heave motion of the floating system (station keeping at zero speed) but the methodology can be extended to other DOFs.

Predictability Window

Wu [6] described that for a given remote sensing point distance d, there is a maximum prediction time window τ_p in which the prediction is reliable, and a minimum time history τ required to ensure sufficient information is acquired for prediction.

Let $S(\omega)$ be the power spectrum obtained from the measurement $\eta^m(t)$. We compute the minimum and maximum frequency, ω_{min} and ω_{max} respectively, such that 90% of the total energy is covered. Once ω_{min} and ω_{max} are obtained, their corresponding group velocity, $c_{g,max}$ and $c_{g,min}$, are computed. The prediction window τ_p and time history window τ are given by the following expression (see [6] for details):

$$\tau_p = \frac{d}{c_{g,max}} \tag{5}$$

$$\tau = \frac{d}{c_{g,min}} \tag{6}$$

3 Simulation

In this paper, we investigate two sea-states based on JONSWAP spectrum (Table 1) and the predictability of a vessel heave at zero speed:

Table 1. Parameters of simulated sea-state based on JONSWAP spectrum.

Heading level	Low sea-state	High sea-state	Units
Significant Wave Height	2.0	5	[m]
Period	8.0	12.0	[s]
Gamma	3.3	3.3	[-]

The reference wave field, $\eta^{ref}(x, t)$ is generated by using the following expression

$$\eta^{ref}(x, t) = \sum_{i=1}^{I} A_i^{ref} \cos(\omega_i t - k_i x + \phi_i^{ref}), \tag{7}$$

where A_i^{ref} and ϕ_i^{ref} are the amplitude and phase obtained by discretizing the prescribed JONSWAP spectrum. The spectrum is discretized into $I = 161$ components, with angular frequency ω_i uniformly distributed between 0–4 rad/s. The reference surface elevation at the floating system location, $\eta^{v,ref}(t)$ and the reference supply vessel [7] heave motion, $\xi^{ref}(t)$ are therefore given as the followings:

$$\eta^{v,ref}(t) = \eta^{ref}(0, t) = \sum_{i=1}^{I} A_i^{ref} \cos(\omega_i t + \phi_i^{ref}), \tag{8}$$

$$\xi^{ref}(t) = \sum_{i=1}^{I} R^q(\omega_i) A_i^{ref} \cos\left[\omega_i t + \phi_i^{ref} + \epsilon^q(\omega_i)\right] \quad (9)$$

The simulated sensing measurements, $\eta^m(t)$ is taken to be the surface elevation of the reference wave-fields at the point location, i.e. $\eta^m(t) = \eta^{ref}(-d, t)$. The distance d is set to 1000 m, a similar range an X-band radar can capture. The current time, t_0 is set to 0, such that $t < 0$ indicates the past, while $t > 0$ indicates the future. Following the methodology presented in Sect. 2.2, one can compute the prediction of floating system motion, $\xi(t)$. The operation limits of the floating system is set to 0.6 m, with an operation window of 20 s. The operation is only permitted if $|\xi(t)| < 0.6$ m throughout the operation window.

Let us first consider the scenario of the lower sea-state in Table 1. It is a relatively benign sea-state and typical operations may take place. However, when several wave components, which are around the peak of the spectrum, occur nearly in phase with one another, the constructive interference will produce a group of higher waves. Moreover, if all the wave components are aligned, a sudden increase in the surface elevation may occur. To simulate the latter event, we adjust the phase of each component and generate a focused wave at the floating system location. Based on the spectrum and the remote sensing location, τ and τ_p are computed to be 316 s and 132 s, respectively. Figure 2 shows the wave profile captured at 1000 m upstream from the floating system location. After performing the computation described in Sect. 2.2, the predicted surface elevation at the same location is obtained and shown in Fig. 3. The reference wave is based on the prescribe spectrum and phase, i.e. A_i^{ref} and ϕ_i^{ref}. Its corresponding vessel motion will be taken as reference. The predicted wave and vessel motion are based on the FFT coefficients of the measured time history between $t_0 - \tau$ to t_0. As the measured time history has only τ seconds, the coefficients obtained from FFT will not be the same as A_i^{ref} and ϕ_i^{ref}, thus causing difference between the predicted and the reference. Using the predicted wave elevation and its frequency domain information, the heave response is calculated using the motion RAO of the floating system, shown in Fig. 4.

As shown in the Fig. 4, our prediction is able to predict the sudden increase of the heave motion induced by the focused wave. This can potentially prevent possible risk and loss or damage when the operation is based on only phase-averaged wave information, such as H_s and T_p.

Next, we move on to study the scenario of a higher sea-state, as listed in Table 1. Operations will be halted for such sea-state, as it has a higher chance that the motion will exceed operation limit. However, if the phase-resolved wave-fields can be predicted, it is possible to utilize the time interval in which the floating system motion is low and within the operation limit. To illustrate this, we simulate the high sea-state and predict the floating system motion. τ and τ_p are computed to be 209 s and 86 s, respectively. Figure 5 shows the measured surface elevation at the upstream remote sensing location, while Fig. 6 shows the predicted surface elevation at the floating system location. The motion of the system is shown in Fig. 7, which shows the motion may go up to 2 m, which is well above the operation limit. However, there are periods where the motion of the system lies within the operation limits for a short duration which could be sufficient depending on the operation. Instead of cutting off the operation at the "rule of thumb" sea-state, the operation envelop of the floating system can significantly be increased.

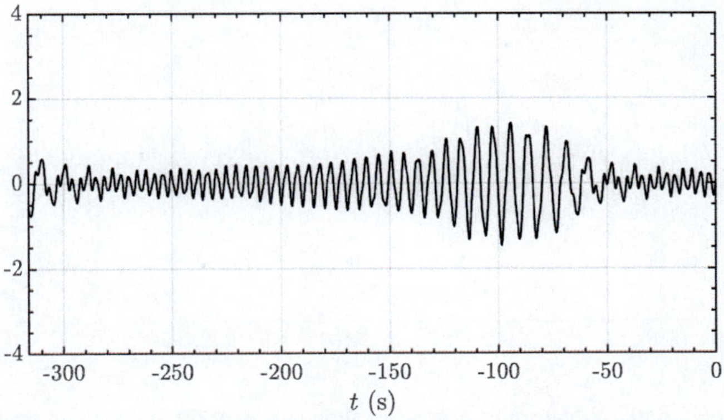

Fig. 2. Low Sea-State – Measured surface elevation at $d = 1000$ m upstream from the floating system location. Minimum time history of $\tau = 316$ s is required.

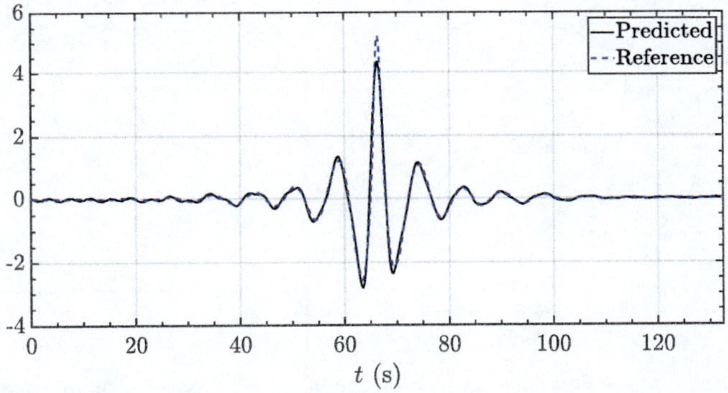

Fig. 3. Low Sea-State – Predicted surface elevation at the floating system location. Reliable prediction is achieved for prediction time window of $\tau_p = 132$ s, beyond which the prediction is not reliable.

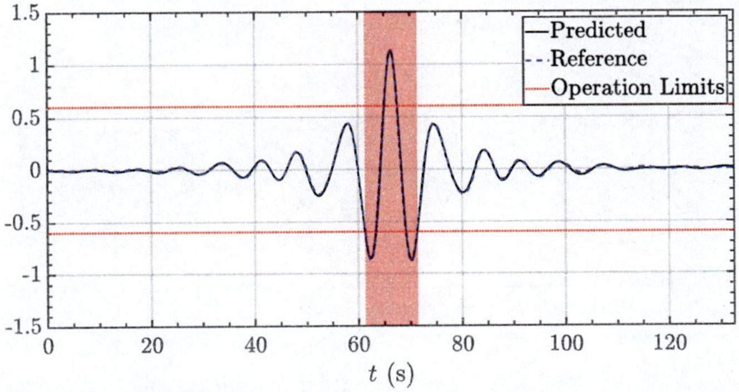

Fig. 4. Low Sea-State – Predicted heave motion of the floating system. The red zone shows the interval where the motion exceeds the operation limits.

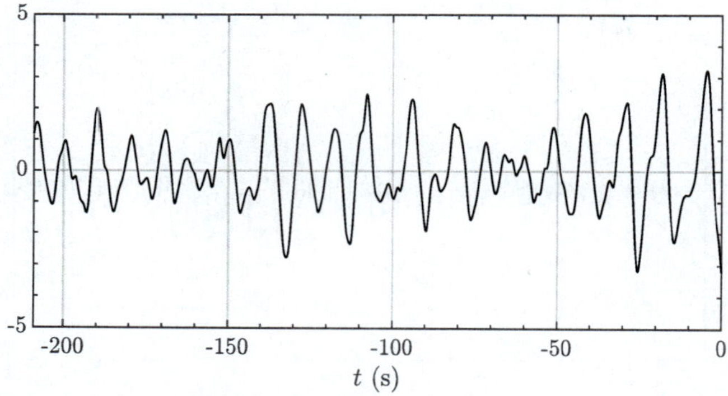

Fig. 5. High Sea-State – Measured surface elevation at $d = 1000$ m upstream from the floating system location. Minimum time history of $\tau = 209$ s is required.

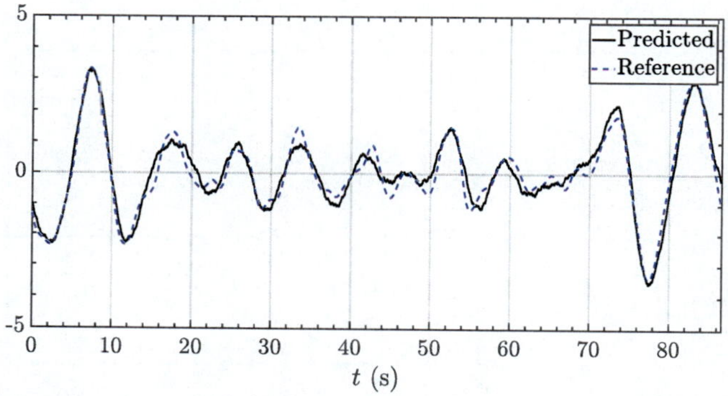

Fig. 6. High Sea-State – Predicted surface elevation at the floating system location. Reliable prediction is achieved for prediction time window of $\tau_p = 86$ s, beyond which the prediction is not reliable.

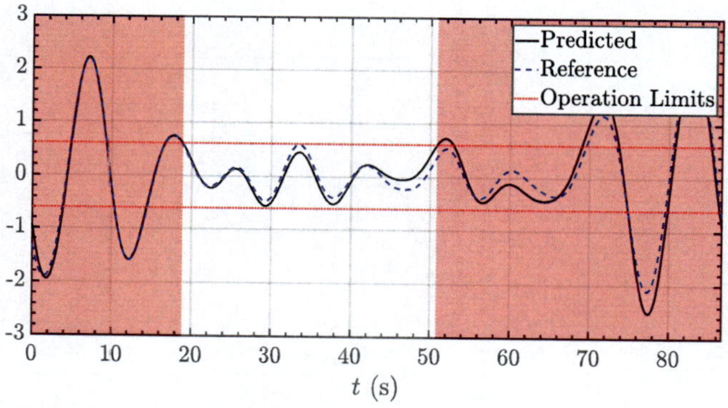

Fig. 7. High Sea-State – Predicted heave motion of the floating system. Red zone indicates time interval where the motion of floating system exceeds the operation limits, and white zone where the available operation window is longer than 20 s.

As future work, this procedure will be tested in a wave basin using remote sensing techniques and demonstrated in the field using an X-band radar. The methodology can also be enhanced to include non-linear wave evolution model, effects of wind, current and hydrodynamic interaction with other nearby vessels or structures. Vessels with forward speed will also be considered.

4 Conclusion

In this paper, the measurement of the incident phase-resolved wave-fields at a distance away from a floating system has been simulated. The incident wave-fields can be used to predict the response of the floating system ahead of time taken for the wave-fields to evolve into the floater location. The concept via simulations of evolving wave-fields and the induced response of a generic floater has been demonstrated. It is shown that the floating system response, and subsequent exceedance of the operational limits, can be predicted in advance. This methodology can help to reduce uncertainties related to the operating environment and the system response, and forms a building block for operability decision support systems.

Acknowledgement. This research is supported by A*STAR under its RIE 2020 Industry Alignment Fund (Grant No: 172 19 00089).

References

1. Clauss, G.N.F., Kosleck, S., Testa, D.: Critical situations of vessel operations in short crested seas: forecast and decision support system. In: International Conference on Offshore Mechanics and Arctic Engineering, vol. 43444, pp. 319–332 (2009)
2. Desmars, N., Pérignon, Y., Ducrozet, G., Guérin, C.A., Grilli, S.T., Ferrant, P.: Phase-resolved reconstruction algorithm and deterministic prediction of nonlinear ocean waves from spatio-temporal optical measurements. In: International Conference on Offshore Mechanics and Arctic Engineering, vol. 51272, p. V07BT06A054 (2018)
3. Naaijen, P., Wijaya, A.P.: Phase resolved wave prediction from synthetic radar images. In: International Conference on Offshore Mechanics and Arctic Engineering, vol. 45509, p. V08AT06A045. American Society of Mechanical Engineers (2014)
4. Hilmer, T., Thornhill, E.: Deterministic wave predictions from the WaMoS II. In: OCEANS 2014 TAIPEI, pp. 1–8. IEEE (2014)
5. Kusters, J.G., Connell, B.S.H., Milewski, W.M., Vinciullo, V.J., van Dijk, R.: Wave characterization and timing using Doppler radar – update on the FutureWaves™ wave and vessel motion forecasting system. In: OCEANS 2019 SEATTLE, pp. 1–9. IEEE (2019)
6. Wu, G.: Direct simulation and deterministic prediction of large-scale nonlinear ocean wave-field. Doctoral dissertation, Massachusetts Institute of Technology (2004)
7. Fossen, T.I., Perez, T.: Marine Systems Simulator (MSS) (2004). https://github.com/cybergalactic/MSS

Neural Network Modeling of Shear Wave Velocity of Macau Soils Using SPT and CPT Data

Z. Zhou and T. M. H. Lok(✉)

University of Macau, Macau, China
mhlok@um.edu.mo

Abstract. Shear wave velocity of soil is an important parameter for earthquake analysis and civil engineering designs. Correlations between shear wave velocity and SPT-N value based on the simple power-law regression model were provided by previous studies. However, due to the inherent uncertainties of the SPT-N values, estimation of shear wave velocity may be improved by incorporating other site investigation data. In this study, artificial neural network models were used for the correlation analysis between the shear wave velocity with SPT (Standard Penetration Test) and CPT (Cone Penetration Test) data, based on recent field test data obtained from the site investigation program of the Macau major transportation project. The analysis results of this study indicated that incorporating the additional soil parameters in the correlation model would improve the prediction performance. When combined with SPT and CPT to form neural network models, a better prediction would be obtained than that using SPT or CPT alone.

Keywords: SPT-N · CPT · Shear wave velocity · Artificial neural network · Nonlinear regression

1 Introduction

The shear wave velocity (V_s) is a fundamental parameter to determine the dynamic properties of soils. It is also helpful in the evaluation of foundation stiffness, liquefaction potential, earthquake site response, soil stratigraphy, soil density, site classification, and foundation settlements [1].

Standard Penetration Test (SPT) is a basic soil field test that is commonly used to indicate the relative density and compressibility of granular soils. It is also commonly applied to estimate the liquefaction potential of saturated granular soils for earthquake design. In the past half-century, a lot of researchers tend to use regression statistical analysis to establish the relationship between the shear wave velocity and SPT-N value worldwide [2].

D. V. K. Huynh et al. (Eds.): VSOE 2021, LNCE 208, pp. 164–171, 2022.
https://doi.org/10.1007/978-981-16-7735-9_16

The motivation of this research is because the direct measurement of shear wave velocity needs to deal with the high cost and the lack of qualified operators to perform the field test. Previous researches were mostly on using statistical regression methods to provide correlation equations that can predict shear wave velocity in a specific site condition.

The correlations between shear wave velocity and SPT-N values have considerable dispersions, which may be due to the test error of shear wave velocity and SPT-N values, geotechnical and geological conditions, and the limitation of regression statistical analysis. To reduce the test error and statistical error, other parameters should be added. Excepting the SPT-N value, it is well known that the cone penetration of soil may also have a relationship with shear wave velocity.

In this study, to improve the accuracy of shear wave velocity prediction, the SPT-N and CPT-qc were taken into consideration together to determine the shear wave velocity by BP neural network. The recent field test data obtained from the site investigation program of a major transportation project at Macau was used for this study, and the comparison between different models is depicted in this paper.

2 Back Propagation (BP) Neural Network

The BP neural network as shown in Fig. 1 was adopted in this study. The algorithm is composed of two processes, the forward propagation of the signal and the backward propagation of the error. In the forward propagation, the input samples enter the network from the input layer L_1 and pass through the hidden layer L2 to the output layer L_3. After the forward propagation of signal and the backward propagation of error, the adjustment of weights and thresholds is repeated until the preset number of training times, or the error is reduced to an allowable level. The process of constant adjustment of weights and thresholds is the training process of the network.

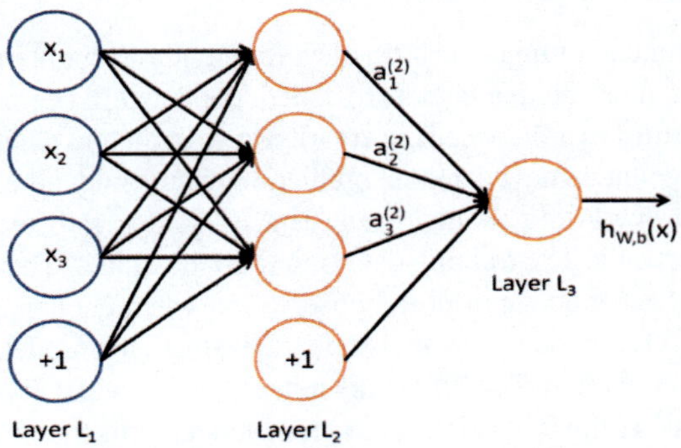

Fig. 1. BP neural network with one hidden layer

3 Case Study

The site investigation data about the shear wave velocity, CPT, and SPT-N values were obtained from a major transportation project at Macau. There are five boreholes contained in this project, the corresponding locations were shown in Fig. 2. The depths of five boreholes are around 18m to 66m. Down-hole seismic test and seismic-cone penetration test (S-CPT) were performed for the measurement of shear wave velocity. A total of 35 pairs of the values of shear wave velocity, SPT-N, and cone resistance values were taken at similar depths from the database of the project.

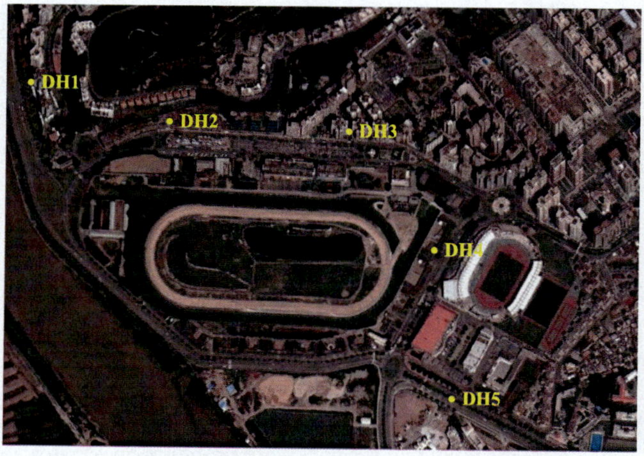

Fig. 2. The corresponding locations of five boreholes for the project

4 Analysis Results

In the field of machine learning, overfitting is a frequently encountered problem. In the BP algorithm, when more hidden layers are added, the network becomes more complex and can make the fitting error between observation and prediction data smaller. However, when the model encounters new data, the prediction results may be very terrible. In this study, to avoid the overfitting problem, the data is divided into two parts, including training data and test data. The training data is to train the model. The test data is mainly to evaluate whether the model can fit the observation when encountering new data. In this study, three different models are analyzed. To evaluate the model, the coefficient of determination (R^2) is adopted. The R^2 mainly reflects the degree of fit between prediction and observation data. If the R^2 is larger, the predicted data fits the observed data better. Finally, by comparing the R^2 among the three models, the best model will be chosen.

4.1 The Model with SPT and Vs

Firstly, the correlation model between SPT and Vs is built, and the SPT-N value is used as the only input to predict the Vs. By changing the number of neurons in the hidden layer and the hidden layers, the error in training data and test data. By comparing the error, the best structure of the BP model is got and shown in Fig. 3.

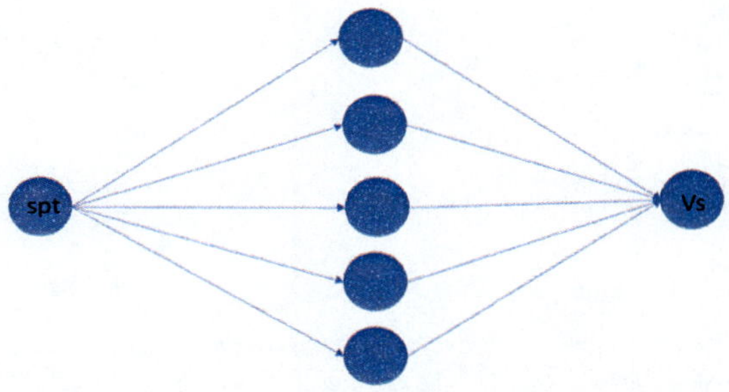

Fig. 3. The best structure of the BP model between SPT and Vs

When this model was trained in the learning process, the training loss became smaller and smaller. Simultaneously, validation loss also became smaller and smaller. The results of these two loss-reducing curves are shown in Fig. 4.

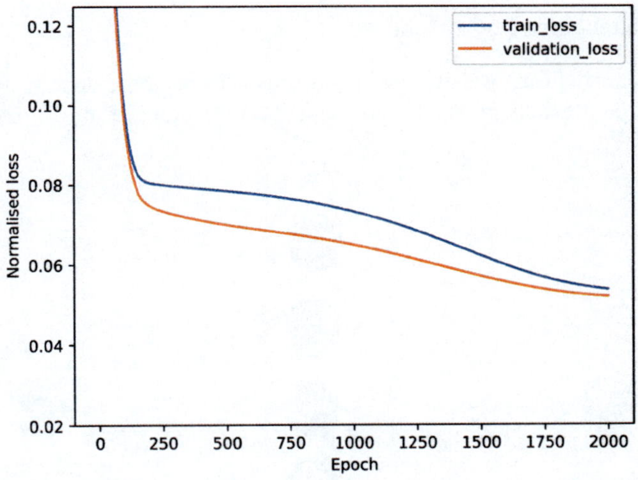

Fig. 4. The loss reducing curve versus the training epoch for SPT and Vs model

From the above figure, both the training and testing losses decrease very quickly and then become stable. Finally, these two lines are very close, which means this model can get a balance between underfitting and overfitting. The prediction by this model and observation results are also shown in Fig. 5. The coefficient of determination (R^2) is 0.352.

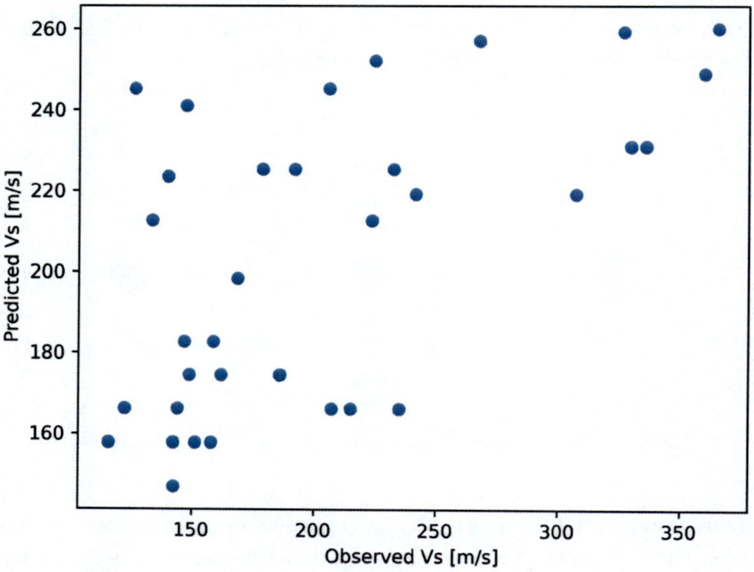

Fig. 5. The fitting results between observation and prediction by SPT

4.2 The Model Between CPT and Vs

Secondly, the model for Vs is obtained using the CPT-cone resistance as the only input. Using the same previous procedure, the same best structure of the BP model was got and shown in Fig. 6.

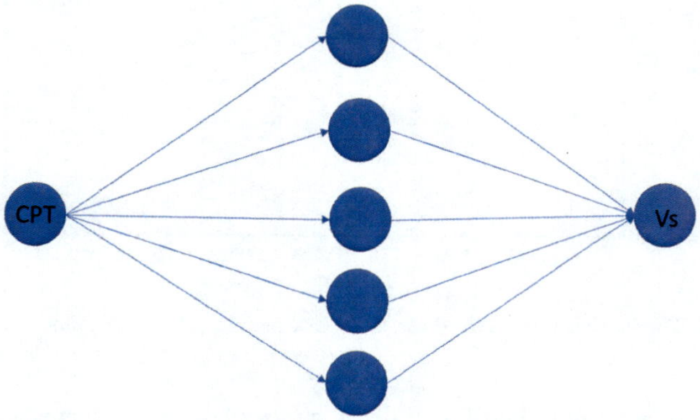

Fig. 6. The best structure of the BP model between CPT and Vs

When the best model was trained in the learning process, similar phenomena are happening with the previous model. The results of these two loss-reducing curves are shown in Fig. 7.

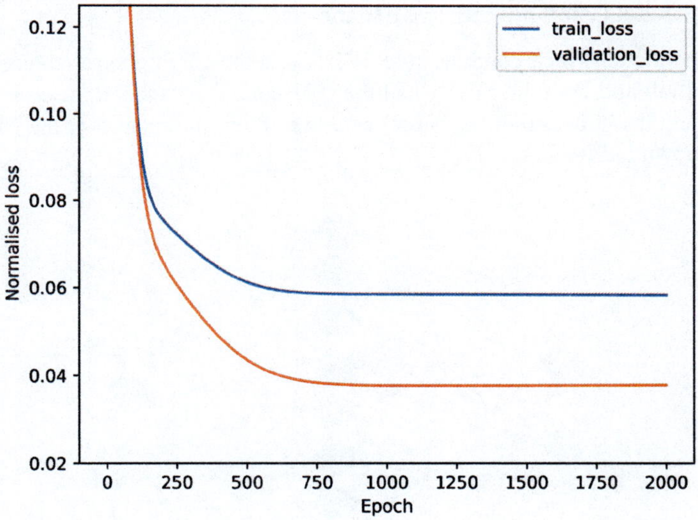

Fig. 7. The loss reducing curve versus the training epoch for CPT and Vs model

From the previous figure, both the training and validation losses decrease and become stable very quickly, but the testing error is smaller than the training error, which means this model has little underfitting problems. The prediction by this model and observation data are also shown in Fig. 8. The coefficient of determination (R^2) is 0.339.

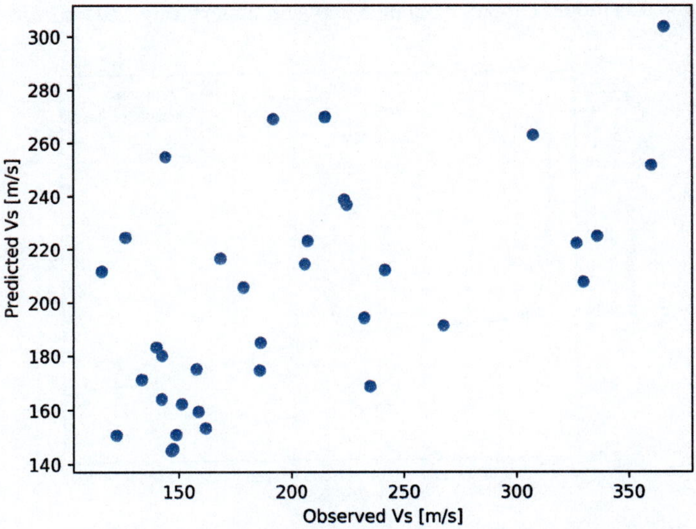

Fig. 8. The fitting results between observation and prediction with CPT and Vs

4.3 The Model Combining SPT, CPT, and Vs

Finally, the model which combines the SPT-N and the CPT-cone resistance to predict the Vs is built and both the SPT-N and the CPT-cone resistance are used as the input. By changing the structure of the neural network, the best structure of the BP model is got and shown in Fig. 9.

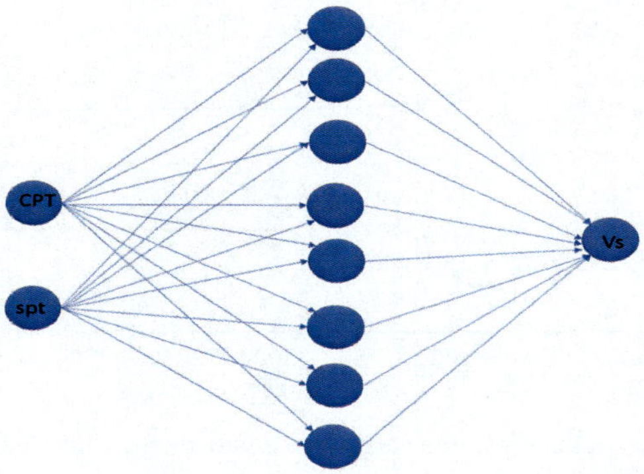

Fig. 9. The best structure of the BP model combining SPT, CPT, and Vs

The results of the two loss-reducing curves for this model are shown in Fig. 10. From this figure, both the training and validation losses decrease very quickly at the early epoch and then become stable, but the testing error is smaller than the training error, which means this model has little underfitting problems. Finally, these two lines are very close, which means this model can get a balance between underfitting and overfitting.

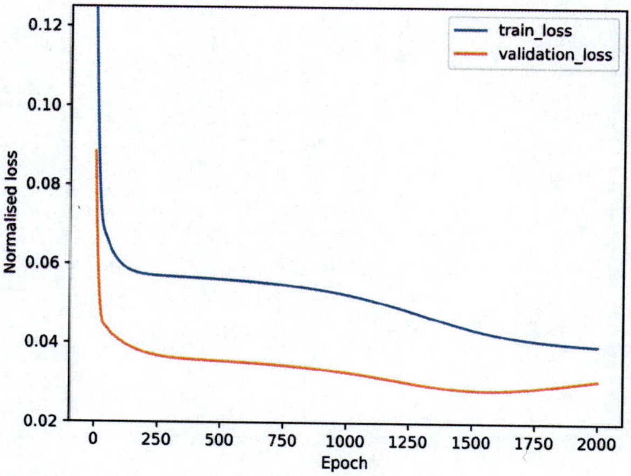

Fig. 10. The loss reducing curve versus the training epoch for combining model

The prediction using this model and observation data are also shown in Fig. 11. The coefficient of determination (R^2) is 0.538.

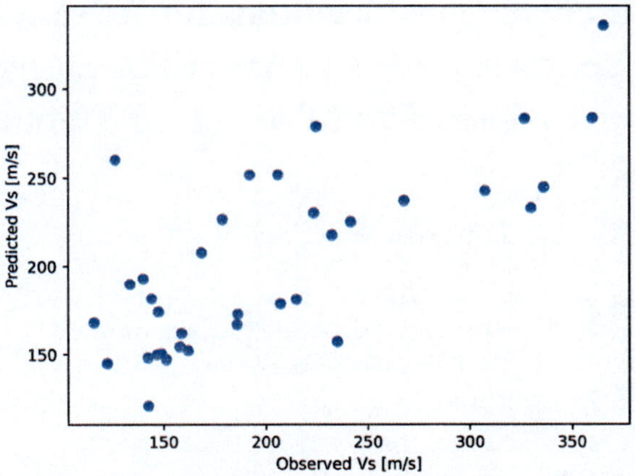

Fig. 11. The fitting results between observation and prediction for the combining model

5 Conclusion

In this study, the BP algorithm is chosen to analyze the database of a major transportation project in Macau. Using this method, the correlation models between SPT-N, CPT, and shear wave velocity are established. By comparison of three models, the model combining the CPT and SPT-N can fit the observed data best.

According to the results, it can be found that the BP method is a powerful method to do nonlinear regression. In traditional statistical analysis, the relationship between SPT-N and Vs are considered. However, according to the above analysis, the CPT also has a relationship with Vs. When combining SPT-N and CPT, the prediction can improve a lot.

References

1. Richart, F.E., Hall, J.R., Woods, R.D.: Vibrations of Soils and Foundations, p. 414. Prentice-Hall, Englewood Cliffs (1970)
2. Marto, A., Soon, T.C., Kasim, F.: A correlation of shear wave velocity and standard penetration resistance. Electron. J. Geotech. Eng. **13**, 463–471 (2013)
3. NG, K.F.: Correlation analysis of shear wave velocity of macau soils using nonlinear curve fitting and genetic programming. Final Year Project Report, University of Macau (2015)

Practical Considerations for Advanced Laboratory Tests to Assist Pile Design for Offshore Wind Projects in Vietnam

Minh Nguyen[1(✉)] and Man Bui[2]

[1] ODE Ltd., London, UK
minh.nguyen@ode-ltd.co.uk
[2] Soil Analysis Services, Dubai, UAE

Abstract. Recently, a significant number of new offshore wind projects has been proposed and some are under development in Vietnam and in Asia Pacific. For Vietnam, such move is logical as the country has abundant potential in offshore wind energy, as well as being very experience in the O&G developments since 1970s and particularly keen on promoting renewable energy. In fact, it is generally expected that although offshore wind is a relatively new in Vietnam, its development may be faster, lower in cost and less troublesome in comparison with other nearby Asian countries.

This paper gives an overview of the differences in obtaining the soil mechanical (stress-strength-stiffness) parameters in the foundation design between an Oil and Gas (O&G) platform jacket and an Offshore Wind Turbine (OWT) piled foundation (jacket pile or monopile) in accordance with accepted offshore standards (e.g. DNVGL-ST-0126 and DNVGL-ST-0437). It then describes the practical considerations which should be given to Advanced Laboratory Tests (ALT), especially cyclic testing strategy to reflect the extra demands given in the cyclic and fatigue behavior of the WTG piled foundation. The scope and content of the ALT strategy should be tailored to each OWT project, including the need for site-specific data, project budget and schedule.

The practical considerations for ALTs are well acknowledged by WTG design engineers but will need to be adapt into Vietnam condition. In addition, the existing capability and capacity constraints of accredited labs are known as other challenges, and these are opportunities for future study and co-operation.

Keywords: Laboratory testing · Site investigation · Offshore wind project · Cyclic tests · Site characterisation

1 Introduction

It has been estimated that Vietnam has a significant capacity of more than 100 GW in offshore wind energy resource, placing the country in the top-ranking potential areas for Offshore Wind Turbine (OWT) projects. The Vietnamese government is currently promoting and supporting renewable energy developments, especially offshore wind. It

D. V. K. Huynh et al. (Eds.): VSOE 2021, LNCE 208, pp. 172–180, 2022.
https://doi.org/10.1007/978-981-16-7735-9_17

is expected that its offshore wind industry may develop quickly due to the success and extensive experience gained from the offshore Oil & Gas industry since 1970s.

In Vietnam, offshore projects are regulated at the national level with references to international offshore codes. It is anticipated that international standards and best practices, such as [1] and [2], will be of popular usage. To obtain a project certification at any stage (e.g. concept, feasibility, FEED, detailed design), the OWTs need to be evaluated for compliance with site-specific conditions. Consequently, the laboratory (lab) testing data, consisting of both standard and advanced lab tests (ALTs), becomes one of the critical inputs within the site dataset throughout the project's design life cycles. The ALT strategy is often a challenging issue for the engineer due to varying technical requirements from different required limit states. Moreover, its scope is subjected to both schedule and cost constraints. Furthermore, each project has many Wind Turbine Generators (WTGs) spreading over a very wide area and potentially employing several types of supporting foundation, introducing additional level of complexity. With these unique characteristics, the ALT strategy for any OWT project differ distinctly with the ones applied for a typical O&G platform.

Due to paper length limitation, the discussions and recommendations in this paper are purposely concentrated on the development of an ALT strategy applicable for WTG piled foundations (including monopile). The ALTs identified herein are specifically limited to the tests for determination of the stress-strength-stiffness characteristics from boring samples. These ALTs shall be carried out by accredited geotechnical lab, requiring full integration and supervision from the design and lab testing teams.

2 ALT Strategy for WTG Piled/Monopile Foundations

2.1 Main Considerations for the Design of Piled/Monopile Foundation

Offshore O&G platforms and WTGs are both subjected to challenging and varying wind and wave loading conditions during their installation and operational life. The offshore wind industry standards have been developed using earlier experience gained from the offshore O&G industry, such as the American Petroleum Industry (API) codes. There are, however, several distinct differences in the design for piles or monopile to support the WTG foundations as follow:

- Unlike an O&G platform, WTG structure is often considered as non-habitable and with shorter design life, and to be of lower Offshore Safety Class
- There are many more WTGs to be considered (via means of grouping/clustering)
- WTG loading is characterised by lower vertical load and higher horizontal and moment load compared to O&G platform, and this trend accelerates with increasing turbine size
- During its operation, the WTG rotates at different speeds/frequencies and therefore there are specific requirements to the system eigen-frequency range
- The jacket (four-leg) or tripod (three-leg) of WTG foundation is generally designed with less piles and offer limited means of loading redistribution. In contrast, O&G platform jacket provides more flexibility on load redistribution

- The monopile supporting the WTG is generally stiff and short, and its working behaviour is widely known to be significantly different from the long slender pile sizes normally employed under the offshore platform (see also [3–6])
- Lateral loading capacity of monopile at Ultimate Limit State (ULS) is often a controlling factor in pile size and pile length.
- Fatigue and cyclic degradation of pile capacity are more critical in WTG foundation
- It is important to establish/estimate the initial soil stiffness and subsequent changes for system frequency check and for the Fatigue Limit State (FLS).
- The WTG supporting structure must satisfy tighter limits for allowable deformations and permanent rotations in Serviceability Limit State (SLS).

2.2 ALTs for the Design of WTG Piled/Monopile Foundation

The discussions are specifically for the group of ALTs described in Table 1, with the details about test procedures and technical specifications can be found in [2] and the international codes for marine offshore site investigation [7].

Table 1. Identifications of the advanced lab tests (ALTs) required for WTG piled foundation

ALT group	Descriptions and objectives	Required testing equipment
ALT-1a	Compressibility/stiffness and consolidation characteristics, including intrinsic parameters	Incremental Oedometer & CRS Oedometer
ALT-1b	Static shear strength, stress-strain-response backbone curve and failure envelope	Triaxial (TX) & Direct Simple Shear (DSS)
ALT-1c	Static strength tests with small strain measurements (initial pseudo-elastic stiffness moduli)	TX with local strain sensors Bender Elements (BEs)
ALT-1d	Interface (soil-pile and soil-soil) strength tests	Torsional Ring Shear Bishop Ring Shear
ALT-2	Dynamic tests for small strain shear modulus, damping ratio, stiffness degradation	Resonant Column (RC) & BEs
ALT-3	Cyclic tests, strength and stiffness dependency w.r.t number of cyclic load and strain levels	Cyclic Triaxial (CTX) & Cyclic Simple Shear (CDSS)

2.3 ALT Strategy for WTG Piled/Monopile Foundation

The ALTs are often employed in conjunction with and as supporting evidence for in-situ tests. For example, tests in group ALT-1b are used to calibrate N_{kt} factor for PCPTs (piezocones). Another example is using ALT-1c & ALT-2 for checking the V_S profile obtained from using P-S suspension logging or seismic CPT. The tests in groups ALT-1c & ALT-2 are essential to define the damping ratio, small strain stiffness, and stiffness degradation. For all the required design limits (ULS, FLS, SLS) a combination of ALT-1a, ALT-1b, ALT-1c & ALT-3 help to create the static responses and the cyclic-dependent characteristics of the soils.

One important aspect of the ALT strategy is to identify a site-specific and design-phase dependent scope of ALTs to enable the derivations of realistic and well-defined soil parameters with reasonable degrees of confidence. The strategy should consider the previously described design characteristics of the WTG piled/monopile foundation and make use of the ground model interpreted. This requires closely collaboration between the design engineer and the site investigation lead engineer. Nowadays, more OWT projects are planned at sites that are outside the known database and experience of the Vietnam O&G industry. For example, silty soils with high mica content and prone to breakage are abundant in many parts of Asia Pacific. Another consideration is natural hazard such as typhoon, high sea current or geohazard like mega mobilized sand-waves. Due to paper's length constraint, such additional site-specific considerations for the ALT strategy are outside the scope of this paper.

ALT strategy becomes even more critical in special conditions, such as highly complex and irregular cyclic loadings, or dealing with intermediate soils (partial drainage), or where the design is pushed to the limit of existing technology. Suitable ratio of tests for ALT-1b (reference static strength tests) versus ALT-3 (cyclic tests) are essential. ALT is also important where previous database or soil type is not available, and to increase to confidence of the correlations to similar soil units/lithologies.

All in all, ALTs are expected as compulsory requirements by certifiers. An early and suitable (often small in relative cost) investment will return cost-efficient foundation solutions, bringing significant savings in overall foundation installation cost, helping to reduce project insurance premium, and limiting potential delays.

3 Establishing Backbone Stiffness Curve from a Combination of Various ALT Techniques

It is well established that geomaterials exhibit non-linear stress strain behaviour (Fig. 1, Fig. 2, and Fig. 3). In principle, The design of WTG piles requires stiffness degradation curves (G/G_{max} or G/G_0 versus shear strain, see Fig. 3) and damping curves.

The establishment of the backbone curve of soils to use with cyclic tests is achieved via using groups ALT-1b, ALT-1c, ALT-2 and ALT-3. Further detailed recommendations can be found in [8, 12]. Figure 3 shown in example of a backbone stiffness curve, drawn using a combination of various techniques: i.e. resonant column test (ALT-2) for strain level between 10-4% and 0.05%, triaxial test with local strain measurement (ALT-1c) for strain from 10-3% to 1–2% and conventional triaxial test (ALT-1b) for strains larger than 0.1%.

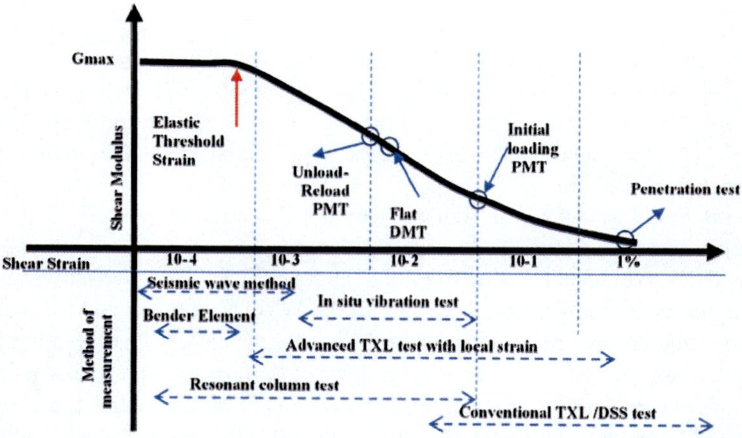

Fig. 1. None-linear stress-strain behavior of soil and testing techniques applied.

Figure 2 shows an example that due to system compliance, the conventional test with external strain measurement (ALT-1b group) yield significantly lower values of Young' modulus in comparison with test in which small strain sensors are employed (ALT-1c). In Fig. 2, the increase in stiffness at 0.05% local axial strain could be attributed to the closing of rock fissures/factures, which can be observed using local strain gauges but not external LVDTs.

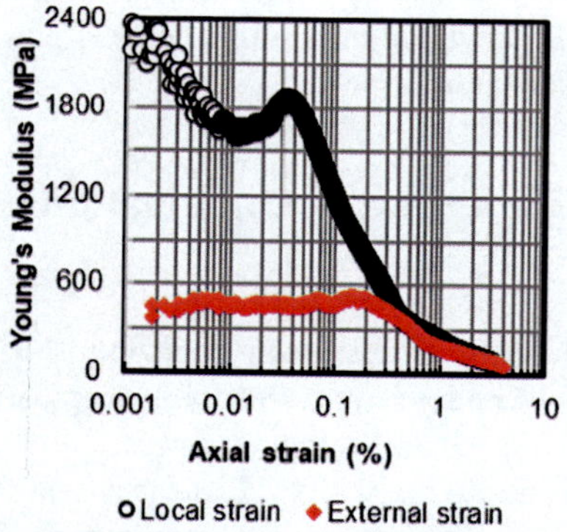

Fig. 2. Young's modulus for marine conglomerate (from ALT-1b/1c tests).

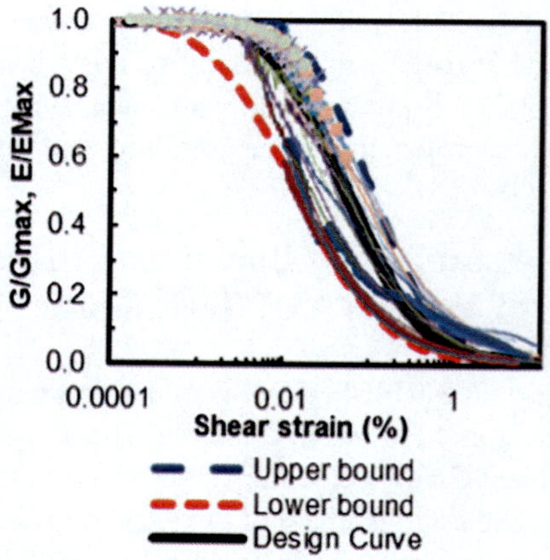

Fig. 3. G/Go curves established from ALT-2 (RC) and ALT-1b (TX) tests

4 Considerations to Cyclic Tests (ALT-3 Group)

Cyclic load causes pore water pressure built up, resulting in effective stress loss and soil strength and stiffness degradation, hence reduction in pile resistant capacity. Drained cyclic loading can also be important in the evaluation of soil stiffness changes. At relative (to vertical load) low loading levels, the WTG foundation is subjected to large number of cycles that create strain accumulation, which can affect the serviceability limit. Large number of loading cycles can also cause excessive permanent foundation settlement which affect the performance of the WTG as illustrated in Fig. 4.

Sample Preparation. Generally, sample preparation shall be carried out in accordance with testing standards (e.g. [7, 10, 11]). For sand it is important that the sand dry density should be the same of that in-situ, which can be estimated from CPT data [9].

Sample Saturation. It should be noted that 'saturation' process is not just simply adding water into the sample. Extra care should be made to avoid sample swelling (sample disturbance) during saturation, especially for overconsolidated clay. Therefore, dry dock method as per ASTM standard [10] or constant moisture content method [11] is recommended for saturation of cohesive soil. The maximum back pressure and minimum B-value should be identified depending on the actual soils.

Sample Consolidation. The consolidation stress path should take in to account the stress history, and/or pile driving effect such as densification and particle crushing for sand [12]. To minimize the effects of sample disturbance, recompression/reconsolidation methods [13] are recommended. For example, due to sample disturbance, undrained shear strength of a stiff clay sample with yielding stress ratio (YSR) of 3.0 could be significant underestimated up to 50%; this can be overcome by consolidating the sample to 2.5 times the overburden effective stress, σ'_{v0}, then reduce the effective stress to σ'_{v0} prior to shearing. Ideally for overconsolidated clay, two phase K_0 consolidation stress path can also be employed, though the cost and testing time will significantly increase. It should be noted that some heavily consolidated clay soils with K_0 greater than one may lead to one-way cyclic loading mode instead of the two-way loading assumption. For sands, pre-drained shear stage may also be considered.

Cyclic Load Amplitude. Offshore WTGs are often subjected to a large number of repeated load cycles at extremely high amplitudes of wave, wind, and earthquake, etc. Cyclic stress amplitude shall be decided based on:

- The results of the structural analysis under the extreme wind/ wave load applied to the structures.
- The peak ground motion due to earthquake, if the windfarm is influenced by an active seismic zone in vicinity.
- The depth of the soil samples, e.g. the cyclic stress amplitude decreases with depth
- The design amplitude in relevant to the pile length for sense-check if the pile behaviour (in pile interaction diagram chart) are in stable zone or meta-stable zone.

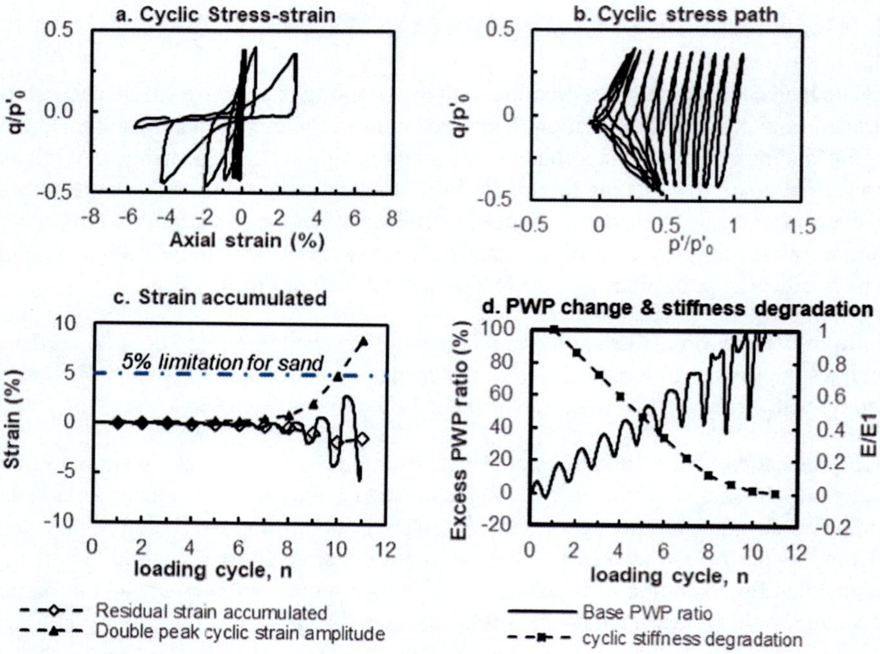

Fig. 4. Cyclic test and liquefaction of a calcareous sand at cyclic stress ratio of 0.20, $n_f = 10$

Loading Frequency. Equally important, the test rate effects shall be well considered. Rate of PWP built up depend on the loading frequency, which shall be specified as closest as possible to the actual frequency of the load actions. If the data are not available, the recommendations in Table 2 [after 7] can be used for guidance.

Table 2. Cyclic loading information in a CTXL test

Load type	Peak Earthquake	Extreme Hight Wave	Storm Gut (Extreme High Wind speed)
Minimum loading periods	10 s	3 h	10 min
commonly range of loading frequency (*)	1.0–2.0 Hz	0.08–0.18	0.20–0.50
Most common frequency specified	1.0	0.1	0.5
Minimum loading cycles	20	100–2000	100–300
Number of loading cycles specified	1500	2000	1500

Undrained vs Drained Cyclic Test. Cyclic tests can be conducted in either undrained or drained condition, depending on the testing purposes, soil types, and loading conditions, etc. For example, undrained test is likely the only option for liquefaction analysis or for assessment of soil stiffness degradation, while drained tests can be conducted for evaluation of long term foundation settlement under service loads [14] or time effects on pile shaft capacity due to increasing of soil stiffness.

Reference Monotonic Test (ALT-1b). It is also important to carry out the cyclic tests in ALT-3 group with identical procedures of sample preparation, saturation, and consolidation applied to the monotonic tests conducted at similar depths and soil units. A larger number of tests in ALT-1b group are commonly conducted in comparison with the cyclic tests in ALT-3 group to establish the reference shear strength envelopes.

5 Summary

This paper describes some main considerations for the ALT strategy to WTG piled foundation. One important aspect of the ALT strategy is to identify a site-specific and design-phase dependent scope of ALTs to enable the derivations of realistic and well-defined soil parameters with reasonable degrees of confidence. ALT strategy becomes even more critical in special conditions, where the design is pushed to the limit of existing technology. The strategy should consider the previously described design characteristics of the pile for WTG foundation and make use of the ground model/interpreted soil units. This requires due considerations to constraints on schedule, lab resource and technical capabilities. The points raised herein are brief due to paper length limitation, but it is expected to start the necessary interest and considerations of the practical geotechnical engineers and project managers.

The practical considerations are well acknowledged by WTG design engineers but will need to be adapt into Vietnam condition. Moreover, there are also known issues on existing capability and capacity constraints of many laboratories in the country. The Authors are looking forward seeing countermeasures to these challenges via co-operation and further studies.

References

1. DNVGL-ST-0126: Support structures for wind turbines (2018)
2. DNVGL-ST-0437: Loads and site conditions for wind turbines (2016)
3. Byrne, B.W., et al.: PISA design methods for offshore wind turbine monopiles. In: Offshore Technology Conference, Paper number 29373 (2019)
4. Byrne, B.W., et al.: PISA design model for monopiles for offshore wind turbines: application to a stiff glacial clay till. Geotechnique **70**(11), 1–18 (2020)
5. Burd, et al.: PISA design model for monopiles for offshore wind turbines: application to a marine sand. Geotechnique **70**(11), 277 (2020)
6. WAS-XL Webinar series in 2020, organized by the Norwegian Geotechnical Institute (link https://www.ngi.no/eng/Projects/WAS-XL/Events/WAS-XL-Webinar)

7. ISO 19901-8:2014 Petroleum and Natural Gas Industries – Specific Requirements for Offshore Structures –Part 8: Marine Soil Investigations
8. Andersen, K.H, Puech, A.A., Jardine, J.R.: Cyclic resistant geotechnical design and parameter selection for offshore engineering and other applications. In: Proceedings of TC 209 Workshop - 18th ICSMGE, Paris 4, September (2013)
9. Kulhawy, F.H., Mayne, P.H.: Manual on estimating soil properties for foundation design, Report EL-6800 Electric Power Research Institute, EPRI, August (1990)
10. ASTM D3999 / D3999M - 11e1: Standard test methods for the determination of the modulus and damping properties of soils using the cyclic triaxial apparatus (2011)
11. BS 1377: Methods of test for soils for civil engineering purposes, Part 8: Shear strength tests (effective stress) (1990)
12. Jardine, R., Sim, W.W., Aghakouchak, A.: Cyclic triaxial tests to aid offshore pile analysis and design. Geotech. Eng. **166**(2), 111–121 (2013)
13. Bjerrum, L: Problems of soil mechanics and construction on soft clays and structurally unstable soils (collapsible, expansive and others). State of the art report, session 4. In: Proceedings of 8th ICSMFE, Moscow, USRR, vol. 3, pp 109–159 (1973)
14. Shariff, E., Bui, M.: Investigation of long-term settlement of flexible steel oil storage tanks on calcareous sand in UAE. In: Proceedings of VSOE2021 (Under Publication) (2021)

Rock Physics Diagnostic (RPD) as a Value-Added Tool in Site Investigation of the Subsea Characteristics for an Offshore Wind Farm Site in Ca Mau, Vietnam

Pham Huy Giao[1,3](✉), Nguyen Duc Thanh[2], Doan Huy Hien[1], Pham Hong Trang[1], Pham Quy Ngoc[1], Bui Viet Dung[1], Ly Van Chien[2], and Le Cao Minh[2]

[1] Vietnam Petroleum Institute (VPI), Hanoi, Vietnam
giaoph@vpi.pvn.vn
[2] Technical World (TW), Ho Chi Minh City, Vietnam
[3] PetroVietnam University (PVU), Ba Ria, Vietnam

Abstract. The study wind farm site is located near-shore in the Ca Mau province, about 35 km far away from Ca Mau city to the East, about 3 km from Ganh Hao town - Bac Lieu province to the South, with a shallow water depth from 2.5 to 3.5 m. An integrated geotechnical-geophysical survey was conducted at the site. In addition to common geotechnical testing, an emphasis in this study was given on PS seismic logging data analysis by Rock Physics Diagnostic (RPD), a method that is more commonly used in analysis of oil and gas well data. As results the subsea soil profile of five layers up to 80 m deep was well characterized by both geotechnical and elastic parameters. In addition, based on the RPD charts one can have a better idea of sedimentological conditions of the young soils at the study site.

Keywords: Near-shore investigation · Offshore wind farm · Rock Physics Diagnostic (RPD) · Seismic logging

1 Introduction

For many countries "*there is currently limited detailed understanding and data on the national offshore seabed characteristics*" [1], thus greater efforts are needed to improve understanding of the subsoil profile underlying the planned offshore wind energy areas to support development decision-making. While PetroVietnam as the leading national company for oil and gas exploration and production still wonders about its suitable position in the supply chain of renewable energy industry, the private sector has been proactively involved in developing wind farms from small to large scale in offshore area, which used to be the traditional operation space of the former. The question is whether or not a win-win cooperation can be made between a state-based organization like Vietnam Petroleum Institute (VPI) and a private company like Technical World (TW) with their own advantages and disadvantages in the period of energy transition. The objective of

D. V. K. Huynh et al. (Eds.): VSOE 2021, LNCE 208, pp. 181–187, 2022.
https://doi.org/10.1007/978-981-16-7735-9_18

this joint research is to perform a near-shore site investigation and its data analysis with an emphasis on seismic logging data acquired at a shallow water depth at the Ca Mau offshore wind farm site with the help of the rock physics diagnostic (RPD) approach, which is more common for oil well log data analysis.

2 Rock Physics Diagnostic (RPD) Method

"*Rock physics is defined as science exploring quantitative relations between various properties (attributes) of the natural rock*" [2]. It has started as a "velocity-porosity" science with applications mainly in oil and gas exploration, in particular for well log data analysis and static reservoir modeling. It has developed to "velocity-porosity-mineralogy" and then "velocity-porosity-mineralogy-fluid" science. Rock physics studies rocks "*as a holistic object whose various properties (attributes) are extracted from experiments simulating processes such as elastic wav propagation, fluid and electrical transport etc.*" [2]. Below are given a number of fundamental equations used in rock physics modeling (RPM):

$$V_p = \sqrt{\frac{K + \frac{4}{3}G}{\rho_b}} \tag{1}$$

$$V_s = \sqrt{\frac{G}{\rho_b}} \tag{2}$$

$$M = \rho_b {V_p}^2 \tag{3}$$

$$G = \rho_s {V_s}^2 \tag{4}$$

$$K = M - \frac{4}{3}G \tag{5}$$

$$AIp = \rho_b V_p \tag{6}$$

$$\vartheta = \frac{1}{2}\frac{(V_p/V_s)^2 - 2}{(V_p/V_s)^2 - 1} \tag{7}$$

$$\lambda = K - \frac{2}{3}G \tag{8}$$

Where: Vp, Vs are the compressional and shear velocity (m/s); K, M, G are the bulk modulus, compressional modulus and shear modulus (GPa), respectively; AIp is the acoustic impedance (m/s*g/cc); ϑ is Poisson ratio; λ is Lame's constant (GPa).

When building out a rock physics model (RPM) one has to detect various velocity-porosity trends and assign each of them to appropriate depth intervals and depositional sequences. This process is called rock physics diagnostic (RPD). For newly deposited sediments a chart of VP versus porosity (F) with the Reuss elastic bound is very useful in

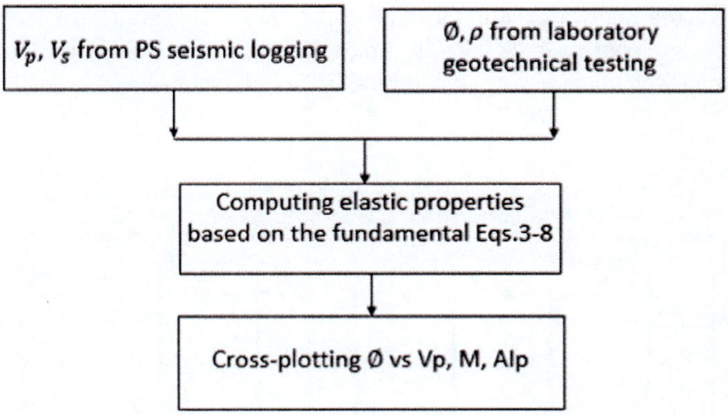

Fig. 1. Flow chart of rock physics diagnostic (RPD).

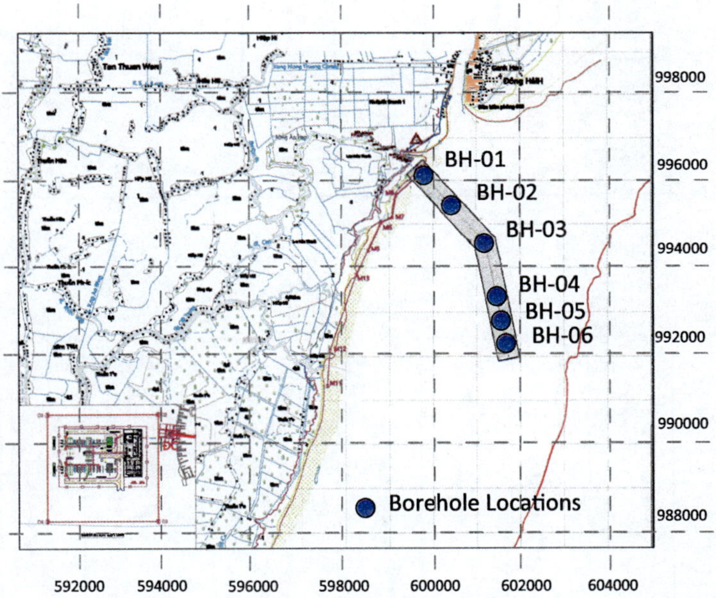

Fig. 2. Study site and seismic borehole locations [4].

deciphering rock texture [3]. For this study the methodology follows the chart as shown in Fig. 1.

The study wind farm site belongs to Dam Doi district, Ca Mau province, with a shallow water depth from 2.5 to 3.5 m. It is about 35 km far away from Ca Mau city to the East, about 3 km from Ganh Hao town - Bac Lieu province to the South as seen in Fig. 2 with the locations of six PS seismic boreholes [4] that were conducted as a part of an integrated geotechnical-geophysical near-shore investigation, which included the drilling and sampling works, laboratory geotechnical testing (grain size analysis, Atterberg limits, water content, bulk density, direct shear test, triaxial test, and consolidation test), in-situ geotechnical testing including CPTU, field vane shear.

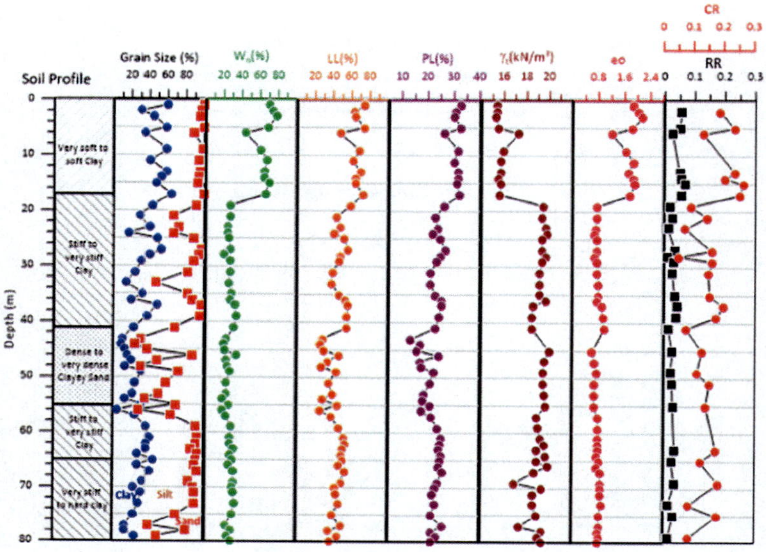

(a)

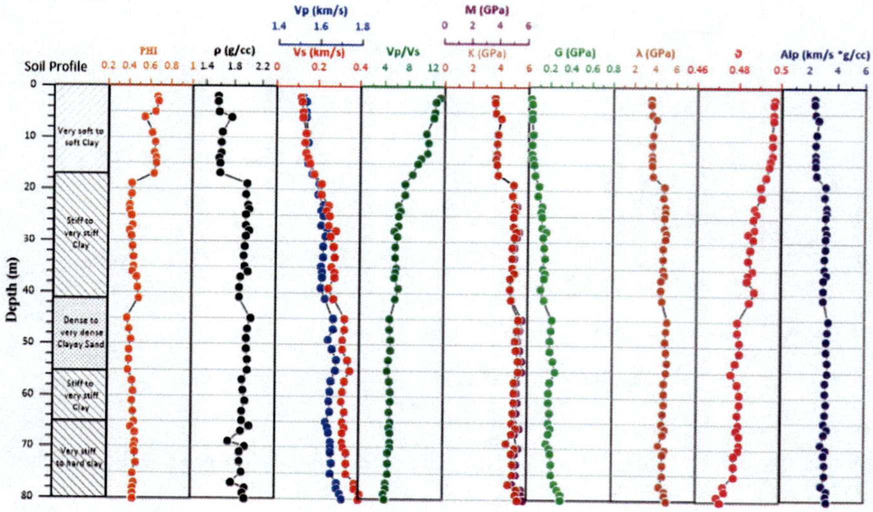

(b)

Fig. 3. (a) The near-shore subsoil profile at the Ca Mau wind farm site and (b) Elastic characteristics of the seabed at the Ca Mau wind farm site.

3 Results and Discussion

Table 1. (a) Geotechnical characteristics of the study site seabed and (b) Elastic characteristics of the study site seabed.

(a)

Layer No	Name	Depth (m)	Geotechnical characteristic analysis									
			Grainsize			LL	PL	W	Bulk density	Void Ratio	CR	RR
			sand (%)	silt (%)	clay (%)	%	%	%	kN/m³	e0		
1	Very soft to soft clay	0-17	3.79	46.24	49.97	65.71	31.02	65.91	15.74	1.81	0.22	0.05
2	Stiff to very stiff clay	17 -41	18.23	48.32	33.45	49.32	23.89	27.46	19.25	0.79	0.13	0.03
3	Dense to very dense clayey sand	41 - 55	50.30	31.50	18.20	34.91	18.81	24.07	19.64	0.69	0.13	0.03
4	Stiff to very stiff clay	55- 64	21.85	48.95	29.19	45.09	23.28	26.23	19.23	0.80	0.18	0.03
5	Very stiff to hard clay	64-80	21.29	51.67	27.03	44.96	23.40	28.94	18.75	0.84	0.13	0.02

(b)

Layer no	Name	Depth (m)	Geophysical characteristics analysis										
			Ø	ρ (g/cc)	Vp (m/s)	Vs (m/s)	Vp/Vs	M (GPa)	G (GPa)	K (GPa)	AIp (km/s *g/cc)	ϑ	λ (GPa)
1	Very soft to soft clay	0-17	0.64	1.61	1538.09	139.81	11.19	3.80	0.03	3.76	2.47	0.50	3.74
2	Stiff to very stiff clay	17 -41	0.44	1.96	1614.03	257.67	6.31	5.11	0.13	4.94	3.17	0.49	4.85
3	Dense to very dense clayey sand	41 - 55	0.42	1.98	1668.10	335.85	4.97	5.51	0.22	5.22	3.31	0.48	5.07
4	Stiff to very stiff clay	55- 64	0.44	1.94	1652.81	323.52	5.11	5.31	0.21	5.03	3.21	0.48	4.90
5	Very stiff to hard clay	64-80	0.46	1.91	1667.68	350.04	4.79	5.32	0.24	5.00	3.19	0.48	4.85

As the results the geotechnical characteristics of the study site seabed are presented in Fig. 3a and Table 1a, while those of elastic characteristics are presented in Fig. 3b and Table 1b, respectively. In this study the whole range of elastic properties such as the ratio between P and S-wave velocities (Vp/Vs), Acoustic impedance (AIp), Poisson's ratio (ϑ), Lame's constant (λ), bulk modulus (K), compressional modulus (M) and shear modulus (G) could be determined using Eqs. 1 to 8.

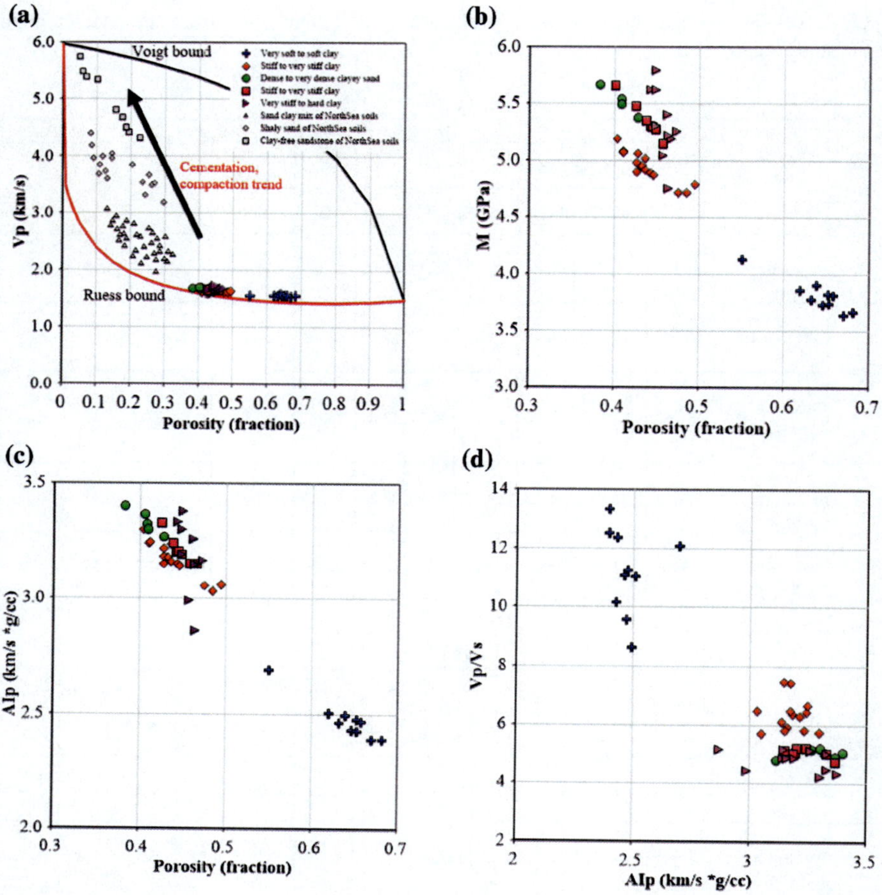

Fig. 4. RPD charts constructed for the study site: (a) Vp vs F in comparison with the North Sea soil data, updated from [4]; (b) M vs. F, (c) AIp vs F; (d) Vp/Vs vs AIp.

Based on the results of an integrated geotechnical and geophysical analysis the subsoil profile up to 80 m deep was completely characterized, consisting of five soil layers, i.e., very soft to soft clay (0–17 m), stiff to very stiff clay (17–41 m), medium dense to very dense clayey sand (41–55 m) having a medium sand of about 2 to 3 m thick in the upper part, stiff to very stiff with two sublayers from 55 to 65 m (stiff fat to lean clay) and 65 to 55–80 m (hard lean clay). Some rock physics diagnostic (RPD) charts were constructed for the study site as shown in Fig. 4a–d. In Fig. 4a the data points of the Ca Mau soils are located close to the curve of Ruess's bound, which indicates of very young sediments, in particular the very soft clay on top. It is interesting to see that the five soil layers identified by geotechnical characteristics could be well defined by elastic characteristics as seen in Fig. 4b–d and Table 1a and b.

4 Concluding Remarks

The seabed characteristics of an offshore wind farm site in Ca Mau, Vietnam, were completely characterized by an integrated geotechnical-geophysical investigation and data analysis. The subsoil profile up to 80 m deep under the sea bottom consists of five well defined layers. Layer 1 of very soft to soft clay on top has highest porosity (0.55–0.7) and Vp/Vs (8.3–14), lowest M (3.5–4 GPa) and AIp (2.3–2.5 km/s*g/cc). In contrast, Layer 5 of very stiff to hard clay is characterized by the highest M (5–6 GPa), AIp (3–3.3 km/s*g/cc), and lowest Vp/Vs (4–5); Layer 4 has a M (5.2–5.7 GPa) higher than that of the layer 2 (4.7–5.2 GPa), whereas Vp/Vs of layer 4 (4.7–5) is smaller than that of the layer 2 (5.8–7.8); The layer 3 has smallest porosity and highest velocity compared to those of the other layers as seen in Table 1b, indicating a mixture of clay-sand. It is first time the RPD method was applied for a shallow seismic logging survey, whose results agree well with the results from geotechnical testing, thus it can be a value-added tool for further nearshore site investigations for offshore wind farms' construction along the coast of Vietnam.

Acknowledgments. Acknowledgments are due to the support from the Vietnam National Foundation for Science and Technology Development (NAFOSTED) under the grant number 105.99-2019.324. Sincere thanks go to the Vietnam Petroleum Institute (VPI) and Technical World (TW) for encouragement to initiate the renewable energy-oriented studies. Thanks are also due to the review panel and two reviewers for their constructive and helpful comments.

References

1. NOWRDC: Research and development road map version 2.0, USA, National offshore wind research and development consortium (2020). https://www.nyserda.ny.gov/All-Programs/Programs/Offshore-Wind/Focus-Areas/Research-and-Development
2. Dovorkin, J.: Rock physics: recent history and advances (2020). https://doi.org/10.5772/intechopen.92161. https://www.intechopen.com/online-first/rock-physics-recent-history-and-advances
3. Avseth, P., Mukerji, T., Mavko, G., Dvokin, J.: Rock-physics diagnostics of depositional texture, diagenetic alterations, and reservoir heterogeneity in high-porosity siliciclastic sediments and rocks—A review of selected models and suggested work flows. Geophysics **75**(5), 7531–7547 (2010)
4. Giao, P.H., Thanh, N.D., Chien, L.V., Minh, L.C., Ngoc, P.Q.: Analysis of near-shore investigation data for a wind farm project in Southern Vietnam. In: The 3rd Asia Pacific Meeting on Near Surface Geoscience, Chiang Mai, Thailand, November 2020 (2020). https://doi.org/10.3997/2214-4609.202071108

Undrained Shear Strength Assessments for Clay of the Offshore Wind Power Projects in Vietnam

Kuanjun Wang[1,2], Mingyuan Wang[1,2], Kanmin Shen[1](✉), Jianhua Lu[1], Ben He[1], and Zhen Guo[3]

[1] Key Laboratory of Far-Shore Wind Power Technology of Zhejiang Province, POWERCHINA Huadong Engineering Corporation Limited, Hangzhou 311122, China

[2] Zhejiang Huadong Construction Engineering Corporation Limited, Hangzhou 310014, China

[3] College of Civil Engineering and Architecture, Zhejiang University, Hangzhou 310058, China

Abstract. As a case study, the CPTU and laboratory tests results Binh Dai offshore wind power projects in Vietnam are analyzed to acquire the accurate strength parameters and interpretative parameters of CPTU. The undrained shear strength s_u derived from UU test and unconfined compression test (Qu) are overly conservative, while s_u from direct quick shear tests (DSq) and consolidated undrained shear tests (CU) are more representative in Binh Dai site. The normalised undrained shear strength $(s_u/\sigma_{v0}')_{NC}$ for the organic clay and stiff clay in are approximately 0.13 and 0.28 obtained from DSq test, respectively. The results from CU test of both organic clay and stiff clay are identical with a value of 0.36, which is relatively higher than the DSq test. By combining SHANSEP and CPTU data, the undrained shear strength of clayey soil can be obtained using the conservative s_u/σ_{v0}' value from DSq test, which show a good agreement with the laboratory test results of organic clay at the top, and reappear the in-situ undrained shear strength of the stiff clay from 20 m to 50 m. Finally, the N_{kt} method is introduced to calibrate the SHANSEP and OCR-to-CPTU interpretation, N_{kt} of 21 is recommended for Binh Dai offshore area.

Keywords: Clay · CPTU · SHANSEP · Normalised undrained shear strength

1 Introduction

In the foundation design of offshore wind power projects, the geotechnical engineering parameters are of great importance. Due to the operation conditions in the marine environment, such as windstorms and waves, the disturbances of borehole sampling in offshore projects are inevitable [1–3] and much higher than the disturbances in onshore projects. The laboratory unconsolidated undrained shear tests (UU) results of the disturbed clay samples might be much lower than the strength of the in-situ clay, which would lead to a huge waste in foundation design if UU test result is adopted for design. Therefore, the piezocone penetration test (CPTU) has become the most important method in offshore geotechnical investigation [4], which can represent the precise soil strength for in-situ soil if appropriate interpretive parameters are adopted [5]. The N_{kt} interpretation method is currently one of the most popular methods for interpreting the undrained

D. V. K. Huynh et al. (Eds.): VSOE 2021, LNCE 208, pp. 188–196, 2022.
https://doi.org/10.1007/978-981-16-7735-9_19

shear strength [6], the key point is how to choose an appropriate empirical coefficient N_{kt}, which has a very wide range [7, 8]. Theoretical equation derived by Modified Cam Clay model [9] or empirical equation of SHANSEP [10] had already established the relationship between normalised undrained shear strength of normally consolidated $(s_u/\sigma_{v0}')_{NC}$ and OCR, which relates directly to the corrected tip resistance q_t from CPTU [11]. However, a lot of tests should be carried out to obtain the normally consolidated $(s_u/\sigma_{v0}')_{NC}$, and appropriate OCR interpretation method should also be employed.

In this paper, CPTU and laboratory test are carried out at Binh Dai offshore area in Vietnam, the empirical parameters of CPTU interpretation had been obtained, which can further improve the strength interpretation of clayey soil in Vietnam in the future.

2 Soil Classification and Physical Properties of Clayey Soil

2.1 Project Site Overview

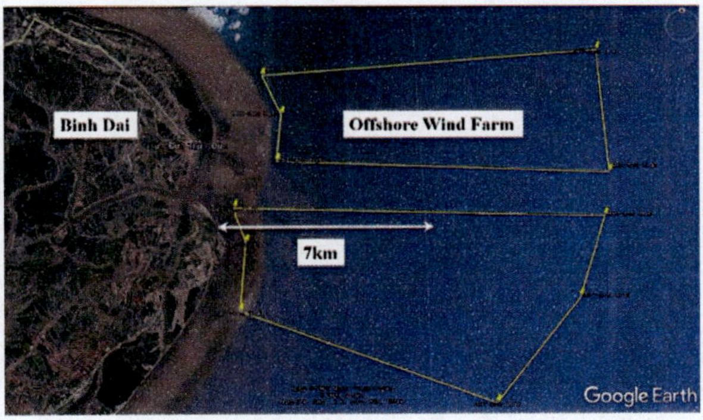

Fig. 1. Overview map showing the location of Binh Dai offshore wind power project

The Binh Dai Wind Power Project is located at Binh Dai District, Ben Tre Province, Vietnam. The district is located on An Hoa islet. The distance from the shore to middle of the site is about 7 km. The water depth ranges from approximately 3 m to 8 m, below mean sea level (MSL). Cohesive soil (clayey soil) samples were taken by thin walled tube sampler according to ASTM D1587, with a diameter of 76 mm (Fig. 1).

2.2 Physical Properties of Clayey Soil in Binh Dai

The Unified Soil Classification System had been adopted to classify the soil type according to ASTM D2487. The particle size distributions and physical properties tests are in accordance with ASTM standards. There are 4 clayey soil layers (Layer 1, 1b, 3 and 5) from the top to 70 m depth in Binh Dai offshore area. The physical properties for the 4 clayey soil layers are summarized in Table 1.

Table 1. Physical mechanical properties for the clayey soil in Binh Dai offshore area

	Unit	1	1b	3	5
		Organic clay	Organic clay	Clay	Clay
Grained size distribution					
Gravel (4.75 mm < d_s < 75 mm)	%	0.0	0.0	0.1	0.0
Sand (0.075 mm < d_s < 4.75 mm)		6.5	6.5	18.7	28.0
Silt (0.002 mm < d_s < 0.075 mm)		56.5	51.6	42.0	44.5
Clay (d_s < 0.002 mm)		37.0	41.9	39.2	27.5
Physical properties					
Moisture contents w	%	49.6	43.0	26.3	1.925
Natural density γ_w	g/cm^3	1.661	1.719	1.914	1.522
Specific gravity G_s		2.656	2.674	2.701	2.700
Void ratio e_0		1.393	1.225	0.783	0.774
Degree of saturation S_r	%	94.6	93.9	90.7	92.4
Liquid limit w_L	%	49.6	49.7	46.1	42.9
Plastic limit w_P		26.7	25.0	22.9	24.0
Plasticity index I_P		22.9	24.7	23.2	18.9
Liquidity Index I_L		1.00	0.73	0.15	0.13

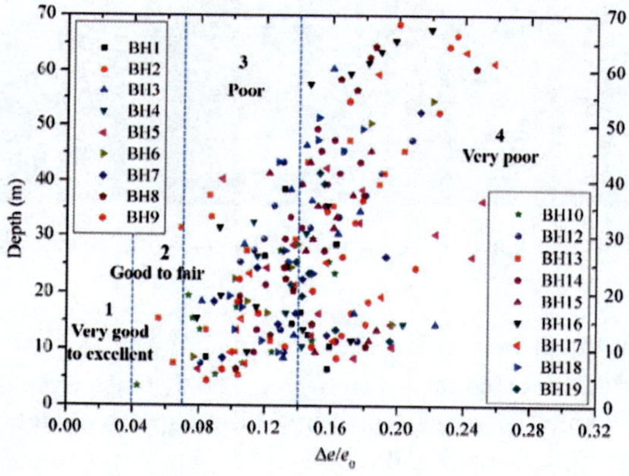

Fig. 2. The Sample quality evaluation of clayey soil in Binh Dai offshore area

3 Mechanical Properties of Clayey Soil in Binh Dai

3.1 Sample Quality Evaluation

Due to the influence of environmental impacts during the offshore drilling works, the samples are greatly disturbed. The soil parameters are systematically influenced by the sample disturbance. Based on previous researches [1, 2], $\Delta e/e_0$ has been adopted to evaluate the sample quality of clayey soil, the evaluation results are shown in Fig. 2.

From the data in Fig. 2, it can be found that most of the soil samples are severely disturbed, the results of UU and Qu tests might not be reliable, which could underestimate the factual soil strength. On account of the sample disturbance, the consolidated undrained triaxial compression test (CU) and direct quick shear (DSq, consolidated quick shear) would be more suitable for the strength assessment, because all the samples experience a consolidation procedure which can reduce the effect of disturbance.

3.2 Undrained Shear Strength from Laboratory Test

Figure 3a presents the results obtained from UU and Qu. For the organic clay from the depth of 0 m to 20 m, an approximately linear correlation is found between the UU and Qu test results and depth, the strength from UU test results are slightly higher than the Qu test. For the clay from 15 m to 70 m, the soil strengths derived by Qu test are larger than the UU test results, where the strengths vary from 33 kPa to 132 kPa in irregular form. However, the UU tests results show a steady increase with the depth. This rather contradictory result may be due to the sample disturbance, and unconfined compression test is not a precise way to obtain the undrained shear strength, and the UU test would underestimate the soil strength when the sample is severely disturbed.

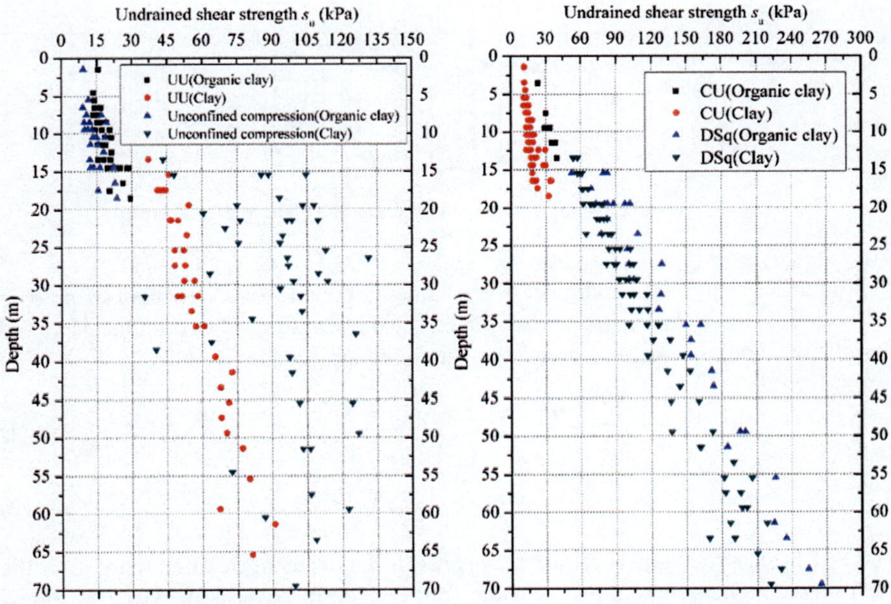

(a) UU and Qu test (b) CU and DSq test

Fig. 3. The undrained shear strength organic clay and clay

Figure 3b provides the CU and DSq tests results of organic clay and clay, the results indicate that there is a clearly positive correlation between the undrained shear strength and the soil depth. For the depth from 15 m to 20 m, there is a significant difference between the organic clay and clay. The soil undrained shear strengths are likely to be soil type dependent, and linear dependent on the soil depth or the vertical stress, which is consistent with the earlier observations [12]. Another possible explanation for these results could be attributed to the different soil states (liquidity).

3.3 Relation Between Normalised Undrained Shear Strength and I_L

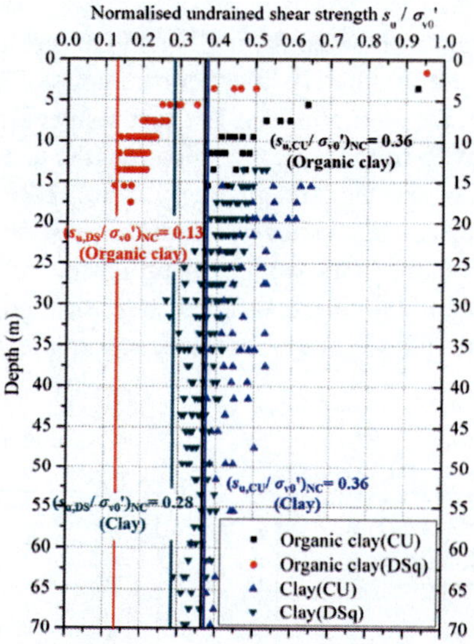

Fig. 4. The relationship between normalised shear strength and depth

To further analyze the undrained shear strength, the effective vertical pressure σ_{v0}' has been adopted to normalise the undrained shear strength s_u. Based on the Modified Cam-Clay model, s_u can be expressed by the following equation.

$$s_u = \frac{M}{2^{\Lambda+1}} \cdot (OCR)^{\Lambda} \cdot \frac{1+2K_0}{3} \cdot \sigma_{v0}' \tag{1}$$

$$\Lambda = (\lambda - \kappa)/\lambda \tag{2}$$

where M is the stress ratio at critical state, OCR is the overconsolidation ratio, K_0 is the earth pressure coefficient at rest, λ is the compression index, κ is the swelling index. Equation (1) can be simplified by

$$\frac{s_u}{\sigma_{v0}'} = a \cdot (OCR)^b \tag{3}$$

where a is the empirical coefficient which is equal to the normalised undrained shear strength $(s_u/\sigma_v')_{NC}$, b is the empirical coefficient varying from 0.5 to 1.0, which is theoretically equal to Λ. Equation (3) is proposed by Ladd and Foott [10], namely the SHANSEP method. The s_u/σ_{v0}' are plotted in Fig. 4 versus the soil depth.

What stands out in Fig. 4 is that s_u/σ_{v0}' of organic clay vary from 0.13 to 0.95 (DSq) and 0.36 to 0.92 (CU). By using Eq. (3), it is easy to discover that the top of organic clay in Binh Dai offshore area is heavily overconsolidated, and the OCR has a sharp fall when the depth increases. For Binh Dai clay at the top layer (15 m to 20 m) have higher s_u/σ_{v0}' than the clay at bottom, the clay at the top is not heavily overconsolidated, but lightly overconsolidated instead. The variation of s_u/σ_{v0}' and OCR with soil depth are in accordance with previous observations. It is possible to hypothesize that the clay at the bottom are more like to be normally consolidated (owing to the consolidation process of CU and DSq tests), the cautious and best estimate of $(s_u/\sigma_{v0}')_{NC}$ are 0.28 from DSq test and 0.36 from CU test respectively, which is in the reasonable range. For organic clay, 0.13 can be chosen as the $(s_u/\sigma_{v0}')_{NC}$ for cautious estimate, 0.36 for best estimate, which is consistent with the value for Binh Dai clay.

3.4 CPTU Interpretation

CPTU is one of the most well-known methods for assessing the strength of soil. Several solutions had been established to interpret the undrained shear strength [11]. The most popularly used is the N_{kt} method, which is expressed as following:

$$s_u = (q_t - \sigma_{v0})/N_{kt} \tag{4}$$

where q_t is the corrected cone resistance, σ_{v0} is the total vertical stress, N_{kt} is the empirical coefficient, which has a very wide range of 7 to 29 [8]. It is hard to choose the suitable N_{kt} value to evaluate the strength of clayey soil.

Lunne et al. [6], propose the following expression relating OCR to q_{net}:

$$OCR = k \cdot (q_t - \sigma_{v0})/\sigma_{v0}' = k \cdot q_{net}/\sigma_{v0}' \tag{5}$$

where k is a variable of value 0.2 to 0.5 [11], which is cautiously assumed as 0.2. Combining the Eq. (5) and Eq. (3), the normalised undrained shear strength can be derived by:

$$s_u/\sigma_{v0}' = a \cdot \left(0.2 \cdot q_{net}/\sigma_{v0}'\right)^b \tag{6}$$

The empirical coefficient a is equal to the $(s_u/\sigma_{v0}')_{NC}$, which are 0.13 and 0.36 respectively for organic clay from DSq test, 0.28 and 0.36 respectively for clay from CU test. And the empirical coefficient b is theoretically equal to $(\lambda-\kappa)/\lambda$ (shown in Table 2), which are 0.8441 and 0.8174 for organic clay and clay respectively.

All 9 CPTU tests are carried out in the relevant borehole locations, each CPT location is close to the BH location with a distance no more than 5 m. One CPTU data of the wind turbine location is selected to further analyse the s_u of the organic clay and clay in Binh Dai. The in-situ undrained shear strength can be determined from the combined SHANSEP and OCR-to-CPTU correlation which are presented in Fig. 5. From the results of organic clay at 9 m to 15 m, the combined SHANSEP and OCR-to-CPTU interpretation by using parameter derived from DSq and CU test are respectively

Table 2. Compression and Swelling index of clayey soil in Binh Dai offshore area

Soil type	Organic clay	Clay
Compression Index λ	0.1867	0.0677
Swelling index κ	0.0291	0.0122
$\Lambda = (\lambda - \kappa) / \lambda$	0.8441	0.8171

consistent with the DSq and CU test results, which has been preliminarily verified to be correct. However, for the clay in 27 m to 44 m, the interpreted results are observed considerable difference from the laboratory test results. A possible explanation for the inconsistence might be the increasing disturbance of the soil sample in the deep layer, which results in the reduction in the laboratory test results. The UU test results are the strong evidence that the sampling disturbance could severely affect the strength of the laboratory test results. The DSq and CU test have much better results which are attributed to the reconsolidation procedures before the shearing.

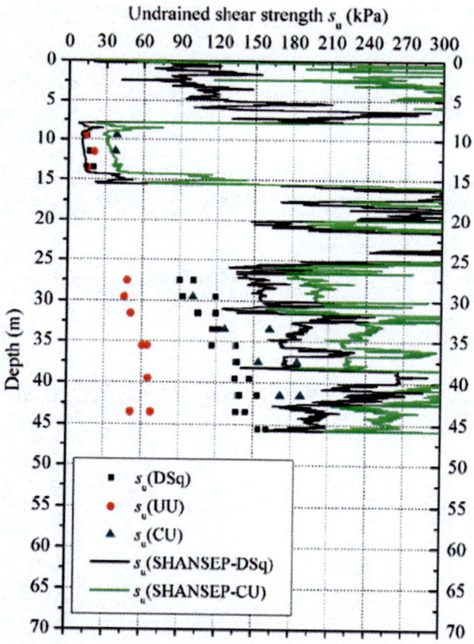

Fig. 5. The undrained shear strength interpretation by SHANSEP and OCR-to-CPTU

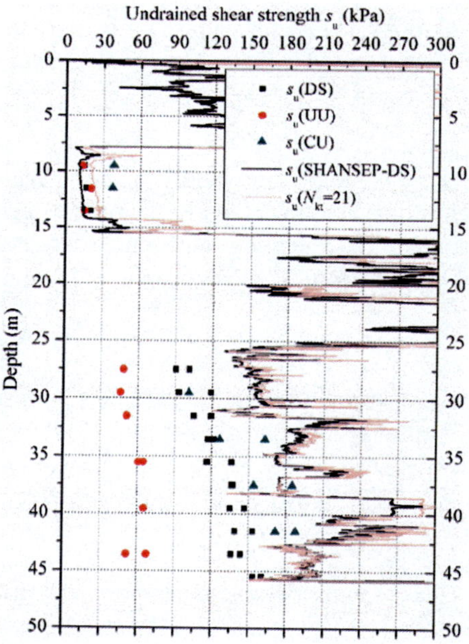

Fig. 6. The undrained shear strength interpretation by N_{kt} method

The SHANSEP and OCR-to-CPTU interpretation method might be slightly complicated for engineering use. For this reason, the N_{kt} method has also been adopted to represent the undrained shear strength by taking the results of SHANSEP method as the reference value. The interpretation results from N_{kt} method, as shown in Fig. 6, are in accordance with the SHANSEP and OCR-to-CPTU interpretation, indicate that the N_{kt} of 21 can be reasonably chosen for interpretation in Binh Dai offshore area.

4 Conclusion

This paper has the analyse the undrained shear strength of clayey soil in Binh Dai offshore area through different methods (DSq, UU, CU, and CPTU). The current data highlight that the sample disturbance of drilling in offshore area is very large. Using triaxial UU test and unconfined compression test (Qu) are overly conservative, while s_u from direct quick shear tests (DSq) and consolidated undrained shear tests (CU) are more representative, attributed to the reconsolidation procedures before the shearing of CU and DSq tests. Furthermore, the normalised undrained shear strength s_u/σ_{v0}' results of organic clay and the stiff clay indicate that the soil at top are heavily overconsolidated, and the s_u/σ_{v0}' for the normally consolidated organic clay and stiff clay are approximately 0.13 and 0.28 obtained from DSq test, respectively. The results from CU test of both organic clay and stiff clay are identical with a value of 0.36, which is relatively higher than the DSq test. By combining SHANSEP and CPTU data, the undrained shear strength of clayey soil can be obtained using the conservative s_u/σ_{v0}' value from DSq test, which show a good agreement with the laboratory test results of organic clay at the top, and reappear the in-situ undrained shear strength of the stiff clay from 20m to 50m. Finally, the N_{kt} method is adopted to calibrate the SHANSEP and OCR-to-CPTU interpretation, the parameter N_{kt} of 21 is recommended for undrained shear strength assessment in Binh Dai offshore area.

Acknowledgement. The authors gratefully acknowledge the supports of the National Natural Science Foundation of China (Grant No. 52108356 and 52101334), Zhejiang Provincial Natural Science Foundation of China (Grant No. LQ20E090001 and LQ19E090002), and Technology project of POWERCHINA Huadong Engineering Corporation (Grant No. KY2020-KC-01).

References

1. Lunne, T., Berre, T., Strandvik, S.: Sample disturbance effects in soft low plastic Norwegian clay. In: Almeida, M., Balkema, A.A. (eds.) Proceedings of the Conference on Recent Developments in Soil and Pavement Mechanics, Rio de Janeiro, Brazil, June 25–27th, Rotterdam, The Netherlands. pp. 81–102 (1997)
2. Lunne, T., Berre, T., Andersen, K.H., Strandvik, S., Sjursen, M.: Effects of sample disturbance and consolidation procedures on measured shear strength of soft marine Norwegian clays. Can. Geotech. J. **43**(7), 726–750 (2006)
3. Lukas, W.G., DeGroot, D.J., DeJong, J.T., Krage, C.P., Zhang, G.: Undrained shear behavior of low-plasticity intermediate soils subjected to simulated tube-sampling disturbance. J. Geotech. Geoenviron. Eng. **145**(1), 04018098 (2019)
4. Lunne, T., Kleven, A.: Role of CPT in North sea foundation engineering. In: Symposium on Cone Penetration Engineering Division, ASCE, pp. 49–75 (1981)
5. Robertson, P.K., Campanella, R.G.: Interpretation of cone penetration tests: part ii: clay. Can. Geotech. J. **20**(4), 734–745 (1983)

6. Lunne, T., Robertson, P.K., Powell, J.J.: Cone Penetration Testing in Geotechnical Practice. CRS Press Taylor & Francis, Boco Raton (2002)
7. Low, H.E., Randolph, M.F.: Strength measurement for near-seabed surface soft soil using manually operated miniature full-flow penetrometer. J. Geotech. Geoenviron. Eng. (ASCE) **136**(11), 1565–1573 (2010)
8. Liu, K., Wang, J.H.: Evaluation on undrained shear strength of marine soil using cone penetration resistance in Zhujiang River mouth basin, South China Sea. Indian J. Geo Marine Sci. **45**(4), 707–718 (2017)
9. Schofield, A., Wroth, P.: Critical State Soil Mechanics. McGraw-hill, New York (1968)
10. Ladd, C.C., Foott, R.: New design procedure for stability of soft soil. J. Geotech. Eng. Divis. ASCE **100**(7), 736–786 (1974)
11. Robertson, P.K., Cabal, K.L.: Guide to cone penetration testing for geotechnical engineering. Gregg Drilling & Testing (2015)
12. Ladd, C.C.: Stability evaluation during staged construction. J. Geotech. Eng. **117**(4), 540–615 (1991)

Design of Offshore Foundations and Structures

Assessment of the Remaining Safety of Pipelines with Corrosion Defects

Long Thanh Do(✉), Quan Hong Mai, and Linh Cao Luong

Faculty of Coastal and Offshore Engineering, Hanoi University of Civil Engineering, 55 Giai Phong Street, Hanoi, Vietnam
{longdt,quanmh,linhlc}@nuce.edu.vn

Abstract. Corrosion is one of the common defects of offshore pipelines and risers. By the time, corrosion defect will reduce the strength of pipelines to limited value, then, a reinforcement solution need to be proposed. The most effective and popular method is installing the Clock Spring. The safety of pile before and after repair, is evaluated by experimental formula accordance to ASME B31G or DNV RP101 standard. However, there are still some disadvantages with using experimental formula, typically as incorrect description of the defect shape on the surface of pile may lead to the deviations of analysis result. This paper presents the results of studying the safety of corrode pile subjected to internal pressure by a numerical simulation in two cases: before and after repair.

Keywords: Internal pressure · Corrosion defects · Numerical simulation · Clock spring · Von Mises stress · FE method

1 Introduction

After a long period of operation under many effects of the marine environment, the steel pipes are corroded and losen metal on the surfaces. This corrosion causes many incidents which affect the entire pipelines and also cause a lot of damages for operators. Therefore, to minimize the risks, corrosion defects survey and inspection are carried out annually as a mandatory regulation. The investigation and inspection of corrosion defects include field work (such as measurement, pig release to determine the position, size, extent of wall corrosion) and work in the room (such as stress analysis, evaluate the remaining working ability of the pipe at the corrosive locations). However, the companies who own the pipelines, still need a more reliable evaluation method to ensure operational safety as well as optimize operating costs before choosing the solution for repair or replacement. One of the most effective repair solutions in order to extend the lives of the corroded pipelines is using Clock spring, a composite wrap repair sleeve and reinforcement system uniquely designed for high-pressure transmission pipelines. The Clock Spring, has been tested and proven up to 8,000 psi and is expected to last 50 years or longer, deeming it a permanent repair by regulators (Fig. 1).

In recent years, many other evaluation methods have been developed mainly based on finite element (FE) studies and burst test results. In this paper, FE method was chosen

D. V. K. Huynh et al. (Eds.): VSOE 2021, LNCE 208, pp. 199–206, 2022.
https://doi.org/10.1007/978-981-16-7735-9_20

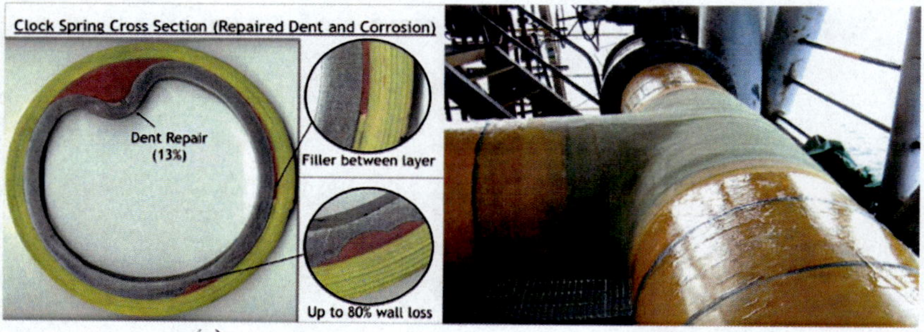

(a) (b)

Fig. 1. (a). The typical Clock Spring cross section, and, (b). A steel pipe with Clock Spring

to model the corrosion defects on the surface of a steel pipe and evaluate allowable limited pressure in two cases: before and after wrapping by Clockspring. Beside, the results from FE model were compared to the failure pressures which were predicted by some conventional methods.

2 Accessment Methods

Semi-empirical Methods: There are two standards, namely ASME B31G and DNV-RP-101, regularly were used to evaluate the remaining strengthen of the corroded pipe. In which, the DNV method can be applied to corrosion subject to axial and bending loads.

ASME B31G is a manual for evaluating the remaining strengthen of corrode pipe. The steps for calculation are shown briefly below:

Step 1: The depth of a corrosion pit is expressed as a percent of the nominal wall thickness of the pipe by:

$$\%P = 100d/t \tag{1}$$

Where:

d: measured maximum depth of the corroded area
t: nominal wall thickness of the pipe

Step 2: Determining the maximum allowable longitudinal extent of the corroded area, LM

$$L_M = 1.12B\sqrt{Dt} \tag{2}$$

D: nominal outside diameter of the pipe

$$B = \sqrt{\left(\frac{d/t}{1.1d/t - 0.15}\right)^2 - 1} \tag{3}$$

Step 3: Determining the factor z:

$$z = L^2/Dt \tag{4}$$

Step 4: Determining the failure pressure S_F:

z ≤ 20,

$$S_F = S_{flow}\left[\frac{1 - 2/3(d/t)}{1 - 2/3(d/t)/M}\right] \tag{5}$$

z > 20,

$$S_F = S_{flow}[1 - d/t] \tag{6}$$

Step 5: Compare S_F to the stress S_o: $S_F \sim S_o{*}SF$

$$S_o = P_oD/2t \tag{7}$$

P_o: Operating pressure.
SF: Safety factor = 1.25.

DNV-RP-F101 gives recommendations to assess corroded pipelines subject to internal pressure, and internal pressure combined with longitudinal compressive stresses. The allowable corroded pipe pressure is determined from the equations below:

Longitudinal corrosion defect, internal pressure loading only:

Step 1: Calculate the failure pressure of the corroded pipe Pf

$$P_f = \frac{2tf_u(1 - d/t)}{(D - t)(1 - \frac{d}{tQ})} \tag{8}$$

Where:

$$Q = \sqrt{1 + 0.31\left(\frac{l}{\sqrt{Dt}}\right)^2} \tag{9}$$

Step 2: Calculate the safe working pressure of the corroded pipe

$$P_{SW} = F.P_f \tag{10}$$

F: Safety factor

Internal pressure and combined compressive loading:

Step 1: Determine the longitudinal stress, at the location of the corrosion defect, from external loads, as axial, bending and temperature loads on the pipe. Calculate the nominal longitudinal elastic stresses in the pipe at the location of the corrosion defect, based on the nominal pipe wall thickness.

$$\sigma_A = \frac{F_X}{\pi(D-t)t} \tag{11}$$

$$\sigma_B = \frac{4M_Y}{\pi(D-t)^2 t} \tag{12}$$

Combined nominal longitudinal stresses:

$$\sigma_L = \sigma_A + \sigma_B \tag{13}$$

Step 2: Determine whether or not it is necessary to consider the effect of the external compressive longitudinal loads on the failure pressure of the single defect:

$$\sigma_L > \sigma_1 \tag{14}$$

Where:

$$\sigma_1 = -0.5 f_u \frac{(1 - d/t)}{\left(1 - \frac{d}{tQ}\right)} \tag{15}$$

Step 3: Calculate the failure pressure of the single corrosion defect under internal pressure only, using the following equation, Ppress

$$P_{press} = \frac{2tf_u(1 - d/t)}{(D-t)(1 - \frac{d}{tQ})} \tag{16}$$

Step 4: Calculate the failure pressure for a longitudinal break, including the correction for the influence of compressive longitudinal stress

$$P_{comp} = \frac{2tf_u}{(D-t)} \frac{(1 - d/t)}{\left(1 - \frac{d}{tQ}\right)} H_1 \tag{17}$$

$$H_1 = \frac{1 + \frac{\sigma_L}{f_u}\frac{1}{A_r}}{1 - \frac{1}{2A_r}\frac{(1-d/t)}{\left(1-\frac{d}{tQ}\right)}} \tag{18}$$

$$A_r = 1 - \frac{d}{t}\theta \tag{19}$$

Step 5: Determine the failure pressure of a single corrosion defect subjected to internal pressure loading combined with compressive longitudinal stresses: Pf = min (Ppress; Pcomp).

Step 6: Calculate the safe working pressure of the corroded pipe, PSW

$$P_{SW} = F.P_f \tag{20}$$

F: Safety factor

FE Methods: The development of computer technology has created conditions for the engineers to solve many complex structural problems on finite element analysis software. This is a numerical method which used to simulate load conditions on a physical system and determine its behavior. Thanks to this method, the number of test samples is reduced because computer simulation allows many models to be simulated quickly and conveniently. Therefore, it saves costs, time and creates more reliable quality designs. In this paper, ANSYS software was used to make a model of defected pipe and analysis stress of pipe subjected to internal high pressure.

3 FE Modeling Cases

Model Data: Three cases were modelled to estimate the failure pressure values, which use the corrosion defect length (L) and the wall thickness (t) as variables. These data and the parameters of material steel and Clockspring are presented in Table 1 and Table 2 below:

Table 1. Case data

Case	Depth of defect	Width of defect	Length of defect	Position
1st	4.15 mm	58 mm	139 mm	Surface/Longitudinal pipe
2nd	5.01 mm	64 mm	1202 mm	Surface/Longitudinal pipe
3rd	4.15 mm	1276 mm	58 mm	Surface/Circumferentinal pipe

Above data was taken from the report of defect accessment of 16A riser pipeline – Bach Ho oil field – Viet Nam - 2016.

Table 2. Parameters of steel material and clockspring

Parameters	Units	Steel pipe	Clock Spring
Material	–	API 5L X52	Polyethylense
Diameter of pipe	mm	406.4	–
Wall thickness	mm	14.3	17.5
SMYS	MPa	359	–
SMTS	MPa	455	25
Modulus	MPa	207000	6900
Poisson factor	–	0.3	0.42
Weight	kG/m^3	7850	950

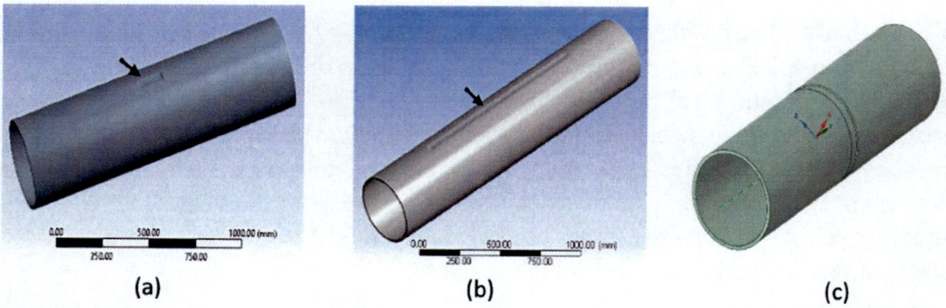

Fig. 2. Pipelines with various defects before wrapped by clockspring (a) Case 1; (b) Case 2; (c) Case 3

In this study, the failure pressure value is considered when Von Mises stress reach maximum stress of the material. Von Mises stress is included of the three principal stress components as below (Fig. 2):

$$\sigma_{VM} = \left(\frac{1}{\sqrt{2}}\right)\sqrt{(\sigma_1 - \sigma_2)^2 + (\sigma_2 - \sigma_3)^2 + (\sigma_3 - \sigma_1)^2} \tag{21}$$

Where:

σ_1 = axial stress acting along the longitudinal direction
σ_2 = hoop stress acting along the circumferential/tangential direction
σ_3 = radial stress acting along the radial direction
σ_{VM} = Von Mises stress.

All boundary conditions such as the constraints and loads are applied to model. The failure pressures were determined by simulating with incremental internal pressure loading, until the von Mises stress is equal to the critical stress of material at failure, $\sigma_{crit} = 600$ Mpa (Fig. 3):

$$\sigma_{crit} = K\varepsilon_{crit}^{n} \tag{22}$$

Where:

- K: hardening coefficient, K = 876 Mpa;

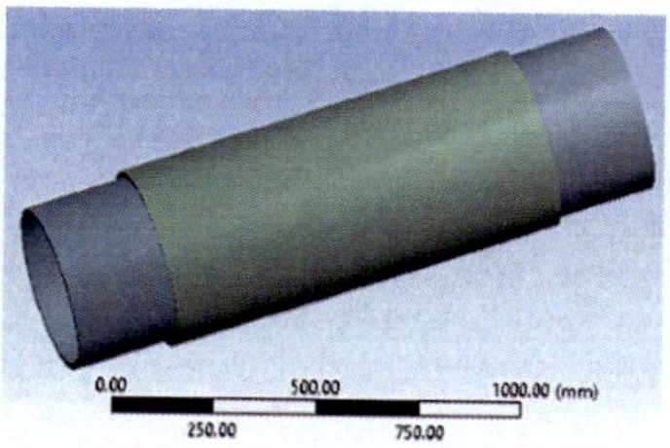

Fig. 3. A model of corroded pipe after wrapped by clockspring

- ε_{crit}: critical strain, $\varepsilon_{crit} = 0.105$;
- n: hardening exponent, n = 0.2;

4 Results and Discussions

4.1 Failure Pressure

The results of failure pressure are shown in Table 3 and Fig. 4.

Table 3. Results of the failure pressure

Case	ANSYS	ASME B31G	DNV RP 101
1st	20.49 MPa	18.14 MPa	18.65 MPa
2nd	15.60 MPa	9.020 MPa	10.32 MPa
3rd	33.80 MPa	20.08 MPa	21.71 MPa

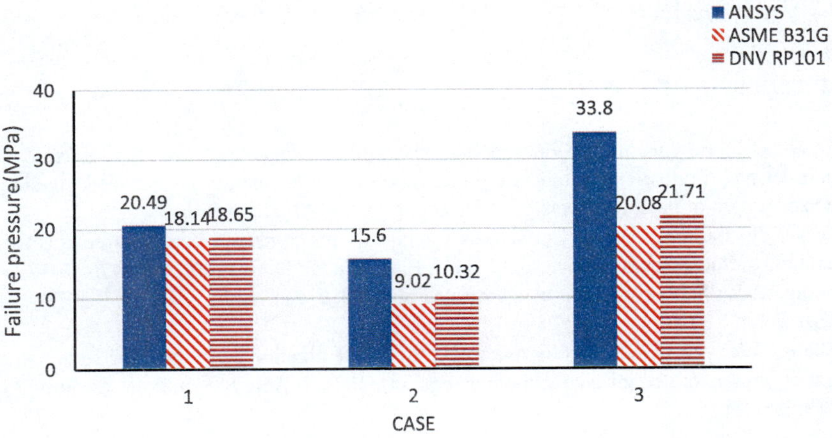

Fig. 4. The chart of failure stress in case 1, 2, 3

4.2 Maximum Von Mises Stress

The results of the Von-mise stress of the pipe before and after wrapping, have change clearly. From the over-the-limit stress value of material (567 Mpa) down to about below 400 Mpa (Table 4).

The case 3 has the corrode defect with circumferential shape. This reason may lead to the result that stress value after wrapping does not decrease significantly but the failure pressure of case 3 is higher than the other cases.

The value of the failure pressure from FE method are higher than the results from ASME B31G and DNV RP101.

Table 4. Results of the maximum Von Mises stress with internal pressure, p = 13.9 MPa

Case	Before repair	After repair
1st	460.48 MPa	384.02 MPa
2nd	567.15 MPa	344.95 MPa
3rd	397.80 MPa	376.20 MPa

5 Conclusions

The FEM method allows to evaluate the stress changes on the corroded pipe before and after wrapping by Clockspring, while the other methods are limited. The analysis results also show the effectiveness of Clockspring coats used for the corroded pipes.

The shape of defects have a important role for the remaining working ability of the corroded pipe.

This study have no the experiments for comparing to the results determined by FE model. This is the scope has to be done in the future for the more reliable assessments and giving the suitable solutions of using Clockspring for corroded pipe.

References

1. Ossai, C.I.: Engineering product development pillar, singapore university of technology and design, "finite element modelling and retained life estimation of corroded pipelines in consideration of burst pressures." Infrastructures **2**, 15 (2017)
2. Terán, G.: Failure pressure estimations for pipes with combined corrosion defects on the external surface: a comparative study. Int. J. Electrochem. Sci. **12**, 10152–10176 (2017)
3. Wang, Y.-L.: State evaluation of a corroded pipeline. J. Marine Eng. Technol. **15**(2), 88–96 (2016)
4. Shang, H.Y.: A methodology for analysis of defective pipeline by introducing stress concentration factor into beam-pipe finite element formulation. Acta Scientiarum Technol. **38**(3), 313–320 (2016)
5. Dick, I., Inegiyemiema, M.: Predicting the structural response of a corroded pipeline using finite element (FE) analysis. Int. J. Sci. Eng. Res. **5**(11) (2014)
6. Karuppanan, S., Aminudin, A.S., Wahab, A.A.: Brust pressure estimation of corroded pipeline with interacting defects using finite element analysis. J. Appl. Sci. **12**(24), 2626–2630 (2012)
7. ASME B31 G -2012, Manual for Determining the Remaining Strength of Corroded Pipelines, the American Society of Mechanical Engineers (2012)
8. DNV RP F101 -2010, Corroded pipelines recommended practice, DNVGL (2010)

Cantilever Secant Pile Constructed in Sand: Numerical Comparative Study and Design Aids

Khaled R. Khater(✉)

Civil Engineering Department, Fayoum University, Fayoum, Egypt
Krk26@fayoum.edu.eg

Abstract. Civil engineering projects include excavation work, therefore some needs retaining structures. Cantilever secant pile wall is an economical supporting system up to 5.0-m depths. The parameters controlling wall tip displacement are the focus of this paper. Three analysis techniques have been arbitrated. They are the conventional method, spring analogy and finite element. Hence, three computer programs have been used, Excel sheet, SPW-2006 and Plaxis-2D. Two soil densities have been considered, i.e. loose and dense sand. Ten wall rigidities have been analyzed covering ranges flexible up to rigid walls. Three excavation depths, i.e. 3.0, 4.0 and 5.0-m were tested to cover the practical range. This work submits beneficial hints about secant piles to assist designers. Also, finite element analysis is recommended to be the fair judge when two designs conflict. A rational procedure using empirical equations has been suggested to upgrade the conventional method to predict wall tip displacement δ. Designer must keep 'δ' $\leq$ 0.8% of the excavation depth, 'h', for safety.

Keywords: Secant pile · Numerical analysis · Design aids

1 Introduction

The manuscript theme is soil retaining structures, with a special emphasis on in cantilever secant pile walls. The wall tip displacement as well the comparison between the methods of analysis that used to handle this problem presents the target of this study. The paper skeleton starts by hints on the literature leads to research questions. Then the research strategy presented by figures, tables, flowchart, the flowchart is the key figure. Then, the results presented, analyzed, discussed, and the conclusions offered.

Worku (2007) studied the use of springs to account for subgrade deformation. Bekdas *et al.* (2020) investigated the analysis of cantilever soldier pile walls in sandy soil based on the finite element and provides an optimal design. Ramadan *et al.* (2018) evaluated the design parameters needed to design contiguous pile supporting shallow excavations in clay. 3D- finite element model is used. Mahesh *et al.* (2019) examined diameter effects on pile capacity, bending moment and wall displacement. The case study involves deep excavation supported by secant piles of reinforced and plain concrete piles. Liao *et al.* (2014) performed comprehensive model tests to study the failure modes of secant faces.

D. V. K. Huynh et al. (Eds.): VSOE 2021, LNCE 208, pp. 207–215, 2022.
https://doi.org/10.1007/978-981-16-7735-9_21

Khater (1991) published a case of sheet pile failures in clay, reported: "*safety factor penetration ratio 20% is insufficient in clayey soil*".

The research statement could be presented by four research questions. Does wall penetration depth have a peak value Q1? Which analysis method is reliable: conventional method, Winkler or finite element, Q2? Is it reasonable to suggest empirical equations to predict the wall tip displacement 'δ', Q3? What is a 'δ' limitation for safe design? Q4. *Satisfactory answer of those questions is the aim and importance of this work.*

2 Methods Employed

Methodology is the strategy that outlines the way in which the research is to be carried out. Here, it is numerical analysis. Methods are the tools chosen to create quality data. Herein, they are the conventional method, Winkler analogy and 2D-FEA.

2.1 Parametric Study

Figure 1 is the variable interactions flowchart. Its basis is simple, 1st level '1-2-3' is the analysis methods. The three doted rectangles present excavation depths. The numbers on the box show 'routes', e.g. '4-7-12-18', is h = 5.0-m, loose soil S1, five wall rigidities I_6: I_{12}. This repeated three times, i.e. (1-4), (2-4) and (3-4). Box 20; explains that every single case has been investigated 10 times by increasing D_O in steps of 10% each, D_O is the minimum penetration depth calculated by the conventional method.

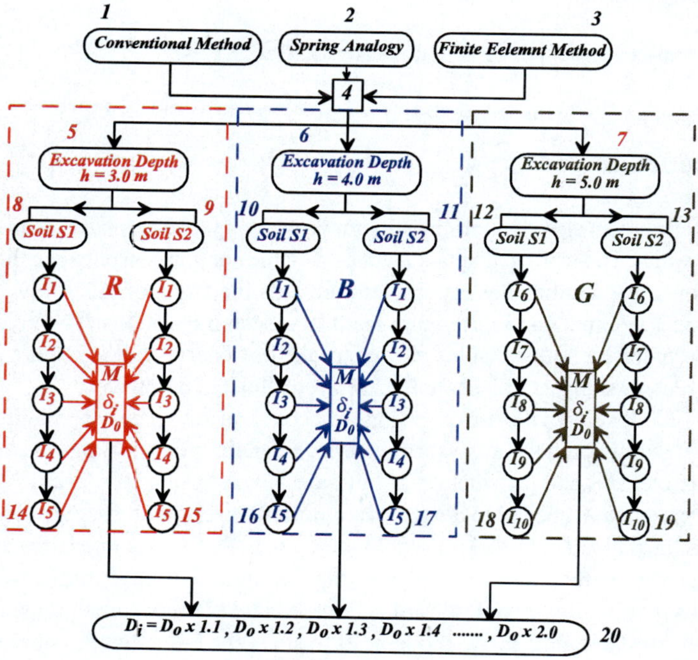

Fig. 1. Flow chart and research algorithm.

Quality data is defined as a set of 'focused-on' data induced by analyzing suggested variables intentionally oriented to achieve the research goal. So, the parametric study technique is appropriate for that achievement. Briefly, two soils have been considered, loose and dense well graded dry sand, S1 and S2 respectively. Three excavation depths "h" have been studied, h = 3.0-m, 4.0-m and 5.0-m. This cover the practical range of cantilever secant pile. Ten wall rigidities "I_n", integer "n" increases from 1 to 10, cover flexibility up to very rigid walls. The above have been analyzed three times, i.e. conventional method, then Winkler analogy and finite element analysis. Figure 1, Table 1 and Table 2 give a detailed description of every single variable used.

Table 1, summarizes the mechanical and physical properties. Even if the soil is dry sand, some parameters such as, γ_{sat}, ψ^o and c are needed for Plaxis-2D, by default. They have been used but never affect the physical meaning or behavior of dry sand.

Table 1. Physical and mechanical properties of the used soil.

Soil type and notations	γ_d	γ_{sat}	ϕ^o	ψ^o	C	μ_S	E_S
SW, loose dry sand, "S1"	14.0	20.0	29	1	1.0	0.30	7500
SW, dense dry sand, "S2"	16.0	20.0	37	7	1.0	0.38	11500

γ_d dry density (kN/m^3) - γ_{sat} saturated density (kN/m^3) - ϕ^o internal friction (deg) - ψ^o dialancy angel (deg) – c cohesion (kN/m^2) - μ_S Poisson's ratio - E_S elasticity modulus (kN/m^2).

Table 2, summarizes the mechanical properties of the wall per meter run. The overlap of RC pile on bentonite pile is kept constant as 0.10-m. The D_{eqv} stands for the equivalent pile diameter induced from neglecting bentonite piles mechanical properties.

Table 2. Physical and mechanical properties of the secant piles.

Secant pile structural parameters	I_n	i	EI	EA	Wt/m'	$D_{P\text{-}Eqv.}$
For excavation depths H = 3.0 and 4.0-m; Soil types: S1 & S2	I_1	0.07490	19260	3433751	7.60	0.25
	I_2	0.08739	28554	3738973	8.28	0.30
	I_3	0.09987	40599	4069631	9.01	0.35
	I_4	0.11236	55733	4414822	9.78	0.40
	I_5	0.12484	74328	4769098	10.56	0.45
For excavation depth H = 5.0-m and Soil type: S1 & S2	I_6	0.14482	112150	5347749	11.84	0.50
	I_7	0.17228	182757	182757	13.64	0.60
	I_8	0.20224	288411	7051280	15.62	0.70
	I_9	0.23096	422127	7913813	17.53	0.80
	I_{10}	0.25964	592019	878001	19.44	0.90

I moment of inertia (m^4) – i radius of gyration (m) – EI flexural rigidity (kN.m^2/m) – W_t wall weigh kN/m') – "E elasticity modulus = 1942^10^4 kN/m^2 - μ_S Poisson's ratio = 0.20"; concrete.

2.2 Methods and Tools

The conventional method formulations described by Das (1987) and incorporated into an Excel sheet is developed to fulfill research needs. Spring analogy used here simulates Elasto-plastic soil response. Beam element models the sheet piles. This behavior has been formulated and runs with 2D noncommercial software 'SPWALL20006'. It has been coded by Prof. Arnold Verruijt for Delft University of Technology (2006). The finite element analysis used is Plaxis-2D. Triangular 15-node elements were used to model the soil with Mohr-Coulomb material model. Plate and interface element were used for the sheet pile modeling. Staged construction procedure was used.

3 Presentation and Analysis of Results

Figure 2 and Fig. 3 demonstrates the case of an excavation depth h = 3.0-m, the other two cases, i.e. h = 4.0-m and 5.0-m having typical trend with different values. Figure 2 presents the FEA results, while Fig. 3 reports the spring analogy, comparison wise. Loose sand S1 and dense S2 have been investigated, with each, five wall rigidities, I_n have been inspected. D_O is the minimum possible penetration depth (no safety factor). D_i is a penetration depth with increasing factor of safety by steps 10% of D_O. Hence, x-axis is the dimensionless ratio D_i/D_O. δ_O is the horizontal displacement of the wall tip corresponding to penetration depth DO. The δ_i is variable displacement of the wall tip at the penetration depth D_i. The y-axis is the dimensionless ratio δ_i/δ_O, too.

3.1 Effective Penetration Depth and Design Hints

Figure 2 and Fig. 3, as I_n or soil density increases, δ_i decreases for the same D_i/D_O ratio. The effect of "I_n" on δ_i is more clear for S1 than for S2. 'δ_i' decreases as D_i/D_O increases, dramatically up to D_i/D_O is 1.50. Then curves get flatter, almost horizontal. It is concluded, for Di/D_O beyond 1.5, no significant reduction of δ_i. Increasing D_O by "design factor to satisfy safety", of 20% appears not to be that generous! The later "point of view" is previously reported in a published case history, Khater (1991).

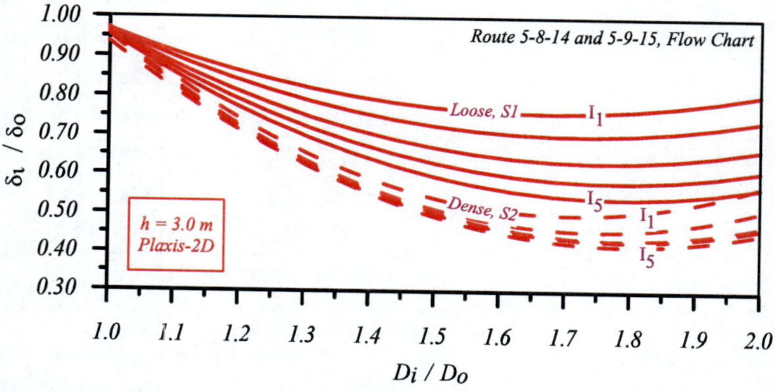

Fig. 2. Effect of penetration depth on wall tip displacement, [Plaxis cases]

Figure 3 shows the results of the spring analogy for the same cases previously studied based on the Plaxis. The curve behaviors are typical in both techniques, but the Winkler approach has numerical values are higher. The numerical deviation is expected due to the lack of soil shear that leads to high values of δ_O, thus forcing all curves to be compacted. The two comments about $D_i/D_O = 1.5$ and $D_i/D_O = 1.2$ are valid herein and through the other two excavation depths. Again, Winkler overestimated 'δ'.

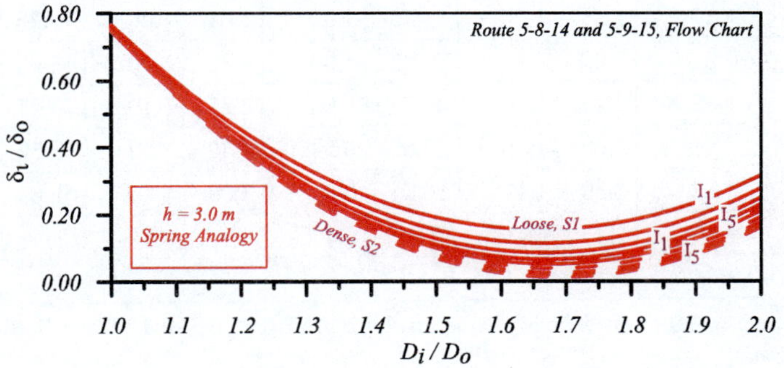

Fig. 3. Effect of penetration depth on wall tip displacement, Winkler cases

Table 3 displays the result variations among the three used methods. Conventional method results are the base of comparison. The ratio (%D_O) definition is D_O computed by spring analogy (or Plaxis) to D_O calculated by the conventional method. The ratio (% BM) definition follows the same (%D_O) concept, and both are decimals. Columns (2) and (6) show the values of (%D_O) for the spring analogy and FEA- Plaxis 2D. The values range from 1.10 to 1.17. Columns (4) and (8) show the values of (%BM), for the spring analogy is 1.0; and for FEA- Plaxis 2D it ranges from 0.51 to 0.68. It is concluded that, the conventional method and the spring analogy underestimate the minimum possible penetration depth $D_{O\text{-}min}$ and overestimate the maximum expecting bending moment BM $_{max}$, in general. This leads to higher values of δ_O predicted by spring analogy, and in comparison reasonable ones for FEA- Plaxis 2D.

Figure 4-a, curves (1) and (2), for all soil densities or wall rigidities, increasing pile penetration depth does not affect maximum bending moment. Figure 4-b, curve (3), (4) and (5), for all soil densities or wall rigidities increases the pile penetration depth reduces the wall tip horizontal displacement. Most of the reduction induces from the increase of wall rigidity. To reduce excessive δ_i it is advised to increase the penetration depth up to 1.5 D_O then increase the rigidity I_n, then change the structural system.

Table 3. Comparison between the three analysis methods

Excavation depth vs. analysis method		Excavation depth[a]							
		h = 3.0 m				h = 4.0-m			
		D_O	% D_O	BM	% BM	D_O	% D_O	BM	% BM
Conventional method	S1	2.92	1.00	51.29	1.00	3.89	1.00	121.57	1.00
	S2	1.96	1.00	31.75	1.00	2.61	1.00	75.25	1.00
Spring analogy	S1	3.20	1.10	51.25	1.00	4.30	1.10	124.5	1.02
	S2	2.15	1.10	31.75	1.00	2.85	1.10	75.25	1.00
FEA plaxis – 2D	S1	3.20	1.10	27.67	0.54	4.30	1.10	71.52	0.59
	S2	2.30	1.17	16.25	0.51	3.00	1.15	46.80	0.62
Notes		(1)	(2)	(3)	(4)	(5)	(6)	(7)	(8)

h and D_O (m), BM (knm/m')

[a]N.B: Case 5, h = 5.0-m gives nearly same results, e.g. FEA % D_O = 1.18 and % BM = 0.68

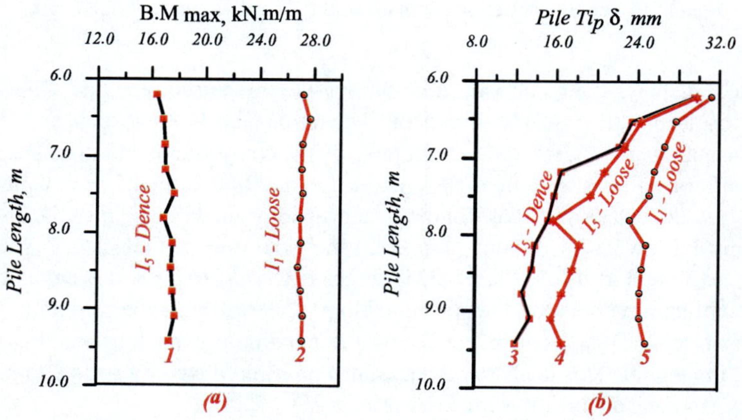

Fig. 4. Hints may enhance the secant pile design.

3.2 Rational Empirical Equations

The variables that affect the displacement of the wall tip "δ_i" could be grouped in three categories. The first one will be ignored, here-in, because it is believed its effect is small, as well as its range of variation, i.e. soil and pile Poisson's ratio.

Second category variables are; soil modulus of elasticity E_S, pile flexural rigidity EI and Rankin's earth pressure coefficient difference ($k_p - k_a$). Increases in the value of these variables reduce the expected δ_i i.e. inverse proportionally. Last category consists of only one parameter, the maximum bending moment M_{max}. Increases in the bending moment increases δ_i, i.e. direct proportionally. The effect of soil density 'γ' is indirectly covered by the existence of E_S, k_p, k_a and M_{max}.

Then, the above parameters have been re-arranged in a common fraction denoted as 'β', Eq. (1). Its denominator is second group variables, while the numerator is $M_{max.}$ It has been assumed, with some confidence, the powers of all variables equal to "1", which is rational and matches the engineering sense.

$$\beta = \frac{M_{max}}{E_S EI\left(k_p - k_a\right)} \tag{1}$$

Consider β is x-axis of and δ/h is its y-axis, Fig. 5. For each excavation depths the input data is fed into Plaxis, run by run, and have been used to calculate β_i, and the corresponding output per each run used to calculate δ_i/h. Then, Fig. 5 graphs have been plotted containing three curves presenting three excavation depths; linear curve fitting is the best fit. Equations (2), (3) and (4) present each curve trend.

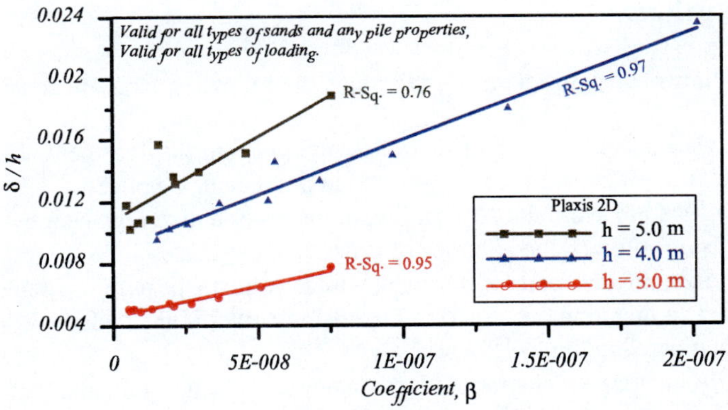

Fig. 5. Prediction of wall tip horizontal displacement, first approximation.

Case of h = 5.m, R-sq = 0.76

$$\delta/\mathrm{h} = 107202.4\,\beta + 0.010815 \tag{2}$$

Case of h = 4.m, R-sq = 0.97

$$\delta/\mathrm{h} = 70791.446\,\beta + 0.00889 \tag{3}$$

Case of h = 3.m, R-sq = 0.95

$$\delta/\mathrm{h} = 38582.394\,\beta + 0.0046 \tag{4}$$

Unit's Key. *The above equations units are:* M_{max} *(kN.m/m),* E_s *(kN/m*^{2}*), EI (kN.m*^{2}*/m),* β *(m/kN),* δ *(m), h (m), the constants "107202.4 - 70791.446 - 38582.394" (kN/m) while "0.010815 - 0.00889 - 0.0046" (dimensionless).*

Those equations present an improvement to the conventional method of design cantilever secant pile walls, as first approximation. The procedures are: (a) calculate penetration depth "D_O" and M_{max} based on the conventional method. (b) Increase D_O by 50% and reduce M_{max} by 35%. (c) Calculate "β", then pick up the equation matches the excavation depth to get δ. At this point, one can acquire reasonable bending moment, safe penetration depth and approximate wall-tip displacement. It could be used for dry sand densities, pile rigidities and any type of loading because the above equations use M_{max}, regardless of loads or soil conditions produced it. This upgrade is limited to homogeneous sand. A look back at Eqs. 2, 3 and 4 says minimum 'δ' occurs if 'β' is "zero". This will never happen unless $M_{max} = 0.0$, not logical, or 'β' denominator is "infinity". The only reason to accept this infinity is the existence of rigid capping beam and corners struts. The capping beam effect is the running research by me and research assistant. Until then, 'δ' is acceptable if it is $\leq$ 5.0-cm, 3.5-cm or 1.5-cm for wall heights 5.0-m, 4.0-m, and 3.0-m respectively. That is to say acceptable limit is $\delta_i \leq 0.8\%$ excavation depth, 'h'.

4 Conclusions

Within the limits and assumptions of this study the following facts are concluded:

1. As the wall penetration depth increases, the wall tip displacement 'δ' decreases. Beyond $50D_O\%$ increase, no weighty 'δ' improvement is noticed.
2. It is suggested to avoid the 20% D_O depth increases and to replace it by 50%.
3. High flexural rigidity and all cases of dense soils decreases δ.
4. Conventional method and spring analogy underestimate penetration depth and overestimate bending moment. Also, spring analogy gives higher values of 'δ' if all the above compared to finite element analysis.
5. Generally, increases penetration depth, does not change the values of maximum bending moment. But, it reduces wall displacement, effectively.
6. To reduce excessive δ_i first increase penetration depth up to 50%, and then increase wall rigidity, if neither insufficient, introduce anchors or struts.
7. The acceptable limit of 'δ_i' is $\leq$ 0.8% 'h', unless stringent values are imposed by adjacent structure. This 'δ_i' is the one that calculated by 2D-FEA model.
8. A rational technique has been suggested using empirical equations to upgrade the capability of conventional method to predict wall tip displacement.

Acknowledgements. Gratitude is expressed to Prof. Dr. M.I Amer, Prof. Dr. Tariq Omran, Prof. Dr. Walid Hammad and Prof. Dr. Youssef Goma for their valuable comments during reviewing this manuscript.

References

Bekdas, G., Arama, Z., Kayabekir, A., Geem, Z.: Optimal design of cantilever soldier pile retaining wall embedded in frictional soils with harmony search algorithm. Appl. Sci. **10**, 3232 (2020). MDPI, 1–17

Brinkgievem, R.: PLAXIS 2D-ver 8.6, Referance and Material model Manuals. Delft University of Technology, The Netherland (2007)

Khater, R. Kh.: Case history: a failure of cantilever sheet pile wall driven in clay. In: Al Azhar 2nd International Conference, vol. III, pp. 110–115, Al Azhar University, Egypt (1991)

Liao, S., Li, W., Fan, K., et al.: Model test on lateral loading performance of secant pile walls. J. Perform. Constr. Facilit. ASCE **28**, 391–401 (2014)

Mahesh, A., Kumar, B.: Effect of Diameter of Secant Pile Wall in Deep Excavation Analysis. Int. J. Recent Technol. Eng. (IJRTE) **7**, 871–876 (2019)

Ramadan, M., Ramadan, E., Khashila, H.: Cantilever contiguous pile for supporting excavation in clay. Geotech. Geol. Eng. **36**(3), 1545–1558 (2018)

Siminar invited Lectures, nine technical papsers: Piles as Retaining Structure. 25th Pragues Geotechnical, Charles University, Albertove 6, Prague (2017)

Verruijt, A.: Offshore soil mechanics. Delft University of Technology, Netherland (2006). Software SPW2006 downloaded. http://geo.verrujt.net

Worku, A.: The use of springs in static of structures to account for short and long tearm subgrade deformations. J. EEA **64**, 11–21 (2007)

Communication and Control for Remotely Operated Underwater Vehicles

Thanh Le Xuan[1](✉), Tuan Phan Anh[2], Duong Tran Khanh[2], Dong Nguyen[2], and Tung Pham Xuan[2]

[1] School of Information and Communication Technology, Hanoi University of Science and Technology, Hanoi, Vietnam
thanh.lexuan@hust.edu.vn
[2] School of Transportation Engineering, Hanoi University of Science and Technology, Hanoi, Vietnam

Abstract. This paper presents a research and development on communication and control for Remotely Operated Underwater Vehicles (ROV). Todays, ROVs are popular used in oceanic research for purposes, such as current and temperature measurement, ocean floor mapping and hydrothermal vent detection. ROVs utilize seafloor mapping, bathymetry, digital cameras, magnetic sensors, and ultrasonic imaging. In this work, a ROV is designed with a torpedo-shaped. The ROV is driven by flexible steering system, including a couple of blade ruder in the horizontal direction and a couple of steering thrusters in the vertical direction. For communication and control the ROV from a Ground Control Station (GCS) on a mothership or on land, a lightweight messaging protocol for communicating with ROV based on developing MAVLink has been used. Communication and control system of the ROV from a GCS are including two steps. The first step, communication between the ROV and a float is conducted via electrical cable technology. The float is used as a wireless access point. The companion computer is designed based on Linux operating system to communicate and transport MAVLink protocol between the main control system of ROV and the float. The second step, communication between the float and a GCS on mothership or on land is conducted via wireless technology. The GCS is setup wireless interface with static IP address, and same router's gateway with the companion computer.

Keywords: Communication · Control · ROV · GCS · Mothership · MAVLink

1 Introduction

Underwater vehicles like Removed Operated Underwater Vehicle (ROV) and Autonomous Underwater Vehicles (AUV) have risen consistently since they were introduced in the 1960s and found their most frequent scientific research use and data collection. Since the 1980s, the offshore oil industry has used ROVs to assist in offshore development, especially as waters became too deep for human divers to go. Underwater vehicle systems face more difficulties in operating in unstructured environments than ground robots, subject to the impact of noise and sensor limitations, leading to difficulty

D. V. K. Huynh et al. (Eds.): VSOE 2021, LNCE 208, pp. 216–221, 2022.
https://doi.org/10.1007/978-981-16-7735-9_22

in control and operation. ROV is a semi-automatic device with human intervention. Thanks to this feature, the vehicles are monitored and controlled to minimize the risk of unexpected incidents. Furthermore, it ensures flexible and stable control in different operating environments. However, communication underwater is limited due to the limitations of electromagnetic radiation in water. Therefore, the research of communication and control system plays an important role in ROV system. Recently, studies are investigating the communication and control for ROV and AUV.

Traditionally, communication is fed through a tether or cable using metallic conductors or fiber optics to transmit and receive data with a single channel RS-232 [1]. It is a practical and low-cost, but low data rate. Due to cable length or cable entanglement limitations, wireless methods have also been investigated to improve ROV mobility. They are based on acoustic and optical communication [2]. However, underwater acoustic communication has many problems such as low data transmission, short-range, and costly [3]. Underwater optical communication has an advanced technology improving the data rate, long-range, but difficult to maintain and costly [4, 5]. This study develops an inexpensive communication and control system using tether and wireless to transmit and receive data between ROV and mothership. The wire communication between ROV and a float is improved using ethernet, which supports local networks, long-distance, and high rate data. The wireless communication between a float and mothership is implemented to increase the long-distance control and ground control system's mobility. Communication and control for system are developed based on MAVLink protocol.

2 Hull Form of the Remotely Operated Underwater Vehicles

The hull form of the Removed Operated Underwater Vehicle in this study is a torpedo-shaped ROV. It is including three main parts and make from steel and aluminum. The fore and the rear of the AUV hull are parabolic shapes, the middle of the AUV hull is a pipe shape. Total length of the ROV hull is 3.2 m, the diameter of the ROV is 0.3 m. The ROV has a camera with LED light system in its fore for recording video under water [6]. Figure 1 shows the main hull form of the ROV.

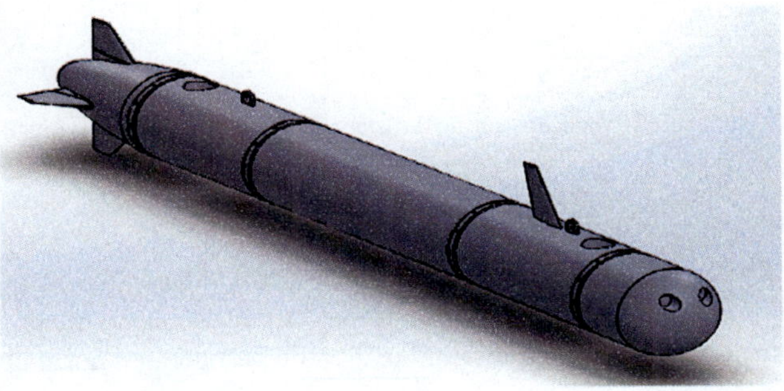

Fig. 1. Hull form of the ROV

3 Propulsive and Steering System

3.1 Propulsive

The Removed Operated Underwater Vehicle uses a main propeller for creating propulsive force for moving forward and backward. It is a fix pitch propeller with four blades. The propeller is made by aluminum. Figure 2 shows the main propulsive propeller of the ROV.

Fig. 2. Propulsive propeller of the ROV

3.2 Steering System

The Removed Operated Underwater Vehicle uses a flexible steering system by combination of blade rudders and thrusters. For horizontal steering, the ROV is driven by a couple of blade rudders that have aerodynamic profiles and located at above and below of the rear of the ROV. Those two blade rudders use the same steering shaft. The steering shaft is located at the symmetrical plane of the ROV. The steering shaft of the blade rudders connect to the ROV hull by ball bearings. The blades rudders are rotated and controlled via steering shaft by a servo motor. The flexible steering system of the ROV is shown in Fig. 3.

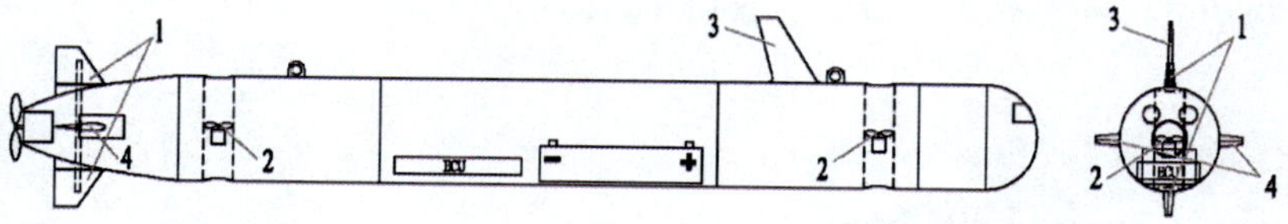

1: Horizontal rudders 2: Vertical steering thrusters 3: Bigger fixed fin 4: Small fixed fins

Fig. 3. Steering system of the ROV

For vertical steering, the ROV use two thrusters. To enhance the ability to generate thrust and torque for steering the ROV, the thrusters are set inside ducts. The thrusters are driven by specialized brushless motors that could be operated normally underwater environment.

In addition, for increasing the stability for the ROV, a set of three fixed fins is designed. They are including a bigger fin with hydrodynamic profile of NACA 0025 is placed on the back of the ROV, two smaller fins placed on left and right at the rear of the ROV.

4 Communication and Control

For communication and control the Removed Operated Underwater Vehicle from a GCS on a mothership or on land, authors have used a lightweight messaging protocol for communicating with ROV based on developing MAVLink.

MAVLink is a transmission protocol developed to communicate between the unmanned systems and mother ship or ground control stations. It supports both point-to-point with retransmission and multicast. It is also equipped with a double checksum to ensure reliability and message integrity. Both the transmitter and the receiver use binary serialization in MAVLink for communication. The transmitter transformed the message into a byte sequence and transmitted them through the network, the receiver deserialization to get the original message. It gains a significant advantage of reducing the size of the message compared to other messages' type [7]. It has two types of protocol headers, which are version 1 and version 2. The significant improvement of version 2 compared to version 1 is a signature at the last of protocol header. It develops the communication security of ensuring that the link is tamper-proof. Figure 4 shows the structure of MAVLink 2 header, which consists of flags, system IDs, components, checksum codes and payload data. The data frames can be from 0 to 255 bytes depends the message type [8].

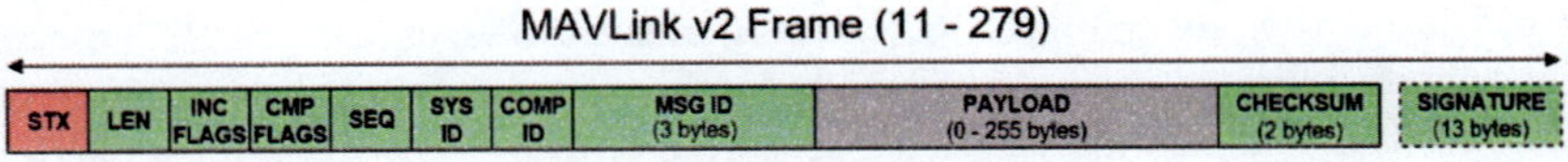

Fig. 4. MAVLink 2 header

The ROV uses both wire and wireless technologies for communication and controlling. The ROV and a float that pulled by the ROV are communicated together via a cable. The float and a GCS on a mothership or on land are communicated via wireless technology. In order to communicate and transport protocols between ROV, the float and the GCS, the network interface is implemented. Figure 5 shows the communication system for the ROV.

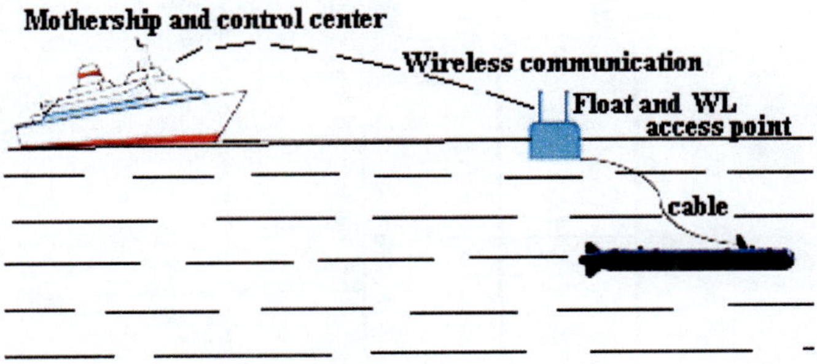

Fig. 5. Communication system

4.1 ROV and Float Communication

The communication between the ROV and the float is conducted via electrical cable technology. It is a cable with 30 m in length. The float is used as a wireless access point. The companion computer is designed based on Linux operating system to communicate and transport MAVLink protocol between the main control system of ROV and the float. The companion computer is functional to forwarding the messages and configure the IP network interface. The companion computer is setup with static IP address.

4.2 Float and Mothership Communication

The communication between the float and the GCS is conducted via wireless technology. The maximum distance for communication between the GCS and the float is 500 m. The float is equipped with wireless router, which is set up with security protocol WPA2 for protection. The GCS on the mother ship or on land is setup wireless interface with static IP address, and same router's gateway with the companion computer.

4.3 Monitoring and Controlling ROV

GCS is developed on the open-source software platform QGroundControl to monitor the system state and control ROV from the GCS through MAVLink messages, as shown in Fig. 6. The main control system on ROV sends the state messages to the GCS. Those messages include information on sensor health, battery status, system status such as angles, velocities. The command messages are sent from GCS to ROV. Those commands are designed based on the system requirements, such as move left, right, forward, backward, or buoyancy control in an emergency. The specific function of command is implemented on the control system on ROV. The command message format is designed on the XML file and generated in C++, and embedded in both GCS, companion computer, and a microcontroller on ROV. On ROV, the ACK (command acknowledgment) is used to get the command results, such as success, failure, or still in progress. Based on this response from ROV, people on mothership can know whether or not ROV implemented that command.

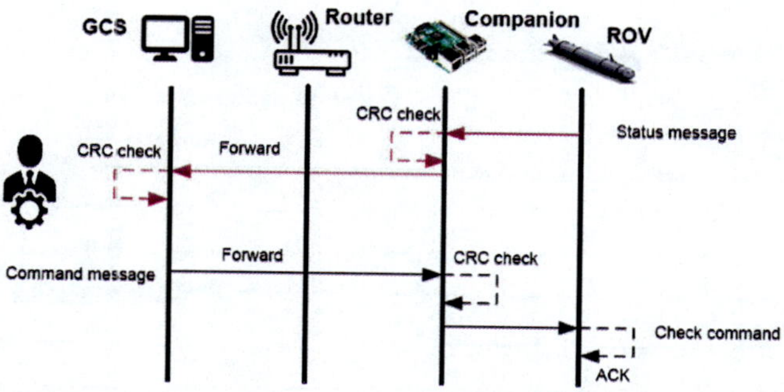

Fig. 6. Diagram of monitoring and controlling ROV

5 Conclusions

In this work, authors have developed a communication and control system for a ROV based on MAVLink. Communication and control system of the ROV from a GCS on a mothership or on land are including two ways. They are including communication from ROV to a float and communication from the float to a GCS. Communication between the ROV and a float is conducted via electrical cable technology. The float is used as a wireless access point. The companion computer is designed based on Linux operating system to communicate and transport MAVLink protocol between the main control system of ROV and the float. Communication between the float and the GCS on a mother ship or on land is conducted via wireless technology. The GCS is setup wireless interface with static IP address, and same router's gateway with the companion computer.

GCS is developed on the open-source software platform QGroundControl to monitor the system state and control ROV from the mothership through MAVLink messages. The main control system on ROV sends the state messages to the GCS. Those messages include information on sensor health, battery status, system status such as angles, velocities. The command messages are sent from GCS to ROV. Those commands are designed based on the system requirements, such as move left, right, forward, backward, or buoyancy control in an emergency.

Acknowledgement. This work is supported by the Ministry of Science and Technology with project code: NĐT.68.RU/19.

References

1. Zain, Z.M., Ahmad, R.B., Arshad, M.R.: Design and development of an RS232-based ROV controller system. In: 2004 IEEE Region 10 Conference TENCON 2004, Chiang Mai, 2004, vol. 4, pp. 487–490 (2004)
2. Farr, N., Bowen, A., Ware, J., Pontbriand, C., Tivey, M.: An integrated, underwater optical/acoustic communications system. In: OCEANS 2010, pp. 1–6. IEEE SYDNEY, Sydney (2010)
3. Kim, G.-J., Park, S.-J.: A wireless remotely operated vehicle using acoustic communication. Marine Technol. Soc. J. **46**, 44–49 (2012)
4. Swathi, P., Prince, S.: Designing issues in design of underwater wireless optical communication system. In: 2014 International Conference on Communication and Signal Processing, Melmaruvathur, pp. 1440–1445 (2014)
5. Kaushal, H., Kaddoum, G.: Underwater optical wireless communication. IEEE Access **4**, 1518–1547 (2016)
6. Anh, T.P., Le, Q., Thanh, H.P.T., Thanh, C.D.: Research and development a remotely operated underwater vehicle. Int. J. Mech. Prod. Eng. Res. Dev. **10**(5), 231–240 (2020). ISSN(P) 2249–6890; ISSN(E) 2249–8001
7. Koubâa, A., Allouch, A., Alajlan, M., Javed, Y., Belghith, A., Khalgui, M.: Micro air vehicle link (MAVlink) in a nutshell: a survey. IEEE Access **7**, 87658–87680 (2019)
8. MAVLink: Common Message Set Specifications. https://mavlink.io/en/messages/common.html. Accessed 15 Nov 2020

Considerations on the Design of Shellfish Reef Against Coastal Erosion

Slobodan B. Mickovski(✉), Alejandro Gonzalez-Ollauri, and Victor Henneton

Glasgow Caledonian University, 70 Cowcaddens Rd, Glasgow G4 0BA, Scotland, UK
Slobodan.Mickovski@gcu.ac.uk

Abstract. The Catterline Bay in Scotland (UK) is regularly subjected to storm surges during spring tide periods which had led to coastal erosion and landslides. As a response to the local population whose material properties and lives are at risk, we have investigated the potential of design and construction of Nature-Based Solutions (NBS) against the geo-hydro-climatic hazards posed by a changing climate. This paper assesses theoretically the wave attenuation capacity of an off-shore shellfish reef, as an NBS, which could be implemented in Catterline. Although less rigid than a traditional, 'grey' solution, this strategy plays on the depth reduction created by its organic structure to break the wave before it reaches the shore. Through numerical modelling using R Studio, we calculated the wave breaking probability (Qb) and the distribution of the wave energy flux (P) from the sea to the shore. Then, we compared two possible scenarios: the current conditions without a shellfish reef and a scenario with a constructed shellfish reef at three different locations and with different dimensions. The modelled NBS showed high wave breaking probability for each wave height trend. Moreover, the reef was shown to be more effective when located furthest from the shore. For maximum wave breaking effectiveness, the reef was shown to require relatively short length when located the furthest from the shore. Finally, based on the design and construction requirements, we offer recommendations about the reefs dimensions, and the shellfish species to implement as well as its management throughout the design life.

Keywords: Shellfish reef · Nature based solutions · Eco-engineering · Sustainability · Wave attenuation

1 Introduction

A Nature-Based Solution (NBS) is defined as a solution to improve the resilience of a site to natural hazards and provide social, environmental, and economic benefits [1]. They aim at mitigating meteorological hazards, which frequency is increasing with climate change [2], by improving the environmental footprint and raising the populations' awareness.

Shellfish reefs have been used as NBS in the past [3], as a solution for water depth reduction created by its organic structure in order to break the wave before it reaches the shore by creating a shallow zone. Therefore, as the major part of the wave energy

D. V. K. Huynh et al. (Eds.): VSOE 2021, LNCE 208, pp. 222–229, 2022.
https://doi.org/10.1007/978-981-16-7735-9_23

stays in the upper layers of the sea [4], the reef position in the water column defines its attenuation capacity. Consequently, the first shellfish must be situated at depth with low to medium wave strength: deep enough to affect the wave strength and shallow enough to avoid movement caused by the wave energy. The shellfish reef is built with time, in layers upon layers of decomposed shellfish. Younger generations of shellfish are growing over their elders which by the time decompose and constitute a bed of organic matter. It is this succession of the decomposition process that expands the reef, strengthens it over time, reduces the water depth over it and increases the friction distance the wave is passing on. The deposition rate differs among species as well as their consistency and resistance to the erosion.

A wave reaching the shellfish reef is usually characterised by its height (H), wavelength (L), period (T), and water depth (h). There are different waves depending on their size and sources, but only wind waves will be studied and discussed in this paper; these are generated by the wind, with a wavelength and strength that increase as the fetch (horizontal distance on which the wind blows) increases. The energy of a wave depends on the wavelength and the water depth and, consequently, three different depth zones can be differentiated [4]: deep-water ($h > L/2$) below which the wave energy is negligible; intermediate zone ($L/20 < h < L/2$); and shallow water zone ($h < L/20$). In the shallow water zone around a shellfish reef, the wave breaks, its velocity decreases and it loses the energy which is dissipated in the surrounding waters in the form of turbulences. Normally, the natural slope of the beach plays this role as the wave goes from the deep-water through to the shallow-water depth zone, but a reef would allow to anticipate this phenomenon by breaking the waves upstream, thus protecting the shore.

The aim of this study is to assess the theoretical wave attenuation capacity of a shellfish reef as an NBS, which would be implemented in Catterline, as well as to recommend optimal reef location, dimensions, shellfish species, and management strategy. The objectives of this study are to assess the wave breaking probability (Qb) and the distribution of the wave energy flux (P) from the sea to the shore through literature study and numerical modelling in R Studio, as well as to compare the effects of the shellfish reef in relation to the reef location within the bay.

2 Methods

2.1 Case Study Site

The Catterline Bay (Fig. 1) had been subjected to storm surges during spring tide periods which had led to important coastal erosion and landslides. As a response, the local community, whose houses are close to the cliffs, is looking actively for protection against coastal erosion. As the waves break permanently on the shore, their action on the beach slope erodes continuously the shore. Originally, 'grey' techniques (e.g. concrete blocks placed upstream or embankments) were used to fix this problem, but the lack of maintenance combined with the detrimental effects of climate change (storm surges, coastal erosion) limited their efficiency.

2.2 Wave Dynamics

Based on [5], the effects of the shellfish reef on the wave force have been assessed numerically by comparing the evolution of the breaking probability (Qb) of the wave and its energy flux (P) within the three depth zones for two scenarios. Scenario 1 was the existing situation without a reef in the bay. Scenario 2 included the reef in the bay, located at three different positions: at 50 m, 100 m, and 150 m from the bay neck (respectively noted as X50, X100, X150). Also, different lengths of the reef in Scenario 2 have been tested to statistically indicate the optimal dimensions. Numerically, the water depth (h) is a variable of the maximum wave height achievable (Hm) which features in the calculation of the breaking probability (Qb, [5]), a ratio varying from 0, for no breaking event, to 1, for a wave breaking at a definite point of depth. Normally, the natural slope of the beach plays this role as the wave goes through the different depth zones. A reef allows to anticipate this phenomenon by breaking the waves upstream.

Wave datasets for Aberdeen (nearest available measuring point, 20 km north of Catterline Bay), and the bathymetry data for Catterline were retrieved from the British Oceanographic Data Centre and analysed to determine the relevant wave and seabed variables. For the wave height, the average, the first, and third quartiles of the significant wave height values on 3-hourly measurements over 3 months were collected, and for each, the total mean calculated to determine respectively the average, the low, and high trend. The same process was applied to the wave crest period to calculate its average only. The wavelength was determined from the annual average wind speed data in Catterline between 2013 and 2016 available from Windfinder, using the Thurman method [6]. The wave propagation through the different seabed depth zones was calculated using the method described in [4] to obtain wave velocity. In our study, the assumption is that the wave breaks in the shallow water zone – a process which induces a complete loss of the wave energy that dissipates in the surrounding waters a form of turbulence [4].

2.3 Reef Dynamics

The wave effects were considered along a 1 m wide, 200 m long corridor situated in the middle of the bay neck, pointing towards the shore (Fig. 1); the wave corridor. Based on the bathymetry data, the bay was 2 m deep at its neck ($x = 0$ m, deepest water), with the depth decreasing linearly to the shore ($x = 200$ m, shallowest depth) with a 1% slope. For the Scenario 2, the reef was first placed at a 100 m seawards from the bay neck and was 2 m wide. Its top surface reached the shallow water depth at the maximum tidal height, corresponding to the most threatening conditions for the shore. Based on the history of hydrometeorological risks on the site, it was assumed that the waves propagated from the open sea, following a straight line to the shore, with no interactions from the sides, and with no reflection or diffraction. The shellfish reef was considered as a straight rectangle positioned across the bay and following the natural curve of the bay (Fig. 1b). The percolation effects and the seabed rugosity were not considered, while a full dissipation of the wave energy through the surrounding water was assumed [4].

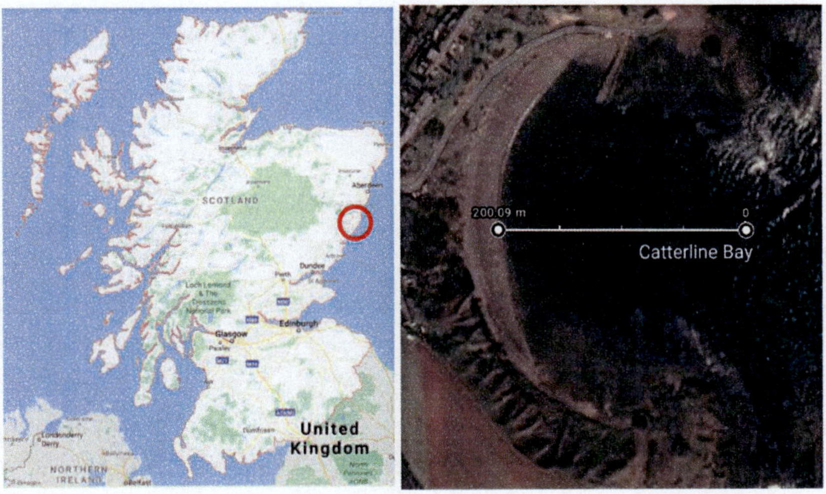

Fig. 1. Location of Catterline Bay, showing the assumed wave corridor and distances from the bay neck (x = 0) to the shore (x = 200 m). Figure lined up on North.

3 Results and Analysis

3.1 Wave Dynamics

The average significant wave height (H) of the site was calculated as 1.85 m (1.2 m at low trend, and 2.34 m at high). The study site had an average wind speed of 13.81 km h^{-1} with a resulting wavelength (L) of 7.32 m, which meant that the shallow-depth water zone started at depth h = 0.37 m (x = 163.4 m). The maximum wave height achievable over the reef (Hm = 0.322 m) gives a breaking probability curve for each reef position. In general, all scenarios have a breaking probability superior to 72% at the beginning (Fig. 2a). The three wave height trends, modelled in R Studio, are mainly dissociated in deeper waters (Qb low = 0.72, Qb average = 0.87, Qb high = 0.92; Fig. 2b-d). Their values then increase slowly as the water gets shallower. Eventually, they join each other at the entrance of the shallow depth zone of the wave (h = 0.366 m at X163, Qb > 0.96).

3.2 Shellfish Reef Efficiency

For each wave height trend calculated in R Studio using the datasets and methods described in Sects. 2.2. and 2.3, the wave was shown to have a very high chance to break over the reef placed at any of the positions considered in Scenario 2 (Qb > 0.96). Table 1 shows the increase in breaking probability for each reef position for every trend compared to the 'no reef' scenario (Scenario 1). As a result, the reef located the furthest from the shore (X50) shows a larger increase in the breaking probability of the wave (+20% of breaking probability in low trend wave height (1.2 m)). On the other hand, the reef closest to the shore (X150) has the smallest increase in this value (+6% in low trend as well).

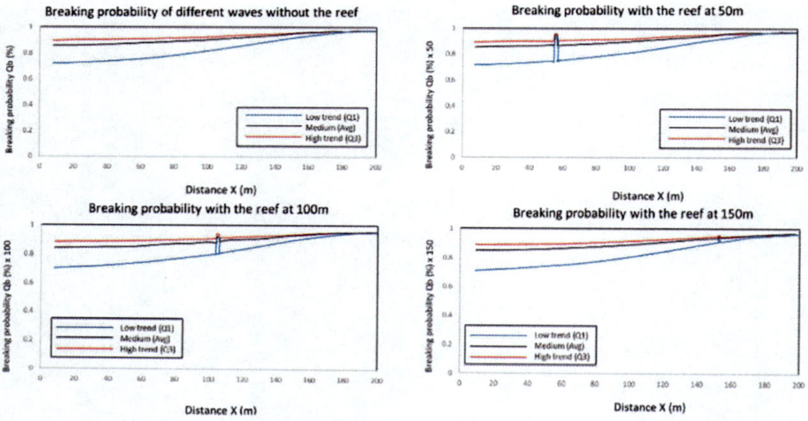

Fig. 2. Breaking probability curves for each reef scenario

However, even though the reef furthest from the shore showed higher efficiency in breaking probability, none of the modelled reef positions showed statistical differences when compared to the 'no reef' scenario ($p > 0.05$). The Kruskal-Wallis test Table 2 also showed that the different trends in wave height have no influence on the statistical impact of the reef.

Since the wave breaking process is affected by the wave velocity decrease due to the friction between the water and the seabed, different reef lengths were tested for each position to increase the wave friction distance until which the reef would show statistical significance (Kruskal test p-value < 0.05) compared to the 'no reef' scenario (Scenario

Table 1. Increase in Qb compared to 'no reef' scenario (%).

Reef position	Increase in Qb compared to 'no reef' scenario (%)		
	Low trend	Average	High trend
X50	20	9	6
X100	12	5	4
X150	2	0	1

Table 2. Kruskal-Wallis test comparison

Reef position	Kruskall-Wallis p-values		
	Low trend	Average	High trend
X50	0.5221	0.5221	0.5221
X100	0.7242	0.7242	0.7242
X150	0.9478	0.9478	0.9478

1). The models showed that the distance between the reef and the shore has an impact on the required reef length to be statistically different: the reef furthest from the shore (X50) would require a minimum reef length of 7 m, the reef in the middle position (X100) would need a reef length of 13 m, whereas the closest (X150) would require a length greater than 50 m (i.e. is not viable for the existing bay geometry).

The 'no reef' scenario modelled in R Studio showed an increasing value of the energy flux (P) as long as the waves are located in its deep or intermediate depth zone ($x = 0$ m to $x = 163$ m; Fig. 3). Then it reaches its maximum before the wave enters shallow waters ($P163 = 58290.35$ Wm^{-1}). Starting from the shallow water zone ($x = 163$ m), the energy flux starts to decrease as the wave is affected by the proximity of the seabed, which reduces its velocity and dissipates its energy with the friction (shoaling effect). The modelled 'no reef' wave finally reaches the shore with the energy that has not been dissipated and thus being delivered to the shore ($P200 = 11413.59$ W.m-1). Our models showed that each reef scenario exhibits a similar progression of the energy flux until the wave encounters the reef. After breaking at the reef, the wave energy is dissipated in the reef surroundings resulting in a drop of the wave energy flux value to 0 $W.m^{-1}$ in the shallow waters between the reef and the shore as classified by [7].

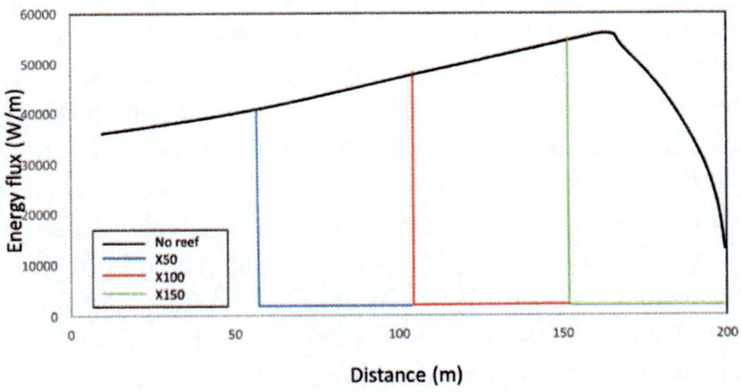

Fig. 3. Wave energy flux (P) scenario comparison.

4 Discussion and Conclusions

This study aimed at theoretical assessment of the efficiency of a shellfish reef to attenuate the wave energy flux to avoid further coastal erosion in Catterline Bay based on readily available data. The datasets were retrieved from open-sourced databases such as meteorological and bathymetric datasets to derive the variables used in the numerical model. A numerical model was created to determine the impact of the reef on the breaking probability of the incoming wave. The energy flux reduction from the waves by the shellfish reef was also modelled. Consequently, this study showed that the efficiency of a reef to attenuate the wave strength in Catterline is affected by its position in the bay as well as its dimensions. The main limitation of this study is that no experiments have been performed to validate the theoretical results due to the feasibility nature of this study and the lack of funding for the local community affected by coastal erosion and instability. Similarly, the calculations considering wave reflection, diffraction and percolation were not carried out but could be added in future studies. The results obtained in this study, however, could be used as a comparison for future both experimental and theoretical studies. The accuracy of the results presented here can be verified by construction and monitoring which should be relevant to the site area and the model used for verification. The relevance to the site area could be taken into account by including the effects of waves propagating from the deep sea to the shore (surge waves) which will be predominant over the wind waves in the nearshore area.

The secondary aim of this paper was to provide guidelines regarding the implementation and management of a shellfish reef on the study site based on a literature study. It does not aim at assessing the suitability of the site to receive a shellfish reef, nor studying the ecology of the advised shellfish species. The feasibility to implement a shellfish reef in Catterline coast must be investigated separately in an environmental study since certain living conditions are required to ensure the shellfish to thrive in the bay. Future studies (feasibility and environmental) should focus on three spheres of interest: the hydrosphere for the hydrological characteristics of a live shellfish reef, the lithosphere for the sedimentation and substrate availability for the shellfish, and the biosphere for ecological assessment of the type and place of the shellfish in its local food chain as well as its interactions with the surrounding flora and fauna. The site must fit all of the criteria within each sphere in order to be potentially suitable for shellfish reef implementation. For example, *Modiolus modiolus*, an ecosystem engineering shellfish species could be used in building the reef based on its current presence in Scotland habitats (67 beds) and its relatively long lifespan (>45 years). Such reef would not only contribute towards the protection of the sea shore but also to increase in knowledge of the ecology of this species as indicated by the OSPAR Commission [8] if a suitable monitoring programme is deployed during and after reef build-up. The implementation of this species could also allow Catterline to be an OSPAR Marine Protected Area. On the other hand, further studies are required with different species to assess the optimum species to implement at a specific site, and it would be recommended to conduct further research on co-habitation possibilities of several species in the reef. The actual management of the shellfish reef which would include management of a potential overgrowth and disturbance of the local food web can be supported by existing tools such as Ecopath with Ecosim model (https://ecopath.org/). Finally, the social acceptance of such a project must be investigated to make sure the project does not oppose the views of the local community and stakeholders. Adapting existing tools such as [9] to gauge the thoughts and views of the relevant stakeholders on the implementation of a shellfish reef as a NBS for coastal protection strategy.

References

1. European Commission. Towards an EU Research and Innovation policy agenda for nature-based solutions & re-naturing cities. Final Report of the Horizon 2020 Expert Group on Nature-Based Solutions and Re-Naturing Cities. Brussels: European Commission (2015)
2. IPPC. Global Warming of 1.5 °C (2018). ISBN 978-92-9169-151-7
3. Scyphers, S.B., Powers, S.P., Heck, K.L., Jr., Byron, D.: Oyster reefs as natural breakwaters mitigate shoreline loss and facilitate fisheries. PLoS **6**, e22396 (2011)

4. World Meteorological Organization: Guide to wave analysis and forecasting. WMO-No. 702 (1998)
5. Battjes, J.A., Janssen, J.P.F.M. Energy loss and set-up due to breaking of random waves (1978)
6. Thurman, H.V.: Introductory Oceanography. Pearson, London (1988)
7. Fernández, J.C., Salcedo-Sanz, S., Gutiérrez, P.A., Alexandre, E., Hervás-Martínez, C.: Significant wave height and energy flux range forecast with machine learning classifiers. Eng. Appl. Artif. Intell. **43**, 115–121 (2015)
8. OSPAR Commission: Background Document for Modiolus modiolus beds (2009)
9. Dalton, T., Jin, D., Thompson, R., Katzanek, A.: Using normative evaluations to plan for and manage shellfish aquaculture development in Rhode Island coastal waters. Mar. Policy **83**, 194–203 (2017)

Design and Construction of Nearshore Wind Turbine Foundations: Some Experiences of APECO and FECON

Tien Dung Nguyen[1](✉), Duc An Ho[1], Nguyen Hai Bui[2], Duc Huyen Pham[1], Huy Dong Ngo[1], and Van Toan Du[3]

[1] FECON Corporation and VNU Vietnam Japan University, Hanoi, Vietnam
nt.dung@vju.ac.vn
[2] Asia Pacific Engineering Consultants (APECO), Hanoi, Vietnam
[3] Vietnam Institute of Seas and Islands, Hanoi, Vietnam

Abstract. Recently, nearshore wind farms along the coast of Vietnam have been designed and constructed increasingly. Different from the job onshore, construction of nearshore wind turbine (WT) foundations exposes to have many difficulties. This paper first presents some key points on selection of foundation types, design codes and geotechnical design for nearshore WT foundations. Some experiences of APECO in design of the foundations are highlighted, especially for elevated pile cap on PHC piles. Finally, key points on design and construction of nearshore WT foundations at a wind farm in Southern Vietnam under construction by FECON are introduced and discussed.

Keywords: Wind turbine foundations · Nearshore · Elevated pile cap · PHC piles

1 Introduction

Following the global goal of sustainable development, Vietnam has actively executed many policies to secure the goal in near future, in which producing electricity from renewable resources (e.g., wind and solar) is one of the most prior tasks. According to the statistics from an official correspondence from the Vietnam Electricity (EVN) to Ministry of Industry and Trade (MOIT), up to the end of March 2021, the total output from twelve operating wind farms in the whole country was about 580 MW, making a portion of nearly 1.0% of the total output (59.0 GW) from different generating sources. Among the twelve projects, two are nearshore ones (Ben Tre and Dong Hai 1). According the Power Development Master Plan VIII (IE 2020), in 2025 and 2030 the rate of power generated from wind energy (for all onshore, near-shore and offshore wind farms) would account for 9% and 12%, respectively, of the total power output generated from all different sources. This indicates that, for the next decade, to meet the plan, many wind farms along the country coast would be constructed.

Compared with onshore areas, nearshore and offshore areas are regarded more productive for the wind farms to be based because of stable wind source and favorable

D. V. K. Huynh et al. (Eds.): VSOE 2021, LNCE 208, pp. 230–237, 2022.
https://doi.org/10.1007/978-981-16-7735-9_24

wind speed. However, construction of foundations of nearshore and offshore wind farms face many more difficulties and challenges compared to that onshore. The difficulties and challenges are especially applied to most Vietnamese contractors, who compete to execute projects at the most budget conditions required. Because of this, the number of nearshore projects completed and being constructed in Vietnam is still limited compared with onshore ones.

This paper first briefly presents some basic considerations in design of nearshore and offshore WT foundations with some experiences of Asia Pacific Engineering Consultants (APECO). The paper then briefly introduces key points in design and construction of nearshore WT foundations of a wind farm in Tra Vinh Province, Vietnam.

2 Design of WT Foundations

2.1 Selection of WT Foundations

The selection of WT foundation types is based mainly on the water depth and the seabed conditions. In term of fixity, WT foundations are classified into two main types: Fixed and floating foundations. Figure 1 shows common types of fixed and floating foundations in practice nowadays.

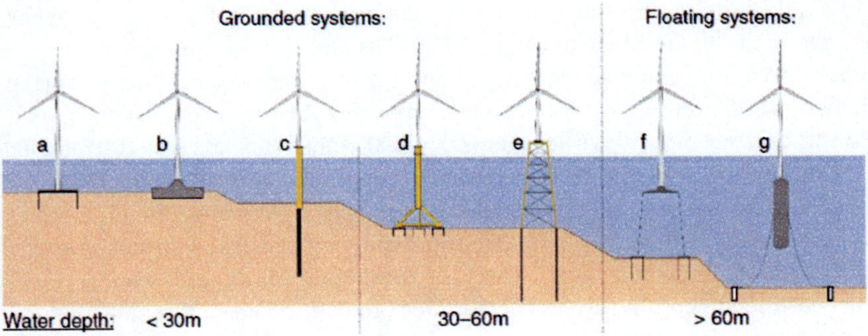

Fig. 1. Types of wind turbine foundations: (a) bucket/suction caisson; (b) gravity-based; (c) monopile; (d) tripod on bucket/suction caisson; (e) jacket/lattice structure; (f) tension leg platform; and (g) spar buoy floating concept (after Bhattacharya 2019).

In general, as illustrated in Fig. 1, the fixed foundation types are applicable up to the maximum water depth of about 60.0 m (Malhotra 2011, Bhattacharya et al. 2017, Bhattacharya 2019). Beyond this depth, the foundations are transited from fixed-bottom structures to floating structures. The floating foundations offer economic and environmental benefits compared with fixed bottom designs due to less-invasive activity on the seabed during the installation.

2.2 Design Codes

In present practice there are design codes, that specify values for safety factors, material parameters, design criterions and other rules and regulations that govern the design.

The design standard IEC 61400-6 is commonly used for design of WT foundation in Vietnam. Besides this, other design codes, such as EN 1997-1, DNVGL-ST-0126, ACI 318-14, are also applied when necessary.

2.3 Geotechnical Design

Fatigue resistance of ground is not normally part of geotechnical design because of the limitation of methods as well as very high cost for laboratory and field testing. Typically, the limit state design (LSD) method is applied in geotechnical design, in which both ultimate limit state (ULS) and Serviceability Limit State (SLS) are considered. Detailed criteria for ULS and SLS that must be taken into account in design are given in IEC 61400-6 (2020).

2.4 Some Experiences of APECO

Up to date, APECO has performed geotechnical design for three nearshore projects in Tra Vinh (TV) Province, Vietnam, as briefly described in Table 1.

Of the three projects, the Dong Hai 1 is now under construction whereas the construction of Trung Nam TV and Hiep Thanh 3 will be commenced by August 2021. As shown in Table 1, at Dong Hai 1 project, one type of foundation (concrete pile cap on PHC piles) is sufficient since the water depth varies from 3.3 to 7.5 m only. For the other two projects, the concrete cap on PHC piles was designed for the WTs at water depths up to 7.5 m, and both steel tripod and concrete cap on steel pipe piles (Fig. 2) were designed for the WTs at larger water depths. For the tripod foundation (Fig. 2(a)), three vertical steel pipe piles (diameter (D) = 2400 mm, thickness (t) = 60 mm, length (L) of about 50 m from the seabed) (i.e., one pile per pod), were designed to support the WT. For the concrete cap on steel pile foundations, eight inclined steel pile piles (D = 1500 mm, t = 30 mm, L = 54 m (from seabed)) were designed to support each TW (Fig. 2(b)). Both foundation types ensure the required bearing capacity under relatively equal construction cost. Analyses for the three projects indicated that steel pipe pile is optimal solution for water depths of larger than 7.5 m because in such water depths the PHC piles would be too slender to support the turbines under extreme loads.

3 A Typical Nearshore Project

3.1 General Information

A typical project discussed in this paper is Tra Vinh No.3 wind farm project, which is located at Truong Long Commune, Duyen Hai District, Tra Vinh Province. The project consists of 12 turbines (WTG01 to WTG12) with a total capacity of 48.0 MW (i.e., 4.0 MW each). The turbines were designed to locate in a straight line with a total length of 3.5 km, as shown in Fig. 3(a). The distance between 2 turbines is approximately 300 m. The distances to the shore from WT01 and WT12 are 0.6 km and 2.5 km, respectively. The foundation works are under construction by FECON, the main contractor for civil works (i.e., CBOP).

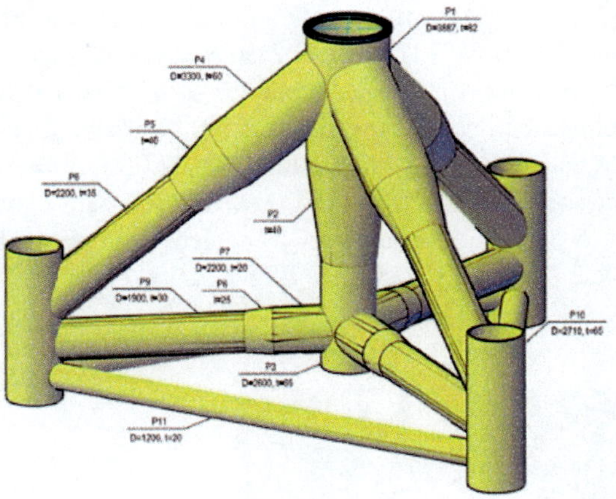

(a)

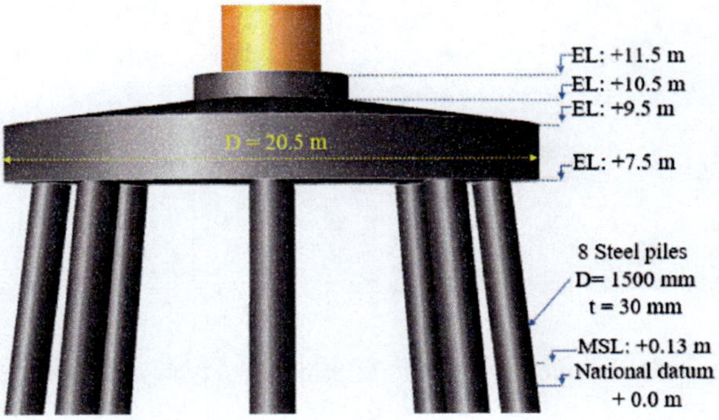

(b)

Fig. 2. Foundation solutions for Trung Nam TV and Hiep Thanh 3 projects at water depths >7.5 m: (a) Tripod foundation; (b) Concrete cap on steel pipe piles

Table 1. Nearshore projects in Tra Vinh province recently designed by APECO

No	Project name	W. depth (m)	Capacity (MW)	No. of WT	Foundation type
1	Dong Hai 1	3.5–7.5	100	25	PHC piles
2	Trung Nam TV	4.0–10.0	300	60	PHC piles, Tripods, Steel piles
3	Hiep Thanh 3	3.0–10.0	210	42	PHC piles, Tripods, Steel piles

The water depth measured from Mean Sea Level (MSL) at turbine locations varies from 2.6 m (WT01) to 3.4 m (WT12). The soil stratum at the site varies rather significantly from turbine to turbine location and is quite complex to describe in detail herein. Broadly, the soil profile (from seabed downward) consists of a fine sand layer (0 – 10.0 m, $N =$ 10 to 25), soft clay mud and sandy clay mud (10.0–22.0 m, $N = 1$ to 4), intercalated layers of clayey and sandy soils (22.0–48 m, $N = 15$–30), hard clayey sand and sandy clay (48.0–60.0 m, $N = 30$ to 60) and then very dense fine sand (60.0–70.0 m, $N =$ 45 to >50). A typical borehole log and the SPT N value at turbine WT12 is shown in Fig. 3(b).

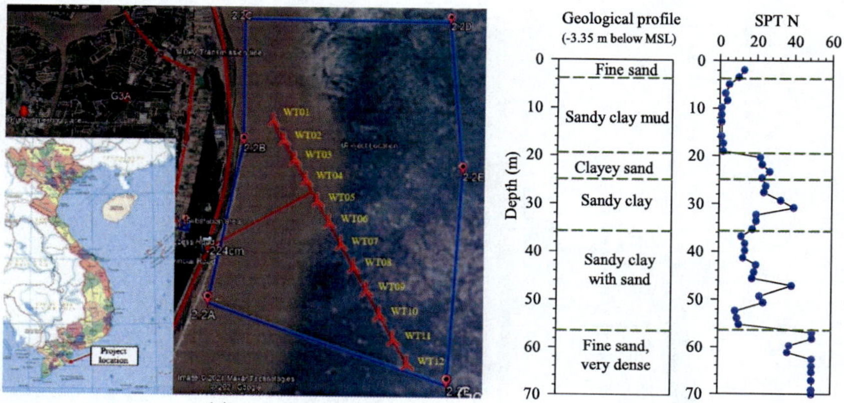

(a) (b)

Fig. 3. (a) Location and plan view of the project; (b) typical soil profile at WT12

3.2 Foundation Design

The foundation design for this project was carried out by FECON and CTE Wind Vietnam (CTV). By considering different criteria, such as water depths, geological conditions and construction cost, the elevated pile-cap foundation type was found most suitable (i.e., economical) and was selected for the turbines at the site. The bottom elevation of the pile caps was determined to accommodate the highest astronomical tide, water rise due to climate variation, storm surges, wave max and a minimum air gap. The key drawbacks of this foundation solution are its extended construction duration and difficulties in construction quality control. Typical design of WT foundation at the site is shown in Fig. 4.

The reinforced concrete pile cap was designed in circular shape with the cap diameter of 21.5m and the pedestal diameter of 6.5 m. The bottom of the pile cap was designed to seat 5.5 m above the MSL. The total thickness of foundation was 4.5m, in which the cap structure was 1.2 m (at the edge) to 2.9 m (at the center) and the pedestal structure was 1.6 m. The structure was placed on 42 Pre-stressed Spun High Strength Concrete (PHC) piles of 800 mm in diameter graded as Type C following the standard JIS A 5373 (2010). The piles were driven in two rings of 21 piles each with radii of 8.55 m (inner

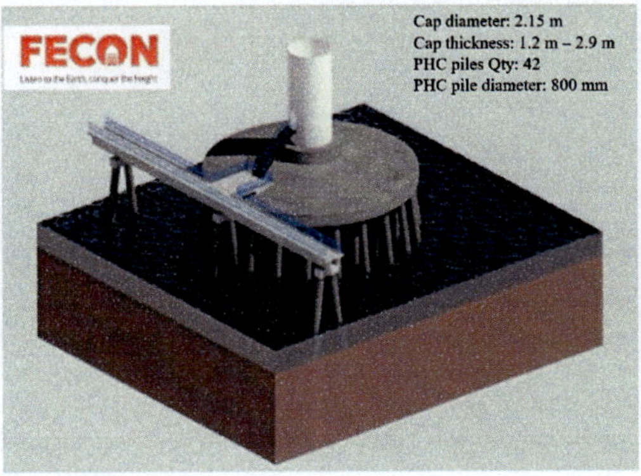

Fig. 4. Typical design of the WT foundation at the project

ring) and 9.85 m (outer ring). The 21 piles of the outer ring were driven oblique with a slope of 7:1 (V:H)) and the 21 piles of the inner ring were vertical ones. The pile length (from seabed) varied from 48–50, that is, the pile toes are socketed into the sandy clay with sand layer.

3.3 Foundation Construction

Most construction activities at the site have been carried out with the support of barges. The PHC piles loaded on a barge are tugged to the site, where the pile driving machine (which also works as a crane at the site) is set on another barge (Fig. 5(a)). For this project, the PHC piles were driven using diesel hammers of 10.0 ton capacity, which can either be hung on the crane's boom end (flying leader) or attached to the Kelly bar of the crane (Fixed leader). For this project, seven flying leader hammers and three fixed leader hammers are simultaneously mobilized. For each WT, the pile driving analyzer (PDA) test (Fig. 5(b)) was carried on two PHC piles to make sure the designed capacity. For the test, the pairs of accelerometers and strain transducers attached oppositely near the pile head were connected to the PDA system placed on the barge. Activities out of the barge (e.g., fixing the transducers) are carried out by engineers with the support of a suspended man basket controlled by the crane.

When all the piles of a turbine are driven to the design depths, the extra part of the pile heads are cut off at the same level of 10 cm above the both of the pile cap. The formwork is then installed on the pile heads (Fig. 6(a)) for the cast-in-place concrete cap. Pieces of steel cages socked into the hollows of the PHC piles are used to secure the firm connectivity between the piles and the cap. For concreting the cap, a concrete batching plant and concrete pumps placed on a barge of 5000 tones are mobilized for the job (Fig. 6(b)).

For the nearshore project with water depth of less than about 7.5 m, the elevated cast-in-place concrete cap on PHC piles was proofed to be the most budget solution. However, construction of this foundation type often deals with many difficulties, such as, low progress in typhoon season; low progress due to the construction of working platform; fragility/damage of oblique PHC piles during driving. Among the factors, the

(a) (b)

Fig. 5. Pile driving at the site: (a) piling work (WT02); (b) PDA test (WT02)

Fig. 6. Formwork and concreting: (a) formwork (WT02); (b) concreting (WT02)

construction progress and quality of the driven piles can be improved significantly by using steel pipe piles. This is because the steel pipe piles are much stronger than PHC piles to overcome the possible damages from driving and they can be produced as a single pile segment to facilitate the piling procedures.

4 Conclusions

This paper presented some experiences of APECO and FECON in recent design and construction of nearshore WT foundations. First, key points on the selection of foundation types, design codes, and geotechnical design are briefly discussed. Design experiences of APECO shows that the foundation type of elevated pile cap on PHC pile is economical solution and technically suitable up to the water depths of about 7.5 m. Beyond this depth other foundation types (e.g., pile cap on steel pipe piles or tripods) should be considered. Then, key points on design and construction of elevated pile cap on PHC piles for WT foundations at a FECON's nearshore project was presented and discussed. The driving PHC piles for this type of foundation faces many difficulties in ensuring construction progress and driven pile quality. Experiences from the construction of the foundations at the site indicates that steel pipe piles are more preferable for this type of foundations.

Acknowledgement. This study was greatly supported by design engineers of APECO and by many design and construction engineers of FECON, who have directly involved in the typical nearshore project.

References

ACI 318-14: Building Code Requirements for Structural Concrete and Commentary. American Concrete Institute, 520 p. (2014)

Bhattacharya, S.: Design of Foundations for Offshore Wind Turbine. Wiley, Hoboken (2019)

Bhattacharya, S., Wang, L., Liu, J., Hong, Y.: Civil engineering challenges associated with design of offshore wind turbines with reference to China. In: Wind Energy Engineering – A Handbook for Onshore and Offshore Wind Turbines, Chap. 13, p. 243–273 (2017)

DNVGL-ST-0126: Support structures for wind turbines. DNV GL, 208 p. (2018)

EN 1997-1: Eurocode 7: Geotechnical design – General rules. European committee for standardization (2004)

IE (Institute of Energy): Power Development Master Plan VIII, Vol. 1 - General statement. Report No. E-542, Institute of Energy, Ministry of Industry and Trade (MOIT) (2021). (In Vietnamese)

IEC 61400-6: Wind energy generation systems - Part 6: Tower and foundation design requirements. International Electrotechnical Commission (2020)

Malhotra, S.: Selection, design and construction of offshore wind turbine foundations. In: Al-Bahadly (ed.) Wind Turbines, Chap. 10, pp. 231–264 (2011)

TCVN 7888: Pretensioned spun concrete piles. Vietnam standards (2014). (in Vietnamese)

Design Considerations for Brushwood Fences Concerning Bathymetry and Fence Locations

Hoang Tung Dao[1,2](✉), Bas Hofland[2], Marcel J. F. Stive[2], Tri Mai[3], and Anh Ngo[4]

[1] Faculty of Marine Science and Island, Hanoi University of Natural Resources and Environments, 41A Phu Dien, Bac Tu Liem, Hanoi, Vietnam
dhtung@hunre.edu.vn

[2] Faculty of Civil Engineering and Geosciences, Delft University of Technology, Stevinweg 1, 2628 CN Delft, The Netherlands
{b.hofland,m.j.f.stive}@tudelft.nl

[3] Faculty of Coastal and Offshore Engineering, Hanoi University of Civil Engineering, 55 Giai Phong, Hai Ba Trung, Hanoi, Vietnam
trimc@nuce.edu.vn

[4] Faculty of Civil Engineering, Thuyloi University, 175 Tay Son, Dong Da, Hanoi, Vietnam
anhntt@tlu.edu.vn

Abstract. Wooden fences are nature-based supporting structures to restore mangroves in the Mekong Delta. The hydraulic functioning of wooden fences was studied in previous studies. However, the role of bathymetry in the dissipation and damping of waves by wooden fences has not been studied yet. Thus, in this study, a numerical approach is used to find the effect of the position of fences and the foreshore bathymetry, including two particular slopes of 1/200 and 1/500, on wave damping due to wooden fences. The results show that the bottom slope significantly influences the dissipation of incoming waves, the so-called pre-dissipation, before damping by the wooden fences. Differences in pre-dissipation occur between fence locations along the cross-shore slopes. The higher pre-dissipation takes place for wooden fences closer to the land, as the depth-limited wave height at the fence reduces. The efficiency in wave damping of wooden fences is also increasing as the freeboard is becoming larger for the fence located closer landward.

Keywords: Wooden fences · Numerical model · SWASH · Bottom slope · Fence locations

1 Introduction

In the Mekong Delta, wooden fences are used as a supportive structure for hard structures to reduce the incoming wave energy and to increase the sedimentation rate inside the downstream basin. In the field, wooden fences contain two to three rows of vertical bamboo poles forming a frame to store horizontal brushwood, such as bamboo branches [1–3]. Previous studies reported that a 50% to 80% reduction of the incoming wave heights was found at field measurements [1–3] and from numerical studies [4]. Due to a lack of understanding of flow over wooden fences, Dao et al. (2020) [5] carried out

D. V. K. Huynh et al. (Eds.): VSOE 2021, LNCE 208, pp. 238–245, 2022.
https://doi.org/10.1007/978-981-16-7735-9_25

experiments to indicate the friction coefficients of wooden fences, i.e., the bulk drag coefficient and the Forchheimer coefficients. The bulk coefficient is the most crucial parameter that is applied in the numerical model.

The Mekong deltaic coast is characterized by a very gentle bathymetry with an average slope from 1/500 to 1/1500 [6, 7], creating a very healthy intertidal area for mangroves to develop up to 1500 m without the presence of sea dikes and/or no human interventions [8]. In reality, the mangrove width reduces by up to about 100 m, due to a rate of erosion of 10 to 50 [6]. The erosion also leads to a steeper slope; for example, the slope in Ganh Hao, Bac Lieu is about 1/200 [8]. In other locations, such as Nha Mat, Bac Lieu, the slope is about 1/500 [9]. Research on observing the presence of wooden fences in front of the remaining mangroves and the role of wooden fences in damping incoming waves on different slopes before reaching the mangroves has not yet been found in any publication.

In this study, the wave damping caused by wooden fences on different slopes and different locations is studied. The outline of this study is as follows. Section 1 is the introduction. The methodology and results are in Sects. 2 and 3, respectively. Finally, Sect. 4 is the conclusions.

2 Methodology

2.1 Bathymetry and Wave Conditions

The intertidal zone of the Mekong Delta is dominated by a tidal range of about 4.0 m with the MSL = 1.95 m, and MHL = 3.95 m [10, 11]. With the gentle slope varying from 1/200 to 1/1500, the intertidal width can be increased up to 1500 m, which provides an area where healthy mangroves are able to grow. In this study, two bottom slopes are used to represent the Mekong deltaic coast. The slopes of 1/200 and 1/500 are found at Nha Mat, Bac Lieu [9], and at Ganh Hao, Bac Lieu [8], respectively.

Wave characteristics are obtained from the study of Tas (2016), based on the wind data at Con Dao station [12] and wave-data from the NOAA wave model (National Oceanic and Atmospheric Administration, 2014). In Table 1, wave conditions at a water depth of 65 m (deep water) with a return period varying from 10 to 50 years are presented. The wave height and wind set-up are obtained from their marginal distributions, but the high correlation that can be expected provides a reasonable and (slightly) safe assumption. From an engineering perspective, the higher return period for a wave condition is normally chosen for a highly safe scenario, directly related to the strength and lifetime of the soft structure. In this study, the wave condition with a return period of 10 years, as given in Table 1, is chosen as wave boundary for the numerical model. The water level is taken as MHL and with a 10-year wind set-up, which amounts to assumed to be five years of this field structure. When designing for a wave condition with a 10-year return period, the probability that this situation will occur during the lifetime is 41%. Hence, this is a condition with a high probability of occurrence, such that the fence should be able to withstand this situation.

Table 1. Extreme wave conditions at a water depth of 65 m obtained at Con Dao and Bach Ho station [7].

Return period (y)	Wave period T_p (s)	Wave height H_s (m)	Wind speed U_{10} (m/s)	Wind set-up (m)
10	**9.4**	**5.1**	**19.7**	**0.6**
20	9.7	5.7	21.9	0.7
25	10.0	5.9	22.6	0.8
40	10.1	6.3	24.1	0.9
50	10.4	6.5	24.8	0.9

2.2 Model Description

The numerical models, SWAN and SWASH, are applied to simulate wave propagation and wave interaction with wooden fences on the Mekong deltaic coast. SWAN is a third-generation wave model and is applied for simulating wave parameters in coastal regions. This model is developed by Delft University of Technology based on the action balance equation [13, 14]. On the other hand, SWASH is a time-domain model applied for simulating non-hydrostatic and free-surface, based on non-linear shallow water equations [15].

Due to the lack of offshore bathymetry in the Mekong Delta, the numerical model profile is assumed to be useful from a distance from the coast about 100 km to a water depth of 50 m. In SWASH, the time-consuming computational requirements for such a long distance are very high, leading to an inconvenient computation of the entire domain. Thus, the combination of SWAN and SWASH is needed to calculate wave propagation from the offshore. Additionally, both the SWAN and SWASH models are validated and efficient to produce wave propagation from the offshore to the nearshore region [15].

The domain of both the SWAN and SWASH models is studied in one-dimensional mode (1D) for which one cross-section perpendicular to the coastline is defined in Fig. 1. The length of the computational domain depends on the bottom slope. In the SWAN model, the computation domain is from z = −45.0 m to z = +6.0 m (Fig. 1, bottom panel) with a total of 108 km long. In Fig. 1, the cross-shore profile in the SWAN model is separated into several bottom slopes with a cell grid of 10.0 m. A standard JONSWAP (JOint North Sea Wave Observation Project) shape spectrum with a wave height and peak period with a return period of 10 years is imposed at the left side of the profile at z = −45.0 m.

The computational domain of the SWASH model is much smaller, but has a much higher resolution than the SWAN model. The domain of SWASH is located from a bottom level, MSL-2 m (x = −2.4 km) to z = +5.0 m (x = +3.0 km) with the grid size of 0.25 m. From MSL (x = 0.0 km), the bottom slope of 1:500, as the base slope, is presented in Fig. 2. All scenarios are tested with a sea dike. The wave boundary is imposed at the offshore boundary, using the spectrum output from the SWAN model at x = −2.4 km (see Fig. 1, bottom panel) at a water depth of 3.95 m under the MHL. The wave characteristics are calculated from variance densities, including H_s = 1.45 m, T_p

= 9.4 s, and d = 3.95 m. For all simulations, one vertical layer is applied, the vertical turbulent mixing of viscosity is set as 3.10^{-4}, and the bottom friction as Manning's roughness is kept the same as in the SWAN model, as 0.038. The further application and boundary conditions are described in [15].

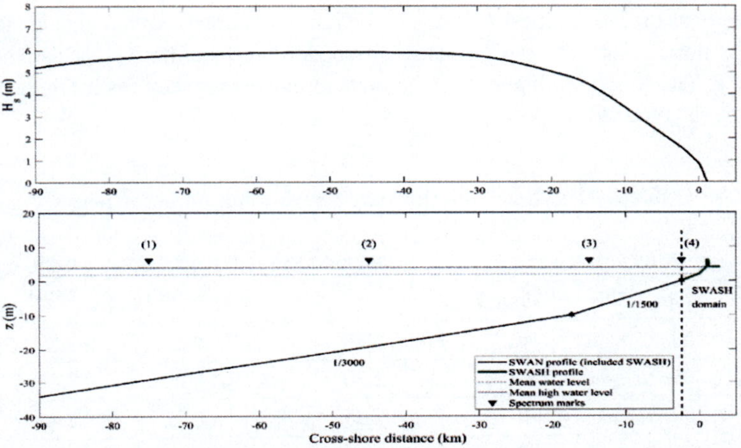

Fig. 1. Wave transformation (top panel) and the computational domain in the SWAN model (bottom panel). The SWASH profile (green line) is defined on the right side of the dash line in the bottom panel.

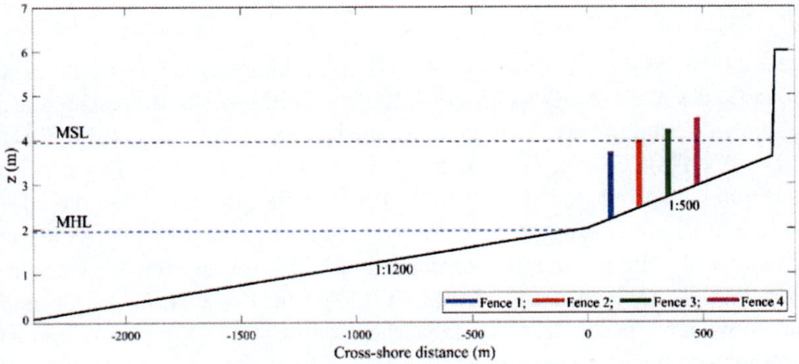

Fig. 2. Computational domain of SWASH in the base slope of 1/500. Wooden fences are located at x_f from 100 m to 500 m.

The vegetation model in SWASH (version 6.01) is applied to simulate wooden fences. The SWASH model was well-validated for the wave-fence interaction aspects [16]. In this model, wave reduction due to an array of stiff cylinders, including horizontal and vertical cylinders, is implemented [17]. For the fence, the brushwood is modelled by horizontal cylinders. The characteristics of a wooden fence are based on Dao et al. (2020) [5], with a branch density of N = 603 cylinders/m^2, a mean branch diameter of

D = 0.02 m, and a bulk drag coefficient of $\overline{C_D} = 2.0$. One fence height of 1.5 m is used for all simulations. One thickness, as B = 1.2 m, is tested at each location. In Fig. 2, several cross-shore locations of the wooden fence from the coastline (at MSL), x_f, are considered from $x_f = 100$ m to $x_f = 475$ m corresponding to water depth at the fence toe (d) that decrease from 1.75 m to 1.00 m. Moreover, the slope of 1:200 is chosen to compare the wave damping with the base slopes. The wooden fences are set at the same water depth as the base slopes. As a result, the fence locations are changed due to shorter profile of slope 1:200. The design cases for wooden fences of both slopes are presented in Table 2. The water depth at each location is remained constant resulting the constant freeboard (R_c) for two slopes.

Table 2. Design scenarios for wave damping due to wooden fences.

Cases	x_f (m)		Water depth d (m)	Freeboard R_C (m)	Bottom level z (m)
	Slope 1:500	Slope 1:200			
1	100	40	1.75	−0.25	2.20
2	225	90	1.50	+0.00	2.45
3	350	140	1.25	+0.25	2.70
4	475	190	1.00	+0.50	2.95

3 Results

The development of the significant wave height on the cross-shore profile in relation to wooden fences at all locations on the 1:200 and 1:500 slopes is presented in Fig. 3. Generally, the incoming wave is more dissipated on the 1:200 slope (Fig. 3b) than on the 1:500 slope (Fig. 3a). In Fig. 3a, the wave height (H_s) at x = 95 m for the first fence position is reduced from 1.03 m to 0.91 m, to 0.78 m, and to 0.66 m on the 1:500 slope for every 125 m. The highest reduction is found at the fence that is located further landwards, mainly due to the effect of depth-induced breaking. For the steeper slopes (1:200) in Fig. 3b, H_s is dissipated by about 0.12 m for every 50 m, corresponding to locations of wooden fences. It is noted that the dissipation of H_s on the steeper slope is also faster than on the gentler slope. For example, x_f at x = 95 m reduces to 0.90 m after 135 m for the 1:200 slope (Fig. 3b), while it needs about 300 m to have a similar value for H_s for the 1:500 slope (Fig. 3a). Due to the fast dissipation of incoming waves, the wave damping at all locations is also different for both slopes. The higher dissipation before contacting the fence, the so-call pre-dissipation, leads to lower transmission waves for all slopes. For the 1:500 slope in Fig. 3a, the transmitted wave heights (H_t) are 0.44 m (x = 110, blue line), 0.38 m (x = 265 m, red line), 0.31 m (x = 360 m, green line) and 0.26 m (x = 485 m, purple line). The H_t for the 1:200 slope are slightly higher corresponding to four locations (Fig. 3b). Interestingly, the bottom slope significantly impacts the pre-dissipation rather than the transmitted wave heights. It is due to H_t being much smaller than the water depth leading to the lower dispersion rate of transmission waves. The results indicate that the closer to wooden fences, the higher the pre-dissipation of waves.

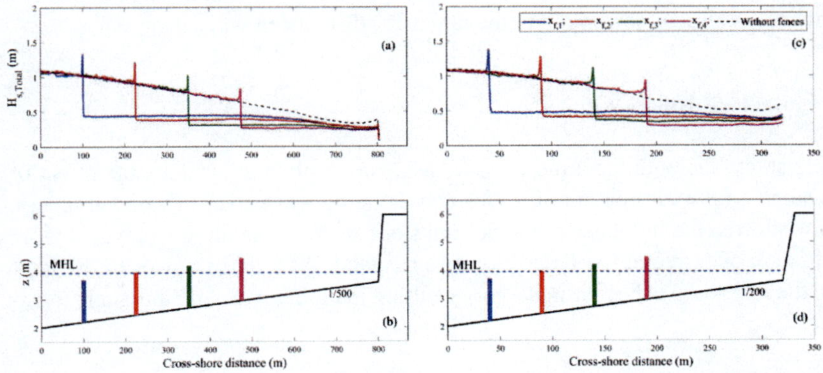

Fig. 3. Wave evolution over wooden fences on the 1:200 and 1:500 slopes at all locations. Wave dissipation without fences (dashed line) is indicated as a reference. Wooden fences at three locations are indicated as the same color as wave heights in the bottom panel.

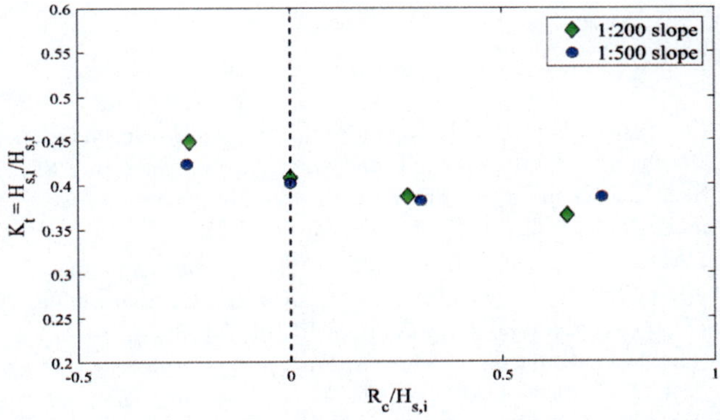

Fig. 4. The relationship between the transmission coefficient (K_t) and relative freeboard (R_c) at two bottom slopes.

As wooden fences move toward the landward side, the fences have a larger freeboard, R_c, due to the bottom slopes (Fig. 2). The increase of freeboard leads to an increase in the wave damping because less wave overtopping occurs. In Fig. 4, the relationship between the transmission coefficient (K_t) and the relative freeboard (R_c) is presented for every fence location and all slopes. As can be seen, the K_t value decreases with the increase of R_c for all slopes which reaches the lowest value, 0.37 (1:200 slope, green diamond) and 0.39 (1:500 slope, blue circle), at R_c of 0.65 and 0.73, respectively. In detail, about 60% wave height can be damped by wooden fences at two last locations, $x_f = 140$ m and 190 m for the steeper slope (Fig. 3c), and $x_f = 350$ m and 475 m for the gentler slope (Fig. 3d). Figure 4 also points out that the transmission wave on the steeper slope (green diamond) is higher when fences are submerged, but it is similar to the gentler slope (blue circle) under emerged conditions. The results suggest that for the fence of a constant height, in this case, the steeper a slope, the larger the wave damping. This adds to the fact that wave height, at the seaward side of the fences that are placed more landward, is already lower due to the larger depth-induced wave damping.

4 Conclusions

In this study, the numerical model, SWASH, was applied to evaluate the effect of bottom slope on pre-dissipation of waves and damping waves due to wooden fences. The results showed the influence of the bottom slope in pre-dissipation when the higher pre-dissipation occurred at the fence closer to the land. As a fence moves to the landward side, the freeboard is becoming larger, resulting in higher wave damping due to wooden fences.

Acknowledgements. The authors gratefully acknowledge the support of the National University of Civil Engineering, Vietnam within the funded project 36-2019/KHXD-TĐ, Delft University of Technology and Hanoi University of Natural Resources and Environments in Vietnam support this research.

References

1. Van Cuong, C., Brown, S., To, H.H., Hockings, M.: Using Melaleuca fences as soft coastal engineering for mangrove restoration in Kien Giang, Vietnam. Ecol. Eng. **81**, 256–265 (2015)
2. Albers, T., San, D.C., Schmitt, K.: Shoreline management guidelines: coastal protection in the lower Mekong delta, vol. 1, pp. 1–124 (2013)
3. Schmitt, K., Albers, T., Pham, T.T., Dinh, S.C.: Site-specific and integrated adaptation to climate change in the coastal mangrove zone of Soc Trang Province, Viet Nam. J. Coast. Conserv. **17**(3), 545–558 (2013). https://doi.org/10.1007/s11852-013-0253-4
4. Dao, T., Stive, M.J.F., Hofland, B., Mai, T.: Wave damping due to wooden fences along mangrove coasts. J. Coast. Res. **34**(6), 1317–1327 (2018). https://doi.org/10.2112/JCOASTRES-D-18-00015.1
5. Dao, H.T., Hofland, B., Stive, M.J.F., Mai, T.: Experimental assessment of the flow resistance of coastal wooden fences. Water **12**(7) (2020). https://doi.org/10.3390/w12071910
6. Phan, L.K., van Thiel de Vries, J.S.M., Stive, M.J.F.: Coastal mangrove squeeze in the Mekong delta. J. Coast. Res. 233–243 (2014). https://doi.org/10.2112/JCOASTRES-D-14-00049.1
7. Tas, S.: Coastal protection in the Mekong Delta: wave load and overtopping of sea dikes as function of their location in the cross-section, for different foreshore geometries. Delft University of Technology (2016)
8. Phan, L.K.: Wave attenuation in coastal mangroves: mangrove squeeze in the Mekong delta. Delft University of Technology (2019)
9. Thieu Quang, T., Mai Trong, L.: Monsoon wave transmission at bamboo fences protecting mangroves in the lower mekong delta. Appl. Ocean Res. **101**, 102259 (2020). https://doi.org/10.1016/j.apor.2020.102259
10. Gagliano, S.M., McIntire, W.G.: Reports on the Mekong river delta. Louisiana State Univ Baton Rouge Coastal Studies Inst (1968)

11. Wolanski, E., Huan, N.N., Nhan, N.H., Thuy, N.N.: Fine-sediment dynamics in the Mekong River estuary, Vietnam. Estuar. Coast. Shelf Sci. **43**(5), 565–582 (1996)
12. Tran, V.L., Nguyen, D.B., Nguyen, V.T.: On the wind pressure zone map of the Vietnam territory. In: Workshop on Regional Harmonization of Wind Loading and Wind environmental Specifications in Asia-Pacific Economies (APEC-WW), pp. 1–7 (2004)
13. Booij, N., Holthuijsen, L.H., Ris, R.C.: The "SWAN" wave model for shallow water. In: Coastal Engineering 1996, pp. 668–676 (1997)
14. Booij, N., Ris, R.C., Holthuijsen, L.H.: A third-generation wave model for coastal regions: 1. model description and validation. J. Geophys. Res. Ocean. **104**(C4), 7649–7666 (1999)
15. Zijlema, M., Stelling, G., Smit, P.: SWASH: An operational public domain code for simulating wave fields and rapidly varied flows in coastal waters. Coast. Eng. **58**(10), 992–1012 (2011)
16. Dao, H.T., Hofland, B., Suzuki, T., Stive, M.J.F., Mai, T., Tuan, L.X.: Numerical and small scale physical modelling of wave transmission by wooden fences. J. Coast. Hydraul. Struct. **1**(4), 1–21 (2021). https://doi.org/10.48438/jchs.2021.0004
17. Suzuki, T., Hu, Z., Kumada, K., Phan, L.K., Zijlema, M.: Non-hydrostatic modeling of drag, inertia and porous effects in wave propagation over dense vegetation fields. Coast. Eng. **149**, 49–64 (2019)

Design of Vibro-Driven Large Diameter Monopiles

Fabian Kirsch[1], Frank Rackwitz[2], Viet Hung Le[2], and Kostas Grivas[1](✉)

[1] GuD Geotechnik und Dynamik Consult GmbH, Darwinstr. 13, 10589 Berlin, Germany
grivas@gudconsult.de

[2] Soil Mechanics and Geotechnical Engineering Division, Technical University of Berlin, Gustav-Meyer-Allee 25, 13355 Berlin, Germany

https://gudconsult.de/, https://www.grundbau.tu-berlin.de/

Abstract. A new trend in Offshore Wind, mainly for environmental reasons (reduced noise emissions), is the installation of piles in predominantly sandy soils with the vibratory method rather than with the traditional impact method. An innovative design methodology for vibro-driven open-ended steel large diameter monopiles is derived and already applied in a North Sea project. This methodology is based on the results of a large test site at Altenwalde (Germany), where the different behavior of fully vibrated monopiles compared to impact driven piles under lateral loading was investigated, and on model scale tests undertaken by TU Berlin, where dependencies incorporating the influence of installation parameters during vibratory driving and their influence on the pile behaviour due to horizontal loading of the pile after installation were determined.

Keywords: Vibratory driving · Vibro-driven monopile · Design monopile · Installation effect

1 Introduction

Large diameter monopiles is a widely used foundation solution for offshore wind turbines and platforms. The main loading situation of these usually open ended steel pipe piles is a cyclic lateral loading. Especially in sandy soils vibratory pile driving is a fast and economic method for the pile installation which has the additional benefit of being considerably more silent that the classical installation of the steel pile by impact driving. As current design methods generally consider impact driven piles, it is important to incorporate the vibratory installation effects into the design method.

During the last years, a number of research activities has been undertaken in order to better understand the effect of the vibratory installation on the lateral bearing behavior of a large diameter monopile, e.g. [1]. Based on the results of in situ tests and scaled model tests, as well as numerical simulations, an analytical procedure is conceptually described in order to take into account the installation effect in the pile design.

D. V. K. Huynh et al. (Eds.): VSOE 2021, LNCE 208, pp. 246–253, 2022.
https://doi.org/10.1007/978-981-16-7735-9_26

2 Experimental Investigation

2.1 Small Scale Model Tests

In this study, small scale model tests are conducted to investigate the cyclic and monotonic behavior of the monopile independent on the installation methods. The tests were performed in a steel container with a base dimensions of 1.70 m by 0.90 m and a height of 1.15 m. On the bottom of the container, a drainage layer of 0.25 m was placed (Fig. 1). The sand was prepared using dry pluviation technique. After the filling, the sand placed in the container was saturated from the bottom through the drainage layer. The model pile had an outer diameter of 20 cm and a wall thickness of 4 mm. Penetration of the pile up to 87 cm was tested. A system of guide rollers above the container ensured vertical penetration of the pile along the glass panel and restricted any tilt. In these experimental model tests, images of the soil movement were continuously recorded and these were then analysed using the method of Digital Image Correlation (DIC) [2]. This was possible by using only a half model, meaning that the test container only represented a halved version of a full container. The pile itself was semi-circular and installed along the glass panel which represented the cut edge of the container. Both pile walls were therefore visible through the glass (Fig. 1). This allowed a direct and continuous visualisation of soil movements during installation and loading without disturbing the soil itself through sensors or soil sampling. More details of the test set up are presented by Labenski et al. [3].

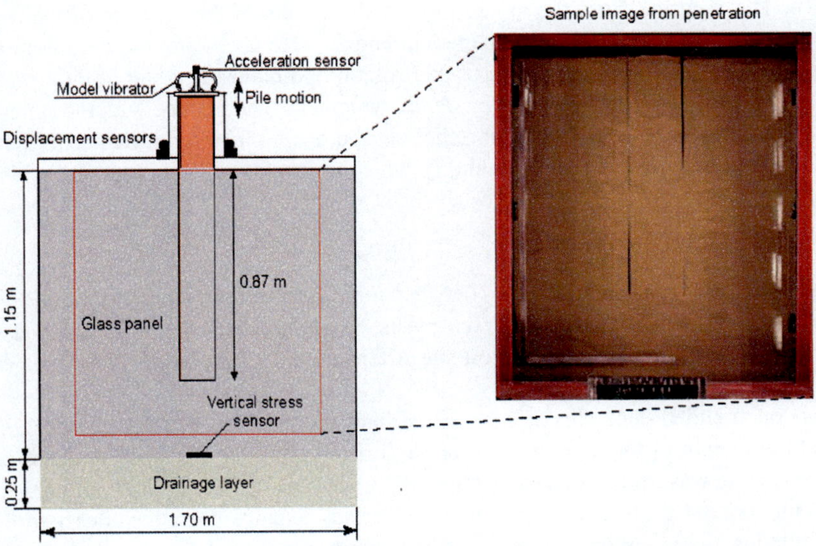

Fig. 1. Schematic view of the test set up with sample image of the glass panel [3]

Figure 2 left shows the deformation behavior of a vibratory pile under static horizontal loading. The lateral load was continuously increased. Unloading–reloading cycles were carried out at lateral loads of 0.9 kN and 1.8 kN. To consider the capacity of the

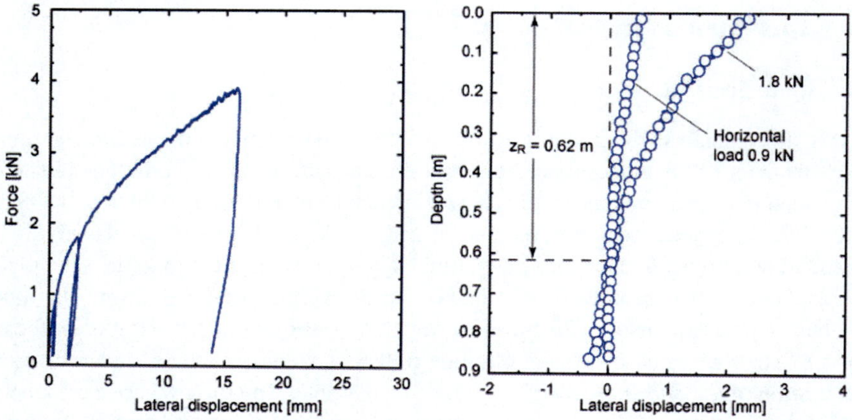

Fig. 2. Lateral load-displacement curves and deflection curves of the vibratory driving pile by dense sand ($I_D = 75\%$)

full-pile, the lateral load of the half-pile test was multiplied by a factor of 2 in the figure. This is valid due to the symmetry conditions of the half-pile model. It was observed that only relatively small lateral deformations occurred. At a lateral load of 1 kN, the lateral displacement equals a value of approximately 1 mm, corresponding to 0.005 D. A lateral displacement of 0.1 D was reached at the maximum lateral load of 4.5 kN.

In Fig. 2 right, the lateral pile-bending curves of two different lateral load levels are shown. These curves from the model test are the results of the analysis of the digital image correlation. Images of the visible cut edges, were taken and the deformation of the pile itself quantified. The depth of pile rotation point, z_R, is equal to 0.62 m. This corresponds to 71% of the penetration depth, which is in accordance with the statement of Klinkvort and Hededal [4], stating that z_R approaches 70% with increasing lateral load, and the investigations of Li et al. [5], who obtained a range of $z_R = 73$ to 85%.

2.2 Large Scale Model Tests

In order to investigate the differences between impact-driven and vibro-driven piles with respect to their lateral capacity, large scale tests were conducted in predominantly dense, non-cohesive, saturated soil in the test site Altenwalde, located on the North Sea coast of Germany near Cuxhaven [6, 7].

Six open-ended steel piles of 4.3 m outer diameter and 21 m length were tested. The penetration depth of the piles reached approximately 18.5 m. The piles were tested in pairs; one pile was impact-driven and the other one vibro-driven.

A number of CPT tests were carried out at various distances from the piles before and after pile installation in order to measure the changes in the relative density of the soil due to installation. Moreover, load-displacement curves of the pile head during testing were recorded and evaluated.

The comparison of the various test results showed that under lateral loading the vibro-driven piles behaved weaker than the impact-driven ones with deflections between 24% and 58% greater, depending on the load magnitude.

A general conclusion of these large scale tests is that the design of vibro-driven monopiles in sand would require the application of reduction factors regarding stiffness. If, however, the installation process could be scheduled in a way that would lead to much less loosening of the surrounding soil, then a similar behavior to impact-driven piles could be expected.

3 Numerical Investigation

3.1 Installation Process

The design approach for vibro-driven monopile required extensive knowledge about the influence of pile penetration on the soil conditions. Additional to the various experimental investigations, numerical simulations were performed. On one side, with the numerical models, different parameters of the soil condition which are not able to be measured in the experimental model can be quantified (such as stress condition). On the other side, the experimental model tests are expensive and time consuming. In the study, only a limit number of tests could be realized. With the numerical models, effects of different parameters could be investigated in details.

The penetration of pile is one of the most complex processes from a soil mechanic point of view. The simulation of this process requires advanced calculation techniques which are able to handle large deformations of the soil due to penetration of the pile. At the same time, sophisticated constitutive model to describe the soil behaviour is necessary. In this study, the pile penetration process using vibratory and impact method was modelled. The theoretical backgrounds and details regarding numerical algorithms and procedures applied by MMALE technique were published in previous works by the research group at TU Berlin [8, 9]. The results used in the study and further details about the boundary conditions of the simulations were presented by Daryaei et al. [10].

3.2 Monopile Under Lateral Loading

For the investigation of the deformation behavior of the monopile under cyclic loading with consideration of the installation effects, different numerical approaches were developed. In the first study, a simple model with a loosening zone directly at the pile shaft and a compaction zone around was applied [11]. Using the results of the experimental model tests, the relative densities in these zones could be estimated. However, in this approach, only the change in density was considered. The influence of the pile penetration on the stress condition was ignored. To overcome this disadvantage, the simulation of the installation process was performed (Sect. 3.1). In a second calculation with a new model, the initial condition of the soil (distribution of the void ratio and stress) was adapted from the previous simulation. This numerical approach is a combination of two separate calculations, where both parameters density and stress condition were taken into account. The calculation concept with these two calculations and their boundary conditions, as well as several results are presented by Le & Rackwitz [12] in detail.

4 Design Approach

The design of a fully vibrated monopile in sandy soil requires the consideration of the installation effects in order to predict the loading behavior. Results of numerical simulations can be utilized to assess the installation effects and to develop and validate the design approach. As described above, numerical simulations are performed to simulate the installation and loading process of a large diameter monopile in order to determine the behavior of the pile under lateral load. Numerical simulations can generally be performed as an independent design approach but, since the computations require considerable effort, they are not suitable for a location specific design at each wind turbine position. Therefore, the here described analytical approach gives the opportunity to provide a relative quick statement of the loading behavior of a fully vibrated large diameter open end steel pipe pile based upon the site conditions, as well as installation and pile parameters.

Since the commonly used lateral soil support in terms of so-called p-y-curves was determined for impact driven piles, these approaches inherently encounter for the installation effect of an impact driving process. One of the most important parameters besides the soil type and the strength of the soil is the stress state, which plays a significant role for the stiffness of the lateral support. The vibratory and the impact driven installation methods are compared on the basis of radial stresses acting on the pile surface and in the vicinity of the pile after the installation. These stresses are determined in numerical simulations and are further utilized to alter the classical p-y-response in order to derive the behavior of a vibrated pile under lateral load.

Numerical simulations allow the determination of radial stresses around a pile installed by vibration or by impact driving. The resulting stresses are obtained at different radial distances to the pile axis. These stresses are put in relation to each other by the calculation of the ratio $\sigma_{vib}/\sigma_{impact}$ (radial stresses after installation) for each radial distance. In order to facilitate the interpretation of the results, the pile is subdivided into a specific number of segments (e.g. 2 m). Subsequently, the ratios $\sigma_{vib}/\sigma_{impact}$ that belong to one segment are averaged, so each segment has one representative value. At this point, the ratios $\sigma_{vib}/\sigma_{impact}$ are available for each segment and at different radial distances to the pile axis. These results represent the influence of the installation by vibration with respect to an impact driven pile for different radial distances and depths. Figure 3 below shows for example the vibration ratios of a 4 m diameter pile with an embedment length of 16 m, for two depths and different distances to the pile axis from a numerical simulation.

Considering that a p-y curve represents the response of the soil at any particular depth, it has to correctly represent the simultaneous response of the soil located at all different radial distances at that same depth. Since the soil closer to the pile is expected to have a greater impact on the lateral response than the soil located further away, a weighting function that decreases with radial distance is required. In order to obtain reasonable weighting functions, a finite element model is used. In this model, a horizontal displacement is applied at the mudline of the pile under consideration and deformation curves are obtained at each segment depths and radial distance. The results of these curves are normalized with the maximum deformation obtained at the outer pile diameter. The

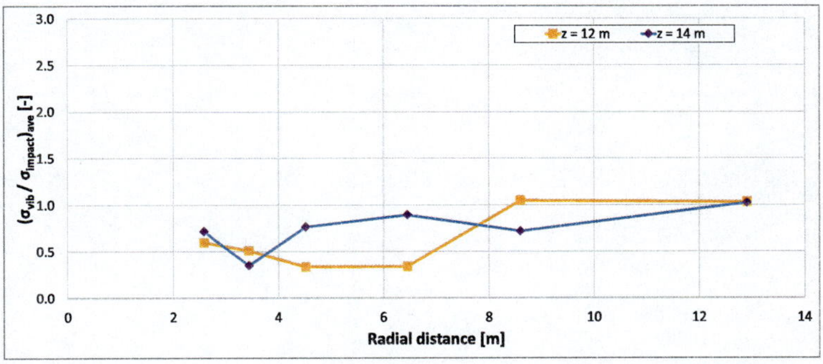

Fig. 3. Vibration ratios of a 4 m diameter pile from numerical simulations

following Fig. 4 shows an example of a weighting function at a specific depth derived from the 3D finite element model of a 4 m diameter monopile in sandy soil.

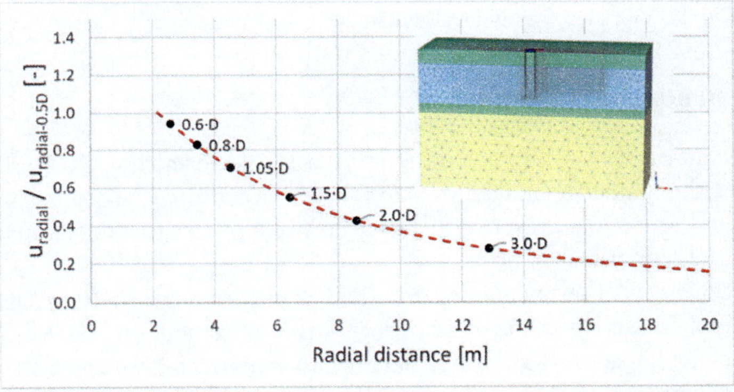

Fig. 4. Example of a weighting function derived from a 3D finite element model

These curves are used as the functions to compute a weighted average of the results of the $\sigma_{vib}/\sigma_{impact}$ ratios at different distances to the pile. By doing so, a vibration factor (VF) is obtained for each depth considered along the pile. The vibration factors represent the influence of the installation by vibration at each depth considered (with respect to a pile installed by impact driving).

For each depth considered in the analyses, (p-y curves)$_{impact}$ are determined according to general offshore practice, possibly with modifications for small strains and large diameters [13]. The (p-y curves)$_{vib}$ are determined by the same procedure, only that the general stress state is adjusted by determining an equivalent depth for each curve in a manner that the depth is multiplied by the corresponding vibration factor. The following Fig. 5 shows exemplary the vibration factors obtained by the prescribed procedure for a 4 m diameter monopile with 18 m embedment length.

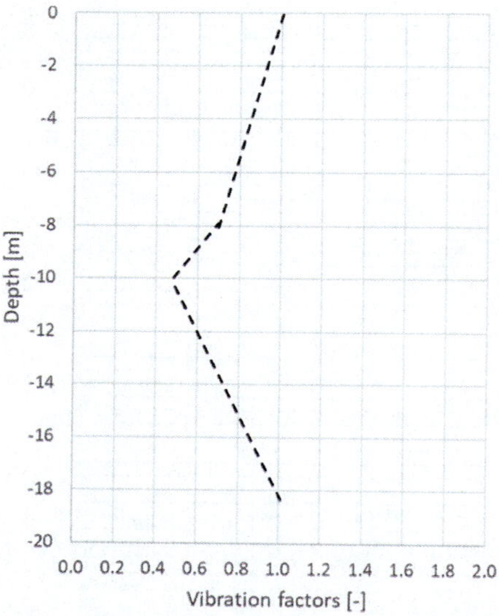

Fig. 5. Vibration factors of a 4m diameter pile from numerical simulations

5 Conclusions

Small scale model tests of vibro-driven piles in sand performed by TU Berlin, showed that only relatively small lateral deformations occur. In addition, the depth of pile rotation point, z_R, was found to correspond to 71% of the pile penetration depth, in accordance with literature references.

Large scale model tests conducted in the test site Altenwalde (Germany) showed that, depending on the load magnitude, vibro-driven piles under lateral loading behave weaker than the impact-driven ones with deflections between 24% and 58% greater.

A design approach for vibro-driven piles in sand under lateral loading is proposed here based on modifications by the so-called vibration factors of the p-y curves determined according to standard offshore practice for impact driven piles. Mathematical functions derived from 3D finite element models with site specific layers are used to compute weighted averages of the results of the $\sigma_{vib}/\sigma_{impact}$ ratios at different distances to the pile, in order to derive such appropriate vibration factors for each depth along the pile.

References

1. Sychla, H., Fischer, J., Stahlmann, J.: Stahlpfähle - vibriert oder gerammt? Stahl im Wasserbau 2013, Fachseminar am 26–27.09.2013 in Braunschweig, Mitteilung des Instituts für Grundbau und Bodenmechanik. Technische Universität Braunschweig, Heft 97, pp. 143–161 (2013)
2. White, D.J.: An investigation into the behaviour of pressed-in piles. Ph.D. thesis, University of Cambridge, UK (2002)
3. Labenski, J., Remspecher, F., Le, V.H., Moormann, C., Rackwitz, F.: Lateral-bearing behaviour of vibratory driven monopiles in different model test set-ups. Int. J. Phys. Model. Geotech., 07 August 2019. https://doi.org/10.1680/jphmg.18.00090

4. Klinkvort, R.T., Hededal, O.: Effect of load eccentricity and stress level on monopile support for offshore wind turbines. Can. Geotech. J. **51**(9), 966–974 (2014)
5. Li, W., Zhu, B., Yang, M.: Static response of monopile to lateral load in over consolidated dense sand. J. Geotech. Geoenviron. Eng. **143**(7), 04017026 (2017)
6. Achmus, M., Schmoor, K.A., Herwig, V., Matlock, B.: Lateral bearing behaviour of vibro- and impact-driven large-diameter piles in dense sand. Geotechnik 43, Heft 3, Berlin (2020)
7. Moormann, C., Kirsch, F., Herwig, V.: Vergleich des axialen und lateralen Tragverhaltens von vibrierten und gerammten Stahlrohrpfählen. 34. Baugrundtagung, Bielefeld, 15 September 2016
8. Bakroon, M., Daryaei, R., Aubram, D., Rackwitz, F.: Investigation of mesh improvement in multimaterial ALE formulations using geotechnical benchmark problems. Int. J. Geomech. **20**(8), 04020114 (2020)
9. Bakroon, M., Daryaei, R., Aubram, D., Rackwitz, F.: Numerical evaluation of buckling in steel pipe piles during vibratory installation. Soil Dyn. Earthq. Eng. **122**, 327–336 (2019)
10. Daryaei, R., Bakroon, M., Aubram, D., Rackwitz, F.: Numerical evaluation of the soil behavior during pipe-pile installation using impact and vibratory driving in sand. Soil Dyn. Earthq. Eng. **134**, 106177 (2020)
11. Le, V.H., Remspecher, F., Rackwitz, F.: Numerical investigation of installation effects on the cyclic behaviour of monopile foundation under horizontal loading. In: Vietnam Symposium on Advances in Offshore Engineering, Hanoi, Vietnam (2018)
12. Le, V.H., Rackwitz, F.: Investigation of the long-term cyclic behaviour of monopile foundation by impact and vibratory installation. In: Vietnam Symposium on Advances in Offshore Engineering, Ho Chi Minh City, Vietnam (2021)
13. Kirsch, F., Richter, Th., Coronel, M.: Geotechnische Aspekte bei der Gründungsbemessung von Offshore-Windenergieanlagen auf Monoplfählen mit sehr großen Durchmessern. Stahlbau Spezial 2014 – Erneuerbare Energien. Ernst & Sohn (2014)

Evaluation of the Wave Damping by a Bamboo Breakwater Applying a Porous Media Model

Nghi Van Vu(✉) and Kien Thanh Trinh

HoChiMinh City University of Transport, 2 Vo Oanh Street,
Binh Thanh District, Ho Chi Minh City, Vietnam
nghi.vu@ut.edu.vn, ttkien@hcmutrans.edu.vn

Abstract. This research uses a numerical model for waves in porous media to evaluate the damped wave effectiveness of bamboo breakwater built in the coastal zone in Mekong delta. The extended Boussinesq model for waves in porous media used in this research was developed by Vu et al. (2018). Internal wave generation technique using a source term addition method is applied to generate waves. The porosity characteristics of the model are matched with the bamboo breakwater from Mai et al. (2019). The numerical results show a very good agreement with the results of Mai et al. (2019) and imply that the effectiveness of the wooden breakwater in dissipating waves is affected by the width and the porosity of the bamboo breakwater and the nonlinearity of the incoming waves.

Keywords: Porous media · Bamboo breakwater · Mekong delta · Wave dissipation · Numerical solution

1 Introduction

Various types of porous breakwaters are being built as pilot projects in coastal zones in Mekong delta in Vietnam (Nguyet-Minh et al. 2019). These porous breakwaters are placed at a distance from 150 m to 200 m away from sea dikes system to dissipate waves energy or reduce waves attacking on sea dikes directly and to trap sediment behind the breakwaters. After being sedimented, the area between the breakwaters and the sea dikes will be covered with mangrove vegetation (Halide et al. 2004; Schmitt et al. 2013; Nguyet-Minh et al. 2019). There are two types of breakwaters being built in Mekong delta: hollow breakwaters and porous breakwaters. The former dissipates the wave energy when waves pass through some hollows fabricated at front and back faces of the breakwater. The empty space inside the breakwater would also help to damp the wave energy. Sizes of the holes are different from each other and depend on specific purposes, e.g., wave heights behind breakwaters, sedimentation (Fig. 1a). A pilot project of this hollow breakwater type was built in Ca Mau province with 180 m in length (Tran et al. 2019). These projects show the efficiency of hollow breakwater in attenuating wave energy. The latter is a porous breakwater. There are two types of porous breakwater being built in Mekong Delta coastal area: reinforced concrete piles filled with rock and bamboo breakwater (Fig. 1b). The reinforced concrete piles filling

D. V. K. Huynh et al. (Eds.): VSOE 2021, LNCE 208, pp. 254–261, 2022.
https://doi.org/10.1007/978-981-16-7735-9_27

with rocks breakwater named CM porous breakwater showed its effect on reducing wave energy and alluvial accumulation (Nguyen and Nguyen 2018; Nguyen and Vu 2019). The bamboo breakwater type or wooden breakwater type was constructed in Thailand (Phansri 2011) and in a pilot project at Bac Lieu province, Vietnam (Schmitt et al. 2013). This is an environment-friendly and effective breakwater in protecting land for mid-slope coastal profile.

Fig. 1. Popular breakwaters built in coastal zone in Mekong delta area: a) hollow breakwater (Tran et al. 2019); b) porous breakwater (Schmitt et al. 2013).

The effectiveness of bamboo breakwater in dissipating wave energy has been studied in various ways such as pilot projects, field experiments, physical experiments and numerical simulations. Halide et al. (2004) evaluated the degree of wave attenuation for several bamboo densities and diameters. In their research, bamboo poles were pushed into the sediment offshore from the coastal mangrove plant. They concluded that the bamboo barrier was an economical method in reducing waves. However, they did not mention the impact of the breakwater porosity.

Dao et al. (2018) and Mai et al. (2019) applied SWASH model to simulate waves propagating through a bamboo breakwater with various conditions of breakwater widths, porosities and incoming waves characteristics. Their numerical results evaluated the effectiveness of bamboo breakwater in damping waves when increasing breakwater width and decreasing the porosity of the breakwater. However, their numerical results did not consider the affection of the nonlinearity of waves on wave transmission coefficients.

Albers et al. (2013), Schmitt et al. (2013), Thieu and Mai (2020) show pilot project results and numerical simulations of waves propagating through bamboo breakwaters, Fig. 2. The field experiment data from these studies show the primary effects of the crest height of the breakwater. Thieu and Mai (2020) applied the IH2-VOF model combined with a volume of fluid method to simulate waves and to track the surface elevation. Turbulence effects were modeled by means of a k-ε model. The numerical results show the influences of the fence parameters (i.e. width, freeboard and porosity) on wave attenuation.

In this research, the extended Boussinesq model for waves in porous media of Vu et al. (2018) is applied to evaluate the damped waves effectiveness of bamboo breakwater due to the width of the breakwater and the porosities of the breakwater also. The research results are well compared with the results from Mai et al. (2019).

The first Section of the paper gives some introduction about some types of breakwater being built in the south west coastal zone of Vietnam, especially the bamboo breakwater.

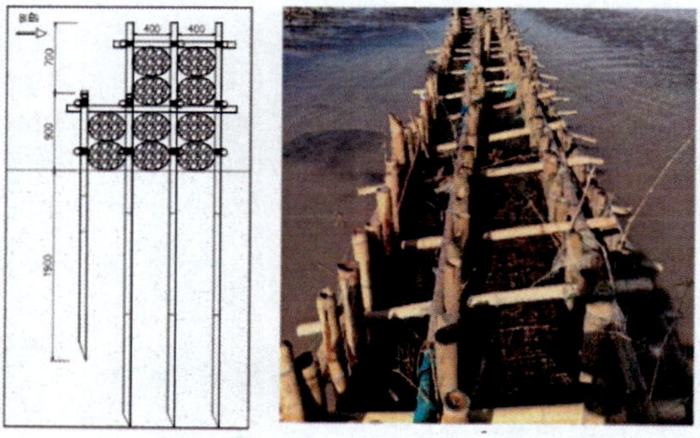

a) b)

Fig. 2. a) Designed cross section of bamboo breakwater; b) bamboo breakwater in reality. Mai and Ngo (2018).

The second Section of the paper introduces the governing equations used in this research. The numerical results from this research are analyzed in Sect. 3. The final Section shows some concluding remarks.

2 Governing Equations for Waves in Porous Media

In this research, the bamboo breakwater is considered as a porous media and is not submerged. Thus, the extended Boussinesq equations for waves in porous media of Vu et al. (2018) is applied. The continuity and momentum equations are given by

$$\frac{\partial \eta}{\partial t} + \nabla \cdot [(h+\eta)\,\mathrm{u}] = 0 \tag{1}$$

$$\begin{aligned}&\left(\beta\frac{\partial}{\partial t}+\alpha\right)\mathrm{u} + g\nabla\eta + \beta\,\mathrm{u}\cdot\nabla\mathrm{u} + \frac{1}{6}\left(\beta\frac{\partial}{\partial t}+\alpha\right)h^2\nabla(\nabla\cdot\mathrm{u}) - \left(\frac{1}{2}+\gamma\right)\times\\ &\left(\beta\frac{\partial}{\partial t}+\alpha\right)h\nabla[\nabla\cdot(h\mathrm{u})] - \gamma\, gh\nabla[\nabla\cdot(h\nabla\eta)] = 0\end{aligned} \tag{2}$$

where $\nabla \equiv \left(\partial/\partial x,\ \partial/\partial y\right)$ is the gradient operator; $\mathrm{u} = (u, v)$ is the depth integrated horizontal velocity; h is the water depth; η is the water surface elevation; γ is the tunning parameter of the extended Boussinesq model, $\gamma = 1/18$; λ is the porosity; α and β are the inertial and drag coefficients, respectively. Porosity is the ratio of the void volume over the total volume. It ranges from 0 to unity, the value of 0 for rigid media and unity for clean water media. The inertial and drag coefficients are given by

$$\alpha = \alpha_l\left(\frac{1-\lambda}{\lambda}\right)^2\frac{\nu}{d^2} + \alpha_t\frac{1-\lambda}{\lambda}\frac{1}{d}|\mathrm{u}| \tag{3}$$

$$\beta = 1 + (1-\lambda)\kappa \tag{4}$$

where α_l are α_t laminar and turbulent drag coefficients, respectively; ν is the dynamic viscosity; d is the gravel diameter and κ is the added mass coefficient.

3 Waves Propagating Through a Bamboo Breakwater

3.1 Model Setup

Mai et al. (2019) applies a non-hydrostatic wave-flow model called SWASH to simulate waves propagating through a bamboo breakwater. It should be noted that SWASH is not a Boussinesq-type wave model (Zijlema et al. 2011). The computational domain can be divided into a fixed number of vertical terrain-following layers which would be complicated. In this work, an extended Boussinesq model is used to compare with the results from Mai et al. (2019). A source term addition method is applied to generate waves in 1 dimensional domain (Fig. 3). The length of the computational domain is 45 m. The bamboo breakwater is placed at 20 m. There are two wave gauges placed in front of and behind the bamboo breakwater with G1 at 19.5 m and G2 at 22.5 m. Two sponge layers are fixed at two ends of the computational domain to avoid re-reflection effects back to the computational domain. The sponge layer is specified by a damping coefficient

$$D_s = \omega \frac{\exp\left[(d/Sp)^{2.5}\right] - 1}{\exp(1) - 1} \tag{5}$$

where ω is the wave angular frequency, d is the distance from the starting point of the sponge layer, Sp is the width of the sponge layer. In this research, Sp is determined as 4 times of a wave length (Vu et al. 2015).

Waves with the height of $H = 0.3$ m are generated in a constant water depth $h = 0.6$ m. The horizontal step size is chosen with $dx = 0.02$ m ($L/dx = 100$, L is the wave length), the time step is chosen in order to make the model stable with a Courant number $Cr < 0.2$. The wave source is placed on the left side of the computational domain.

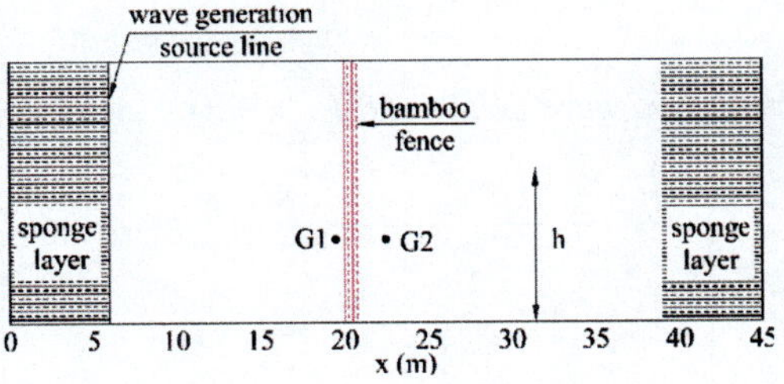

Fig. 3. One-dimensional computational domain.

The widths of the bamboo breakwater are fixed with the research of Mai et al. (2019) as 0.4 m and 0.8 m. The porosity characteristics of the Boussinesq model are matched with SWASH as $\alpha_l = 1100$, $\alpha_t = 1.5$, $\kappa = 0.4$, $\lambda = 0.5,\ 0.9$, $\nu = 1.004 \times 10^{-6}\ \text{m}^2/\text{s}$. The widths and porosities of the bamboo breakwater are combined in four cases which are given in the following Table 1.

Table 1. Cases of simulation.

Cases	Breakwater porosity λ	Breakwater width B_f, m
1	0.5	0.4
2	0.5	0.8
3	0.9	0.4
4	0.9	0.8

3.2 Influence of Porosity and Bamboo Breakwater Width on Transmission Coefficient

Figure 4 shows the comparison between the numerical results of the present work and the numerical results using SWASH of Mai et al. (2019) for 4 cases. The present results are almost the same with Mai et al. (2019)'s results. Figure 4 illustrates the influence of the relative width (B_f/L) and the porosity (λ) of the bamboo breakwater on the transmission coefficient which is given in Eq. (6). When the porosity decreases and the breakwater's width increases, the transmission coefficient decreases. The numerical results present that the influence of the porosity on the transmission coefficient is dominant. When the porosity keeps small, i.e. $\lambda = 0.5$, the breakwater width increases from 0.4 m to 0.8 m, the transmission coefficient reduces largely from 0.9 to 0.55. However, with a greater value of porosity, i.e. $\lambda = 0.9$, when the breakwater width increases with the same amount as the previous case, the transmission coefficient does not reduce much. For field cases, small porosity is conducted by putting bundles of small-size branches inside bamboo breakwaters.

The transmission coefficient K_t in conventional form is given by

$$K_t = \frac{H_t}{H_i} \tag{6}$$

where H_t is the transmission wave height and H_i is the incident wave height.

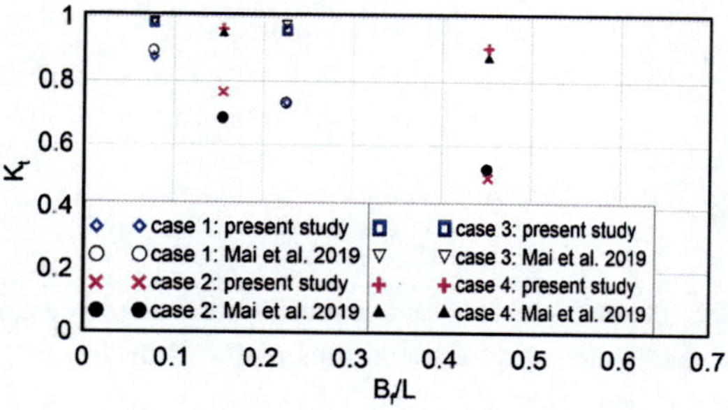

Fig. 4. Comparison between the numerical results of the present research and the results of Mai et al. (2019).

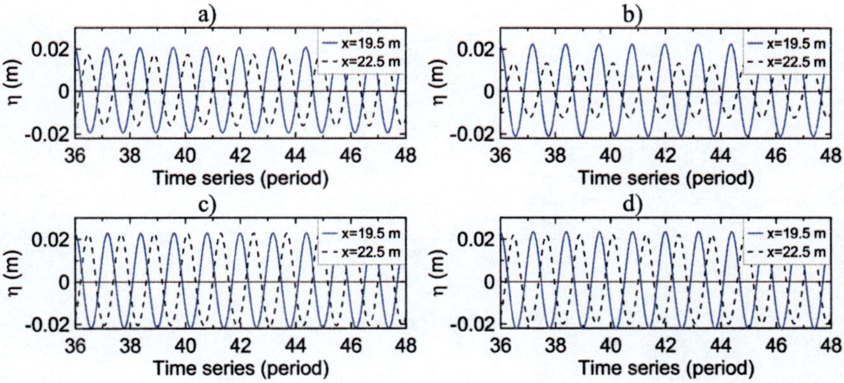

Fig. 5. Time series of water surface elevations at Gauge 1 (solid lines) and Gauge 2 (dashed lines) for 4 cases: a) case 1; b) case 2; c) case 3; d) case 4.

Figure 5 (a, b, c, d) show more detail about the influence of porosity on transmission coefficient when comparing water surface elevations in front of the bamboo breakwater (at gauge G1) with ones behind the bamboo breakwater (at gauge G2) for 4 cases. The water surface elevations are recorded in 12 periods, from the 36th period to the 48th period. Figure 5 (a, b) shows a large reduction of waves behind the breakwater while Fig. 5 (c, d) show a small variance. It should be noted that the porosities in cases 1 and 2 (Fig. 5 a, b) are smaller than those in cases 3 and 4 (Fig. 5 c, d).

3.3 Effectiveness of Wave Nonlinearity on Wave Transmission Coefficient

This part shows the effectiveness of wave nonlinearity (H/h, H is the incident wave height and h is the water depth) on wave height behind the bamboo breakwater in 4 cases. The nonlinearity increases from 0.1 to 0.58. Figure 6 shows the transmission coefficient in a downward trend when increasing the wave nonlinearity. It can be seen clearly that when the porosity is small ($\lambda = 0.5$, cases 1 & 2) the affection of the nonlinearity is large and when the porosity is large ($\lambda = 0.9$, cases 3 & 4) the affection of the nonlinearity is small. Thus, the porosity of the bamboo breakwater is significant. It is interesting to find that the similarity trend of nonlinearity effect on transmission coefficient through porous breakwater can be found in literature (Liu and Wen 1997; Lynett et al. 2000; Vu et al. 2018).

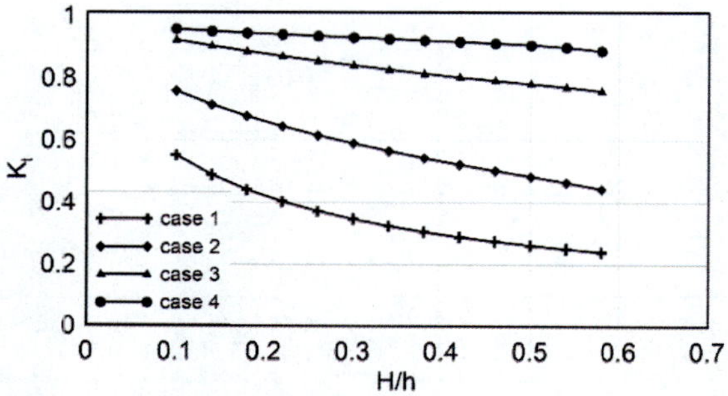

Fig. 6. Effectiveness of wave nonlinearity on wave transmission coefficient.

4 Concluding Remarks

Most of the previous works on bamboo breakwaters have focused on waves attenuating by changing the breakwater width, porosity, crest height and waves nonlinearity. However, the importance of the breakwater porosity was unclear. This work proves the effectiveness of the breakwater width, porosity and waves nonlinearity on waves height behind the bamboo breakwater. It is concluded that the breakwater porosity is the most dominant factor.

The numerical results applying extended Boussinesq equations for waves in porous media of Vu et al. (2018) show a very good agreement with the numerical results from Mai et al. (2019) applying SWASH. The research results prove small porosities give significant effect on dissipating wave energy.

Acknowledgement. "This research is funded by Vietnam National Foundation for Science and Technology Development (NAFOSTED) under grant number 107.03-2019.338".

References

Albers, T., San, D.C., Schmitt, K.: Coastal Protection in the Lower Mekong Delta 124 (2013)

Dao, T., Stive, M.J.F., Hofland, B., Mai, T.: Wave damping due to wooden fences along mangrove coasts. J. Coast. Res. **34**, 1317–1327 (2018). https://doi.org/10.2112/JCOASTRES-D-18-00015.1

Halide, H., Brinkman, R., Ridd, P.: Designing bamboo wave attenuators for mangrove plantations. Indian J. Mar. Sci. **33**, 220–225 (2004)

Liu, P.L.F., Wen, J.: Nonlinear diffusive surface waves in porous media. J. Fluid Mech. **347**, 119–139 (1997). https://doi.org/10.1017/S0022112097006472

Lynett, P.J., Liu, P.L.-F., Losada, I.J., Vidal, C.: Solitary wave interaction with porous breakwaters. J. Waterw. Port Coastal Ocean Eng. **126**, 314–322 (2000). https://doi.org/10.1061/(ASCE)0733-950X(2000)126:6(314)

Mai, C.T., Ngo, T.T.A.: Wave transmission through bamboo fences in Bac Lieu coastal area in the south-west season. J. Water Res. Environ. Eng. **61**, 115–120 (2018). (in Vietnamese)

Mai, T., Dao, T., Ngo, A., Mai, C.: Porosity effects on wave transmission through a bamboo fence. In: Trung Viet, N., Xiping, D., Thanh Tung, T. (eds.) APAC 2019, pp. 1413–1418. Springer, Singapore (2020). https://doi.org/10.1007/978-981-15-0291-0_191

Nguyen, B.C., Nguyen, H.N.: Đánh giá tác động của tuyến kè hiện hữu và giải pháp công trình đề xuất mới đối với chế độ thủy-thạch động lực biển ven bờ mũi cà mau nhằm bảo vệ và phát triển vùng đất mũi. J. Water Res. Sci. Technol. **46**, 44–56 (2018). (in Vietnamese)
Nguyen, T.T.L., Vu, V.N.: Propagation of waves over a submerged porous breakwater in shallow water area. J. Transp. Sci. Technol. **34**, 50–56 (2019). (In Vietnamese)
Nguyet-Minh, N., et al.: Evaluating the effectiveness of existing coastal protection measures in Mekong delta. In: Trung Viet, N., Xiping, D., Thanh Tung, T. (eds.) APAC 2019, pp. 1419–1429. Springer, Singapore (2020). https://doi.org/10.1007/978-981-15-0291-0_192
Phansri, W.: Coastal erosion protection and enhancing sediment deposition by bamboo wall at Samut Songkhram Province, Thailand. Master thesis, Kasetsart University (2011)
Schmitt, K., Albers, T., Pham, T.T., Dinh, S.C.: Site-specific and integrated adaptation to climate change in the coastal mangrove zone of Soc Trang Province, Viet Nam. J. Coast. Conserv. **17**(3), 545–558 (2013). https://doi.org/10.1007/s11852-013-0253-4
Thieu, Q.T., Mai, T.L.: Monsoon wave transmission at bamboo fences protecting mangroves in the lower Mekong delta. Appl. Ocean Res. **101**, 102259 (2020). https://doi.org/10.1016/j.apor.2020.102259
Tran, V.T., Nguyen, H.H., Pham, D.H., Nguyen, D.N., Nguyen, T.T.: Hollow cylinder breakwater for dissipation of wave energy to protect the west coast of Ca Mau Province in Vietnam. In: Proceedings of the 1st Vietnam Symposium on Advances in Offshore Engineering, VSOE 2018, pp. 599–605 (2019). https://doi.org/10.1007/978-981-13-2306-5_85
Vu, V.N., Lee, C., Jung, T.H.: Extended Boussinesq equations for waves in porous media. Coast. Eng. **139**, 85–97 (2018). https://doi.org/10.1016/j.coastaleng.2018.04.023
Vu, V.N., Lee, C., Jung, T.H.: Internal generation of damped waves in linear shallow water equations. Coast. Eng. **104**, 13–25 (2015). https://doi.org/10.1016/j.coastaleng.2015.07.004
Zijlema, M., Stelling, G., Smit, P.: SWASH: an operational public domain code for simulating wave fields and rapidly varied flows in coastal waters. Coast. Eng. **58**, 992–1012 (2011). https://doi.org/10.1016/j.coastaleng.2011.05.015

Study the Influence of Newly Built Constructions on the Wave Field of the Dyke Area from Ben Got to Hoang Chau on Cat Hai Island, Hai Phong, Vietnam

Trung Dung Nguyen(✉) and Tri Mai

Faculty of Coastal and Offshore Engineering, Hanoi University of Civil Engineering, 55 Giai Phong, Hai Ba Trung, Hanoi, Vietnam
dungnt4@nuce.edu.vn

Abstract. Currently, the Cat Hai island in Hai Phong in Vietnam is strongly developing with the goal to become a smart island in the Northern region. The construction of Lach Huyen dikes to protect the Vinfast factory will affect the wave field in the coastal water from Ben Got to Hoang Chau in the island. Therefore, the anchoring of vessels and maritime activities in the area will be affected. This paper is going to study and assess the impact of the newly constructed dikes on wave fields around Cat Hai island using numerical models. The Bouss2D model of the SMS (Surface-water Modeling System) is applied in this study to simulate 2-dimensional wave propagation from offshore to the coastal area. This model is rather new model and has not widely been applied in Vietnam. The results of this study will show the impact of newly built constructions on the regional wave field from Ben Got to Hoang Chau in Cat Hai Island.

Keywords: Bouss2D model · Wave diffraction · Cat Hai island

1 Introduction

Cat Hai is a small island with one side is adjacent to the sea and the other 3 sides are bordered by estuaries and canals (Fig. 1). The area from Ben Got to Hoang Chau is directly influenced by waves from sea side, which is one of the factors that directly affect the accretion, erosion and inundation of the area [1]. There were some studies on wave fields in Cat Hai island area few year ago [2, 3]. However, from 2018 up to now, there are two container ports has been built and operating in Cat Hai island. In addition, there is a 10 km long dike under construction in the island and the Vinfast car factory encroached the sea, the furthest place nearly 1 km which is obviously changing the coastline in the island (see Fig. 2). Currently, there is no study on the effects of this new construction on the local wave field from Ben Got to Hoang Chau. Meanwhile, the population density is very dense behind this dike area.

Nowadays, with the development of science and technology, numerical models have become an indispensable tool for forecasting the coastal wave field. The models mentioned here: Bouss-2D model in the SMS model set [4, 5], MIKE 21 model in MIKE

D. V. K. Huynh et al. (Eds.): VSOE 2021, LNCE 208, pp. 262–269, 2022.
https://doi.org/10.1007/978-981-16-7735-9_28

model set [6, 7], D-WAVE model (SWAN) in the model set DELFT 3D model [8, 9], model EFDC [10], model STWAVE in CEDAS model set [11, 12]. Bouss-2D model is a modern and powerful model in calculating the wave field in the coastal area and has not been applied in Vietnam. This study is going to use the Bouss-2D model to calculate the wave field in the Cat Hai island area for the cases with and without the new constructions/dikes to assess the impact of constructions/dikes on the wave field from Ben Got to Hoang Chau.

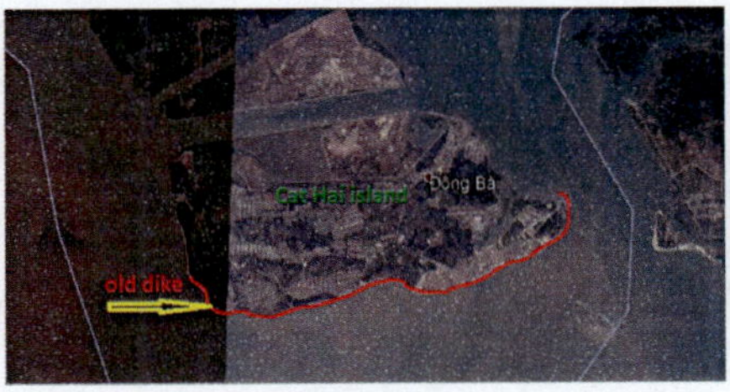

Fig. 1. Image of Cat Hai island location taken in August 2014.

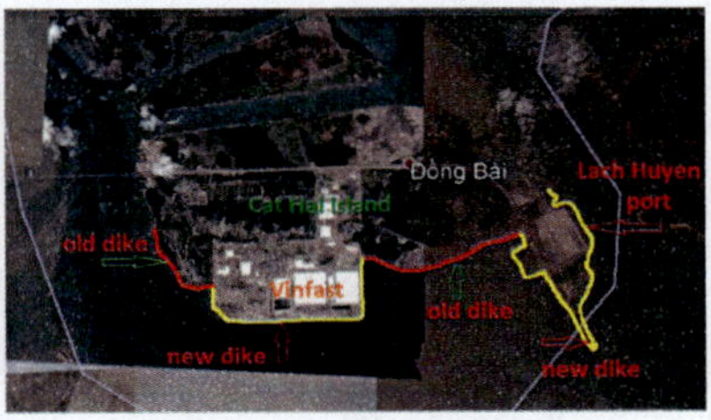

Fig. 2. Construction image of Cat Hai island taken in May 2019.

2 Methodology

2.1 Theoretical Description

BOUSS-2D is based on Boussinesq-type equations derived by Nwogu [13, 14]. The equations are depth-integrated equations for the conservation of mass and momentum for nonlinear waves propagating in shallow and intermediate water depths. For short-period waves, the horizontal velocities are no longer uniform over depth and the pressure is non-hydrostatic. The vertical profile of the flow field is obtained by expanding the velocity potential, Φ, as a Taylor series about an arbitrary elevation; z_α, in the water column. For waves with length, L, much longer than the water depth, h, the series is truncated at second order resulting in a quadratic variation of the velocity potential over depth (See Eq. 1).

$$\Phi(x, z, t) = \phi_\alpha + \mu^2(z_\alpha - z)[\nabla\phi_\alpha.\nabla h] + \frac{\mu^2}{2}\left[(z_\alpha + h)^2 - (z + h)^2\right]\nabla^2\phi_\alpha + \mathrm{O}(\mu^4) \tag{1}$$

Where $\phi_a = \Phi(x, z_\alpha, t)$, $\nabla = (\partial/\partial_x, \partial/\partial_y)$, $\mu = h/L$ is a measure of frequency dispersion. The horizontal and vertical velocities are obtained from the velocity potential as.

$$\begin{aligned} u(x, z, t) = \nabla\Phi &= u_\alpha + (z_\alpha - z)[\nabla(u_\alpha.\nabla h) + (\nabla.u_\alpha)\nabla h] \\ &+ \frac{1}{2}\left[(z_\alpha + h)^2 - (z + h)^2\right]\nabla(\nabla.u_\alpha) \end{aligned} \tag{2}$$

$$w(x,z,t) = \frac{\partial \Phi}{\partial z} = -[u_\alpha.\nabla h + (z+h)\nabla.u_\alpha] \tag{3}$$

Where $u_a = \nabla\Phi|_{z_\alpha}$ is the horizontal velocity at $z = z_\alpha$. Given a vertical profile for the flow field, the continuity and Euler (momentum) equations can be integrated over depth, reducing the three-dimensional problem to two dimensions. For weakly nonlinear waves with height, H, much smaller than the water depth, h, the vertically integrated equations are written in terms of the water-surface elevation $\eta(x,t)$ and velocity $u_\alpha(x,t)$ as;

$$\eta_t + \nabla.u_f = 0 \tag{4}$$

$$\begin{aligned} &u_{\alpha,t} + g\nabla\eta + (u_\alpha.\nabla)u_\alpha + z_\alpha\left[\nabla(u_{\alpha,t}.\nabla h) + (\nabla.u_{\alpha,t})\nabla h\right] \\ &+ \frac{1}{2}\left[(z_\alpha + h)^2 - h^2\right]\nabla(\nabla.u_{\alpha,t}) = 0 \end{aligned} \tag{5}$$

where u_f is the volume flux density given by:

$$\begin{aligned} u_f &= \int_{-h}^{\eta} u dz = (h+\eta)u_\alpha + h\left(z_\alpha + \frac{h}{2}\right)[\nabla(u_\alpha.\nabla h) + (\nabla.u_\alpha)\nabla h] \\ &+ h\left[\frac{(z_\alpha+h)^2}{2} - \frac{h^2}{6}\right]\nabla(\nabla.u_\alpha) \end{aligned} \tag{6}$$

The depth-integrated equations are able to describe the propagation and transformation of irregular multidirectional waves over water of variable depth. The elevation of the velocity variable z_α is a free parameter and is chosen to minimize the differences between the linear dispersion characteristics of the model and the exact dispersion relation for small amplitude waves. The optimal value, $z_\alpha = -0.535h$, is close to middepth.

For near-rupture waves in shallow water, the small nonlinear wave hypotheses are no longer correct. Nwogu [14] further developed these systems of nonlinear equations by presenting nonlinear components as a function of velocity on the free surface. These systems of nonlinear equations are written as follows;

$$\eta_t + \nabla.u_f = 0 \tag{7}$$

$$\begin{aligned} &u_{\alpha,t} + g\nabla\eta + (u_\eta.\nabla)u_\eta + w_\eta\nabla w_\eta + (z_\alpha - \eta)\left[\nabla(u_{\alpha,t}.\nabla h) + (\nabla.u_{\alpha,t})\nabla h\right] \\ &+ \frac{1}{2}\left[(z_\alpha+h)^2 - (h+\eta)^2\right]\nabla(\nabla.u_{\alpha,t}) - \left[(u_{\alpha,t}.\nabla h) + (h+\eta)\nabla.u_{\alpha,t}\right]\nabla\eta \\ &+ \left[\nabla(u_{\alpha,t}.\nabla h) + (\nabla.u_{\alpha,t})\nabla h + (z_\alpha + h)\nabla(\nabla.u_\alpha)\right]z_{\alpha,t} = 0 \end{aligned} \tag{8}$$

where, z_α is now a function of time and is given by $z_\alpha + h = 0.465(h+\eta)$. The volume flux density is given by;

$$u_f = (h+\eta)\left\{u_\alpha + \left[(z_\alpha + h) - \frac{(h+\eta)}{2}\right][\nabla(u_\alpha.\nabla h) + (\nabla.u_\alpha)\nabla h]\right.$$

$$+\left[\frac{(z_\alpha + h)^2}{2} - \frac{(h+\eta)^2}{6}\right]\nabla(\nabla.u_\alpha)\Bigg\} \tag{9}$$

The fully nonlinear equations are able to implicitly model the effects of wave-current interaction. Currents can either be introduced through the boundaries or by explicitly specifying a current field, U.

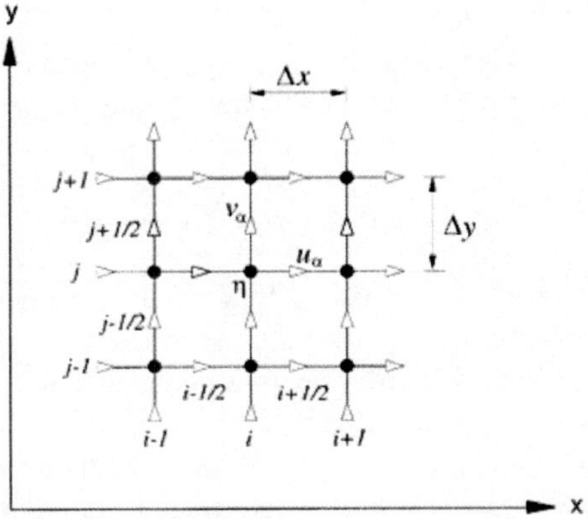

Fig. 3. Computational grid for finite difference scheme.

The weakly and fully nonlinear Boussinesq equations are solved in the time-domain using a finite-difference method. The computational domain is discretized as a rectangular grid with grid sizes Δx and Δy, in the x and y directions, respectively. The equation variables η, u_α, v_α are defined at the grid points in a staggered manner as shown in Fig. 3. The water depth and surface elevation are defined at grid points (i, j), while the velocities are defined half a grid point on either side of the elevation grid points (Fig. 3).

The numerical solution scheme is an implicit Crank-Nicolson scheme with a predictor-corrector method used to provide the initial estimate. The first step in the solution scheme is the predictor step in which values of the variables at an intermediate time-step $t = (n + 1/2)\Delta t$ are determined using known values at $t = n\Delta t$ (Δt is the time-step size). The second step is the corrector step in which predicted values $t = (n + 1/2)\Delta t$ are used to provide an initial estimate of the values at $t = (n + 1)\Delta t$. The last step is an iterative Crank-Nicolson scheme, which is repeated until convergence.

The resulting Crank-Nicolson formulation for the weakly nonlinear form of the mass and momentum equations can be written as:

$$\delta_t \eta = -\delta_x^{(1)} u_f^{n+1/2} - \delta_y^{(1)} v_f^{n+1/2} \tag{10}$$

$$\delta_t(u_\alpha + f_4\delta_x^{(2)} u_\alpha + f_2\delta_{xx}u_\alpha) = -g\delta_x^{(1)}\eta^{n+1/2} - \frac{1}{2}\delta_x^{(1)}\left[\left(\overline{u_\alpha^{n+1/2}}^{x}\right)^2\right]$$
$$-\delta_y^{(1)}\left(\overline{v_\alpha^{n+1/2}}^{x}\overline{u_\alpha^{n+1/2}}^{y}\right) - f_4\left[\delta_y^{(1)}(h)\delta_t\delta_x^{(1)}\left(\overline{v_\alpha^{n+1/2}}^{y}\right) + \delta_x^{(1)}(h)\delta_t\delta_y^{(1)}\left(\overline{v_\alpha^{n+1/2}}^{x}\right)\right] - f_2\delta_t\delta_{xy}v_\alpha^{n+1/2} \tag{11}$$

$$\delta_t(v_\alpha + f_4\delta_y^{(2)} v_\alpha + f_2\delta_{yy} v_\alpha) = -g\delta_y^{(1)}\eta^{n+1/2} - \frac{1}{2}\delta_y^{(1)}\left[\left(\overline{v_\alpha^{n+1/2}}^y\right)^2\right]$$
$$- \delta_x^{(1)}\left(\overline{v_\alpha^{n+1/2}}^x \overline{u_\alpha^{n+1/2}}^y\right) - f_4\left[\delta_y^{(1)}(h)\delta_t\delta_x^{(1)}\left(\overline{u_\alpha^{n+1/2}}^y\right) + \delta_x^{(1)}(h)\delta_t\delta_y^{(1)}\left(\overline{u_\alpha^{n+1/2}}^x\right)\right] - f_2\delta_t\delta_{xy}u_\alpha^{n+1/2} \tag{12}$$

To solve the governing equations, appropriate boundary conditions have to be imposed at the boundaries of the computational domain. The types of boundaries considered in BOUSS-2D include:

Fully reflecting or solid wall boundaries: Along solid wall or fully reflecting boundaries, the horizontal velocity normal to the boundary must be zero over the entire water depth, i.e.,

$$u.n = 0 - h < z < \eta \tag{13}$$

External wave generation boundaries: Along external wave generation boundaries, time-histories of velocities u_α or v_α, and flux densities u_f or v_f corresponding to an incident storm condition are specified. The time-histories may correspond to regular or irregular, unidirectional or multidirectional waves.

Internal wave generation boundaries: in BOUSS-2D with the governing equations modified to allow for the generation of waves inside the computational domain and absorb reflected waves in a damping layer placed behind the generation boundary to prevent a buildup of wave energy inside the domain.

Wave absorption or damping regions: Expressed through coefficients $\mu(x)$ is the damping strength with units of s^{-1}.

Porous structures: The equations for the porous area were obtained by substitution u with u/n, where n is the porosity.

2.2 Numerical Model Setup

The seabed topography is a very important parameter for running the model, the author has collected from the Central Hydrometeorological Forecast Center (Fig. 4). These are the depth data of the coastal and estuarine areas of Hai Phong city, supplemented and updated from the depth measurement data of a number of other project projects that have been done in this area such as the field visits. Surveying the Port and Waterway Consulting Center (TEDI) in 2006, the survey of the Institute of Geography in 2007 and 2008. Then the author used software such as DTM, Google Earth Pro to recalibrate the topography to the real variety. The coastline of Cat Hai in 2015 has been encroached upon compared to 2008 and describes the shoreline of newly constructed works.

The author establishes the offshore boundary at a depth of nearly 40 m, more than 50 km from the shore, corresponding to the size of the calculated grid area 7600 × 54000 m, the calculated grid cell is 10 × 10 m. Deep water wave parameters: for storm level 10, design tidal water level frequency of 5%, storm surge frequency of 20%. Therefore, the deep water wave parameters include: the design water level of + 3.25 m, the average wave height of 5.03 m and the average wave period is 9.88 s [2]. The JONSWAP spectrum wave conditions are used in this simulation study. Boundary

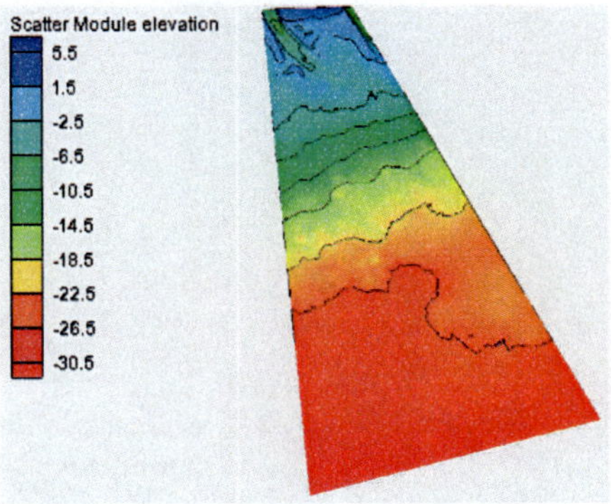

Fig. 4. Seabed topography of the Cat Hai island. Units in m.

conditions applied in this study includes the damping width of 20 m, the damping coefficient of 0.2, the constant chezy roughness type with coefficient of 80 and the porosity friction factors: 2.4 for turbulent, 800 for lamia and the added mass coefficient of 0.8.

3 Results and Discussions

The wave field from Ben Got to Hoang Chau have been simulated without and with the newly constructions in the Cat Hai island and the simulated results have been presented in Fig. 5 and Fig. 6, respectively. It can be seen from Fig. 5 that the wave height varies from 1.2 m to 2.1 m along the coastal area from Ben Got to Hoang Chau for the case without newly constructions. For the case with new constructions (ports, Vinfast dikes), the wave height has been significant increased up to 1.8 m to 3.2 m (see Fig. 6).

Several points, P1–P5, along the coastal line have been chosen to investigate the change of the significant wave heights without and with the new constructions (See Fig. 7 & Fig. 8). It is shown that at point P1 (at the Hoang Chau dike toe) the significant wave height is about from 1.2 m to 1.63 m with no construction in place (Point P1 in Fig. 7). When the new constructions are built, the significant wave height at P1 have been increased to about from 1.6 m to 1.98 m (Point P1 in Fig. 8) and that wave field have significant effects on the activities of the vessels in the area.

Table 1 present the simulated significant wave heights at point P1–P5 along the coastal area from Ben Got to Hoang Chau (Fig. 7 & Fig. 8). As we can see from Table 1 that in general the significant wave heights at points P1–P3 have been increased with the present of the Vinfast dike and other new constructions. On the other hand, the significant wave heights at points P4–P5 are decreased from 1.4 m to 0.82 m with the new constructions in place and this shows that the new constructions play a role as a protection measure for the coastal area from P4 to P5 (Ben Got to Lach Huyen port).

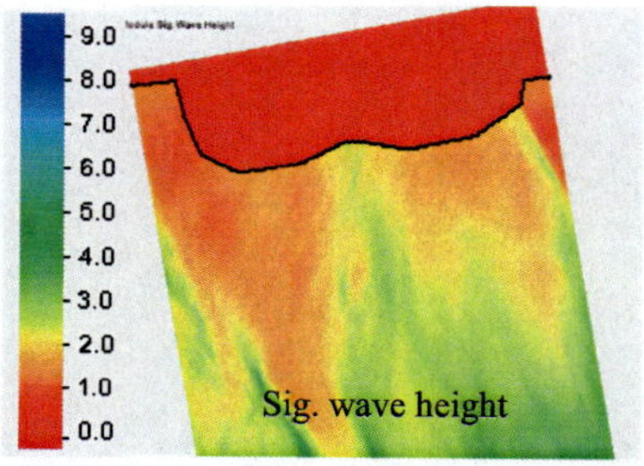

Fig. 5. Results of the local wave field without new construction. Units in m.

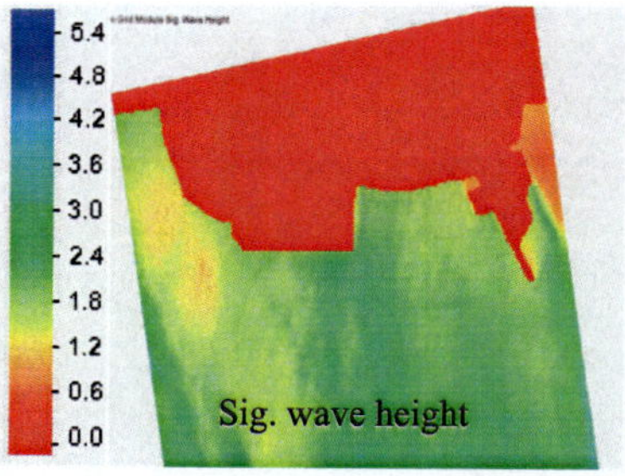

Fig. 6. Results of the local wave field with new construction. Units in m.

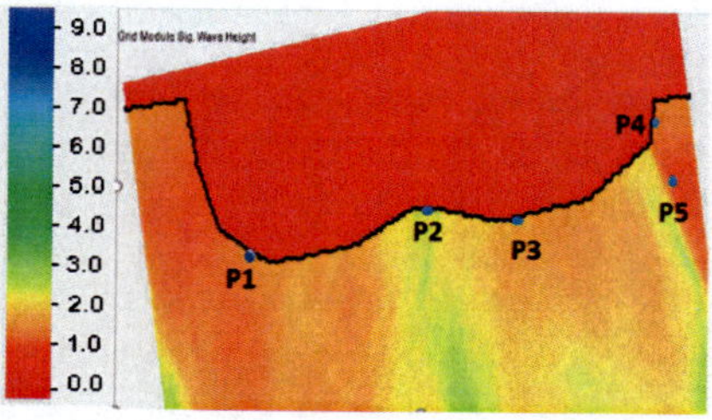

Fig. 7. Wave heights along the coastal area without new constructions. Units in m.

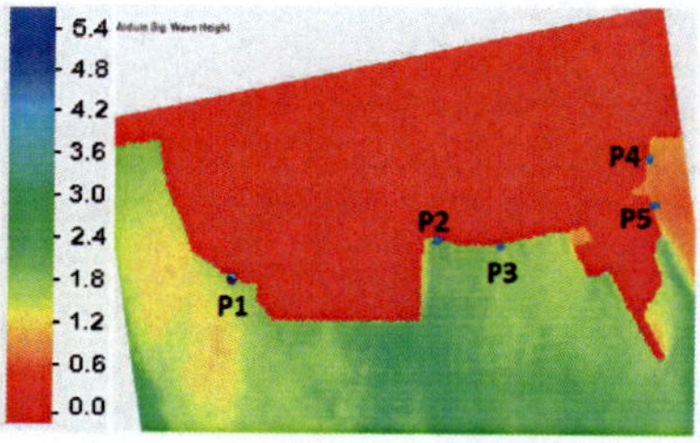

Fig. 8. Wave heights along the coastal area with new constructions. Units in m.

Table 1. Summary of the significant wave heights at points P1-P5.

Significant wave height H_s (m)	P1 (Hoang Chau dike)	P2 (Van Chan-Gia Loc)	P3 (Gia Loc-Ben Got)	P4 (Ben Got dike)	P5 (Lach Huyen port)
Without the new structures	1.20–1.63	1.30–1.90	1.30–2.10	1.30–1.39	1.00–1.40
With the new structures	1.60–1.98	1.67–2.30	1.67–2.21	0.86–0.99	0.67–0.82

4 Conclusions

This paper have applied the modern numerical model BOUSS-2D in the SMS model set to investigate the effects of the newly construtions on the wave field of the coastal area in Cat Hai island in Hai Phong province in Vietnam. The simulation has been carried out for the coastal area from Ben Got to Hoang Chau for the cases without and with the new constructions along the island. The results showed that for the case with the new constructions the significant wave height at the toe of the dikes from Ben Got to

Hoang Chau is higher than that without new constructions about from 0.3 m to 0.4 m. In constrast, in the Lach Huyen port area, the significant wave height is reduced from 0.4 to 0.6 m for the case with the new constructions. The finding of this study may help the community to propose measures to reinforce dikes and to protect the living area behind the dikes in Hoang Chau, Van Chan- Gia Loc, Gia Loc- Ben Got during storms because the effects of the new buildings.

Bouss2D software is open software for free research in a short time of two weeks. The software is easy to install, easy to use interface and with enthusiastic software support team. It is recommended for the students and researchers to use the Bouss2D for their study and research.

Acknowledgements. The authors would like to thanks the National University of Civil Engineering for their support under the funded project code of 108-2019/KHXD.

References

1. Hung, P.K.: Synthesis report on choosing the appropriate solutions for typical sea dykes in Vietnam. Project of the Vietnam Association of Science and Technology (2003–2004) (2005)
2. Dung, N.T.: Study on wave overtopping process and propose design solutions to minimize the impact of overtopping waves for dike section Gia Loc-Van Chan, Cat Hai District, Hai Phong. Master thesis, Truong Dai Construction School, Hanoi (2015)
3. Hang, N.T.T.: Design of a breakwater for coastal protection in Cat Hai - Hai Phong. An engineering graduate project, Thuy Loi University, Hanoi (2010)
4. SMS model's. https://www.aquaveo.com/software/sms-surface-watermodeling-system-introduction
5. SMS User Manual (v13.0)
6. MIKE model's. http://www.dhigroup.com/
7. MIKE 21 spectral waves FM module (2012) – User manual
8. Delft3d's. http://oss.deltares.nl/web/delft3d
9. D-WAVE User manual (2017)
10. Environmental fluid dynamics code (EFDC) (2013) – User manual. http://www.dsintl.biz/
11. CEDAS - Coastal Engineering Design and Analysis System
12. CEDAS model's. http://www.veritechinc.com/products/cedas/CEDASdetails.php
13. Nwogu, O.: Alternative form of Boussinesq equations for nearshore wave propagation. J. Waterw. Port Coast. Ocean Eng. ASCE **119**(6), 618–638 (1993)
14. Nwogu, O.G.: Numerical prediction of breaking waves and currents with a Boussinesq model. In: Proceedings of 25th International Conference on Coastal Engineering, ICCE 1996, Orlando, vol. 4, pp. 4807–4820 (1996)

Estimation of Wave Characteristics from the Measured Pressures on Support Structure of Offshore Wind Turbine

Tri Mai(✉)

Faculty of Coastal and Offshore Engineering, Hanoi University of Civil Engineering, 55 Giai Phong, Hai Ba Trung, Hanoi, Vietnam
trimc@nuce.edu.vn

Abstract. Waves at sea are very important for designing an offshore wind turbine support structure. There are several techniques used to observe waves at sea. These techniques can be in-situ technique and remote-sensing technique. This study is going to present an in-situ technique, in which pressure sensors are mounted on a support structure of offshore wind turbine and they are deployed in the water. Wave characteristics have been recovered from the measured pressures on the structure applying the spectral analysis method. Results show that the significant wave height can be well recovered from the subsurface pressure at the angle of 150, 180 and 210° (the angle of the orientation of the pressure sensor with respect to the incident wave direction). There is slightly effects of the submergent level of the pressure measurement on the recovered wave heights.

Keywords: Wave recovery · Wave pressures · Offshore wind turbine · Physical model

1 Introduction

Wave climate is a vital important in estimation of wave loading on coastal and offshore structures. There are various techniques used to observe waves at sea. These techniques can be divided into two main techniques. The first one is in-situ technique in which instruments are deployed in water such as wave buoys, wave poles, inverted echo-sounders, pressure transducers and current meters. The second is remote-sensing technique, in which instruments are deployed at some distance above the water surface such as imaging techniques and altimetry.

The pressure transducer is one of the in-situ techniques. Traditionally, a pressure transducer can measure wave-induced pressure fluctuations at various depths below the water surface. These fluctuations, in combination with the linear wave theory (Airy theory) can be used to estimate wave characteristics. There have been many studies on application of this technique in estimation of wave parameters since 1947 (Folsom 1947; Grace 1978; Cavaleri 1980; Bishop and Donelan 1987; Mai and Schlurmann 2012; Clamond 2013; Oliveras et al. 2012; Kogelbauer 2015). Grace (1978) found that the Airy theory can be used to predict individual wave height from concurrent pressure variations

D. V. K. Huynh et al. (Eds.): VSOE 2021, LNCE 208, pp. 270–277, 2022.
https://doi.org/10.1007/978-981-16-7735-9_29

at or near the sea floor with an empirical correction factor is included. Disturbance of the orbital motion of water particles under waves on the wave measurements by pressure transducers was found by Cavaleri (1980) and a measurement system has been devised to avoid the effect of dynamic pressure by consisting of a water filled plastic sheath surrounding the captive part of the transducer. His result shows that the dynamic effect of water motion has been cancelled and the pressure transducers were picking up just the pressure field due to the presence of waves. Gabriel et al. (1986) investigated the effects of currents on interpretation of sub-surface pressure spectra. Their study shows that the Doppler effect should be taken into account when recovering surface elevation spectra from subsurface pressure measurement taken in the presence of currents, particularly when the pressure transducer is deployed close to the sea bed. Their study indicates that there is better agreement between the measured water elevation spectra and the one recovered from the pressure measurement at level closer to the water surface. Clamond (2013) derived some new exact relations for the recovery of the steady free surface wave profile from pressure measurements at the flat sea bed, with the introduction of the holomorphic function which yields a reformulation of the problem into a much simplified form. Oliveras et al. (2012) presented a new nonlocal nonlinear equation relating the pressure and the surface elevation, and the solvability properties of this nonlocal equation are rigorously analyzed using the Implicit Function Theorem. They derived various new approximate formulas and these approximations are compared with the nonlocal equation by using numerical and experimental data. The nonlocal equation consistently outperforms its different approximations. Kogelbauer (2015) obtained a relation between the bottom pressure and the free water surface, expressed in terms of Fourier coefficients by using a conformal hodograph transform and the governing equations. Application of the relation in Kogelbauer (2015) to experiment has not been made yet.

This study is going to apply the linear transfer function to recover wave heights from concurrent pressure measurements at various levels below mean water level. The pressure transducers were mounted on a mono-pile to measure pressure distribution around as well as along the mono-pile under wave condition. A polar coordinate has been applied to obtain transfer function which also depends on the location of pressure measurement on the mono-pile. Significant wave heights are recovered from pressure measurements using frequency domains. The theoretical and experimental results are then compared.

2 Methodology

2.1 Physical Model

A small scale 1:40 physical model was constructed in the new wave basin in Hannover, Germany to examine different multi-directional waves interacting with a large mono-pile structure (the diameter of the mono-pile is large in comparing with wave length). The wave basin has dimensions of 40 m wide and 24 m long and the basin can be filled up to 1 m of water depth for experiment. The 3D wave maker has 72 individual wave paddles and each wave paddle has 0.40 m wide and 1.8 m high. Maximum stroke of each paddle is ±0.6 m. The tested mono-pile has its diameter of 0.3 m and height of 0.9 m (see Fig. 1a).

The tested wave condition (UI4) was a unidirectional irregular wave with an underlying JONSWAP spectrum. The experimented wave characteristics had the significant wave height of 0.12 m and the peak wave period of 2.0 s. The diffraction parameter, kr_0, was 0.22 in which k is the wave number and r_0 is the radius of the mono-pile. The tests were performed within a water depth of 0.6 m. The pressures upon the surface of the mono-pile were measured by applying pressure sensors mounted at various levels (z/h = −0.07 to −0.77; z is the submergence of pressure sensor, h is the water depth in the basin) on the surface of the structure. Water elevations were measured synchronously next to the structure (see Fig. 1b).

(a) (b)

Fig. 1. (a) Location of pressure transducers on the mono-pile (Note units in cm); (b) Configuration of physical model in the 3D-wave basin.

2.2 Theoretical Aspects

2.2.1 Subsurface Pressure Under Water Wave (Bishop and Donelan 1987)

The pressure beneath a progressive wave can be expressed from Bernoulli's theorem:

$$p(t) = -\rho g z + \rho\frac{\partial\phi}{\partial t} - \frac{1}{2}\rho\left(u^2 + w^2\right) + p_s + \rho\gamma(t) \tag{1}$$

in which, p is the total pressure, ρ is the water density, g is the gravity acceleration, z is the depth of submergence of the pressure transducer (measured positive upward from the mean water level), ϕ is the velocity potential, u is the horizontal velocity, w is the vertical velocity, p_s is the atmospheric pressure at the surface and $\gamma(t)$ is a function of time (see Bishop and Donelan 1987).

The first three terms on the right hand side of Eq. 1 are the hydrostatic pressure (This hydrostatic pressure is independent of the presence of the wave), the pressure due to the passage of the wave form (or the hydrodynamic pressure due to the wave and therefore this term represents the wave-induced pressure) and the pressure due to the local kinetic energy. We are normally interested in the hydrodynamic terms:

$$p(t) = \rho\frac{\partial\phi}{\partial t} - \frac{1}{2}\rho\left(u^2 + w^2\right) + \rho\gamma(t) \tag{2}$$

According to the Airy wave theory, pressure due to the passage of a wave form, at a level z under the water surface, can be estimated by:

$$p(t) = \rho \frac{\partial \phi}{\partial t} = \rho g \frac{H}{2} \frac{\cosh k(d+z)}{\cosh kd} \cos(kx - \omega t) \tag{3}$$

in which, p is the pressure at a level z below still water level, ρ is the water density, g is the gravitational acceleration, H is the wave height, d is the water depth, k is the wave number, ω is the angular frequency and $(kx - \omega t)$ is the phase of wave.

The water elevation can be estimated by using Airy theory as:

$$\eta(t) = \frac{H}{2} \cos(kx - \omega t) \tag{4}$$

Substitute Eq. 4 into Eq. 3 then:

$$p(t) = \rho g \frac{\cosh k(d+z)}{\cosh kd} \eta(t) \tag{5}$$

Therefore, the water fluctuation can be estimated from the fluctuation of pressure:

$$\eta(t) = \frac{1}{\rho g} \frac{\cosh kd}{\cosh k(d+z)} p(t) \tag{6}$$

Since pressures upon a vertical mono-pile have been measured, the water fluctuation can also be estimated by these pressures via the linear theory function which will be presented in the following sections.

2.2.2 Pressures upon a Vertical Circular Cylinder (Sumer and Fredsøe 2006)

From the linear feature of potential flow, the total potential function (ϕ) can be written as the sum of two potential functions:

$$\phi = \phi_i + \phi_s \tag{7}$$

in which, ϕ_i is the potential function of the undisturbed incident wave and ϕ_s is the potential function of the scattered (reflected and diffracted) wave by presence of the vertical mono-pile.

In the case of a wave passing a small vertical mono-pile ($D/L < 0.2$, D is the diameter of the mono-pile and L is the wave length), only potential function of the undisturbed incident wave is taken into account.

$$\phi = \phi_i \tag{8}$$

Therefore, according to MacCamy & Fuchs (Sumer and Fredsøe 2006) the dynamic pressure exerts on the surface of the mono-pile can be expressed as a function of the time, frequency, submerge level and the angle of orientation of the point on the surface of the mono-pile with respect to the incident wave direction:

$$p(t, f, z, \theta) = \frac{\rho g H}{2} \frac{\cosh k(h+z)}{\cosh(kh)} \cdot \left[J_0(kr_0) + \sum_{p=1}^{\infty} 2 i^p J_p(kr_0) \cos(p\theta) \right] \cdot e^{-i\omega t} \tag{9}$$

in which, ρ is the water density, g is the gravity acceleration, H is the wave height, k is the wave number, h is the water depth, z is the depth of pressure sensor's submergence which is measured positive upward from still water level, r_O is the radius of the mono-pile and θ is the angle of orientation of the pressure sensor with respect to the incident wave direction such as definition in Fig. 2. $J_O(kr_O)$ and $J_p(kr_O)$ are the first and second kind of Bessel functions, respectively.

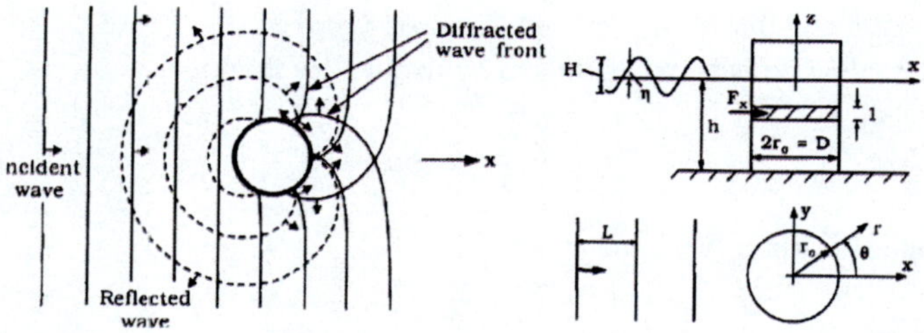

Fig. 2. Sketch of the incident, diffracted and reflected wave fronts for a vertical circular cylinder (Sumer and Fredsøe 2006).

In spectral analysis, the transfer function, $TF(f, z, \theta)$ for the dynamic pressure at any point on the surface of the structure can be expressed as a function of the frequency f, submergence level z and angle θ (Sumer and Fredsøe 2006):

$$TF(f, z, \theta) = real\left\{\frac{\cosh k(d+z)}{\cosh(kd)} \cdot \left[J_O(kr_O) + \sum_{p=1}^{\infty} 2i^p J_p(kr_O)\cos(p\theta)\right]\right\} \tag{10}$$

Spectral density of the dynamic pressure at any point on the surface of the mono-pile can be expressed as:

$$S_{pp}(f, z, \theta) = |\rho \cdot g \cdot TF(f, z, \theta)|^2 S^p_{\eta\eta}(f, z, \theta) \tag{11}$$

in which, $S_{pp}(f, z, \theta)$ is the spectral density of the measured dynamic pressures upon the mono-pile and $S^p_{\eta\eta}(f, z, \theta)$ is the predicted spectral density of the water surface elevation from the measured pressure spectral density.

An empirical factor, $N(f, z, \theta)$, is introduced to account for the differences between the measured water surface wave and the one predicted from dynamic pressure. The empirical factor is expressed in frequency domain as:

$$N(f, z, \theta) = \sqrt{\frac{S^m_{\eta\eta}(f)}{S^p_{\eta\eta}(f, z, \theta)}} \tag{12}$$

in which, $S^m_{\eta\eta}(f)$ is the spectral density of the measured water surface elevation.

3 Results and Discussions

The experimental data of the tested wave UI4 are applied to recover the wave heights from subsurface pressures on the monopile. The tested wave spectrum from wave gauge

WP1 and the spectra of concurrent pressures upon the mono-pile at various levels ($|z/h|$ = 0.01 to 0.76) on the surface of the mono-pile and at an angle θ, are presented in Fig. 3.

Figure 4 presents the wave spectra recovered from pressures measured at a submergence level $|z/h| = 0.095$ on the complete circumference ($\theta = 0°$ to 360° in step of 30°) of the mono-pile under the tested wave UI4 (two top plots). It can be seen from Fig. 4 that the wave spectral densities recovered from pressures at angles $\theta = 150°$, 180° and 210° have the same peak as the measured one from wave gauge WP1. The recovered wave spectral densities at these three angles are lower than the measured one in the low frequency range ($f < f_{peak}$) and are higher than the measurement in the high frequency range ($f > f_{peak}$). In the higher frequency range ($f > 1.25 f_{peak}$) these recovered spectral densities are much higher than the measurement.

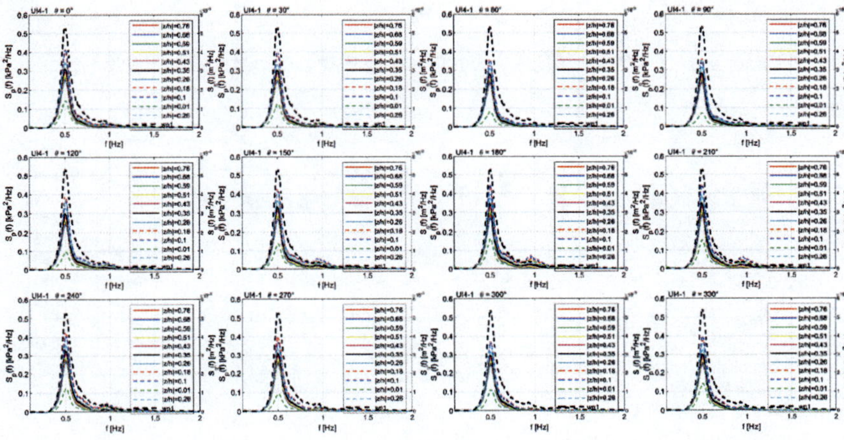

Fig. 3. The tested wave spectrum and concurrent pressures upon the mono-pile.

The theoretical transfer functions *TF(f)* at the submergence level $|z/h| = 0.095$ and on the complete circumference ($\theta = 0°$ to 360° in step of 30°) of the mono-pile are presented in two middle plots in Fig. 4. As it has been mentioned in Sect. 2 that $TF(f, z, \theta) = TF(f, z, 180° - \theta)$ because we get the real part of Eq. 10. It is shown that the higher submergence level z the higher transfer function values at all angles $\theta = 0°$ to 360° (Fig. 4). Two last plots of Fig. 4 present the empirical factors *N(f)* from all pressure data at the submergence level $|z/h| = 0.095$. In the low frequency range ($f < 0.43$ Hz), the empirical factors of all pressure data are very sensitive values. In frequency range $f > 0.43$ Hz the empirical factor trend lines are quite smooth and they vary gradual up to frequency of about 0.83 Hz then they have been reduced dramatically in frequency range of 0.83 Hz to 0.95 Hz. In the frequency range $f > 0.95$ Hz, the empirical factors from most of pressures are nearly horizontal developments. The empirical factors from pressure data at position $\theta = 150°$, 180° and 210° are quite similar and in frequency range 0.43 Hz $< f <$ 0.83 Hz the empirical factors from those positions θ vary from 0.85 to 1.2. The empirical factors from pressures at different submergence level and at a position θ are quite similar to each other. Their difference is about of 10% therefore the submergence level ($|z/h| > 0.095$) has not affected much to the empirical factor *N(f)*.

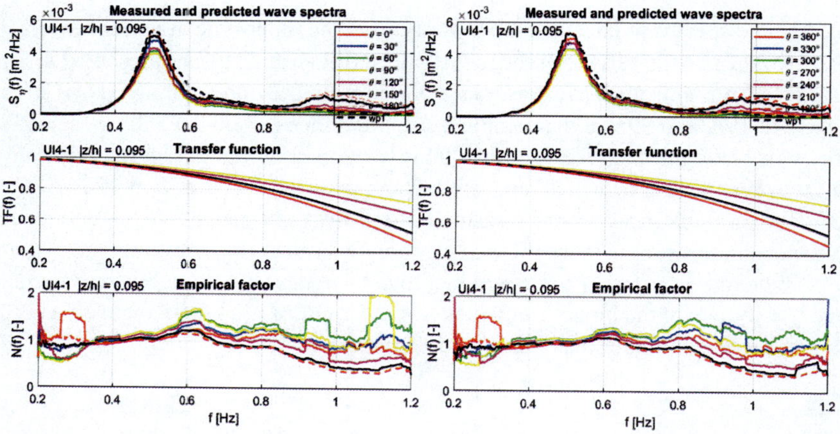

Fig. 4. $S_\eta(f)$, $TF(f)$ and $N(f)$ at different angles θ and at level $|z/h| = 0.095$ (P9).

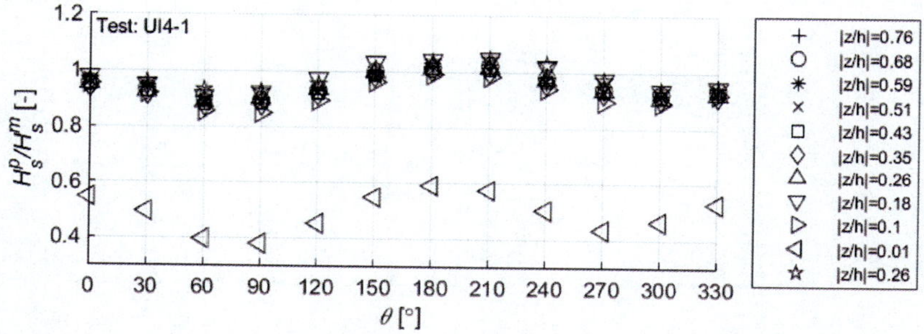

Fig. 5. Ratio of predicted and measured significant wave heights.

The ratios of the recovered significant wave heights H_s^p and the measured one H_s^m are presented in Fig. 5 for the tested wave UI4. As we can see from Fig. 5 that the values from pressures at level $|z/h| = 0.012$ (mostly at the water surface) are much lower than the values from the higher levels ($|z/h| = 0.095$ to 0.76). Therefore, the pressure data at the level $|z/h| = 0.012$ cannot be used to recover the wave heights. The significant wave heights recovered from pressure records at different submergence levels have small differences. Figure 5 shows the maximum difference of about 10% for the tested wave UI4. Also from Fig. 5, the recovered significant wave heights H_s^p from pressure records at position $\theta = 150°$, 180° and 210° are mostly similar to the measured/tested significant wave heights H_s^m. Therefore, the ratios of H_s^p/H_s^m at those positions θ are nearly unity. It can be seen that the significant wave height can be recovered quite well from pressure records at positions $\theta = 150°$, 180° and 210° (at all submergence levels $|z/h| > 0.095$), in comparing with the measured one (H_s^m).

4 Conclusions

This study has applied the linear transfer function to recover the significant wave heights from the tested subsurface pressures under waves. The physical model was carried out in a wave basin in Hannover, Germany. The results show that the significant wave height can be recovered from the tested pressures upon a small vertical mono-pile at the angle $\theta = 150°$, $180°$ and $210°$ (θ is the angle of orientation of the pressure sensor with respect to the incident wave direction) which give very good results in comparing with the measured one. In addition, it is found that the recovered significant wave heights have not been affected much by the submergence level ($|z/h| > 0.095$) of the pressure measurement.

Acknowledgements. The author gratefully acknowledges the support of the German Federal Environment Ministry (BMU) within the funded project "GIGAWIND alpha ventus – LUH" (BMU code 0325032/A).

References

Bishop, C.T., Donelan, M.A.: Measuring waves with pressure transducers. Coast. Eng. **11**, 309–328 (1987)

Cavaleri, L.: Wave measurement using pressure transducer. Oceanol. Acta **3**(3), 339–346 (1980)

Clamond, D.: New exact relations for easy recovery of steady wave profiles from bottom pressure measurements. J. Fluid Mech. **726**, 547–558 (2013)

Folsom, R.G.: Subsurface pressures due to oscillatory waves. Trans. Am. Geophys. Union **28**(6), 875–881 (1947)

Gabriel, D.W., Hedges, T.S.: Effects of currents on interpretation of subsurface pressure spectra. Coast. Eng. **10**(4), 309–323 (1986). https://doi.org/10.1016/0378-3839(86)90018-9

Grace, R.A.: Surface wave heights from pressure records. Coastal Eng. **2**, 55–67 (1978)

Kogelbauer, F.: Recovery of the wave profile for irrotational periodic water waves from pressure measurements. Nonlinear Anal. Real World Appl. **22**, 219–224 (2015)

Mai, T.C., Schlurmann, T.: Wave heights recovery from subsurface pressures upon a small vertical cylinder. Coast. Eng. Proc. 1(**33**), (2012). Posters 16, Santander, Spain. https://doi.org/10.9753/icce.v33.posters.16

Oliveras, K.L., Vasan, V., Deconinck, B., Henderson, D.: Recovering the water-wave profile from pressure measurements. SIAM J. Appl. Math. **72**(3), 897–918 (2012)

Sumer, B.M., Fredsøe, J.: Hydrodynamics Around Cylindrical Structures (Revised Edition). World Scientific Publishing Co. Pte. Ltd., Denmark (2006)

Evolving Soil-Conductor Stiffness Due to Multiple-Episode Cyclic Loading

Mariajose Guevara[1](✉), James Doherty[1], Phillip Watson[1], and David White[1,2]

[1] University of Western Australia, Perth, Australia
mariajose.guevaracastillo@research.uwa.edu.au
[2] University of Southampton, Southampton, UK

Abstract. Offshore drilling operations typically involve a wellhead/casing system connected to a floating vessel, with metocean forcing on the vessel translating to lateral movement at the wellhead. The conductor is the outer-most casing and must be checked against fatigue due to repeated cycling, which can create 'hot-spots' of high accumulated damage. Assessing this involves soil-structure interaction analysis, in which the soil-conductor lateral behaviour is modelled as non-linear springs called p-y curves.

Current industry practice often only accounts for stiffness degradation due to cyclic loading, excluding load history effects and/or potential regain in stiffness resulting from pore-pressure dissipation after cycling.

This paper compares results from centrifuge testing of a rigid length of conductor installed in reconstituted samples of carbonate silt and kaolin clay, subject to sequences of cyclic lateral displacement. The main aspects studied are the effect of previous cycling (load history), the effect of consolidation between episodes of undrained loading, and the impact of one-way vs two-way loading.

Keywords: Conductor · p-y curves · Centrifuge testing · Secant stiffness · Fatigue

1 Introduction

Conductors are a critical part of drilling riser operations, and are susceptible to fatigue because of the continuous motion of the system. Safe design depends on the assessment of fatigue utilisation, which is performed with interaction analyses where the soil-conductor behaviour is modelled as a series of non-linear springs, characterised by their secant stiffness. This stiffness will influence the distribution of bending moment along the length of the conductor and the location of the hot-spot, where the risk of fatigue damage is greatest.

Recently proposed approaches for soil-conductor lateral behaviour analysis [1] assume that the secant stiffness reaches an amplitude-dependent steady state behaviour after a few hundred cycles of loading. This method, which will be referred to as the "steady state approach" in this paper, is based on centrifuge tests on kaolin and Gulf of Mexico Clay.

D. V. K. Huynh et al. (Eds.): VSOE 2021, LNCE 208, pp. 278–285, 2022.
https://doi.org/10.1007/978-981-16-7735-9_30

However, recent centrifuge tests on carbonate silt [2] have shown that the secant stiffness increases with time from an undrained fully remoulded minimum (steady-state), consistent with the behaviour observed previously by multiple authors [3, 4]. In addition, the undrained fully remoulded stiffness and subsequent increase depend on the highest previous load experienced by the soil [2, 5] and the number of cycles applied.

The results presented in this paper aim to expand on previous work by examining fully remoulded stiffness and its load-history dependence. The tests presented here are part of a broader campaign aimed at improving fatigue design of subsea well conductors and casings.

2 Centrifuge Test Description

The tests were conducted in the 1.8 m radius beam centrifuge at the National Geotechnical Centrifuge Facility, University of Western Australia, on samples of reconstituted carbonate silt and kaolin clay. Samples were prepared from slurry that was into a strongbox of internal dimensions of 650 mm long, 390 mm wide and 325 mm height (with a sand base drain). The moisture content (w) of the slurries was 140% for the carbonate silt and 130% for the kaolin clay. After positioning it in the centrifuge, the sample was spun for ~5 days at an acceleration of 40 g until fully consolidated.

A miniature T-bar penetrometer (5 mm in diameter and 20 mm in length [6]) was used to track changes in soil strength with time. T-bar tests were performed at a penetration rate of 3 mm/s and included a cyclic stage to investigate remoulding. Typical measured strength gradients were 1.65 and 1.05 kPa/m, with sensitivities of 3.3 and 2.40, for the carbonate silt and kaolin clay, respectively. Values of consolidation coefficient have been reported by other authors as $c_h \approx 1$ m^2/year [7] for the carbonate silt and 3.3 m^2/year [8] for the kaolin clay ($\sigma'_{v0} = 15$ kPa).

It is important to note that, while the carbonate silt was recovered from an offshore site, its response does not necessarily replicate in-situ soil behaviour. This stems from the reconstitution process used in the laboratory, which cannot replicate the natural deposition process – often resulting in lower sensitivity in the model than in the field. However, it is noted that the strength and sensitivity of the carbonate silt are similar to properties seen in deep water Gulf of Mexico clay, hence making these results directly relevant for conductors in such soils.

A schematic of the test setup used is shown in [2], and briefly described here. The stainless-steel model pile had an outer diameter (D) of 19.5 mm and a wall thickness (t) of 0.95 mm (D/t = 20.5), representing a prototype conductor of 780 mm (30 in.) and 38 mm (1.5 in.) in diameter and wall thickness, respectively. The horizontal load H applied was measured using a bending leg, with two independent lasers measuring the displacement and rotation of the pile. Although the pile was practically rigid for the range of loads applied, there was a very small amount of compliance in the connection of the pile and the bending leg. This was accounted for by developing a relationship between the applied load and pile rotation, which was then used to determine the pile lateral displacement at 50% of the penetration depth.

The piles were installed in flight to an embedment depth of 4.5D or 6D (depending on the test) into 8 different locations within the sample. The test locations were separated by at least 6D from each other and from the walls of the box to avoid interaction effects. After installation, the lateral loading sequence imposed on the pile was preceded by a stationary period of 2.15 h to allow the dissipation of the pore pressures generated during the installation process.

The tests discussed in this paper comprised four episodes of displacement controlled cyclic motion at the actuator level, followed by a stationary period to allow for pore pressure dissipation. The duration of the stationary period was assessed based on the solution proposed by [9] for consolidation around a laterally loaded pile, and set to achieve approximately 75% dissipation. This solution was also applied to determine the time frame to consider an episode of cycling as (effectively) undrained, by restricting its duration to the time that would result in no more than 25% consolidation. Each episode consisted of an arrangement of five packets of increasing and decreasing lateral displacement amplitude (Δy), shown in Table 1, and were applied to both carbonate silt and kaolin clay. In addition to the tests in Table 1, a monotonic test was performed to determine the lateral capacity (H_{ult}) for each soil, for comparison to cyclic testing.

Table 1. Test type and description

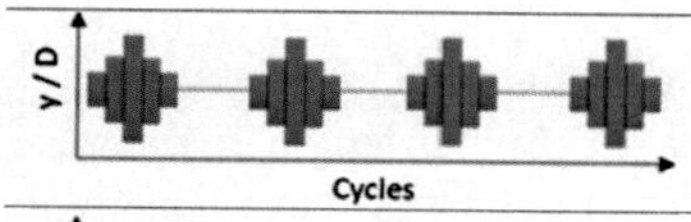	**Test A:** Amplitude of each 100 cycles packet (Δy/D)					
	CS	0.04	0.08	0.16	0.08	0.04
	KC	0.05	0.10	0.20	0.10	0.05
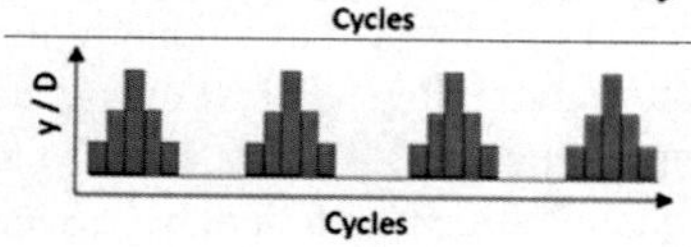	**Test B:** Amplitude of each 200 cycles packet (Δy/D)					
	CS	0.04	0.08	0.16	0.08	0.04
	KC	0.05	0.10	0.20	0.10	0.05

3 Results and Discussion

The results are reported in terms of a normalised secant stiffness (K), which has been defined in this paper as:

$$K = \frac{\Delta H / H_{ult}}{\Delta y / D} \tag{1}$$

Where ΔH is the cyclic range in horizontal load for a particular cycle, representing the integrated response of the soil pressure along the pile length; H_{ult} is the measured lateral load capacity from the monotonic test; Δy is the difference between the peak lateral displacement and the trough lateral displacement at mid embedment depth of the conductor per cycle (range of displacement), and D is its diameter.

A plot of the horizontal load and displacement at the actuator level, for the first two episodes of test type 1 in the carbonate silt with 4.5D embedment, is shown in Fig. 1a. The first and last cycle of packets 2, 3 and 4 are highlighted in red, green and magenta, respectively, for the first episode of the test. The load against displacement at mid embedment depth of the pile is shown in Fig. 1b for the highlighted cycles, with an indication of how ΔH and Δy are obtained for Eq. (1), for the first cycle of the 3rd packet (green continuous line on the plot).

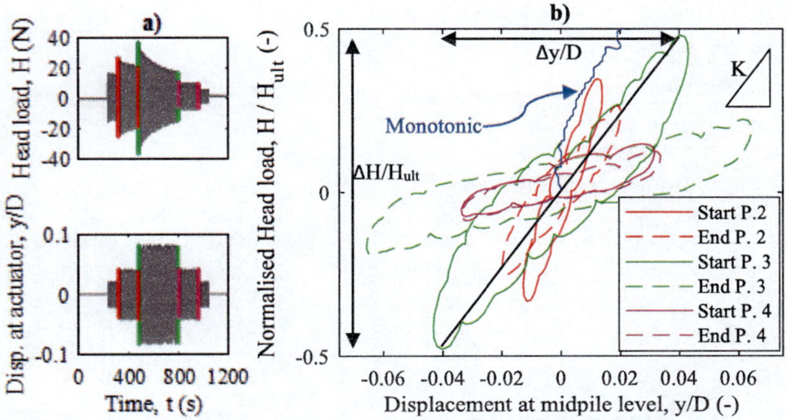

Fig. 1. Data from test Type 1 at 4.5 D embedment depth: a) Head load and displacement at the actuator level vs time (model scale); b) Head load vs displacement at mid-embedment level for the start and end cycle of the 2nd, 3rd and 4th packets of loading, and the monotonic push test (model scale).

The Effect of Previous Cycling on the Stiffness During an Undrained Event

The secant stiffness K vs cycle number for all the packets of episode 1, 2, 3 and 4, from test type A (two-way loading) and type B (one-way loading) at 6D embedment depth, can be observed in yellow, red, green and blue, respectively, in Fig. 2, for both carbonate silt and kaolin clay. As described by [2], previous cycles of high amplitude displacement have an effect on the subsequent lesser amplitude cycles within an undrained episode of loading. At the end of the third packet of the first episode on the carbonate silt, and second packet for the kaolin clay, the subsequent stiffness for smaller amplitudes was fully remoulded and exhibited a value comparable to that of a large amplitude cycling (~0.1 D). Despite higher sensitivity, it appears from these plots that the carbonate silt shows a more resilient behaviour than the kaolin clay: the stiffness fully degraded for the silt when cycled at Δy/D = 0.16, whereas the stiffness of the kaolin fully degraded when cycled at Δy/D = 0.10.

One-Way Cycling vs Two-Way Cycling

A comparison of the secant stiffness from the final cycle of each packet of episode 1 and episode 4 for both test type A and test type B is plotted against half the cyclic amplitude range (Δy/2D) in Fig. 3 for both soils. Not much difference is observed between one-way (offset) loading and two-way loading tests on both soils, except for very small displacement amplitudes (Δy) during the increasing phase of episode 1 of the tests – although it is difficult to conclude on with certainty whether this is a real effect, due to challenges in measuring the response at such small movements.

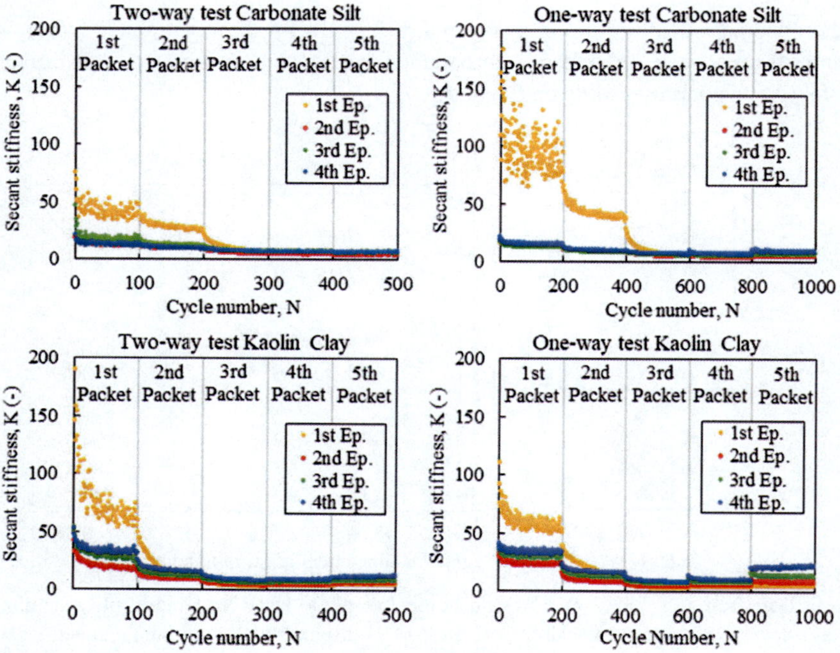

Fig. 2. Secant Stiffness against cycle number for test type A (two-way cycling) and test type B (one-way cycling) on carbonate silt and kaolin clay.

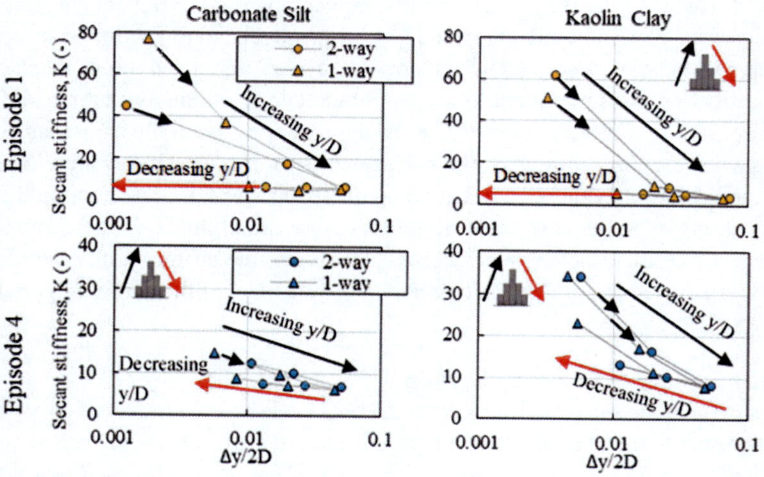

Fig. 3. Secant stiffness of the final cycle in each packet against episode amplitude for test type A (two-way) and test type B (one-way) on carbonate silt and kaolin clay, for first and final episodes.

Regain in Strength After Undrained Cycling Episodes and Consolidation Periods

The undrained secant stiffness (K) of the final cycle of each packet during the increasing amplitude phase of all the episodes for test type A (two-way) on carbonate silt and kaolin clay can be observed in Fig. 4a, plotted against the amplitude of cycling. After the first loading episode in the carbonate silt, the stiffness of the subsequent episodes follows a distinctive path, in which the stiffness values are approximately half of those in the first episode. However, the stiffness path for the kaolin clay progressively increases with each episode of cycling and consolidation after the fully remoulding process of the first episode.

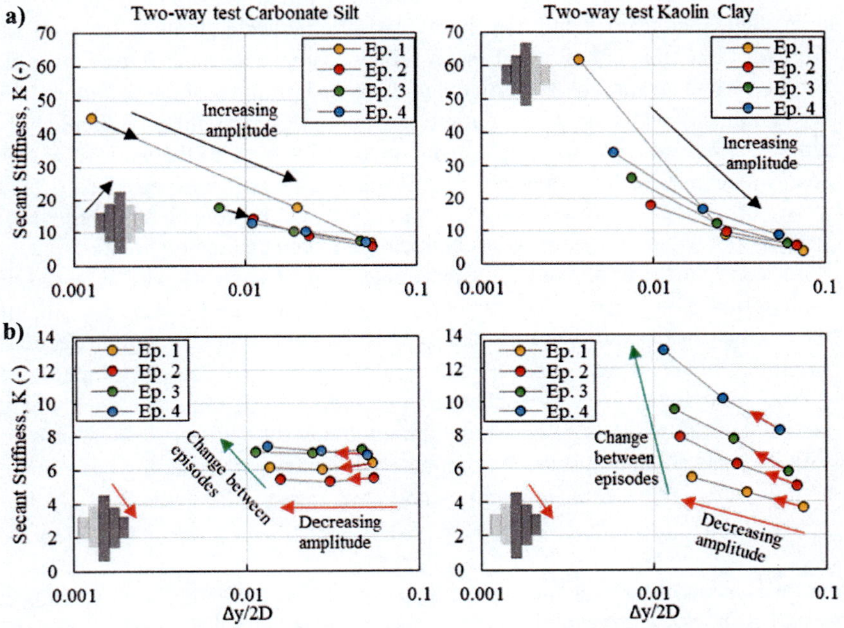

Fig. 4. Secant stiffness against amplitude per episode of undrained cyclic loading of two-way cyclic amplitude tests on carbonate silt and kaolin clay on a) The increasing cyclic amplitude phase, and b) the decreasing cyclic amplitude phase.

The undrained secant stiffness (K) against amplitude per episode of undrained cyclic loading for the decreasing amplitude phase of test type A on carbonate silt and kaolin clay is plotted in Fig. 4b against the amplitude of cycling. It is observed that, after the first episode of loading, the stiffness decreases further on the second episode and shows a recovery at the last episode, in which the stiffness value is 1.5 times the minimum value. Conversely, the kaolin clay shows an increase in stiffness with the decrease in cyclic amplitude, and a progressive recovery after each episode. It is also observed that the gradient of the path of the stiffness becomes steeper for each subsequent episode.

Interestingly, the kaolin clay exhibits more degradation than the silt on the largest amplitude packet of the first episode of cycling, but recovers up to 2 times its minimum value.

4 Conclusions

1. It was verified that one-way and two-way cycling do not yield significantly different secant stiffness values, which agrees with results reported from previous studies. Differences were observed for very small displacements on the increasing amplitude phase of the tests, but more tests are needed to confirm if this is a real phenomenon. This means that a small offset in displacements (<0.1D), produced by drifting of the vessel, could be analysed as if the vessel was positioned on top of the conductor, which is common practice when performing wellhead systems fatigue life assessments.
2. The soil-conductor stiffness on the decreasing amplitude phase of each event, is lower than on the increasing amplitude phase. Instead of returning to its steady state value from the increasing amplitude phase, K remained constant as the amplitude reduced for the carbonate silt, being set by the previous maximum sustained amplitude. For the kaolin clay on the decreasing amplitude phase, K approximately doubled as the amplitude reduced by a factor of 10, but remained about half the value of K from a steady state approach. This means that, if the conductor has experienced sufficient cycles of a larger amplitude loading, its stiffness will be considerably reduced and the hot-spot will migrate down the conductor, influencing the fatigue life estimation.
3. A recovery in the undrained fully remoulded secant stiffness was observed after each period of pore pressure dissipation, where K increased to 1.5 times its minimum value for the carbonate silt and 2 times for the kaolin clay, on the decreasing amplitude phase of each episode. This stiffening effect of the consolidation after cycling partially removes the reduction in stiffness of unloading relative to loading, and implies that assuming a post-large-amplitude secant stiffness for smaller cycling amplitudes is not appropriate. An adequate soil-conductor modelling has to include both the effect of previous remoulding and pore pressure dissipation.

Acknowledgements. This work was supported by the ARC Industrial Transformation Research Hub for Offshore Floating Facilities which is funded by the Australian Research Council, Shell Australia, Woodside Energy, Bureau Veritas and Lloyd's Register (IH140100012). The third and fourth authors also acknowledge support from Shell Australia via the Shell Chair in Offshore Engineering at UWA.

References

1. Zakeri, A., Sturm, H., Jeanjean, P.: Validation and extension of soil response framework for fatigue analysis of offshore wells and piles. In: Offshore Technology Conference, p. 26 (2019)
2. Guevara, M., Doherty, J., Watson, P., White, D.: Key features impacting soil-conductor lateral behaviour as illustrated by centrifuge tests. In: Proceedings of 4th International Symposium Front Offshore Geotech, ISFOG 2020, Austin, pp. 1–10 (2020)
3. Zhang, C., White, D., Randolph, M.: Centrifuge modeling of the cyclic lateral response of a rigid pile in soft clay. J. Geotech. Geoenviron. Eng. **137**, 717–729 (2011)

4. Zhou, Z., O'Loughlin, C.D., White, D.J., Stanier, S.A.: Improvements in plate anchor capacity due to cyclic and maintained loads combined with consolidation. Géotechnique **70**, 732–749 (2019)
5. Doherty, J., White, D., Watson, P., Grime, A.: Life cycle changes in p-y stiffness for a conductor pile installed in carbonate silt. In: Randolph, M.F., Doan, D.H., Tang, A.M., Bui, M., Dinh, V.N. (eds.) Proceedings of the 1st Vietnam Symposium on Advances in Offshore Engineering: Energy and Geotechnics, pp. 362–368. Springer, Singapore (2019). https://doi.org/10.1007/978-981-13-2306-5_50
6. Stewart, D.P., Randolph, M.F.: A new site investigation tool for the centrifuge. In: International Conference on Centrifuge 1991, Balkema, Rotterland, pp. 531–538 (1991)
7. Chow, S.H., O'Loughlin, C.D., Zhou, Z., White, D.J., Randolph, M.F.: Penetrometer testing in a calcareous silt to explore changes in soil strength. Géotechnique **70**, 1160–1173 (2020)
8. Richardson, M.D., O'Loughlin, C.D., Randolph, M.F., Gaudin, C.: Setup following installation of dynamic anchors in normally consolidated clay. J. Geotech. Geoenviron. Eng. **135**, 487–496 (2009)
9. Osman, A.S., Randolph, M.F.: Analytical solution for the consolidation around a laterally loaded pile. Int. J. Geomech. **12**, 199–208 (2012)

Experimental Investigation of Wave Scattering Around a Large Vertical Circular Cylinder

Tri Mai(✉)

Faculty of Coastal and Offshore Engineering, Hanoi University of Civil Engineering, 55 Giai Phong, Hai Ba Trung, Hanoi, Vietnam
trimc@nuce.edu.vn

Abstract. Wave-structure interaction is one of the most interesting studying for an offshore structure. This paper is going to present an experimental investigation of wave scattering around a large vertical circular cylinder in a 3D wave basin to understand how the scattered wave field is affected by a large cylinder and wave steepness. The existing phase-based harmonic separation method has been applied to extract the interger harmonic components. It is found that the amplitude of the second and third harmonic components at the bow and around the cylinder is significantly increased as the wave steepness is increased. In practical design of offshore structures, the second, third and fourth harmonics should be taken in assessment of springing- and ringing-type structural responses and the air gap.

Keywords: Wave-structure interaction · Diffracted waves · Circular cylinder · Focused waves · Physical model

1 Introduction

Wave-induced load components at integer harmonics of the dominant linear input wave frequency can excite high frequency resonant response for floating offshore structures (e.g. floating offshore wind turbine, floating wave energy converter, and floating platforms more generally) and also for bottom-fixed offshore structures (e.g. gravity-based structures - GBS). There may be a nonlinear transfer of energy to a higher-harmonic response of the structure owing to nonlinear wave-wave interaction effects and nonlinear wave-structure interaction effects. Therefore, waves with the incident spectral energy at peak frequency (f_p) can in principle excite structural responses at multiples of the linear peak frequency ($2f_p$, $3f_p$, $4f_p\ldots$). These higher-harmonic frequencies are known to cause highly intense nonlinear structural behaviours called *springing* (at double frequency) and *ringing* (at triple frequency), which were first observed in a model test of the Hutton platform which was operated in the UK North Sea from 1984–2001 [1]. The second-order excitation at the double frequency dominates for springing, while the higher-order (3rd and 4th order) frequencies trigger the ringing of gravity-based platforms and tension leg platforms, which is a transient elastic response [2, 3].

Fitzgerald et al. [4] studied higher harmonic diffracted wave fields around a surface-piercing column due to a focused wave group using NewWave theory (Tromans et al.

D. V. K. Huynh et al. (Eds.): VSOE 2021, LNCE 208, pp. 286–293, 2022.
https://doi.org/10.1007/978-981-16-7735-9_31

1991). They simulated wave-structure interaction of a focused wave group (with $kA = 0.1$, where k is the wavenumber corresponding to the spectral peak energy period, and A is the total amplitude of the linear harmonic) and a 0.25 m diameter cylinder, using a fully nonlinear higher-order BEM potential flow model. Their simulation showed the second and third harmonics of the total and scattered wave fields at locations upstream and downstream of the model. Their general phase-based harmonic separation method [4] is applied to decompose the local wave field in this work.

The importance of these scattered waves and the associated loads depends on the application. One of the principal loading components on floating bodies is the slow drift force - the 2nd order difference frequency loading - as this is a major loading component in the design of the spread moorings, together with current and wind. Double and higher frequency harmonics of force are less important here. However, another design condition for floating bodies is the occurrence and severity of loading arising from green water on deck [5–7]. The onset of green water will be significantly influenced by 2nd order difference and sum (and higher harmonic) contributions to local surface elevation, as negative freeboard is required for green water to appear on the deck.

In the present work, those higher harmonic components of the scattered wave are experimentally investigated to get a better understanding of how wave steepness effects on the local scattered wave field, may affect local and global loads, which may cause damage to or failure of offshore structures.

2 Methodology

2.1 Physical Model

The experimental work was carried out in the Ocean Basin at University of Plymouth's COAST Lab. A water depth of 2.93 m was used for this experiment. The truncated cylinder was made of aluminium. The tested model had the height of 0.3 m, and diameter of 0.3 m (Fig. 1). The draft was 0.15 m for all test cases. The cylinder was rigidly fixed to the gantry (which spans the width of the Ocean Basin).

The input waves were focused wave groups generated using the NewWave methodology [8] with an underlying JONSWAP spectrum ($\gamma = 3.3$), focusing at the bow of the models. Hence, the input wave group is a representation of the average shape of the largest (linear) waves in a random sea-state with a JONSWAP spectrum. The local wave steepness were 0.13 and 0.18, and the incident wave angle was 0° which corresponds to a head sea. The ratio of body size over wavelength (corresponding to T_{p}) is about 0.09 for the tested cylinder. The local wave field around the model were measured by 6 resistance wave gauges, as shown in Fig. 2, with a sampling frequency of 128 Hz [9, 10].

2.2 Phase-Based Harmonic Separation Method

In this study the extended phase-based harmonic separation method (Fitzgerald et al. 2014) is applied to extract the linear and higher-order harmonic components of the free-surface elevation by applying simple linear combinations of the relevant time histories.

Fig. 1. The tested truncated cylinder in the Ocean Basin.

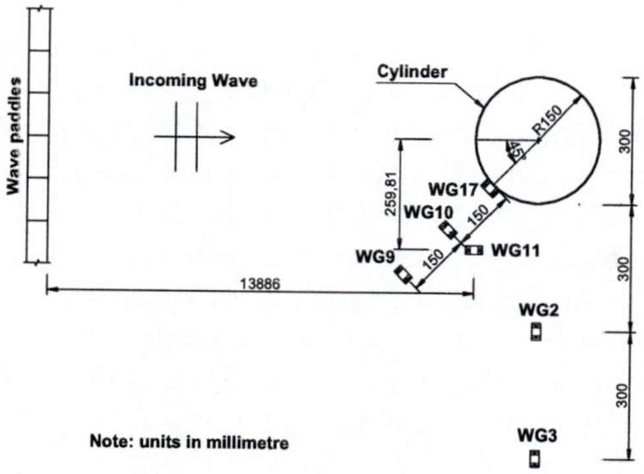

Fig. 2. Layout of wave gauges around the tested model.

This method requires the data from four incident focused wave groups that are each exactly 90° out of phase. There is then minimal post-processing applied to extract the linear, second-, third- and fourth-harmonics [9, 10].

An incident wave group that has amplitude A and relative phase θ can be expressed as the classic Stokes perturbation expansion [11], as follows

$$\begin{aligned}\eta(A,\theta) &= B_{11}A\cos\theta + A^2(B_{20} + B_{22}\cos 2\theta) + A^3(B_{31}\cos\theta + B_{33}\cos 3\theta)\\ &+ A^4(B_{40} + B_{42}\cos 2\theta + B_{44}\cos 4\theta) + O\left(A^5\right)\end{aligned} \quad (1)$$

where B_{ij} are the coefficients in Fourier series for $\eta(A,\theta)$; i is the amplitude content order; and j is the frequency content order. Equation (1) can be rewritten in a more compact form as:

$$\eta(A,\theta) = \eta_{11} + (\eta_{20} + \eta_{22}) + (\eta_{31} + \eta_{33}) + (\eta_{40} + \eta_{42} + \eta_{44}) + O\left(A^5\right) \quad (2)$$

where, η_{ij} are the jth-order harmonic components, $\eta_{ij} = A^i\, B_{ij}\cos(j\theta)$. If i and j are identical, then η_{ij} are the jth-order harmonic sum, e.g. the first-order sum η_{11}. On the other hand, if i and j are different, then η_{ij} are the jth-order harmonic difference, e.g. the term η_{31} is at the first-harmonic in frequency but 3rd order (cubic) in input wave amplitude. Henceforth, the subscript i is referred to as the (amplitude) order and j as the harmonic.

By considering four focused wave groups generated from the same wave amplitude components but with the phase of each Fourier component shifted 0°, 90°, 180° and 270° it is possible to obtain the four time histories of free-surface elevation η^0, η^{90}, η^{180} and η^{270} respectively. The linear combinations of these time histories and the Hilbert transforms of the 90° and 270° free-surface elevation time histories are applied to extract the linear and the first three superharmonics (2nd, 3rd and 4th order); these are important for springing and ringing and are presented in Eqs. (3)–(6). A more detailed explanation of the separation method can be found in [4].

$$\frac{\eta^0 + H(\eta^{90}) - \eta^{180} - H(\eta^{270})}{4} = \eta_{11} + \eta_{31} \tag{3}$$

$$\frac{\eta^0 - \eta^{90} + \eta^{180} - \eta^{270}}{4} = \eta_{22} + \eta_{42} \tag{4}$$

$$\frac{\eta^0 - H(\eta^{90}) - \eta^{180} + H(\eta^{270})}{4} = \eta_{33} \tag{5}$$

$$\frac{\eta^0 + \eta^{90} + \eta^{180} + \eta^{270}}{4} = \eta_{20} + \eta_{40} + \eta_{44} \tag{6}$$

where H is the harmonic conjugate of the signal. Also note that due to the relatively weak nonlinearity of the difference components compared with the sum components, for example η_{31} compared with η_{11}, the difference components are likely to be negligible.

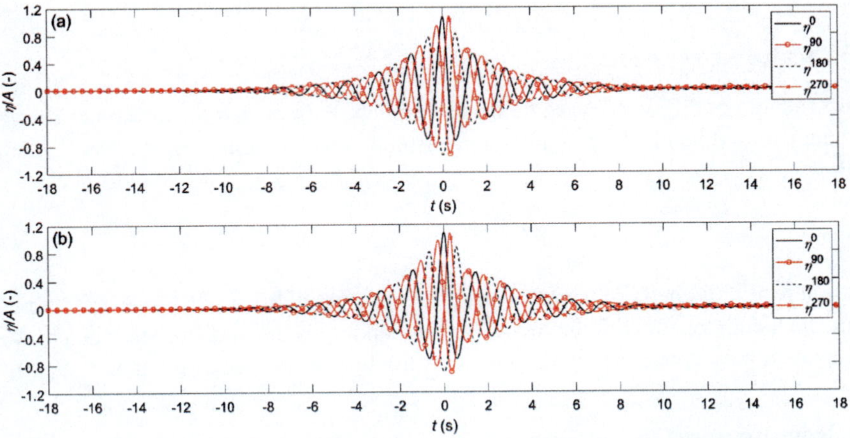

Fig. 3. Wave profiles at the focus location (without cylinder): (a) $kA = 0.13$; (b) $kA = 0.18$.

Time histories of the free-surface elevation η^0, η^{90}, η^{180} and η^{270} at the focus location (wave gauge WG11) are shown in Fig. 3 for the tested wave groups with $kA = 0.13$ and 0.18. The vertical axis is the dimensionless free-surface elevation (η/A), where η is the free-surface elevation and A is the linear amplitude at the focus location and time. The horizontal axis is the time scale with the focus time at $t = 0$ s.

Applying the linear combinations presented in Eq. (3)–(6) for η^0, η^{90}, η^{180} and η^{270} in Fig. 3, the linear and the next three higher harmonic components have been obtained and presented in terms of their separated time histories (Fig. 4). Minimal post-processing (Fourier bandpassing) has been applied to remove the leakage of the linear component in the higher harmonics [9, 10]. Similar to the finding by [9], there was a double frequency error wave train off the wave paddles since only linear wave theory was used to create the waves. This can be seen at around $t = +13$ s for the second harmonic presented in Fig. 4b.

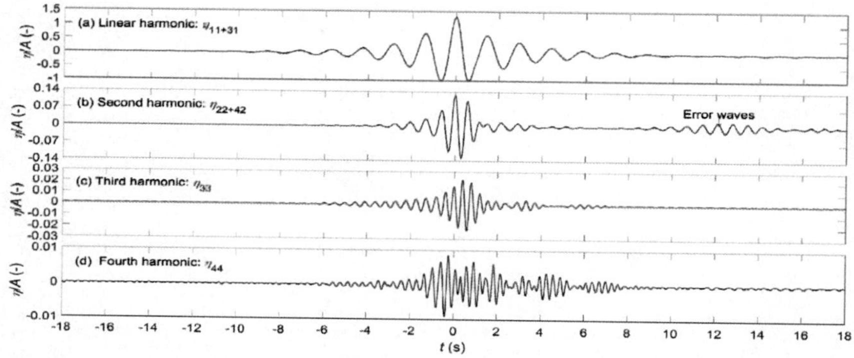

Fig. 4. Time histories of the separated components at the focus location (without cylinder, $kA = 0.18$). Note the different vertical scales.

2.3 Scattered Wave Field

The scattered or diffracted wave field can simply be extracted as the difference between the undisturbed incident wave and the measured wave in the presence of the model as follows

$$\eta_{ij}^{dif} = \eta_{ij}^{Model} - \eta_{ij}, \tag{7}$$

where η_{ij}^{dif} is the $(i,j)^{th}$ component of the scattered/diffracted wave, η_{ij}^{Model} is the equivalent component with the model present, and η_{ij} is the undisturbed incident wave component measured at the same gauge location in the absence of the model [9, 10].

3 Results and Discussions

The effect of wave steepness on the scattering is reported for the truncated cylinder with the focused wave groups of two different steepnesses ($kA = 0.13$ and 0.18) and five gauge locations: at the bow (WG17), near to the bow (WG10) and to the side (WG2). The non-dimensional ratio D/L_p of the tested waves varies from 0.09 to 0.11 (with the linear component), from 0.37 to 0.43 (with the second-harmonic), from 0.83 to 0.98 (with the third-harmonic) and from 1.48 to 1.74 (with the fourth-harmonic). Due to

$D/L_p < 0.2$, the diffraction of the linear component is much weaker than those of the higher harmonic components (second-, third- and fourth-harmonics) which have $D/L_p > 0.2$.

As found in [9], there is no evidence of significant cumulative evolution beyond linear as the wave propagates from the paddle to the position of the model. Therefore, we can treat the incident linear components as identical in shape, simply with an amplitude scaling.

Applying the phase-based separation method presented in Sect. 2, the filtered time histories of the linear and second, third and fourth harmonics of the separated components are examined. Only the more interesting higher harmonics are presented in this paper to examine the effect of wave steepness on the wave-structure interaction, because the linear component simply scaled with wave steepness. The filtered time histories of the higher harmonics of the scattered wave fields at locations WG17, 10 & 2 are presented in Fig. 5 to Fig. 7, respectively. In general, as would be expected, the higher harmonics are seen to increase as the wave steepness is increased from $kA = 0.13$ (solid line) to $kA = 0.18$ (dashed line). There is considerable amplification of the second and third harmonics at all locations (see Fig. 5a&b, Fig. 6a&b, Fig. 7a&b). At the bow, the second and third harmonics are significantly increased as increasing the wave steepness and this may due to effects of the wave run-up on the cylinder (Fig. 5a&b). The fourth harmonic component near the bow (Fig. 6c) and alongside (Fig. 7c) is slightly higher than that at the bow of the model (Fig. 5c).

The second-, third- and fourth-order frequency components obtained from this physical model test should be taken into account to assess wave loading for offshore structure design. The assessment of load components that might produce springing- and ringing-type structural responses in fixed and taut moored structures should be done in practical design. Usually, numerical modelling approaches are based on the linear theory and cannot predict these strongly nonlinear effects [12] and therefore the high order or fully nonlinear approaches should be taken. For designing the air gap and location of acoomodation in offshore structures, the effects of the superharmonics (up to fourth-order) on crest elevation should also be considered.

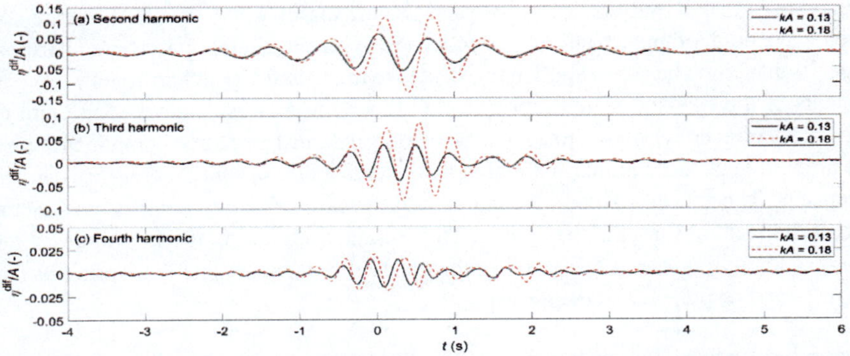

Fig. 5. Time histories of the scattered waves at the bow of the cylinder (WG17). Note the different vertical scales.

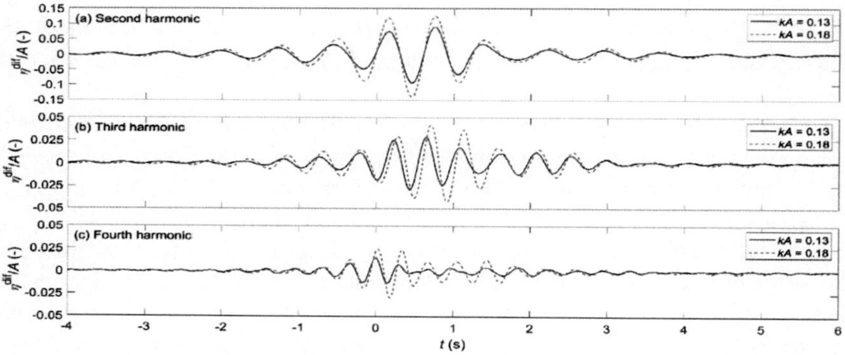

Fig. 6. Time histories of the scattered waves near the bow of the cylinder (WG10). Note the different vertical scales.

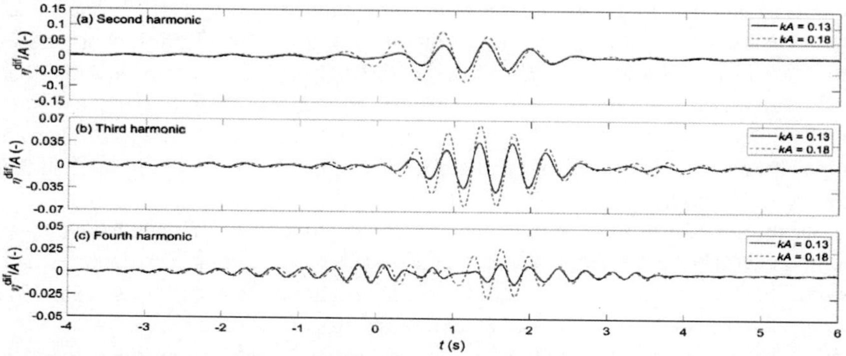

Fig. 7. Time histories of the scattered waves alongside the cylinder (WG2). Note the different vertical scales.

4 Conclusions

This experimental study have investigated the effects of wave steepness on the scattered wave field around a truncated cylinder. An existing general phase-based harmonic separation method has been applied to extract the linear and higher harmonic wave components of the free-surface elevation around the cylinder. Results show that there is a significant effect of wave steepness on the second and third harmonic components where the amplitude of those higher hamonic components increases with increasing the wave steepnees. The fourth harmonic component at the bow of the cylinder is smaller than that at the locations around the cyliner. The second, third and fourth harmonics should be considered in practical design of offshore structure for assessment of springing- and ringing-type responses and the air gap.

Acknowledgements. This study is a part of the FROTH (Fundamentals and Reliability of Offshore sTructure Hydrodynamics) project supported by the Engineering and Physical Science Research Council (EPSRC Grant EP/J012866/1).

References

1. Mercier, J.A.: Evolution of tension leg platform technology. In: Proceedings of the 3rd International Conference Behaviour of Offshore Structures, Massachusetts Institute of Technology, Cambridge, MA. (1982)
2. Faltinsen, O.M., Newman, J.N., Vinje, T.: Nonlinear wave loads on a slender vertical cylinder. J. Fluid Mech. **289**, 179–198 (1995)
3. Faltinsen, O.M.: Hydrodynamics of marine and offshore structures. J. Hydrodyn. **26**(6), 835–847 (2014). https://doi.org/10.1016/S1001-6058(14)60092-5
4. Fitzgerald, C.J., Taylor, P.H., Taylor, R.E., Grice, J., Zang, J.: Phase manipulation and the harmonic components of ringing forces on a surface-piercing column. Proc. R. Soc. Lond. A Math. Phys. Eng. Sci. **470**, 20130847 (2014). https://doi.org/10.1098/rspa.2013.0847
5. Barcellona, M., Landrini, M., Greco, M., Faltinsen, O.M.: An experimental investigation on bow water shipping. J. Ship Res. **47**(4), 327–346 (2003)
6. Greco, M., Colicchio, G., Faltinsen, O.M.: Shipping of water on a two-dimensional structure. Part 2. J. Fluid Mech. **581**, 371 (2007)
7. Schoenberg, T., Rainey, R.C.T.: A hydrodynamic model of green water incidents. Appl. Ocean Res. **24**(5), 299–307 (2002)
8. Tromans, P.S., Anaturk, A.R., Hagemeijer, P.: A new model for the kinematics of large ocean waves - application as a design wave. In: Proceedings of the 1st International Offshore and Polar Engineering Conference, pp. 64–71, Edinburgh, U.K., 11–16 August (1991)
9. Mai, T., Greaves, D., Raby, A., Taylor, P.H.: Physical modelling of wave scattering around fixed FPSO-shaped bodies. Appl. Ocean Res. **61**, 115–129 (2016)
10. Mai, T.C.: On the role of aeration, elasticity and wave-structure interaction on hydrodynamic impact loading (Doctoral dissertation, University of Plymouth) (2017)
11. Fenton, J.D.: A fifth-order Stokes theory for steady waves. J. Waterw. Port Coast. Ocean Eng. ASCE **111**, 216–234 (1985)
12. Veritas, D.N.: Environmental conditions and environmental loads. Recommended Practice DNV-RP-C205 (2010)

Experimental Study on the Flexural Performance of RC Beam Strengthened with CFRP Sheet

Phi Long Nguyen(✉), Ngoc Gianh Pham, and Trung Hieu Nguyen

Ho Chi Minh City University of Transport, No 2 Vo Oanh St., Binh Thanh Dist., HCMC, Vietnam
long.nguyen@ut.edu.vn

Abstract. The main goal of this work is to experimentally evaluate the influence of anchorage length of carbon fiber reinforced polymer (CFRP) on the mechanical performance of concrete beam strengthened with CFRP sheet. Twelve reinforced concrete beams are strengthened in flexure using CFRP sheet at four different cases of anchorage length using wet-layout technique. The standard anchorage length is calculated based on Niedermeier's formula. The experimental result shows that the ultimate load of CFRP reinforced concrete beam increases up to 43% as the anchorage length of CFRP sheet increases up to 3 times of standard one. The failure mode changes from ductile to brittle mode as the increase of anchorage length due to the improvement in stiffness of beam segment bonded with CFRP sheet. From this result, the numerical model for flexural behavior of the CFRP strengthened reinforced concrete beam can also be developed for parametric studies and then engineering design.

Keywords: Experimental study · RC beams · Strengthening · CFRP · Anchorage length

1 Introduction

Upgrading the existing civil engineering infrastructure is becoming one of great importance for over recent decades. Deterioration of these structure may be attributed to ageing, environmentally induced degradation, poor initial design and/or construction, and lack of maintenance. The increasing decay combined with the need for upgrading to meet more stringent design requirements demand for structure renewal throughout the world. Recent decade, among available options, the use of polymer reinforced with carbon fibre (Carbon Fibre Reinforced Polymer - CFRP) is becoming popular in civil engineering. This material is used to reinforcing concrete or steel structure thanks to its advantages in high tensile/weight ratio, good properties in corrosion resistance and fatigue. This can be applied for not only near shore structures but also offshore structures such as oil rig, base caisson,… In the literature, there are several reports on mechanical performance of RC beam reinforced with CFRP which shows the increase in load-carrying capacity within the adequate bond strength [1]. The configurations of CFRP-concrete

D. V. K. Huynh et al. (Eds.): VSOE 2021, LNCE 208, pp. 294–301, 2022.
https://doi.org/10.1007/978-981-16-7735-9_32

bond also have been intensively studied by many authors and synthesized by Kang et. al. [2]. Several guidelines by organizations including ACI Committee 440, ISIS-Canada, and CFRR A-Japan are under way of standardizing the use of carbon fiber reinforced polymer (CFRP) in civil engineering [3]. However, to confirm the analytical models, several extended experiments need to be conducted despite of its expensive cost. This paper presents the results of experimental studies concerning the flexural strengthen of RC beams by the externally bonding of CFRP sheet to the tension region of the beams under mechanical load. The objective is to contribute to the standardization of anchorage length in strengthening techniques. The ultimate results are then discussed and validated with available analytical models in the literature.

2 Experimental Approach

To study the efficiency of flexural strengthening solution, the three point bending test or four point bending test can be conducted for its mechanical behavior. These tests can help characterize the behaviour of bending sections in real structure such as simply supported beam or even continuous beam. In this study, the four point bending system on simply supported beam was used to study the flexural performance of CFRP RC beam at different anchorage lengths. The specimen and testing procedure are detailed in the following subsections.

2.1 Specimens

A total of twelve beams were tested. All beams have identical rectangular cross-sections and the same size: 150 mm × x 200 mm × 2200 mm (Fig. 1). During the test, the mechanical load and displacement at middle bottom of each beam will be recorded. Three beams were used as referenced beam and other nine beams were strengthened in flexure using one layer of externally bonded CFRP sheet at three different anchorage lengths (L_b, Fig. 1). The minimum anchorage length is calculated as function proposed by Niedermeier's formula (1) and then adopted by FIB bulletin 14 [3]. It is because, among the proposed formula, the one by Niedermeier results in the maximum length. Base on the properties of constituent material, the minimum anchorage length in this study was estimated at 132 mm, and then selected as 150 mm. Other two case of anchorage lengths are multiplied from the minimum anchorage length. The main flexural reinforcement consisted of two 10 mmdeformed bars with sectional area 157 mm^2 (steel ratio was 0.524%), two 10 mmround bars were used as compression reinforcement. Shear reinforcement consisted of 6 mm diameter round steel stirrup spaced at 150 mm (Fig. 1). This was to prevent the shear failure and separate the flexural behaviour from shear behaviour. The number of CFRP layer is selected as one layer to prevent the failure mode of debonding at adhesive of at the concrete substrate.

$$l_{b,max} = c_2 \sqrt{\frac{E_f t_f}{f_{ck} f_{ctm}}} \text{ [mm]} \quad (1)$$

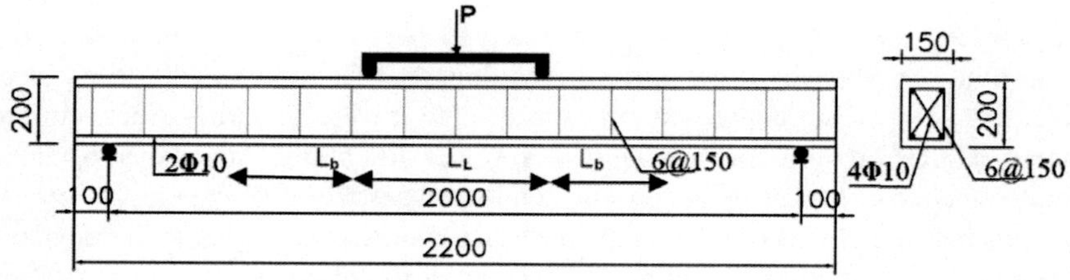

Fig. 1. Details of test beams (dimensions in mm)

In which: E_f: modulus of elasticity of FRP (MPa); t_f: thickness of FRP (mm); f_{ck}: characteristic value of the concrete compressive strength (MPa); f_{ctm}: mean value of the concrete tensile strength (MPa); c_2: coefficient of material obtained through calibration with test results (1.44).

2.2 Materials

Concrete was designed and prepared with grade of compression of C25 according to Eurocode standard. The composition was presented in Table 1. Six cylinder concrete specimens were made at the time of casting and cured for 28 days before being tested for compression strength. The steel strength was 365 MPa according to the material specification. The CFRP material was uni direction type (Fig. 2), fiber weigh was 200 g/m2, sheet thickness was 0.111 mm. Its tensile strength was 3400 MPa, its modulus was 230 GPa. The CFRP sheet was bonded to concrete with an epoxy with tensile strength was 30 MPa, flexural strength was 40 MPa, and compression modulus was 1500 MPa (according to product specifications provided by supplier). The CFRP sheet was bonded to concrete beam using wet layout technique after the preparation of the bonding surface (Fig. 3).

Table 1. Concrete mixing ratio (for 1 m^3 of concrete)

Cement (kg)	Water (kg)	Gravel (kg)	Sand (kg)
336	180	1262	622

2.3 Test Procedures

All specimens were tested in four-point bending over 2 m simple span in a 30 kN test frame. Loads were applied by a hydraulic cylinder fixed to the strong frame during these test procedures. The load level was maintained by a controlling system. The displacement at middle bottom of the beam was measured by an LVDT (LVDT Controlab can measure displacement up to 25 mm). During the test, the load level and middle-beam displacement were recorded. The two load points were offset 250 mm from the top mid-span of the beam (L_L = 500 mm, Fig. 1). An initial cyclic load between 1–2 kN was applied and

Fig. 2. Carbon fiber reinforced polymer (CFRP) sheet

Fig. 3. Four point bending test

released for five times to stabilize the testing system. The mechanical load was then increased monotonically until the failure of the beam. During the test, the development of crack was marked and the corresponding load level was noted. The details of tests undertaken in this study are presented in Table 2.

Table 2. Summary of tests

No	Test group	Dimension, mm			Anchorage length, L_b, mm	Number of test
		b	h	L		
1	B.REF	150	200	2200	0	3
2	B.LB15	150	200	2200	150	3
3	B.LB45	150	200	2200	450	3
4	B.LB75	150	200	2200	750	3

3 Results and Discussion

3.1 Experimental Results

Table 3 displays the ultimate strength obtained from four tested groups and Fig. 4 shows the Mechanical load - displacement diagram of two typical tested groups: non-strengthened and fully strengthened by CFRP sheet. Table 3 shows that the mechanical performance of RC beam increases as it is reinforced with CFRP and also with the increase of the length of CFRP in extend bonding region. For the reference RC beams, the failure loads are about 41.5 kN to 45.5 kN; RC beams with anchorage length of 150 mm, the failure loads increase to about 49.3 kN to 53.1 kN; with the anchorage length 450 mm, the failure loads range from 54.1 kN to 64.5 kN; and at the anchorage length of 750 mm, the range of failure loads develop up to 63.4 kN–65.1 kN (Table 3). Figure 4a shows the Mechanical load - displacement diagram of non-strengthened RC beam and Fig. 4b shows the Mechanical load - displacement diagram of CFRP strengthened RC beam with anchorage length at 750 mm. These two diagrams display similar stiffness before the longitudinal steel yielding, while beyond this, the stiffness of the beams strengthened with CFRP is greater than that of the non-strengthened beam. They indicate that, by externally bonding CFRP to tension region, the mechanical performance of RC beam increases in both terms of stiffness and ultimate capacity. Before the yielding point, beams which are strengthened with CFRP experience less crack than that of reference beams, and its deformation is smaller. The failure of all the CFRP strengthened RC beams is caused by the tensile rupture of CFRP layer, subsequently follow by the rupture of tensile steel bar.

Table 3. Summary of test results

No	Specimen	Failure load, kN	No	Specimen	Failure load, kN
1	B.REF.01	45	7	B.LB45.01	64.4
2	B.REF.02	41.5	8	B.LB45.02	54.1
3	B.REF.03	45.5	9	B.LB45.03	64.48
4	B.LB15.01	53.1	10	B.LB75.01	63.4
5	B.LB15.02	53	11	B.LB75.02	65.1
6	B.LB15.03	49.32	12	B.LB75.03	59.6

3.2 Discussions

Figure 5 shows the development of ultimate load of RC beam as the increase of CFRP anchorage length from 0 to 750 mm from the loading point. It shows that, as the RC beams are strengthened with CFRP, the failure load increases from 44 kN (reference beam) to 51.18 kN (at $L_b = 150$ mm) which correspond to 18% of load development; to 61.66 kN (at $L_b = 450$ mm) which correspond to 40% of load development; and to 62.7 kN (at $L_b = 750$ mm) which corresponds to 43% of load development. This shows that with the minimum anchorage length of CFRP as proposed by Niedermeier's formula (1) (adopted by FIB bulletin 14 [3]), the mechanical performance of RC beam slightly increases while at the 5 times than the proposed length, the mechanical performance of RC beam increases significantly.

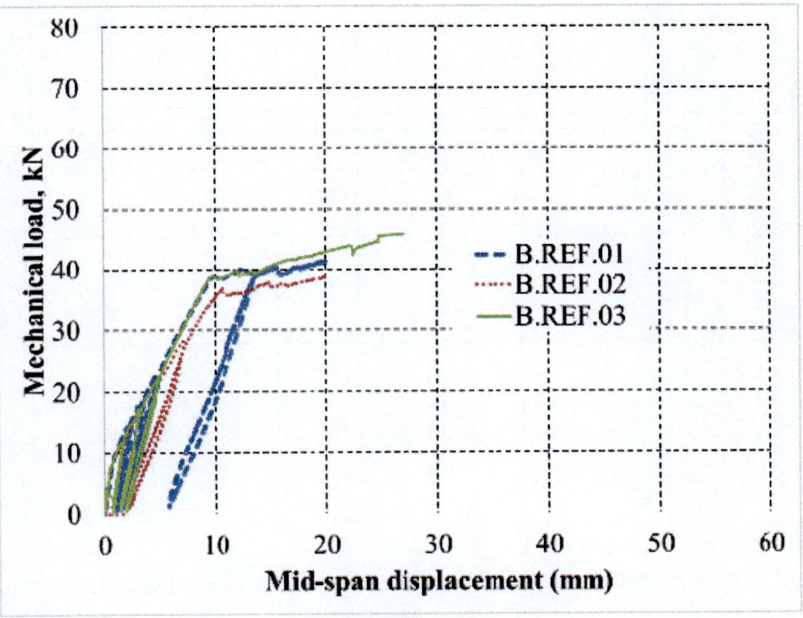

a) Reference beams

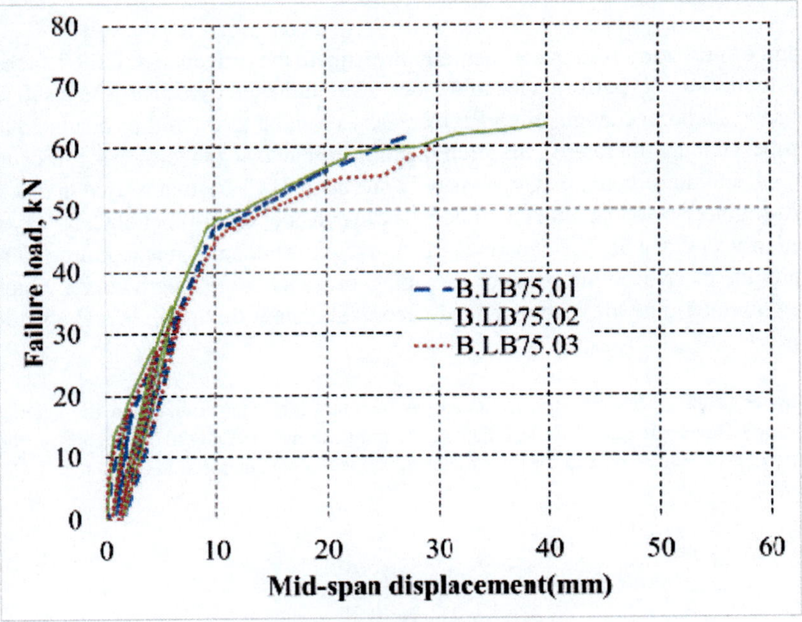

b) Fully CFRP strengthened beam

Fig. 4. Mechanical load- displacement diagram of two typical tested groups

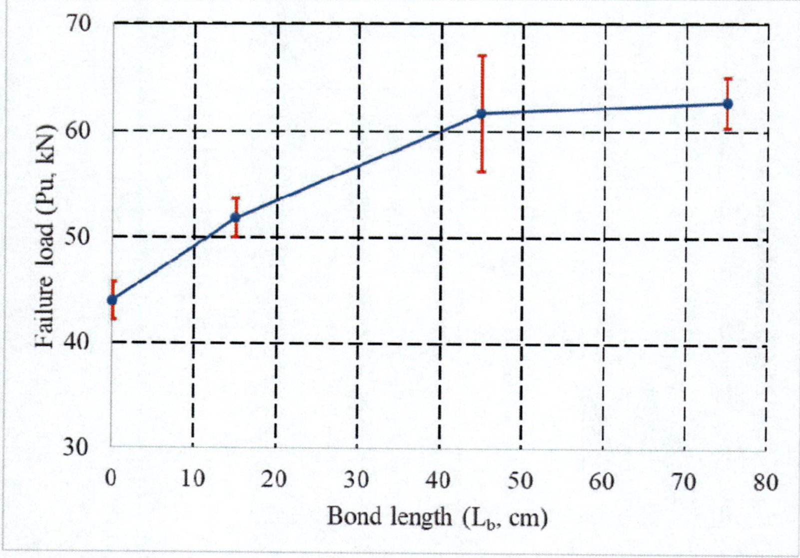

Fig. 5. The development of ultimate load of RC beams as the increase of anchorage length.

4 Conclusion

The aim of this study is to experimentally investigate the influence of CFRP anchorage length to mechanical performance of RC beam strengthened by CFRP. The result shows that, as the anchorage length of CFRP increase to 3 and 5 times of the calculated length (as proposed in the literature), the strengthening efficiency can increase significantly up to 43%. It should be noted that, there is only one layer of CFRP that is used in this study. In civil engineering application, it is necessary to consider the number of CFRP layer that is used to strengthen the RC concrete structures. In further study, testing more elements will provide more accurate strength gained against anchorage length curve, which can be used to determine the recommended anchorage length that provides reliable design results.

Acknowledgment. This research is funded by Vietnam National Foundation for Science and Technology Development (NAFOSTED) under grant number 107.01-2020.32 and by Phu Bac Construction & Trade Co., Ltd for their material and technical supports.

References

1. Wenwei, W., Guo, L.: Experimental study and analysis of RC beams strengthened with CFRP laminates under sustaining load. Int. J. Solids Struct. **43**, 1372–1387 (2006). https://doi.org/10.1016/j.ijsolstr.2005.03.076
2. Kang, T.H.K., Howell, J., Kim, S., Lee, D.J.: A state-of-the-art review on debonding failures of frp laminates externally adhered to concrete. Int. J. Concr. Struct. Mater. **6**, 123–34 (2012). https://doi.org/10.1007/s40069-012-0012-1
3. International Federation for Structural Concrete, editor. Externally bonded FRP reinforcement for RC structures: technical report on the design and use of externally bonded fibre reinforced polymer reinforcement (FRP EBR) for reinforced concrete structures. Lausanne: fib (2001)

Investigation of Scour Around Two Side-by-Side Piles with Different Spacing Ratios in Live-Bed

Amir Bordbar(✉), Soroosh Sharifi, and Hassan Hemida

University of Birmingham, Birmingham B15 2TT, UK

Abstract. In this study, a developed model is employed for numerically investigating scour around a group of two identical side-by-side piles. The model benefits from a coupled approach between a hydrodynamic and a morphodynamic model which is successfully validated for predicting scour around a single circular pile. The hydrodynamic model solves unsteady Reynolds-Averaged Navier-Stokes (RANS) equations in combination with the k-ω-SST turbulence closure model. The morphodynamic model entails bedload and suspended load sediment transport models for non-cohesive sand. The morphological evolution of the bed due to the sediment transport fluxes is expressed by a means of a mass-balance equation. The model is run for two side-by-side piles with three different spacing ratios $G/D = 1$, 2, and 3, where G represents the distance between the piles and D is the pile diameter, under steady-current flow and live-bed conditions. The results of the investigation reveal an inverse correlation between the spacing ratio and the scour hole depth adjacent to piles. The results are generalized and an equation is obtained for equilibrium scour hole depth as a function of the spacing ratio.

Keywords: Scour · Side-by-side piles · CFD · Sediment transport

1 Introduction

The flow around riverine structures is characterized by strong 3D turbulent motions. In the case of bridge piers, these turbulent motions may wash off part of the sediments around the piers and cause local scour which may lead to bridge failures. A study over 30 years in the United States [1] concluded that 60% of total bridge failure cases were due to local scour. Hence, protecting bridge piers from scour damage can lead to significant savings in maintenance and repair costs which requires a thorough knowledge of scouring characteristics.

There are variety of elements that may affect the local scouring around piers. These factors generally include foundation geometry, fluid and flow parameters, and bed material parameters. In this study, the effect of foundation geometry in a group of two identical side-by-side circular piles with different spacing ratios on flow behavior and local scour is studied. In the literature, few experimental and numerical studies have investigated the impact of a group of side-by-side piles on scouring. Among them, Ataie-Ashtiani and Beheshti [2] experimentally investigated local scour around pile groups including two to eight circular piles with different arrangements and spacing ratios in clear-water

D. V. K. Huynh et al. (Eds.): VSOE 2021, LNCE 208, pp. 302–309, 2022.
https://doi.org/10.1007/978-981-16-7735-9_33

steady current conditions. In total, 112 tests were carried out. It was found that scour around a pile group can be different from that of a single pile depending on pile spacing. In their study, for the case of two side-by-side piles, the maximum scour depth happened for the $G/D \approx 0.25$, where G represents the distance between the piles and D is the pile diameter. In this case, the maximum scour depth was 50% more than that for a single pile. In another work, Kim et al. [3] numerically investigated the progression of the scour patterns around two adjacent circular cylinders in side-by-side and tandem arrangements with diverse spacing ratios between 0.25 to 4 in clear-water condition. Large Eddy Simulation (LES) approach was employed to solve the Navier-Stokes equations which was followed by a Lagrangian approach for sediment transport. They validated their model against the experimental data from Khosronejad et al. [4] for scour around a single circular pile. From their results, they found out that in the case of two side-by-side cylinders when $G/D \geq 1.5$, the local scour patterns are similar to those of a single cylinder. Zhang et al. [5] ran a set of numerical simulations to study local scour around a different arrangement of a group of up to three adjacent piles under steady current flow in clear-water conditions. Flow-3D was applied to investigate the local scour. The software used the volume of fluid (VOF) method to capture the water-air interface. Reynolds Averaged Navier-Stokes (RANS) equations in combination with the RNG k-ε turbulence closure was applied to solve the hydrodynamic equations. Sediment transport was modeled by both bedload and suspended load models. The model was validated against experimental data from Khosronejad et al. [4] for the scour around a single circular pile. The simulations were conducted for circular piles with $D = 1.5$ m and $G/D = 1, 2$ and 3. It was observed that with increase in the spacing ratio the pile group effect decreases, as for $G/D = 3$, there was almost no interaction between the piles effects. Vaghefi et al. [6] experimentally studied the scour pattern around three side-by-side and tandem arrangements of circular cylinders in a 180° bend under clear-water and steady current flow conditions. Circular cylinders with constant $G/D = 2$ were placed at the positions of 60°, 90° and 120° of the bend. In their study, the maximum scouring depth occurred in the side-by-side arrangement at the position of 90° of the bend and around the pier closer to the outer bank. Qi et al. [7] made a series of experimental tests of local scour at twin piles under combined current and wave. Two pairs of piles with diameters D = 12 cm and 8 cm at four different flow skew angles (α_s = 0°, 30°, 60°, 90°) with different $0 \leq G/D \leq 3$ were considered. It is found that with increase in G/D the pile group effects gradually diminished. In their work, the maximum pile group effect on the scour depth was found to occur for $G/D = 0$ in side-by-side arrangement. An empirical equation was established to estimate the maximum scour depth around twin piles based on G/D and α_s.

In the pertinent literature, most of the studies were done in clear-water condition and to the author's best knowledge, no numerical investigation on local scour around a pile groups has been carried out in live-bed condition. Therefore, this study aims to bridge this gap with studying the flow behavior and progression of local scour pattern around two identical side-by-side circular piles with different $G/D = 1, 2$ and 3 in steady current and live-bed conditions.

For this purpose, a coupled hydrodynamic and morphodynamic model is implemented in the framework of OpenFOAM®. The model only solves the water-phase around the piles. Hence, the computational domain is limited to the water-air interface on top. The model is validated against the experimental data of Roulund et al. [8] for scouring around a circular pile under steady current flow in live-bed condition. In this study, the similar flow and bed material condition as Roulund's scour test is used in all simulations.

2 Numerical Method

2.1 Governing Equation of the Flow

In this study, the flow field is governed by incompressible unsteady RANS equations. The Reynolds stress terms in equations are modeled by the k-ω SST turbulence closure. The k-ω SST is chosen as it has a good capability in predicting adverse pressure gradients and separating flow. The computational domain and boundaries for a case of two side-by-side piles are presented in Fig. 1a. Figure 1b shows the dimensions of the computational domain in the xy plane, while the height of the domain is equal to the water depth. The flow and turbulence properties need to be quantified in the boundaries of the computational domain. At the walls, which contain a smooth pile surface and the rough sandy bed, no-slip boundary condition is considered while the near-wall treatment of Cebeci and Bradshaw [9] is imposed on the center of the first layer of mesh cells closest to the wall. This near-wall treatment is valid in the logarithmic region of the boundary layer and can be adapted for both smooth and rough walls to calculate friction velocity u_f. The top boundary of the domain is satisfied with the frictionless slip wall as the model does not consider the water-air interface. The sides are treated with symmetry boundary condition whereas the effects of the side walls on the approaching flow are ignored. In all simulations, the inlet flow is supplied with the approaching flow velocity profile from Roulund et al. [8] and pressure is treated with the Neumann (wall-normal zero gradient) condition. At the outlet, pressure is treated with Dirichlet condition and equal to zero while the all other flow properties are set as zero normal gradient.

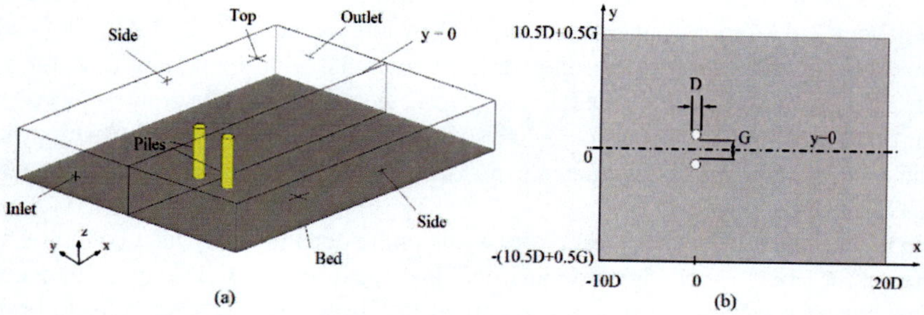

Fig. 1. (a) 3D view of the computational domain and boundaries and (b) top view for a group of two side-by-side piles.

2.2 Sediment Transport and Morphodynamic Models

The sediment transport contains bedload and suspended load. Bedload includes a transfer of sediments that roll, slide or jump along the bed in a thin layer very close to the bed. In contrast, Suspended load includes sediment particles which transfer in the flow over the bedload without continuous contact with the bed and is dominated by fluid turbulence. These two sediment transport modes are separated by a hypothetical layer called the reference level with the height of Δ from the bed. In this study, the bedload rate for each computational grid, q_B, is calculated using a similar approach used by Roulund et al. [8] for a 2D bed. The suspended load is modeled solving an advection-diffusion equation to fulfill the case of a non-uniform and transient flow around structures:

$$\frac{\partial c}{\partial t} + \nabla.\left[(\vec{u} + \vec{w}_s)c\right] = \nabla.[D_s \nabla c] + S_\Delta \tag{1}$$

where c is the value of fractional volumetric suspended sediment concentration, $\vec{u}$ and $\vec{w}_s$ are the fluid velocity and sediment fall velocity vectors, respectively, D_s is the sediment diffusivity coefficient and $S_\Delta = E_\Delta - D_\Delta$ is the source/sink term for the entrainment E_Δ and deposition rate D_Δ. In the present work, sediment fall velocity is calculated using the method proposed by Soulsby [10]. The entrainment rate, E_Δ, is obtained from:

$$E_\Delta = c_\Delta w_s \tag{2}$$

where c_Δ is the reference level concentration of suspended load which is calculated as suggested by Rijn [11], where Δ is adopted based on the mean particle diameter d_{50} and equal to $4d_{50}$. The deposition rate D_Δ is obtained from the approach applied by Liu and García [12]:

$$D_\Delta = c_b w_s \tag{3}$$

where c_b is the sediment concentration at Δ, which here, is equal to the sediment concentration in the center of the first layer of cells closest to the bed. The results of the sediment transport models are reflected as a deformation on the bed boundary of the computational domain. This is achieved by solving a mass-balance equation on a 2D mesh according to the bed boundary of the 3D domain.

3 Model Validation

The model is validated against data collected from a physical scour test around a circular pile under steady current flow provided in Roulund et al. [8]. The experiment was carried out in a 10 m length and 4 m width flume. A circular pile with 0.1 m diameter was mounted in a sandy bed at 6.6. m downstream of the inlet section. The depth average of the flow for a test condition was 0.46 m/s with the water depth of 0.4 m. The sandy bed was covered with sand with a mean particle diameter $d_{50} = 0.26$ mm.

For the numerical modeling, a computational domain with 30D length and 20D width with a circular pile placed in the symmetry plane and at 10D downstream of the inlet was considered. The domain height was equal to water depth and equal to 4*D*. As a result of mesh dependency test, a generated mesh with around 400,000 cells was chosen. The thickness of the first layer of the grids closest to the bed was 2 mm to satisfy the requirements of the near-wall treatment method and the reference level height from the bed. Figure 2 depicts the progression of the scour hole in terms of scour depth in front and back of the pile which is compared with the experimental and simulation results of Roulund et al. [8]. It is found that the maximum discrepancy in the prediction of maximum scour hole from the present simulation results and Roulund's experimental data was less than 5% during the evolution process.

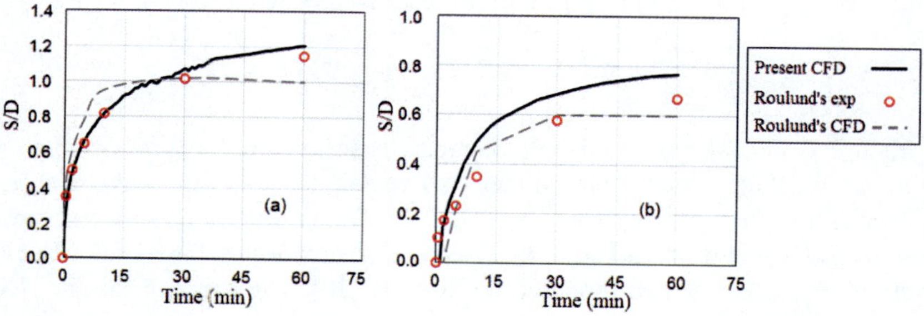

Fig. 2. The evolution of scour hole in terms of scour depth (a) in front and (b) in back of the circular pile.

4 Results

Two sets of simulations were conducted to investigate the flow behavior and formation of scour hole around a group of two circular side-by-side piles under steady current flow. One was on a rigid bed and the other one was on a mobile bed. A similar geometry (as shown in Fig. 1) and generated mesh was used for each case in both series of tests. The generated meshes included around 600,000 grids.

The first series of simulations were run to evaluate the influence of the changes in the spacing ratio in the local flow field and bed shear stress on the rigid rough bed. Figure 3 shows the contours of flow field in $x = 0$ plane. It is seen that the flow velocity between the piles highly depending on the G/D, while for $G/D \geq 3$ the effect of the pile group in the flow field is small.

The results of the bed shear stress distribution in a dimensionless form for different arrangements of two side-by-side piles are shown in Fig. 4. The dimensionless bed shear stress (BSS) is τ/τ_∞ where τ and τ_∞ are values of local and undisturbed bed shear stress, respectively. The results reveal that a decrease in the value of G/D significantly raising the BSS between the piles which may result in deeper scour hole in the area.

The second series of simulations were conducted on the mobile sandy bed to assess the progression of scour hole pattern adjacent to two side-by-side piles under current flow. Figure 5 demonstrates the results of the scour hole pattern around piles at $t = 30$ min of physical time which took around 3 months of simulation time on a 2.3 GHz CPU

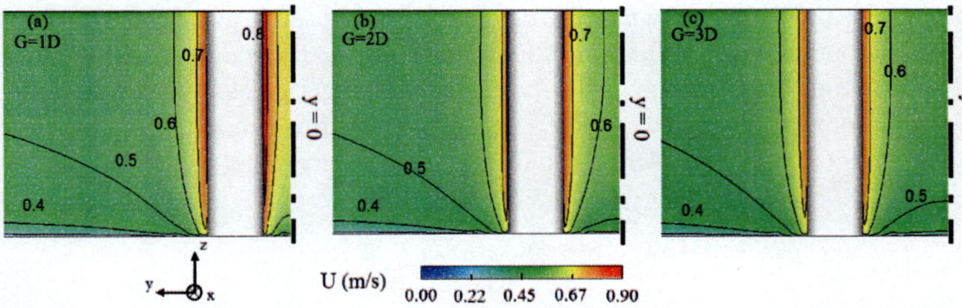

Fig. 3. Flow field adjacent to the two side-by-side piles in plane $x = 0$ with (a) G = 1D, (b) G = 2D, and (c) G = 3D.

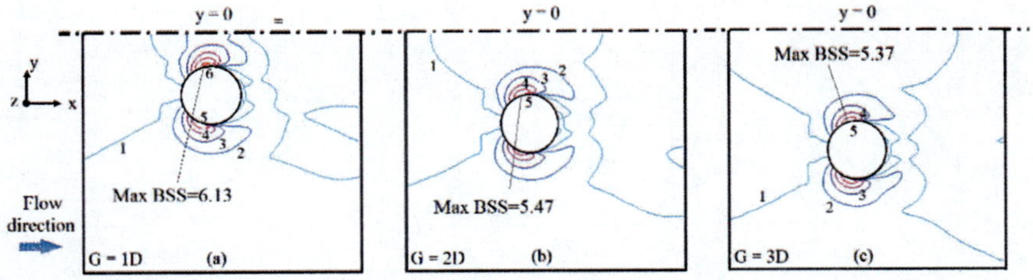

Fig. 4. Bed shear stress distribution around the two side-by-side piles for (a) G = 1D, (b) G = 2D, and (c) G = 3D.

with 8 MB RAM. The results indicate notable impacts of changes in G/D on scour pattern and maximum scour depth around the piles. The effects mitigate with increase in G/D, as for the case $G = 3D$ the depth in inner and outer sides of the pile group is roughly similar to that of a single pile.

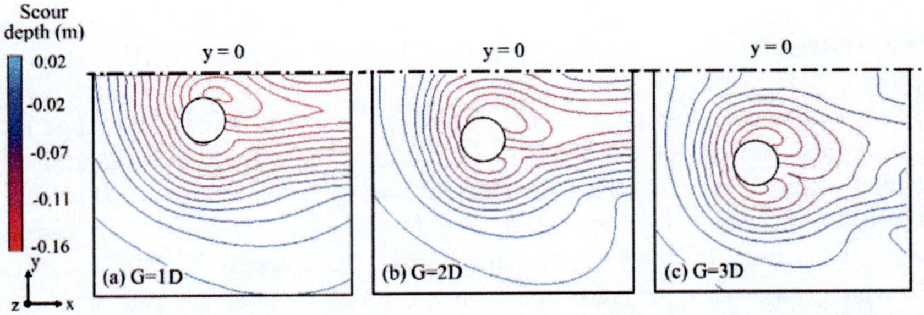

Fig. 5. Scour patterns around two side-by-side piles at t = 30 min for (a) G = 1D, (b) G = 2D, and (c) G = 3D.

It was assumed that the scour development reaches the equilibrium condition after a first one hour of physical time for all cases similar to that of the single pile. Hence, Fig. 6a presents the evolution of dimensionless scour hole depth S/D in time for different spacing ratios of two pile groups and single pile until the equilibrium condition. The equilibrium depth for pile groups was obtained by elongation of the scour depth curves for the first 30 min with maintaining the ratio with that of a single pile. Figure 6b displays values of dimensionless equilibrium scour hole depth S_m/D for different arrangements of two pile groups. It was assumed that the pile group effects are completely negligible in $G/D = 5$. Hence, S_m/D equal to that of a single pile was considered for two side-by-side piles with $G/D = 5$. A curve was fitted for the set of data with a formula as following:

$$\frac{S_m}{D} = 1.186 + 1.6exp\left(-0.92\left(\frac{G}{D}\right)\right), \quad \frac{G}{D} \geq 1. \tag{4}$$

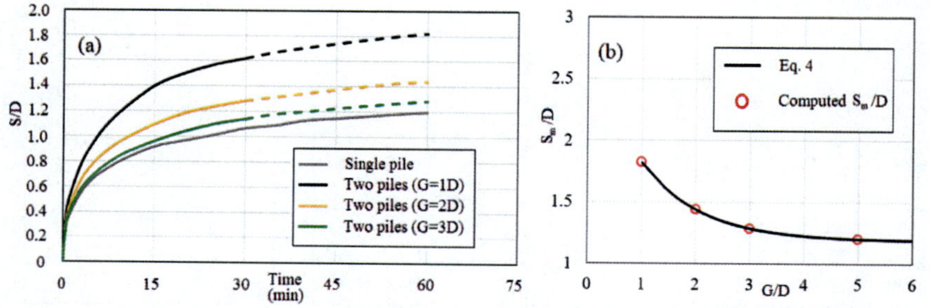

Fig. 6. (a) evolution of scour depth for two side-by-side piles and a single pile and (b) equilibrium scour depth for two side-by-side piles with different spacing ratios.

Sumer et al. [13] inferred that the equilibrium scour depth around a single pile under current flow and live-bed condition was only proportional to pile diameter D. With considering this, Eq. 4 can be generalized in case of live-bed condition for two identical circular side-by-side piles, however more investigation is needed to understand the impacts of the other factors.

5 Conclusion

In the present study, a numerical model for the prediction of flow behavior and scour pattern around pile groups is implemented. The results for the flow field, bed shear stress and scour pattern for two identical side-by-side piles for different G/D values are provided. The results depict that for $G/D \geq 3$ the pile group effect gradually vanishes. An equation is fitted for the computed equilibrium scour depths for different spacing ratios which can be used for prediction of maximum scour depth adjacent to two side-by-side piles under current steady flow and live-bed condition. The equation is valid for $G/D \geq 1$, as the physics of the problem may change for lower spacing ratios.

References

1. Briaud, J.-L., Ting, F.C.K., Chen, H.C., Gudavalli, R., Perugu, S., Wei, G.: SRICOS: prediction of scour rate in cohesive soils at bridge piers. J. Geotechn. Geoenviron. Eng. **125**(4), 237–246 (1999)

2. Ataie-Ashtiani, B., Beheshti, A.A.: Experimental investigation of clear-water local scour at pile groups. J. Hydraul. Eng. **132**(10), 1100–1104 (2006)
3. Kim, H., Nabi, M., Kimura, I., Shimizu, Y.: Numerical investigation of local scour at two adjacent cylinders. Adv. Water Resour. **70**, 131–147 (2014)
4. Khosronejad, A., Kang, S., Sotiropoulos, F.: Experimental and computational investigation of local scour around bridge piers. Adv. Water Resour. **37**, 73–85 (2012)
5. Zhang, Q., Zhou, X.L., Wang, J.H.: Numerical investigation of local scour around three adjacent piles with different arrangements under current. J. Ocean Eng. **142**, 625–638 (2017)
6. Vaghefi, M., Motlagh, M.J.T.N., Hashemi, S.S., Moradi, S.: Experimental study of bed topography variations due to placement of a triad series of vertical piers at different positions in a 180 bend. J. Arabian J. Geosci. **11**(5), 102 (2018)
7. Qi, W.G., Li, Y.X., Xu, K., Gao, F.P.: Physical modelling of local scour at twin piles under combined waves and current. J. Coast. Eng. **143**, 63–75 (2019)
8. Roulund, A., Sumer, B.M., Fredsøe, J., Michelsen, J.: Numerical and experimental investigation of flow and scour around a circular pile. J. Fluid Mech. **534**, 351–401 (2005)
9. Cebeci, T., Bradshaw, P.: Momentum Transfer in Boundary Layers, p. 407. Hemisphere Publishing Corp, Washington, DC, McGraw-Hill Book Co., New York1977 (1977)
10. Soulsby, R.: Dynamics of marine sands: a manual for practical applications: Thomas Telford (1997)
11. Rijn, L.C.V.: Sediment transport, part II: suspended load transport. J. Hydraul. Eng. **110**(11), 1613–1641 (1984)
12. Liu, X., García, M.H.: Three-dimensional numerical model with free water surface and mesh deformation for local sediment scour. J. Waterw. Port Coast. Ocean Eng. **134**(4), 203–217 (2008)
13. Sumer, B.M., Christiansen, N., Fredsøe, J.: Influence of cross section on wave scour around piles. J. Waterw. Port Coast. Ocean Eng. **119**(5), 477–495 (1993)

Limitations of Helical Screw Anchors for Offshore Structures

Alan J. Lutenegger(✉)

University of Massachusetts, Amherst, Ma 01003, USA
lutenegg@ecs.umass.edu

Abstract. Helical anchors have been used since the mid 1800's to secure vessels and structures offshore. There has recently been renewed interest in this technology for anchoring offshore floating and submerged structures such as large floating wind generators and submerged current or tidal generators. Helical anchors may offer a number of distinct advantages over other types of anchors for these applications. They can be designed with a round or square central shaft and can have a single helical plate or multiple helical plates depending on the design load requirements. They may be removed easily if needed, as in the case of temporary works or decommisioning, they have a low carbon footprint, and installation generally produces low noise and vibration levels. However, screw anchors may also have limitations that need to be clearly recognized in order to make efficient and cost-effective use of the technology. Estimating the required installation torque for different geometries at specific sites may be challenging. Potential limitations of using helical anchors offshore are discussed.

Keywords: Helical · Anchors · Offshore

1 Introduction

1.1 Helical Anchors - Background

Anchorage of offshore structures, such as wind turbines, drilling platforms and other structures has been a subject of intense research in the past decade as a search for reliable, cost effective and environmentally sensitive anchorage systems has intensified. There are a number of options for engineers to choose from for providing anchorage for offshore structures, including drag anchors, driven pile anchors, torpedo anchors and helical anchors. Whereas early offshore applications of helical anchors were largely based on trial-and-error, like many other areas of Civil Engineering, by contrast, modern geotechnical engineering is based on principles of soil mechanics with some attention to experience. Helical anchors have been used since the mid 1800's in shallow offshore applications resist tensile loads as mooring anchors, partial support for offshore lighthouses and ocean front piers around the world [1]. In fact, the first "screw piles" were installed offshore to support an iron lighthouse [2].

The use of helical anchors for offshore structures was a common global construction practice that extended through the end of the 19th century. Most helical plates were on the

D. V. K. Huynh et al. (Eds.): VSOE 2021, LNCE 208, pp. 310–316, 2022.
https://doi.org/10.1007/978-981-16-7735-9_34

order of 2.5 ft. to 4 ft. in diameter and were largely installed by hand without applying a downward "crowd" force. Helical piles were installed both vertically and on a batter. A decline in the use of screw-piles for both offshore and onshore structures occurred after 1900 as other deep foundation technologies emerged and became available, especially for shallow water support of ocean front piers and onshore bridges. However, Since the 1980s, this technology has reemerged as a viable cost effective foundation system to provide resistance for both compression and tension loads and now may have potential for anchoring offshore structures.

Using large scale helical elements would be a return to earlier offshore applications in the mid 20th century. For example, Morgan [3] described the use of double-helix and triple-helix screw piles with helical plates of 7 and 8 ft. diameter attached to 3 ft. diameter shafts to support a jetty in Scotland. Spacing of the helical piles was 16 ft. on center. Likewise, Wilson [4] described the installation of single-helix screw piles with 8 ft. diameter helical plates on 3 ft. diameter shafts to support the jetty at Newport. In both cases installation was performed using electric powered capstans attached to the top of the pile shaft and power was monitored to estimate installation torque.

Engineers should have a clear understanding of not only the advantages but the limitations of any technology. Advantages of using helical piles and anchors to resist loads have been described extensively elsewhere. In this paper some potential limitations for using helical anchors to resist tension loading in offshore applications are discussed.

2 Potential Limitations of Helical Anchors for Offshore Applications

In the past twenty years extensive field research on the installation and performance of helical piles and anchors has improved the understanding of their behavior and led to development of appropriate design methods. However, use of helical anchors offshore may present some unique limitations relative to typical land use. These include:

Estimating Load Capacity Prior to Deployment
Estimating Installation Torque
Providing High Quality Installation
Load Testing
Estimating Installed Load Capacity from Installation Torque

2.1 Estimating Load Capacity Prior to Deployment

As with all foundation design, the geotechnical engineer has the challenge of obtaining sufficient and relevant site information from a site to use is selecting soil properties in that design. The combination of test borings, sampling and laboratory testing and in situ testing needed to characterize a site are as important when considering the use of helical piles and anchors as in any other project. Some current practice for smaller scale onshore projects often rely on the installation contractor to provide a design for anchors based on installation torque and an appropriate safety factor. Occasionally this is done by performing trial installations at the site in order to select the most appropriate geometry for the project needs. However this will likely not do for large-scale projects, either onshore or offshore.

It was recognized very early that the load capacity of a screw pile depended on a number of factors. As noted by Mitchell [2]:

> "Whether this broad spiral flange, or "Ground Screw," as it may be termed, be applied to the foot of a pile to support a superincumbent weight, or be employed as a mooring to resist an upward strain, its holding power entirely depends upon the area of its disc, the nature of the ground into which it is inserted, and the depth to which it is forced beneath the surface. The proper area of the screw should, in every case, be determined by the nature of the ground in which it is to be placed, and which must be ascertained by previous experiment."

For tension loading, as in the case of screw-moorings, Mitchell [2] suggested that the holding power could be calculated assuming an appropriate failure surface in the form of a frustrum of a cone. However, this type of failure generally would only describe the behavior of very shallow installations, with depth to plate diameter ratios on the order of 1.5 to 3, which historically was used for moorings and lighthouses.

Most installations on land result in a "deep" anchor installation as compared to a "shallow" anchor installation. For different soils, the transition from "shallow" to "deep" is related to the soil type and strength and the helical plate diameter to embedment ratio. Shallow anchor installations are now relatively uncommon since deep installations develop higher capacity at about the same cost. Future offshore applications would be more cost effective if the helical plate is installed beyond about 6 plate diameters, depending on the specific anchor geometry, e.g., shaft to plate diameter ratio.

Depending on the site stratigraphy and site restrictions, shear strength of individual layers, stiffness parameters and in the case of clays, soil sensitivity is needed for design. Some understanding of site variability is also needed so that large scale economics may be realized by using helical elements that are fabricated as the same geometry for an individual project. The engineer will be able to make initial estimates of installation torque requirements, load capacity and load-displacement behavior provided that sufficient information is available from the site investigation. In order to size the appropriate equipment needed for installation the estimate of required installation torque for any given anchor geometry will be essential.

Measurement of installation parameters, including incremental installation torque, anchor advance, rate of advance, inclination and rotation speed should be used to evaluate the installation quality. Based on these data the engineer can then adjust the initial design as appropriate to the site and installation so that modifications to anchor geometry or installation can be made quickly in order to obtain the desired capacity. It is suggested that some trial helical installations should be performed as a part of the site investigation using smaller scale helical elements. This would provide the engineer with an indication of the required torque for installation and also can be used to evaluate subsurface variability.

Since the 1980s, design methods for helical piles and anchors have followed traditional foundation approaches, i.e., Total Stress Analysis (TSA) for clays and Effective Stress Analysis (ESA) for sands. Details on specific design methods can be found elsewhere. Single-helix anchors derive capacity from plate bearing and shaft resistance, as appropriate and use traditional bearing capacity theory for design. Capacity of multi-helix anchors has been described using failure conditions assumed to be appropriate for the anchor geometry; either Individual Plate (IP) Bearing (Independent Plate Mechanism) or Perimeter Shear (PS) (Interacting Plate Mechanism) depending on the helix spacing. The transition from PS to IP takes place as the spacing between consecutive plates increases and occurs at a relative helix spacing of about 2.5 for clays and 3 for sands [5–7].

As helical plates are scaled up in diameter for offshore anchoring to sizes on the order of 10 ft. or more, the pitch of the plate will likely be increased to 12 in. or more. Using a larger pitch reduces installation time, helps the plate advance and reduces the potential for installation disturbance.

2.2 Estimating Installation Torque

Because of the high cost of mobilization and operation, the prediction of required installation torque of helical anchors offshore becomes much more important as compared to helical anchors being installed on land where equipment modifications can be made easily. Several models have been suggested for estimating installation torque for a given geometry anchor in both fine-grained and coarse-grained soils. A number of these models require input from site specific cone penetration tests (CPT) and generally are based only on cone tip resistance [8–10].

Further development of torque prediction models might also include use of local sleeve friction from CPT measurements or the use of direct torque measurements from the Standard Penetration Test, using the SPT-Torque procedure. There have been only limited results published comparing predictive models to measurements of torque on large scale helical piles in the field. As larger diameter shafts are becoming increasingly common, where torque may be dominated by pipe shaft resistance, the relative contributions of pipe shaft and helical blade to both installation torque and load capacity will need to be considered and validated.

2.3 Providing High Quality Offshore Installation

Early offshore installation of helical anchors was performed using an anchored barge or ship to allow workmen to rotate a capstan. The anchored barge produced the required reaction. The reaction to the installation torque is provided by multiple anchors extending from the ship. Most of the early work was performed by hand until the early part of the 20th century when steam power was introduced for installation. Figure 1 shows an early arrangement for installing helical anchors for mooring ships.

Installation of helical anchors in offshore environments has two principle challenges: 1) providing sufficient torque to perform installation; and 2) providing sufficient reaction to the installation torque. It may be possible to develop sufficient reaction using multiple anchorages for vessels that might be deployed for offshore installation.

Another option may be to use parallel twin helical anchors installed simultaneously using a special frame and pairs of anchors having standard and reverse pitch helical plates. This would effectively counterbalance the installation torque eliminating the need for extensive vessel anchorage. This is occasionally done when installing helical anchors

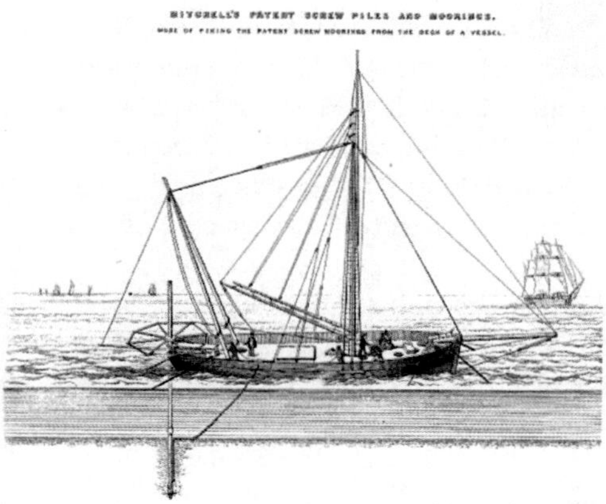

Fig. 1. Using a capstan and anchored ship to install a helical anchor offshore (circa 1860).

on opposite sides of a pipeline for resisting uplift forces. An example of this approach was suggested in the 1870s for installing helical anchors using a steam engine mounted to a railway carriage.

It has now been well established that the behavior of helical anchors depends on the quality of the installation. Detailed monitoring of the installation is a requirement in order to assess the quality of the installation and evaluate the influence of installation quality on predicted behavior. High quality manufacturing should provide a plate that is formed as a continuous helical screw section with a uniform pitch. Ideally, the helical plate will advance one pitch length for each full revolution of rotation. For example, for a helical pitch of 6 in., just two full rotations of the anchor are needed to advance the plate 1 ft. to obtain ideal installation. This ideal installation of a helical anchor is referred to as "pitch-matched" installation, in that the advance of the helical blade is one pitch distance for each full 360° revolution. This produces the least disturbance to the soil. Rotation of the helical plate without advance of the plate pitch produces progressively more disturbance to the soil which reduces load capacity.

Lutenegger et al. [11] suggested defining the Installation Disturbance Factor (IDF) as the ratio of the number of plate rotations per 1 ft. of advance to the number of rotations for perfect installation:

$$\text{IDF} = (\text{observed no. of rotations/ft.})/(\text{perfect no. of rotations/ft.}) \quad (1)$$

By recording the number of rotations for each advance interval, the IDF can be determined for each anchor installation and then be used to adjust the expected load capacity to account for installation quality as needed. Pitch-matched installation results in IDF equal to 1. Observations have shown that most installations on land give values of IDF typically in the range of 1.0 to about 1.5 for high quality installation practice but could be as high as 2 to 4 for very poor quality installation [11]. In clays, higher values of IDF indicated more disturbance and more remolding leading to lower anchor capacity. The IDF is particularly important for tension loading since the anchor (tension) capacity is derived from the behavior of the soil immediately above the plate through which the plate has already passed. Even with perfect installation the soil experiences some disturbance and a reduction in shear strength from the installation.

Skempton [12] suggested reducing the undisturbed undrained shear strength in clays to account for installation disturbance. The suggestion to use a strength between undisturbed and fully remolded means that it would be important to evaluate sensitivity as part of the site investigation if anchors will be installed in clay. An initial framework for considering the influence of sensitivity and IDF on available shear strength in clays was presented by Lutenegger et al. [11]. Aging after installation may alleviate some installation issues. Installation effects in sands are still not fully understood. How does installation influence sand strength and stiffness and ultimate load capacity? In many cases, the behavior of helical anchors may be directly related to the quality of installation.

2.4 Load Testing

Load tests are an important component in the design processes to validate site specific design and provide the engineer with the opportunity to modify the design for site specific conditions. For offshore helical anchors, conventional static tension tests will be cost prohibitive however there is a possibility that dynamic tests can provide a reliable means for rapid and economic testing. Several examples have shown that high strain dynamic impact testing of helical piles can provide accurate estimates of capacity [13, 14].

2.5 Limitations of Empirical Torque-To-Capacity Correlations

Manufacturers of helical piles and anchors have suggested for many years that there is an empirical correlation between installation torque and ultimate capacity of a helical piles or anchors; i.e., $Q_{ult} = TK_T$. The premise is that both the ultimate capacity, Q_{ult}, and the installation torque, T, are a function of the specific geometry of the pile/anchor and soil strength and are related through a correlation factor, K_T. The expectation of the engineer is that as installation torque increases, ultimate capacity increases proportionally; however this is not strictly the case.

Unfortunately, the value of K_T is not a constant and many reported comparisons between installation torque and capacity show wide variations. Both T and Q_{ult} may be influenced by a number of factors that can make any direct correlation between torque and capacity tenuous [16]. An important factor that can influence the measured installation torque but have little to no effect on the ultimate capacity is the downforce applied to assist the advance. Installation torque and capacity are developed from the resistance provided by both the helical plates and the central pipe shaft. The distribution of load between the shaft and helical plate depends on specific anchor geometry, e.g., diameter and length of shaft and diameter of helical plate, as well as the relative load level. A further complication is that T is measured during installation while Q_{ult} is measured at some time after installation after the soil has had some time to adjust to the installation effects.

3 Summary

Helical anchors offer great potential over other traditional piles and anchors for support of structures in offshore environments. While there may be challenges with their use, they offer the following potential advantages: ability to resist both tension and compression loads; verification of load capacity through monitoring of installation torque; evaluation of installation quality by monitoring other installation parameters; removal and reuse for temporary structures; rapid installation; immediate use; and suitability to a wide range of ground conditions. Potential disadvantages of helical anchors in offshore work include: ability to reliably predict installation torque in both sands and clays; ability to incorporate installation disturbance in estimating ultimate load capacity; and difficulties with deployment to maintain proper alignment and to monitor required installation parameters.

References

1. Lutenegger, A.J.: Historical development of iron screw-pile foundations: 1836–1900. Int. J. Hist. Eng. Technol. **81**(1), 108–128 (2011)
2. Mitchell, A.: On submarine foundations: particularly screw-pile and moorings. Civ. Eng. Archit. J. **12**, 35–40 (1849)
3. Morgan, H.D.: The design of wharves on soft ground. J. Inst. Civ. Eng. **22**, 5–25 (1944)
4. Wilson, G.: The bearing capacity of screw piles and screwcrete cylinders. J. Inst. Civ. Eng. Lond. **34**(5), 4–73 (1950)
5. Lutenegger, A.J.: Cylindrical shear or plate bearing? – Uplift behavior of multi-helix screw anchors in clay. In: Contemporary Issues in Deep Foundations, ASCE, pp. 456–463 (2009)
6. Lutenegger, A.J.: Behavior of multi-helix screw anchors in sand. In: Proceedings of the 14th Pan American Conference on Geotechnical Engineering, p. 6 (2011)
7. Merifield, R.S.: The ultimate uplift capacity of multi-plate helical type anchors in clay. J. Geotechn. Geoenviron. Eng. ASCE **137**(7), 704–716 (2011)
8. Gavin, K., Doherty, P., Spagnoli, G.: Prediction of installation torque resistance of large diameter helical piles in dense sand. In: Proceedings of the 1st International Geotechnical Symposium on Helical Foundations (2013)
9. Al-Baghdadi, T.A., et al.: CPT based design procedure for installation torque prediction for screw piles installed in sand. In: Proceedings of the 8th International Conference on Offshore Site Investigations and Geotechnics, no. 1, pp. 346–353 (2017)
10. Spagnoli, G.: A CPT based model to predict installation torque of helical piles in sand. Mar. Georesour. Geotechnol. **35**(4), 578–585 (2017)
11. Lutenegger, A.J., Erikson, J., Williams, N.: Evaluating installation disturbance of helical anchors in clay from field vane tests. In: Proceedings of the 39th Annual Deep Foundations Institute Conference (2014)
12. Skempton, A.W.: Discussion of the bearing capacity of screw piles and screwcrete cylinders. J. Inst. Civ. Eng. Lond. **34**(5), 76–81(1950)
13. Cannon, J.G.: The application of high strain dynamic pile testing to screwed steel piles. In: Proceedings of the 6th International Conference on the Application of Stress-Wave Theory to Piles, pp. 393–398 (2000)
14. Beim, J., Luna, S.C.: Results of dynamic and static load tests on helical piles in the varved clay of Massachusetts. Deep Found. Inst. J. **6**(1), 58–66 (2012)
15. Lutenegger, A.J.: Factors affecting torque correlations for screw-piles and helical anchors. In: Proceedings of the 1st International Geotechnical Symposium on Helical Foundations, pp. 211–224 (2013)

Mudmat Foundation Stability on Very Soft Clay - Coupling Analytical and Numerical Analysis for a Platform in Vietnam Water

Tien Dat Chu(✉) and Van Hoan Vu

PTSC Mechanical & Construction, PetroVietnam Technical Services Corporation (PTSC), No 31, 30/4 Road, Ward 9, Vung Tau City, Vietnam
{datct1,vuhoan}@ptsc.com.vn

Abstract. Mudmats are widely used in the oil and gas industry to temporarily support jacket before and during pile installation phase. This paper presents a case study on mudmat foundation stability which has taken on a geotechnical challenge when the mudmat foundation is seated on a 13.8 m thick layer of very soft clays. The mudmat stability is firstly verified using conventional procedures recommended by the industry-specific codes (ISO, API). The challenge of this analytical approach consists of consideration of mudmat stability when the degree of undrained strength heterogeneity exceeds a typical limit value of ten (10). In this case, the study of Salencon and Matar (1982) is used to extend the application range of the conventional approach. Secondarily, a numerical analysis is carried out using PLAXIS 3D Foundation. The numerical results show relative agreement in comparison with the analytical approach. Combining analytical and numerical analysis allows confirming the mudmat foundation stability. The study of Salencon and Matar (1982) seems to be a reliable reference to determine the foundation shape factor for the seabed with a high degree of undrained strength heterogeneity.

Keywords: Mudmat stability · Very soft clay · Coupling analysis

1 Introduction

In the construction of a fixed offshore platform, when pre-piling and suction buckets are not chosen, mudmats are widely used to provide temporary stability for the fixed platform during the installation phase. The design of the mudmat foundation is commonly considered as a shallow foundation. Loads acting on mudmats can be complex and tridimensional due to environmental loads, operational loads, and jacket self-weight.

The conventional design approach, as described in ISO [1], DNV [2] and API [3] focuses on the linear superposition of applied vertical, horizontal, and moment loadings. These ultimate loadings are determined based on the classical bearing capacity equation for the failure of a vertically loaded strip foundation on uniform soil. Semi-empirical factors then are introduced to account for foundation shape, load inclination and eccentricity, heterogeneity in soil strength profile, and embedment. According to

D. V. K. Huynh et al. (Eds.): VSOE 2021, LNCE 208, pp. 317–324, 2022.
https://doi.org/10.1007/978-981-16-7735-9_35

[4], the conventional approach will under-predict the capacity for non-uniform shear strength profiles, even for simple eccentricity with no lateral load.

An alternative approach that aims to represent ultimate limit states under combined loading consists of using interaction diagrams or failure envelopes. Failure envelopes can be expressed in planes of the vertical, moment, or horizontal load (V-H, V-M, H-M) or as a three-dimensional surface (V-H-M). The failure envelope approach has been developed for shallow foundations of various geometries including strip, circular, and rectangular [5–11]. Details of various important contributions were summarized in [4, 12]. Therefore, most of the available closed-form expression of failure envelope for the undrained condition were provided for a strength heterogeneity up to 10. For a larger degree of strength heterogeneity, closed-form expressions of failure envelope were provided in [8], but the foundation was limited to a circular shape.

This paper presents a case in which mudmat lies on very soft clay with complex combined loading. Firstly, an analytical approach is used to verify the mudmat stability according to the API standard [3]. The very soft clayey layers challenge the analytical approach with a degree of strength heterogeneity larger than 10. Subsequently, a numerical study is also performed to compare with the analytical results to confirm the mudmat configuration.

2 Project Description and Geotechnical Data

The project is a moderately sized gas development, located in the gas fields offshore southern Vietnam. The platform model is shown in Fig. 1. This platform is configured as an unmanned minimum facilities wellhead platform.

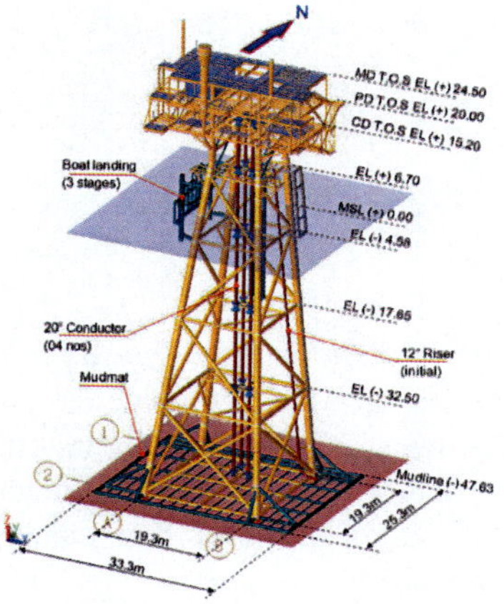

Fig. 1. Overview of the platform and timber mudmat

Due to Transportation & Installation (T&I) method constrained by the client, a mudmat foundation is chosen to support the jacket before and during the pile driving phase. The results of a preliminary design show that the necessary dimension of the mudmat is 33.3 m long and 25.3 m wide. The seastates expected at the proposed window are given in Table 1. The mudmat stability is checked with minimal and maximal water depth for all steps of the pile driving phase. The up-ending of the jacket is assisted by two buoyancy tanks. Therefore, the on-bottom stability was verified with loadings before and after the removal of two buoyancy tanks. The on-bottom stability was analyzed with 221 load combinations following [3].

Table 1. On-bottom stability seastates

Parameter	Value	Unity
Maximum wave height	3.0	m
Wave period	6.0	s
Current velocity	0.5	m/s
Number of load combinations due to environmental loadings and T&I sequence	221	[-]

The geotechnical survey was carried out by a qualified contractor. The investigation comprised of one 150-m sampling borehole and one 150-m continuous Piezocone Penetration Test (PCPT) borehole. The clayey soil is classified as very soft to soft from the seafloor to 13.80 m below the seafloor, overlaying firm to stiff clays to 27.10 m. For depths greater than 27.10 m, the clays are generally very stiff. The undrained shear strength, S_u increases respectively from 2 kPa at the seafloor to 25 kPa at 13.80 m below the seafloor (Table 2).

Table 2. Geotechnical data

Stratum	Depth [m]		Soil type	Unit weight [kN/m^3]	Sensitivity, S_t [-]	S_u [kPa]	E_u kPa
1	Top	Bottom					
	0.00	3.00	Clay	15.60	3.0	2 to 8	160 to 640
	3.00	13.00	Clay	15.60	3.0	8 to 22	640 to 1760
2	13.00	13.80	Clay	19.00	3.0	25	2000
3	13.80	16.00	Clay	18.80	2.0	45 to 75	7200 to 12000
	16.00	20.00	Clay	18.80	2.0	75	12000
	20.00	24.50	Clay	18.80	2.0	75 to 80	12000 to 12800

(continued)

Table 2. (*continued*)

Stratum	Depth [m]		Soil type	Unit weight [kN/m^3]	Sensitivity, S_t [-]	S_u [kPa]	E_u kPa
	24.50	27.10	Clay	18.80	2.0	80 to 85	12800 to 13600
4	27.10	40.60	Clay	18.30 to 17.30	4.0	110	17600
5	40.60	42.50	Clay	19.50	2.0	110 to 225	17600 to 36000

To perform numerical analysis, undrained Young's modulus is necessary as an input parameter. This parameter was not provided in the geotechnical data and was derived from correlations available in the literature. Among various correlations, the one proposed by Duncan and Buchignani (1976) reported in [13] was used to estimate undrained Young's modulus of soils. Stiffness ratio E_u/S_u was described as a function of overconsolidation ratio (OCR) and plasticity index (I_P). For a conservative assumption, the lower bound of the stiffness ratio was used such as $E_u/S_u = 80$ for the three first layers below the seafloor and $E_u/S_u = 160$ for other layers. Adopted values of the undrained Young's modulus are given in Table 2.

3 Analytical Approach

The mudmat stability was firstly verified according to the industry-specific codes API 2GEO [3] and API RP 2A-WSD [14]. Following the soil data provided in Sect. 2, the seabed was classified as a saturated clayey profile. Mudmat was checked with the undrained bearing capacity-linearly increasing shear strength. The vertical allowable load under undrained conditions is given as follows:

$$Q_d = F\left(s_{u0}N_c + \frac{\kappa B'}{4}\right)K_c A'$$

where Q_d is the vertical allowable load, F is the correction factor given as a function of $\kappa B'/s_{u0}$, κ is the rate of increase of undrained shear strength with depth, S_{u0} is the undrained shear strength of cohesive soil, N_C is a dimensionless constant equal to 5.14, B' is the minimum effective lateral mudmat dimension, A' is the effective area of the foundation, K_C is the correction factor that accounts for load inclination (i_C), footing shape (s_C), depth of embedment (d_C), inclination of the base (b_C), and inclination of the seafloor surface (g_C).

The correction factor (K_C) and foundation shape factor (s_C) are determined by the following equations in [3]:

$$K_C = 1 + s_C + d_C - i_C - b_C - g_C$$

$$s_C = s_{CV}(1 - 2i_C)(B'/L')$$

While other components of K_C (except s_C) can be calculated by the formulas given in [3], the foundation shape correction factor, s_C is only given as a function of $\kappa B'/s_{u0}$ for the range $0 \leq \kappa B'/s_{u0} \leq 10$. However, the degree of strength heterogeneity for the present project, $\kappa B'/s_{u0}$ varies between 10.73 and 19.47, is larger than the range proposed by API 2GEO [3]. Therefore, the shape factor cannot be directly determined with this code. This shape factor can be determined for large soil strength heterogeneity basing on the study of Salencon and Matar [15]. Figure 2 shows the diagram reproduced from [15], which was used to determine the circular foundation shape factor.

The sliding stability was checked following API 2GEO code [3] for the undrained condition in the following equation:

$$H_d = s_{u0}A$$

where A is the surface area of the mudmat. The overturning stability was also verified by calculating the ratio between the restoring moment and the overturning one around x-direction (long edge of the mudmat) and y-direction (short edge of the mudmat).

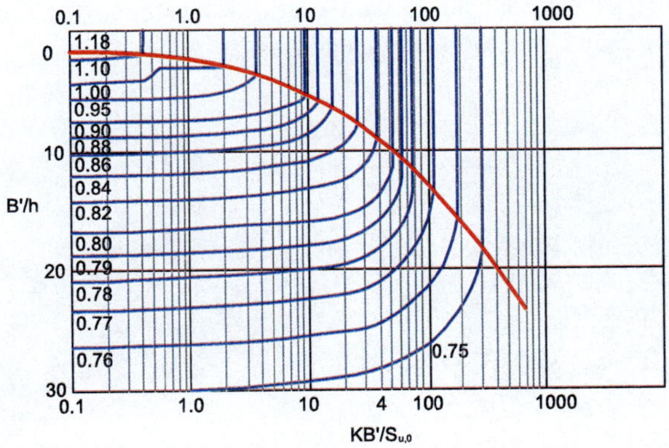

Fig. 2. Diagram for determination of s_{CV} reproduced from [15]

For each load case summarized in Table 1, a pair of the upper-bound and the lower-bound cases was considered. The lower-bound corresponds to the lower value of the correction factor F (for fully smooth interface condition) and the upper one corresponds to the higher value of the correction factor F (for fully rough interface condition) [3]. The most critical cases in terms of factor of safety for both the lower and upper-bound conditions are shown in Table 3.

Table 3 shows that the factors of safety (FoS) determined by the analytical study satisfy the requirements of industry-specific codes [3, 14]. However, due to the high degree of undrained strength heterogeneity, the foundation shape factor cannot be directly determined by the industry-specific codes. Consequently, the analytical results may have some uncertainties. Therefore, these results need to be confirmed by other approaches such as numerical analysis. The most critical cases 1013, 2006, and 2122 are chosen for the numerical analysis.

Table 3. Factor of safety (FoS) of the on-bottom check following API [3]

Load case	Minimal FoS					
	Lower bound			Upper bound		
	Vertical	Sliding	Overturning	Vertical	Sliding	Overturning
1013	2.02	2.37	2.23	2.41	2.37	2.23
2006	3.48	2.17	8.21	4.18	2.17	8.21
2122	1.79	2.28	2.71	2.15	2.28	2.71

4 Numerical Approach

The numerical model was carried out using PLAXIS 3D Foundation. This finite element model is shown in Fig. 3. The mudmat was modeled in the following manner: (i) beams represent the mudmat horizontal bracings and (ii) floors (shells) represent the mudmat timber. The structural element unit weight was set as zero to eliminate additional weights to the mudmat. In this finite element model, 23496 of 15-noded tetrahedral elements with 62281 nodes were generated. The Poisson's ratio was taken as 0.49 for all clayey soil layers as common practice in total stress analyses.

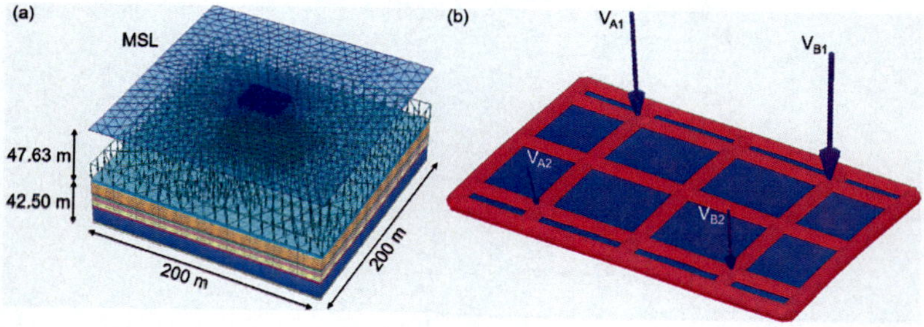

Fig. 3. Mudmat model in PLAXIS 3D Foundation

The loading at four legs of the mudmat was derived from the three selected load cases (2122, 2006 and 1013) and summarized in Table 4. These forces were then converted to the system coordinates of PLAXIS 3D Foundation.

In this numerical analysis, the mudmat settlement was first considered. The mudmat settlement for load case 2122 is shown for row 1 in the x-direction (Fig. 4a) and row B in the y-direction (Fig. 4b). The maximal settlement of the mudmat was about 116.29 mm for this load case. According to the difference of vertical loading acting on 4 legs, differential settlements were correspondingly observed for the mudmat base. The maximal differential settlement for three selected cases along mudmat diagonals was about 88 mm, equivalent to an inclination of 0.12° of the mudmat. The differential settlements of the mudmat seem small and shall be accounted for in the T&I analysis.

To compare with the analytical analysis, the factor of safety (FoS) was determined for the load cases 1013, 2006, and 2122 using the phi/c reduction procedure in PLAXIS

Table 4. Loading on four jacket legs

Load case	Leg point	Loading (kN)		
		Vx	Vy	Vz
1013	A1	0.7		−1716.4
	A2	0.7	177.9	−511.1
	B1			−1536.5
	B2		177.9	−331.1
2006	A1		−135.8	−1340.7
	A2			−1504.6
	B1	−139.0	−135.8	−846.7
	B2	−139.0		−1010.7
2122	A1	129.6		−2010.2
	A2	129.6	131.6	−837.1
	B1			−2487.2
	B2		131.6	−1314.1

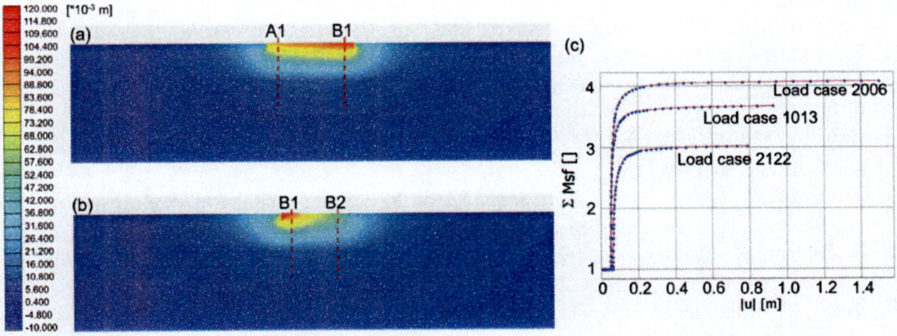

Fig. 4. Mudmat settlement for the load case 2122 (a) Row 1 in x-direction; (b) Row B in y-direction; (c) FoS determined for the three load cases 1013, 2006 and 2122

3D Foundation. The results were shown in Fig. 4c. In these numerical analyses, the FoS was the combined FoS for simultaneous vertical, horizontal, and moment loading. In contrast, the FoS determined in the analytical approach was separately determined for vertical loading, horizontal sliding, and moment overturning. Therefore, the comparison in terms of FoS between the analytical and numerical analysis is relative and essentially in the aspect of the FoS magnitude.

In some cases, the analytical approach gives smaller values of FoS than the numerical method. These discrepancies of the FoS may be explained by the difference in terms of the FoS calculation method and by the under-prediction of the bearing capacity in the analytical approach [4]. In general, a relative agreement between the analytical and numerical analysis was obtained. The numerical results have helped to confirm that the given mudmat satisfies the requirements mentioned in [3, 14].

5 Conclusions

In this work, the presence of very soft clayey layers under the mudmat leads to a high degree of undrained strength heterogeneity that challenges the conventional design procedure basing on the industry-specific codes. To confirm the mudmat foundation stability, a coupled study including an analytical and a numerical study was carried out. The factors of safety determined by the two approaches were in relatively good agreement, essentially in terms of magnitude. The results of these two approaches confirmed the satisfaction of the proposed mudmat according to the requirements specified in the standard API 2GEO. When the degree of undrained strength heterogeneity of the seabed is high, the study of Salencon and Matar (1982) seems to be a helpful reference to determine the foundation shape factor and the undrained bearing capacity.

References

1. ISO, ISO 19901-4: Petroleum and natural gas industries - Specific requirements for offshore structures - Part 4: Geotechnical and foundation design considerations. 2003: Geneva, Switzerland
2. DNVGL, DNVGL-RP-C212: Offshore soil mechanics and geotechnical engineering (2017)
3. API, ANSI/API RP 2GEO: Geotechnical and foundation design considerations (2014)
4. Randolph, M.F., Gourvenec, S.: Offshore Geotechnical Engineering. 2011, 2 Park Square, Milton Park, Abingdon, Oxon OX14 4RN, Spon Press - an imprint of Taylor & Francis, UK (2011)
5. Bransby, M.F., Randolph, M.F.: Combined loading of skirted foundations. Géotechnique **48**(5), 637–655 (1998)
6. Taiebat, H.A., Carter, J.P.: Numerical studies of the bearing capacity of shallow foundations on cohesive soil subjected to combined loading. Géotechnique **50**(4), 409–418 (2000)
7. Gourvenec, S.: Failure envelopes for offshore shallow foundation under general loading. Géotechnique **57**(9), 715–727 (2007)
8. Vulpe, C., Gourvenec, S., Power, M.: A generalized failure envelope for undrained capacity of circular shallow foundations under general loading. Géotech. Lett. **4**, 187–196 (2014)
9. Feng, X., Randolph, M.F., Gourvenec, S., Wallerand, R.: Design approach for rectangular mudmats under fully three-dimensional loading. Géotechnique **64**(1), 51–63 (2014)
10. Suryasentana, S.K., Dunne, H.P., Martin, C.M., Burd, H.J., Byrne, B.W., Shonberg, A.: Assessment of numerical procedures for determining shallow foundation failure envelopes. Géotechnique **70**(1), 60–70 (2020)
11. Salgado, R., Lyamin, A.V., Sloan, S.W., Yu, H.S.: Two- and three-dimensional bearing capacity of foundations in clay. Géotechnique **54**(5), 297–306 (2004)
12. Randolph, M.F.: Offshore geotechnics - the challenges of deepwater soft sediments. In: Geotechnical Engineering State of the Art and Practice, pp. 241–271 (2012)
13. Lunne, T., Robertson, P.K., Powell, J.J.M.: Cone penetration testing in geotehnical practice. Blackie, London (1997)
14. API, API Recommended Practice 2A-WSD: Planning, Designing, and Constructing Fixed Offshore Platforms - Working Stress Design, 22nd Edition (2014)
15. Salencon, J., Matar, M.: Capacité portante des foundations superficielle circulaires. J. de Mécanique théorique et appliqué **1**(2), 237–267 (1982)

New Practical Approach to Assess the Robustness and Redundancy of Offshore Structures: Application in Life Extension of Fixed Steel Offshore Platforms

Binh Nguyen Thanh[1](✉) and Hung Nguyen Manh[2]

[1] DNV, Bærum, Norway
thanh.binh.nguyen@dnv.com
[2] PTSC Mechanical and Construction, Vũng Tàu, Vietnam
hungnm1@ptsc.com.vn

Abstract. Life extension of fixed offshore platform structures is a growing demand in oil and gas industry worldwide. Although the industry has been developing various theories, models and guidelines for structural integrity assessment of ageing structures, it remains challenging for operators in real-life practice. This paper presents a new practical approach for assessment of residual strength of ageing fixed offshore platform structures. Numerous structural models with damaged members and joints can be randomly selected for structural assessment in strength limit state using industry recognized program suites such as USFOS or SACS. There will be a massive workload for structural engineers to implement the structural assessment and verify the analysis results in conventional approach. However, a new approach is proposed in this paper aiming to simplify computational process by developing a new framework and an integrated software is developed for assessment of analysis results. The new approach will also assist Operator's management in a quick decision-making process if damage is detected on certain members or joints of the structures. The framework is then applied to an actual ageing jacket as a case study and recommendations are drawn accordingly. In conclusion of this paper, applicability, validity and limitations of the new approach are discussed.

Keywords: New approach · Steel structure · Life extension · Fixed offshore platform · Structural integrity · Framework · Recommended practice · Nonlinear assessment · Fracture mechanics · Material ductility · Fracture toughness

1 Historical Background

Offshore structures are designed in accordance with internationally recognized standards such as API RP 2A [1], ISO 19900 [3], DNVGL-OS-C101/C201 or NORSOK N-001 [4]. These standards define requirements for fulfillment of different limit states:

D. V. K. Huynh et al. (Eds.): VSOE 2021, LNCE 208, pp. 325–332, 2022.
https://doi.org/10.1007/978-981-16-7735-9_36

- The Ultimate Limit State (ULS);
- The Fatigue Limit State (FLS);
- The Serviceability Limit State (SLS);
- The Accidental Limit State (ALS).

It is worth noting that it is not the intention of the authors to provide guidelines for comprehensive structural analysis, but it is intended to highlight the differences in design conditions between South East Asia (SEA) seas and other harsh environmental regions such as North Sea or North Atlantic Sea with important findings that structural failure in ULS is dominant for fixed offshore structures in SEA seas rather than FLS which is highly attractive to many researchers and engineers in structural integrity assessment of ageing offshore structures. These findings will serve as basis for development of a new practical approach for structural integrity assessment of ageing fixed offshore structures in SEA seas with a target to answer two major questions of the structure owners as listed below:

- Will the damaged structure collapse?
- Is it possible to extend the structure design life with such damaged condition?

It is intention of this paper to help the owner/operator answer above questions by discussion about a computational framework that helps to determine the ultimate capacity of damaged structures by non-linear finite element analysis method.

2 New Approach

This section is intended to present a modified approach that suits environmental conditions in SEA and specifically fixed offshore platforms in Vietnams' seas.

2.1 Robustness and Redundancy of Offshore Structures

API RP 2SIM [2] proposed a term namely "reserve strength ratio (RSR)": A measure of the ultimate load carrying capacity of a platform, defined as the ratio of the base shear at ultimate capacity to the base shear derived from a 100-year return period weather event.

The detail of terms applied in collapse process of a redundant structure with significant plastic strength can be seen on Fig. 1.

Furthermore, API RP 2SIM recommends that the RSR should be at least 1.6 for platform type L-1 (manned and high consequence of failure, [1]) and 0.8 for platform type L-3 (unmanned and low consequence of failure, [1]). In fact, due to probabilistic nature of 100-year return period events, there would be an existing relationship between RSR and extreme design event [7] that is re-produced in Fig. 2.

Due to the absence of required RSR for structural assessment in Vietnams' seas which is normally considered as benign environment (Region C). In this paper, it is conservative to adopt curve-A for metocean design condition in Vietnam. The Author's experience in RSR assessment of ageing offshore structures in Vietnam is highlighted below:

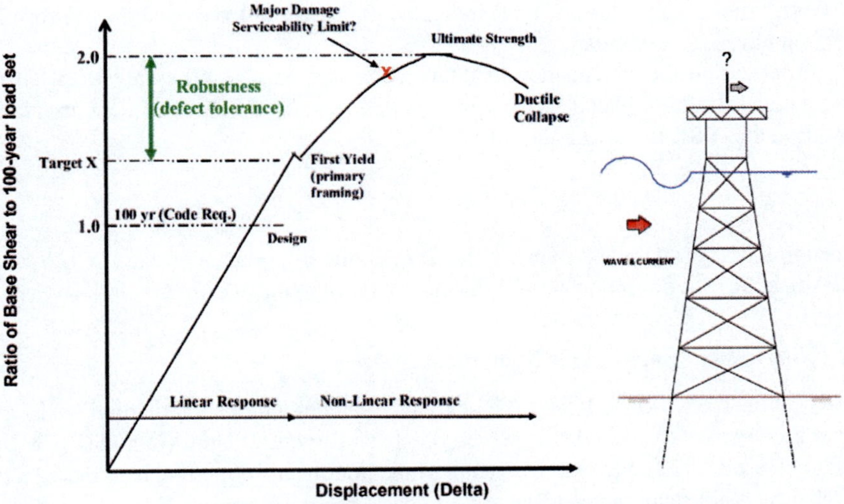

Fig. 1. Basic concepts in progressive collapse analysis

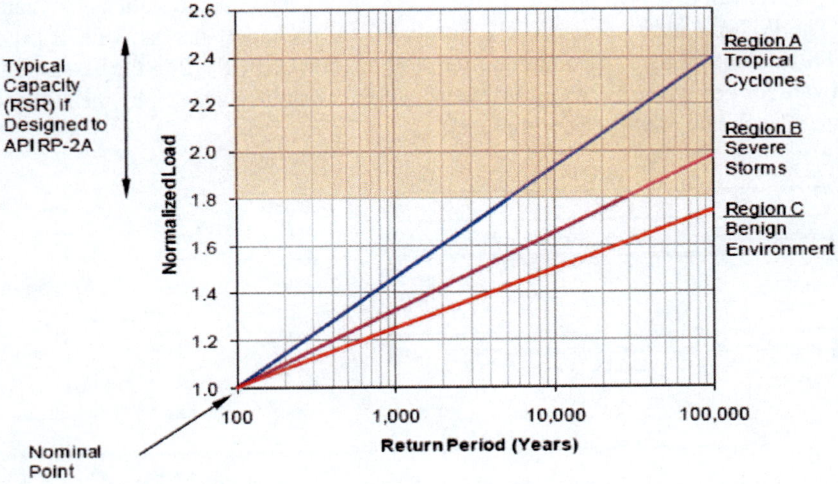

Fig. 2. Typical environmental hazard curves for different regions

- Vietsovpertro's offshore structure type of MSP has the RSR of 3.2, i.e. the platform will globally collapse in a 100,000-year wave event, approximately. This platform has originally been designed to Russian standard.
- Hoan Vu JOC's Ca Ngu Vang WHP has minimum RSR of 1.7 that yields a global collapse in a 5,000-year wave event, approximately. This platform has been designed in accordance with API RP 2A.
- Hoang Long JOC's H1 WHP has minimum RSR of 2.4 that yields a global collapse event in a 100,000-year wave event, approximately. This platform has been designed in accordance with API RP 2A.

Above numerical values of RSR indicated that existing fixed offshore platforms in Vietnam have high redundancy.

To determine the redundancy of damaged structure, it is necessary to determine the Residual Strength Factor (RIF) which is defined as ratio between RSR in damage condition and RSR in intact condition:

$$RIF = \frac{RSR_{damage}}{RSR_{intact}} \tag{1}$$

High RIF (normally less than 1.0) means that the importance of damage is low and vice versa. This definition is helpful in assessment of damage effect.

2.2 Non-linear Space Frame Structure Analysis

To determine the RSR and RIF, it is essential to conduct nonlinear structural analysis of space frame structures in ULS in which Idealised Structural Unit Method (ISUM) [8] should be used. The fundamental concept of the nonlinear FE method is to model a real member by a physical member that allows coarse mesh of joints.

The theory of space frame nonlinear analysis can be briefly described in Fig. 3 for hinge formulation in a beam subject to bending moment. Reference is made to [5, 10] and [11] for full details of nonlinear analysis of space frame in both static and dynamic.

The ISUM is used in commercial non-linear FE packages such as USFOS [5] and ALPS/ISUM [9]. There are some common opinions that USFOS [12] is the best available software for non-linear FE analysis. Hence, USFOS had been chosen by the authors to further development of the new approach.

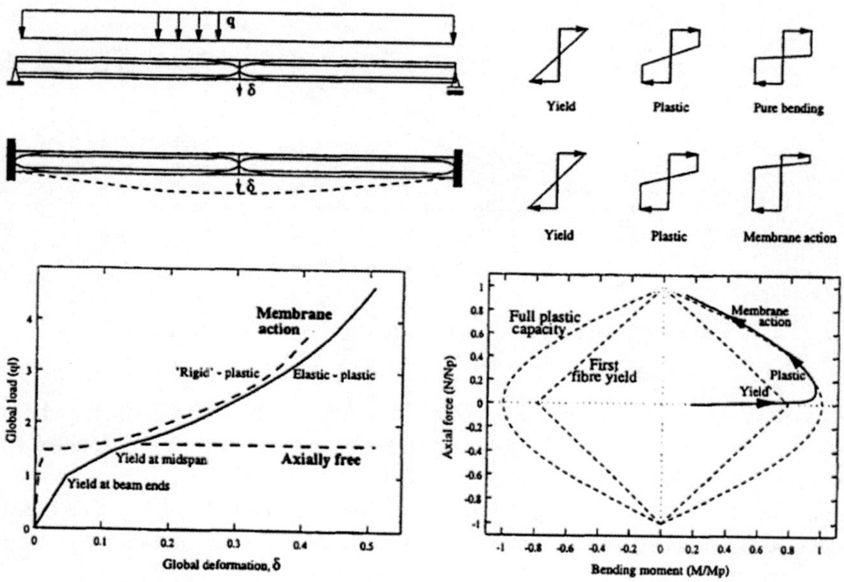

Fig. 3. Yield Hinges, Beam Bending and Membrane Effects

2.3 Automation of USFOS Programs

The center of the new approach is launching of a computer programing toolset namely AutoRIFS® (Fig. 4) that helps to determine RSRs/RIFs in a manner that it is less time consuming by running USFOS in batch-mode (both Windows and Linus platforms) and can be automatically and remotely controlled by the batch files which have been programmed by the authors.

After running non-linear analyses by using USFOS in batch-mode, a huge-size database will be produced. AutoRIFS® can help structural engineers to access into the database subfolders and obtain important data, summarize then export results to office suites such as Microsoft Excel in smart manner.

Normally, non-linear analyses are very time-consuming, but AutoRIFs will effectively assist engineers' practice and maximize usage of commercial software. Hence, the non-linear analyses can be run as much as possible to yield enough database of ultimate structural capacity taking into account massive damage scenarios.

3 Application Case Study

Redundancy analyses have been carried out for a 4-legged platform with analysis information tabulated in Table 1.

Table 1. Basic parameters of a case study for determining ultimate structural capacity in damage scenarios

Parameter	Description	Platform USFOS model
Number of elements, (2 nodes)	2587	Note: Group ID 2 consist of 2 Elements: 258 and 426[a]
Number of nodal points	1866	
Number of wave/current load cases	7 (0°, 20°, 30°, 90°, 120°, 160° and 180°)	
Number of permanent load cases	2 (Gravity - Dead Weight - Equipment Load + Buoyancy Load)	
Number of element-group removals	176	
Total analysis cases	$7 \times (176 + 1) = 1239$	
Time consumed	About 3 days (4,336.5 min)	
Results file size of each non-linear analysis	Approx. 219.8 MB	
Project folder size	266 GB	

[a]RSR of 2.969 for Intact Analysis with Wave Load Case No.4, continued to remove Group ID 2 will give RIF of 0.968, i.e. the analysed member group is deemed less important.

The number of non-linear analysis in redundancy assessment will be increased incredibly due to combinatorial number of removed members. Table 2 presents approximately required computer resource for various combination removals.

Table 2. Required computer resources for various removal scenarios

Number of Group ID removed	Number of USFOS analysis Case	Approx. storage size (GB)	Approx. analysis time (days)
0	7	1.5	0.02
1	1,232	264.4	2.99
2	215,600	46,278.2	524.03
3	37,514,400	8,052,407.3	91,180.83
4	6,489,991,200	1,393,066,470.5	15,774,284.17
5	1,116,278,486,400	239,607,432,920.6	2,713,176,876.67[b]

[b]The fastest supercomputer (Fugaku with a LINPACK benchmark score of 415 PFLOPS) may have to spend 1.35 days to complete these analyses of 5-member groups removal.

With huge data amount yielded from non-linear analyses, author's software will be used to access this database with a few efforts to export important data to Microsoft Excel sheets (See Fig. 4) (Fig. 5).

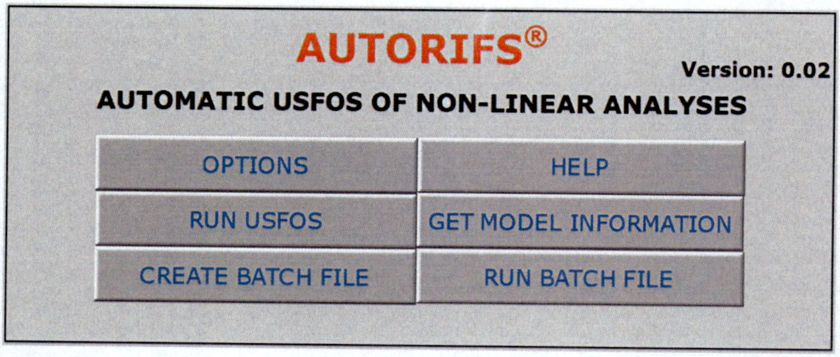

Fig. 4. AutoRIFS® interface

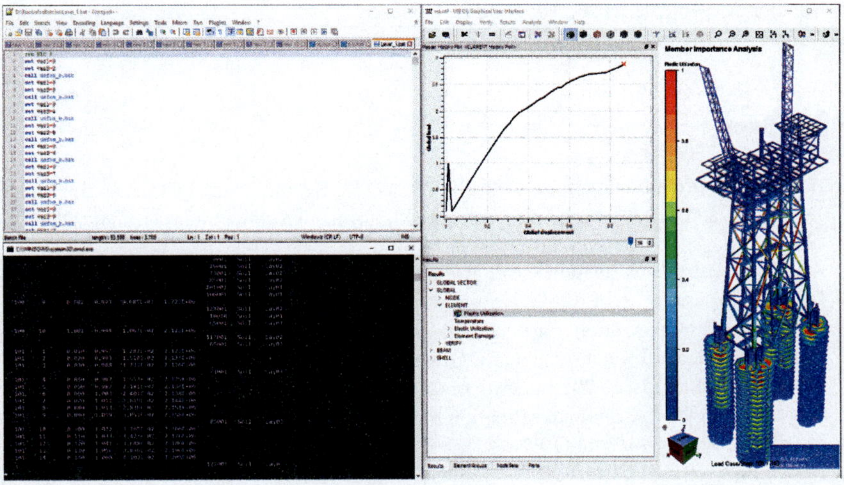

Fig. 5. Arrangement of analysis for case 1_177 (WLC 4 and remove Group ID 2)

Create USFOS Batch Files for 1 Group Removal | Get Peak Loads from USFOS Result Files

Analysis No.	WLC	Group ID	Folder Name	Adjusted RSR	RIF
Case 1_172	3	169	WLCID_3__GroupID_169	2.521	0.504
Case 1_173	3	170	WLCID_3__GroupID_170	1.716	0.343
Case 1_174	3	171	WLCID_3__GroupID_171	2.407	0.481
Case 1_175	3	172	WLCID_3__GroupID_172	2.356	0.471
Case 1_176	3	173	WLCID_3__GroupID_173	2.351	0.470
Case 1_177	4	2	WLCID_4__GroupID_2	2.875	0.968
Case 1_178	4	3	WLCID_4__GroupID_3	2.953	0.995
Case 1_179	4	4	WLCID_4__GroupID_4	3.016	1.016

Fig. 6. Analysis results presented in a smart manner

4 Conclusion

This technical paper is intended to present a new practical approach for determining ultimate structural capacity in different damage scenarios in order to assist offshore structure owners/operators in structural integrity management. The following important conclusions are given accordingly:

- Because of warm temperature in tropical regions, materials are likely to have high ductility properties in fracture;
- Existing offshore platforms has high redundancy; hence, it is of high probability that when one member begins yielding, it may not result in global collapse of the entire platform;
- Progressive ultimate collapse is the most governing concern in assessment of ageing offshore fixed structures in SEA seas rather than fatigue failure;
- The software AutoRIFS® could help to optimize usage of non-linear F.E analysis software by summarizing huge databases with less efforts so as to help the offshore platform owner/operator utilize the resources efficiently.

Further developments such as required RSR by regulators, boat impact, fire/explosion analysis are being continued by the authors.

References

1. American Petroleum Institute: API RP 2A WSD, Recommended Practice for Planning, Designing, and Constructing Fixed Offshore Platforms - Working Stress Design, 22nd Edition (2014)
2. American Petroleum Institute: API RP 2SIM, Structural Integrity Management of Fixed Offshore Structures, 1st Edition, November 2014
3. International Organization for Standardization: ISO 19900, Petroleum and Natural Gas Industries - General Requirements for Offshore Structures (2019)
4. NORSOK Standards: N-001 Integrity of Offshore Structures (2020)
5. SINTEF: USFOS - A Computer Program for Progressive Collapse Analysis of Steel Offshore Structures. Theory Manual (1993)
6. Det Norske Veritas: Classification Note No. 30.6 Structural Reliability Analysis of Marine Structures, 1 July 1992
7. Amer, M.Y.: Development of Structural Regional Acceptance Criteria for Existing Fixed Offshore Structures in the Arabian Gulf (2010)
8. Ueda, Y., Rashed, S.M.H.: An ultimate transverse strength analysis of ship structure. J. Soc. Naval Architect. Japan, Tokyo. **136**, 309–324 (1974). (in Japanese)
9. Paik, J.K., Thayamballi, A.K.: Ultimate Limit State Design of Steel-Plated Structures. 1st Edition (2013)
10. DNV- SINTEF- BOMEL: Ultiguide Phase 2 - Best Practice Guidelines for Use of Non-linear Analysis Methods in Documentation of ULSs for Jacket Type Offshore Structures, April 1999
11. SINTEF-DNV: Ultiguide Phase 2 - Evaluation Report on Dynamic Effects (1998)
12. USFOS webpage. https://www.usfos.no/

Numerical Investigation on Bearing Capacity of Rigid Footing on Sandy Soils Under Eccentrically Inclined Load

N. Quang Pham[1](✉) and Satoru Ohtsuka[2]

[1] Faculty of Transportation Mechanical Engineering, The University of Danang-University of Science and Technology, 54 Nguyen Luong Bang Street, Danang City, Vietnam

[2] Department of Civil and Environmental Engineering, Nagaoka University of Technology, 1603-1 Kamitomioka, Nagaoka, Niigata 940-2188, Japan

Abstract. The objective of this study was to evaluate the bearing capacity of a rigid footing on uniform sandy soil under eccentrically inclined load, using the rigid plastic finite element method (RPFEM). The footing was considered to be rigid and rough at the contact plane, as it most often is in reality. The study thoroughly considered the effect of the soil properties on the load inclination factor i_γ to investigate the validity of the current design methods. In particular, the effect of the horizontal load in two opposite directions on the ultimate bearing capacity and the limit load space in the *V-H-M* space were clarified. Through a series of numerical analyses, new equations were proposed for the load inclination factor i_γ, and the limit load space in the *V-H-M* space, by taking into account the direction of the horizontal load.

Keywords: Limit load space · Rigid footing · Eccentric-inclined load

1 Introduction

The bearing capacity of a rigid footing subjected to an inclined central load is commonly calculated by the bearing capacity formula for sandy soil with no embedment which is expressed as follows:

$$V = \frac{1}{2} i_\gamma \gamma B^2 N_\gamma \tag{1}$$

where N_γ is the bearing capacity factor, and B, γ are the width of the footing (M) and the soil unit weight (kN/m^3). The load inclination factor i_γ is the function of the internal friction angle of the soil, ϕ and the inclination angle of the load, α. In practice, rigid footings are often subjected to eccentric-inclined coupled loads. The stability of a rigid footing on the free surface of soils under complex loading is of practical interest. Meyerhof [4], Hansen [2], and Vesic [8] conducted a series of model tests to propose empirical and semi-empirical inclination factors, as shown in Table 1. However, the applicability of these inclination factors has not been clarified due to the lack of a

D. V. K. Huynh et al. (Eds.): VSOE 2021, LNCE 208, pp. 333–341, 2022.
https://doi.org/10.1007/978-981-16-7735-9_37

systematic analysis based on a reliable stability method. Therefore, the validity of the current design methods needs to be verified by applying load inclination factor formulas for i_γ. Although many works have been performed to investigate the ultimate bearing capacity under eccentrically inclined loads, a comprehensive understanding of both the failure mechanism of the footing-soil system and the limit load space in the vertical load - horizontal load – moment (*V-H-M*) space has not yet been established.

Table 1. Summary of load inclination factor i_γ for bearing capacity of footing.

Meyerhof (1963)	Hansen (1970)	Vesic (1975)	Loukidis (2008)
$i_\gamma = \left(1 - \frac{\alpha}{\phi}\right)^2$	$i_\gamma = (1 - 0.7\tan\alpha)^5$	$i_\gamma = (1 - \tan\alpha)^3$	$i_\gamma = \left(1 - 0.94\frac{\tan\alpha}{\tan\phi}\right)^{(1.5\tan\phi+0.4)^2}$

In this study, the rigid plastic finite element method (RPFEM) is employed to determine the ultimate bearing capacity of the footing placed on a uniform layer of sandy soil subjected to an eccentric vertical load. New constitutive equations are introduced to simulate the solid element and the interface element based on the upper bound theorem in the limit analysis. The objective of this study is to determine the limit load space in the *V-H-M* space and the failure mechanism of the footing-soil system. Focus is placed here on the ultimate bearing capacity of a footing against the action of eccentric and inclined loads taking into account the direction of the horizontal load. The sign convention for the eccentrically inclined loading in this study, based on the one suggested by Butterfield et al. [1], is shown in Fig. 1.

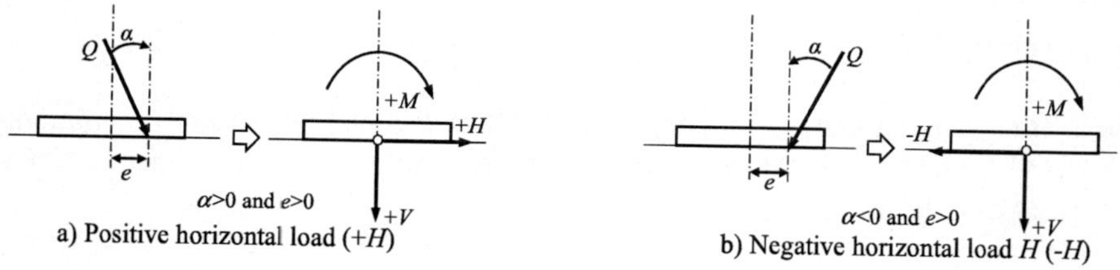

Fig. 1. Sign convention for positive and negative combinations of *H* and *M*

The combined loading on the rigid footing can be represented by a resultant load Q at eccentricity e and inclination angle α which is divided into three statically equivalent loads, namely, V, H, and M. Moment load M is positive when acting clockwise and horizontal load H is positive in the positive direction of the x-axis. Inclination angle α is positive when acting counterclockwise. For these purposes, load inclination factor i_γ is reported in the results as a function of the friction angle of the soil, while the limit load space in the *V-H-M* space are reported as equations of an ellipse depending on the positive and negative directions of the horizontal loads.

2 Rigid Plastic Constitutive Equation for Finite Element Method

2.1 Rigid Plastic Constitutive Equation for Solid Elements

Tamura et al. [7] developed the rigid plastic constitutive equation for frictional material. The Drucker-Prager's type yield function is expressed as follows.

$$f(\boldsymbol{\sigma}) = aI_1 + \sqrt{J_2} - b = 0 \tag{2}$$

where $I_I$is the first invariant of stress σ_{ij} and $I_1 = tr(\sigma_{ij})$ in which extension stress is defined positive. J_2 is the second invariant of shear stress s_{ij} and $J_2 = \frac{1}{2}s_{ij}s_{ij}$ and the coefficients a and b are the material constants corresponding to internal friction angle and cohesion under the plane strain condition. The strain rate $\dot{\boldsymbol{\varepsilon}}$, which is purely plastic component should satisfy the volumetric constraint condition as follow:

$$h(\dot{\boldsymbol{\varepsilon}}) = \dot{\varepsilon}_v - \frac{3a}{\sqrt{3a^2 + \frac{1}{2}}}\dot{e} = \dot{\varepsilon}_v - \eta\dot{e} = 0 \tag{3}$$

where $\dot{\varepsilon}_v$ and $\dot{e}$ indicate the volumetric strain rate and the norm of strain rate, respectively. The parameter η is a coefficient determined by Eq. (2). The study derived their constitutive equation by introducing the constraint condition into the constitutive equation on the strain rate directly with the use of penalty method. The stress-strain rate relation for the Drucker-Prager's yield function is expressed as follow:

$$\boldsymbol{\sigma} = \frac{b}{\sqrt{3a^2 + \frac{1}{2}}}\frac{\dot{\boldsymbol{\varepsilon}}}{\dot{e}} + \kappa(\dot{\varepsilon}_w - \eta\dot{e})\left(\boldsymbol{I} - \eta\frac{\dot{\boldsymbol{\varepsilon}}}{\dot{e}}\right) \tag{4}$$

where K is a penalty constant. FEM with this constitutive equation provides the equivalent equation of the upper bound theorem in plasticity so that this method is called as RPFEM in this study. It is noted property of this constitutive equation that the relationship between stress and strain rate is specified. At the limit state of the soil-footing system, since the displacement field of the system is indeterminate for the applied load, the norm of the strain rate in the failure area becomes indeterminate, too in the system. Stress is determined for normalized strain rate using its norm.

2.2 Rigid Plastic Constitutive Equation for Interface Elements

At the contact plane where the displacement velocity is discontinuous, the stress at the limit state is expressed by the following coulomb yield function:

$$f(\mathbf{t}) = |t_s| - c_s + t_n\tan\phi_s = 0 \tag{5}$$

where, t_s and t_n are the shear and normal components of the stress vector at the contact plane. ϕ_s and c_s are the material parameters corresponding to the internal friction angle and the cohesion at the contact plane.

The interface element of zero thickness is introduced into the contact plane between two bodies in Fig. 2(a). $\Delta\dot{\mathbf{u}}$ is the vector of the relative displacement velocity at the discontinuous line in displacement velocity, and $\Delta\dot{\mathbf{u}} = \dot{\mathbf{u}}^{+} - \dot{\mathbf{u}}^{-}$ is shown in Fig. 2(b).

The volumetric constraint condition is expressed as follows:

$$h(\Delta\dot{\mathbf{u}}) = |\Delta\dot{u}_s|\tan\phi_s - \Delta\dot{u}_n = \left(\frac{\Delta\dot{u}_s}{|\Delta\dot{u}_s|}\cdot\tan\phi_s - 1\right)\begin{pmatrix}\Delta\dot{u}_s\\ \Delta\dot{u}_n\end{pmatrix} = \mathbf{a}\cdot\Delta\dot{\mathbf{u}} = 0 \tag{6}$$

where, $\Delta\ \dot{u}_n$ is a component of relative displacement velocity normal to the discontinuous line; and $\Delta\ \dot{u}_s$ is a tangential component.

The stress and relative displacement velocity relation for the interface element is expressed by directly considering the volumetric constraint condition with a penalty constant ξ (Pham et al. 2020) [6] as follows.

$$\mathbf{t} = \frac{c_s}{\cos\phi_s\left(1+\tan^2\phi_s\right)}\frac{\Delta\dot{\mathbf{u}}}{\|\Delta\dot{\mathbf{u}}\|} + \xi(\mathbf{a}\cdot\Delta\dot{\mathbf{u}})\mathbf{a} \tag{7}$$

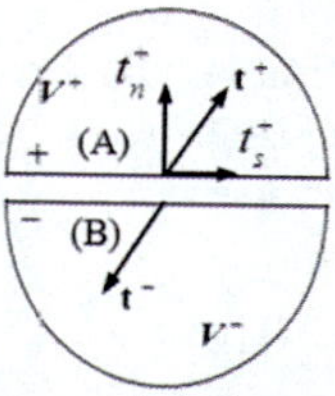

(a) Stress conditions on interface surface

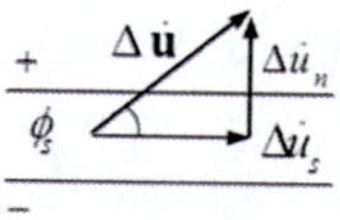

(b) Line of motion on interface surface

Fig. 2. Stress vector and vector of relative displacement velocity

3 Ultimate Bearing Capacity for Inclined Central Load

As mentioned earlier, a two-dimensional model is introduced to investigate the behavior of a footing under an inclined load. The footing was assumed to be perfectly rough such that the friction angle of the interface elements between the footing and the soil would be taken as equal to the friction angle of the soil. This study examines the effect of the internal friction angle of the soil on the load inclination factor i_γ and the failure envelope in the *V-H* plane in the case of a concentric load. Figures 3 shows the obtained result on the load inclination factor i_γ for various internal friction angles of the soils in the case of an inclined central load. The figure demonstrates that the load inclination factor i_γ depends greatly on the value of the internal friction angle, although Hansen [2] and Vesic [8] proposed the inclination factors that do not depend on the value of the internal friction angle. Since differences are found among the obtained results, this study positively proposes the inclination factor i_γ as a function of ϕ, as seen in Loukidis et al. [3].

$$i_\gamma = \left(1 - 0.76\frac{\tan\alpha}{\tan\phi}\right)^{(1.7\tan\phi+0.4)^2} \tag{8}$$

$$\frac{H}{V_{ult}} = \frac{\tan\phi}{0.76}\frac{V}{V_{ult}}\left[1 - \left(\frac{V}{V_{ult}}\right)^{1/(1.7\tan\phi+0.4)^2}\right] \tag{9}$$

Figure 4 shows the results of the failure envelopes in the *V-H* plane obtained for case studies of $\phi = 30°$, $35°$, and $40°$. It shows that the sizes and shapes of the failure envelopes in the *V-H* plane seem to be dependent on the value of ϕ. these findings are well in agreement the results reported by Loukidis et al. [3]. From Eq. (8), the failure envelope in the *V-H* plane is derived based on the results computed with the RPFEM, as seen in Fig. 4. Moreover, the obtained results for the strain rate distributions of the footing under an inclined central load on sandy soil of $\phi = 30°$ at load inclination angles of $\alpha = 10°$ and $20°$ are shown in Fig. 5. Figure 5(a) indicates that the failure mode of the footing has an asymmetric shape and becomes largely one-sided as load inclination angle α increases, as seen in Fig. 5(b). The observations of the failure mechanisms agree well with those in the results of Loukidis et al. (2008) [3] using the finite element method. Moreover, the ultimate bearing capacity of the rigid footing was obtained at about $Q = 2232$ kN/m for an inclination angle of $\alpha = 10°$, and was $Q = 1077$ kN/m for an inclination angle of $\alpha = 20°$. The difference in ultimate bearing capacities due to the increase in the load inclination angle α is seen to be very large. This is because the vertical and horizontal extents of the failure mechanism decrease with the increasing load inclination α, which corresponds to the smaller bearing capacity. It indicates that the RPFEM can provide reasonable predictions of the bearing capacity and the failure mode of the rigid footing on sandy soil under the inclined central load.

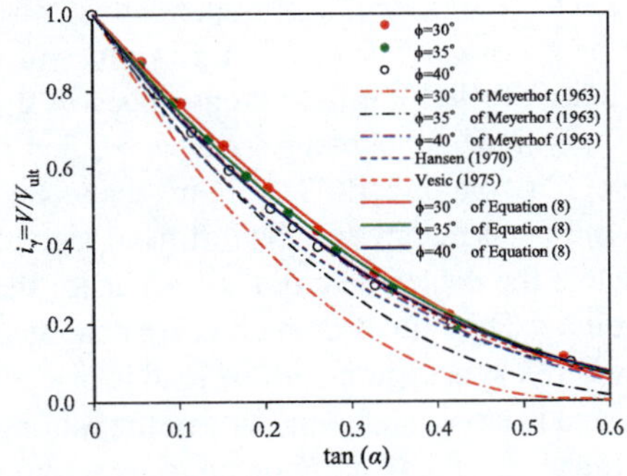

Fig. 3. Effect of internal friction angle of soil on load inclination factor i_γ in sandy soil

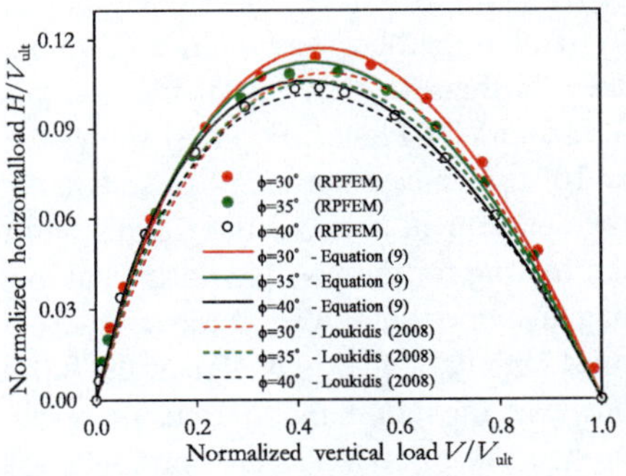

Fig. 4. Effect of internal friction angle on failure envelopes in *V-H* plane in sandy soil

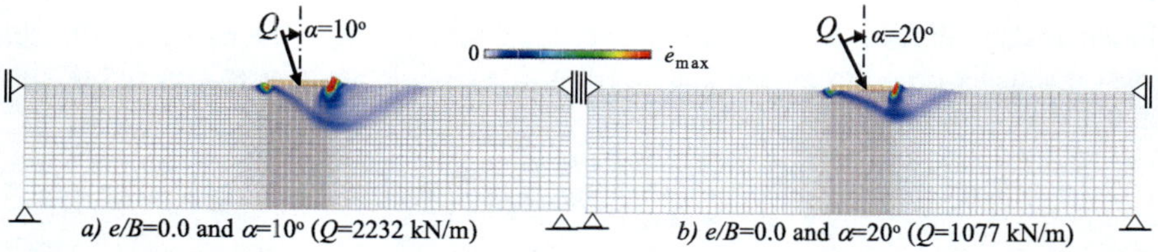

Fig. 5. Deformation diagrams of footing-soil against inclined central load in case of sandy soil $\phi = 30°$

4 Ultimate Bearing Capacity for Eccentrically Inclined Load

In case of the eccentrically inclined load, the contact stress between the footing and the soil needs to be properly computed by considering the detachment of them. Pham et al. [5] reported that the application of an interface element to the footing-soil system was effective for analyzing the distribution of the contact stress and the failure mechanism of the sandy soils. In this study, a series of analyses are conducted for the eccentric-inclined loaded footing in the case of sandy soil of $\phi = 30°$ taking account into the direction of the horizontal load, by using the interface element. in particular, the study investigates the effect of two different loading paths on the failure envelope in *V-H* and *V-M* planes, namely, the normalized eccentricity *E/B* and the inclination angle tan(α) in which they vary stepwise from 0.0 to 0.3.

Figures 6 and 7 Show the failure envelopes in the *V-H* and *V-M* planes for various values of eccentricity length *e* and inclination angle tan (α). It is seen that, when the eccentricity length *e* and tan(α) Increase, the size of the failure envelopes in the *V-H* and *V-M* planes decrease for both the positive and negative horizontal loads. This is because the bearing capacity of the rigid footing decreases with an increase in the eccentricity *e* and the inclination angle α regardless of the direction of the horizontal loads. It can be observed from the figures that the shapes of the failure envelopes for the positive horizontal load are presented as nearly symmetric shapes, while those for the negative horizontal load are presented as asymmetric shapes. The maximum values of H/V_{ult} and M/BV_{ult} for the positive horizontal load are seen to be smaller than those of the negative horizontal load corresponding to each value of *e*, tan(α). The difference is caused by the fact that the direction of the horizontal load affected the failure envelopes in the *V-H* and *V-M* Planes. Figure 8 shows the failure modes obtained for $e/B = 0.3$ and $\alpha = 10°$ in two cases of the positive and negative horizontal loads. The figure shows that the detachment between the footing and the soil occurred around the left-hand side of the footing for the positive horizontal load, while the detachment did not occur for the negative horizontal load at the large eccentricity *e*. The failure domain is concentrated most largely on the right edge of the footing where the eccentric vertical load is placed. however, the failure mode forms more deeply and is larger along one side of the footing in the same direction as the horizontal load. It can be seen that giving consideration to the direction of the horizontal load leads to an increase in bearing capacity *Q* in the case of the negative horizontal load. The rigid footing can support a higher applied load at the negative horizontal load.

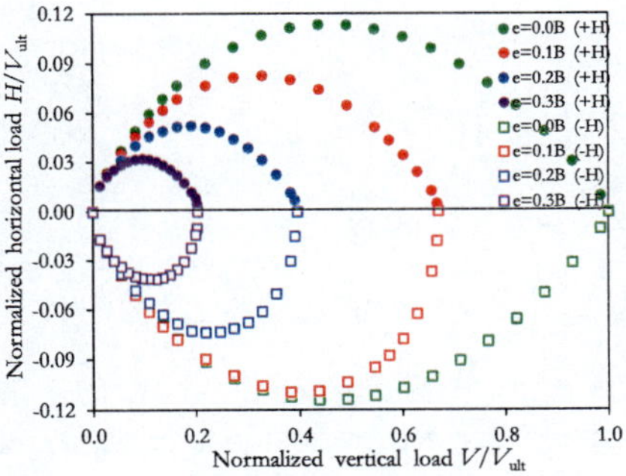

Fig. 6. Failure envelopes in *V-H* plane in case of sandy soil $\phi = 30°$

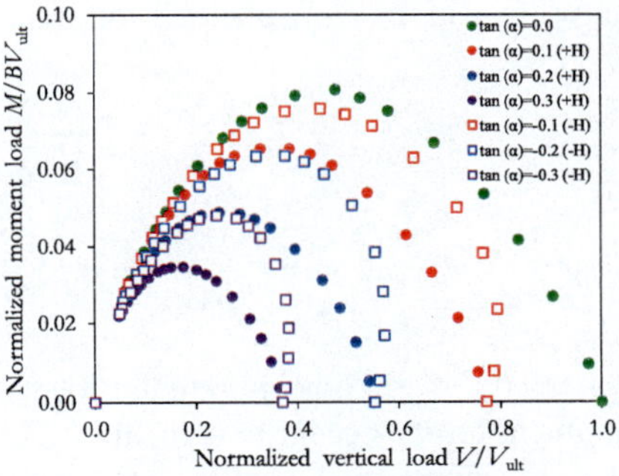

Fig. 7. Failure envelopes in *V-M* plane in case of sandy soil $\phi = 30°$

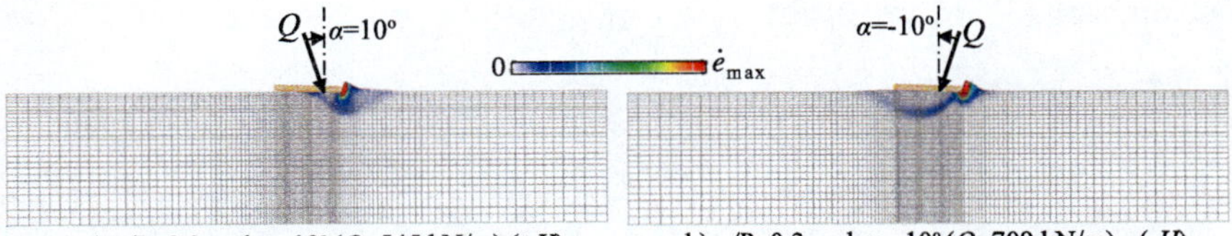

a) *e/B*=0.3 and *α*=10° (*Q*=545 kN/m) (+*H*)   b) *e/B*=0.3 and *α*=-10° (*Q*=709 kN/m) - (-*H*)

Fig. 8. Deformation diagrams of footing-soil against eccentric inclined load taking into account direction of horizontal load (*e/B* = 0.3, sandy soil of $\phi = 30°$)

5 Limit Load Space of a Rigid Footing in *V-H-M* Space

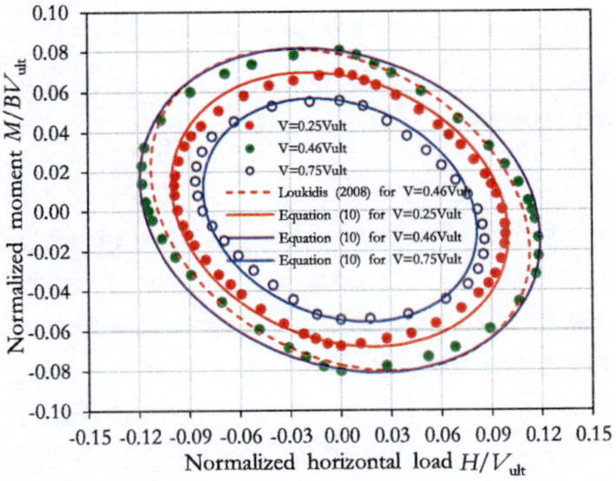

Fig. 9. Failure envelopes in *H-M* plane in case of sandy soil of $\phi = 30°$

The limit load space in the *V-H-M* space for the effect of the eccentric-inclined coupled loads have relatively complex geometries. Figure 9 shows the failure envelopes in the *H-M* plane under the eccentric-inclined coupled loads at $V/V_{ult} = 0.25$, 0.46, and 0.75. It can be seen that the sizes and shapes of the failure envelopes are dependent on the level of normalized vertical load V/V_{ult}. Moreover, these diagrams clearly show that the rigid footing can support a higher applied load with the combination of the negative horizontal and positive moment loads. This is because the direction of the horizontal load affects the failure envelopes in the *H-M* plane. Most of the past works model the failure envelopes in the *H-M* plane by considering the equation for an ellipse by Buttterfield et al. [1]. This study proposes $C = 0.195$ for sandy soil from the computed failure envelopes in the *H-M* plane as follows:

$$\left(\frac{H}{H_o}\right)^2 + \left(\frac{M}{M_o}\right)^2 + 2C\left(\frac{H}{H_o}\right)\left(\frac{M}{M_o}\right) = 1 \tag{10}$$

$$\frac{M}{BV_{\mathrm{ult}}} = 0.55\frac{V}{V_{\mathrm{ult}}}\left(1 - \left(\frac{V}{V_{\mathrm{ult}}}\right)^{0.49}\right) \tag{11}$$

where H_o is the limit horizontal load for the inclined central load and M_o, the limit moment for the eccentric vertical load under the prescribed vertical load of V_o. H_o and M_o are provided by Eqs. (9) and (11) by inserting V_o into V. There, Eq. (11) was proposed by Pham et al. [5] for the eccentric vertical load in case of sandy soil. According to the work of Loukidis et al. [3], parameter C is equal to 0.267, and H_o and M_o are given by his equations. The failure envelope by RPFEM is in good accordance with that by Loukidis et al. [3] at $V_o = 0.46\ V_{\mathrm{ult}}$, as seen in Fig. 9.

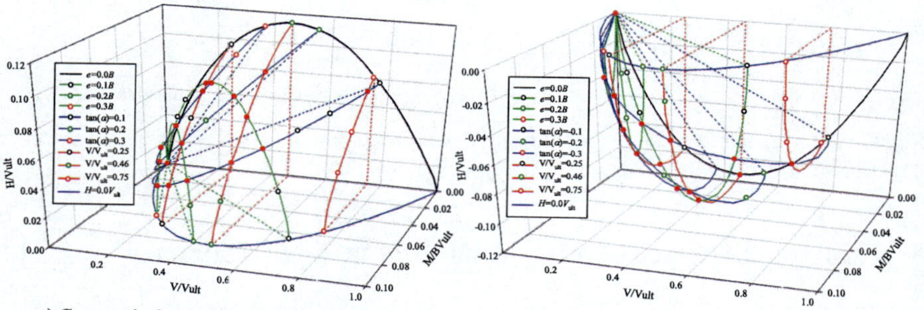

a) Case study for positive horizontal load (+*H*) b) Case study for negative horizontal load (-*H*)

Fig. 10. Limit load space of *V-H-M* for various values of eccentricity length *e*, inclination angle tan (α) and normalized vertical load V/V_{ult} on sandy soil ($\phi = 30°$)

In addition, once the limit load space in the *V-H-M* space of a rigid footing under an eccentrically inclined load is proposed, the question is raised as to whether it can be applied for the different loading paths. Almost none of the previous studies considered the uniqueness of the limit load space. The present study widely investigates the intersecting points in the three-dimensional failure envelopes in the *V-H-M* space with three different computation conditions, namely, the eccentricity length *e*, the inclination angle tan(α), and the normalized vertical load V/V_{ult}. A three-dimensional image of the limit load space under an eccentrically inclined load is presented in Fig. 10. The representations of the failure envelopes in the *V-H-M* space are shown as contour plots. The bold red plots denote the intersecting points calculated from the RPFEM. The figure shows some intersecting points among the contour plots of $e/B = 0.1$, 0.2, and 0.3, of tan (α) = 0.1, 0.2, and 0.3, and of $V/V_{\mathrm{ult}} = 0.25$, 0.46, and 0.75. It can be seen that the limit load space in the *V-H-M* space for the different loading paths are almost coincidental for the same limit surface. It can be concluded that the limit load space in the *V-H-M* space is unique for each value of ϕ.

6 Conclusions

The Conclusions of This Study Are as Follows:

(1) The numerical results by the rigid plastic FEM for inclined central load indicated the load inclination factor i_γ in the ultimate bearing capacity formula changed with the internal friction angle ϕ in case of sandy soil. Based on the analyzed results, new equation was proposed for the load inclination factor.
(2) Under eccentrically inclined load, the ultimate bearing capacity of footing was properly estimated by employing the interface element between the footing and the soils. In the computed results, the collapse mode changed by the direction in horizontal load even though the eccentricity length for the applied load was same.
(3) The ultimate bearing capacity formula was newly proposed for eccentrically inclined load. The limit load space in the normalized *V-H-M* spaces was proved to be unique for various loading paths.

References

1. Butterfield, R., Houlsby, G.T., Gottardi, G.: Standardized sign conventions and notation for generally loaded foundations. Géotechnique **47**(5), 1051–1054 (1997)
2. Hansen, J.B.: A revised and extended formula for bearing capacity. Danish Geotechn. Inst. Bull. No. **28**, 5–11 (1970)
3. Loukidis, D., Chakraborty, T., Salgado, R.: Bearing capacity of strip footings on purely frictional soil under eccentric and inclined loads. Can Geotech J. **45**(6), 768–787 (2008)
4. Meyerhof, G.G.: Some recent research on the bearing capacity of foundations. Can Geotech J. **1**(1), 16–26 (1963)
5. Pham, Q.N., Ohtsuka, S., Isobe, K., Fukumoto, Y., Hoshina, T.: Ultimate bearing capacity of rigid footing under eccentric vertical load. Soils Found. **59**(6), 1980–1991 (2019)
6. Pham, Q.N., Ohtsuka, S., Isobe, K., Fukumoto, Y.: Limit load space of rigid footing under eccentrically inclined load. Soils Found. **60**(4), 811–814 (2020)
7. Tamura, T., Kobayashi, S., Sumi, T.: Rigid plastic finite element method for frictional materials. Soils Found. **27**(3), 1–12 (1987)
8. Vesic, A.S.: Analysis of ultimate loads of shallow foundations. J. Soil Mech. Found. Division. ASCE **99**(1), 45–73 (1973)

Numerical and Experimental Investigation of Wave Loading on a Monopile Support Structure of Offshore Wind Turbine

Linh Cao Luong(✉) and Tri Mai

Faculty of Coastal and Offshore Engineering, Hanoi University of Civil Engineering, Hai Ba Trung, 55 Giai Phong, Hanoi, Vietnam
{linhlc,trimc}@nuce.edu.vn

Abstract. In recent years, the renewable energy is highly developing and therefore more offshore wind farms have been built in the world and Vietnam. The offshore wind farms have been usually installed in shallow water and the monopile is one of the most structures used for offshore wind turbine support structures. Wave load is an important aspect in designing of the offshore wind turbine support structures. This paper is going to investigate the wave loading on a monopile support structure of the offshore wind turbines by numerical and experimental methods. Wave loading on the monopile has been simulated by ANSYS Fluent software and the fifth-order Stokes wave theory is applied in the model. The free surface of waves is tracked by using the volume of fluid (VOF) technique. The numerical and experimental data have been analysed and compared. Then, discussions and conclusions will be given in the paper for about the wave loading on a monopile support structure of offshore wind turbines.

Keywords: Wave load · Monopile · Offshore wind turbine · Physical model · ANSYS FLUENT

1 Introduction

Currently, the majority of wind power plants in the world are located on land [1]. Offshore wind power plants, however, have already begun to be built and will soon develop with a clear advantage in the potential for wind power at sea over land. Several types of wind turbine support structures are being used and studied for offshore wind power projects as gravity-based structure, monopile, tripod, jacket, spar, tension-leg platform, or semi-submersible. Most of the current offshore wind power projects are located in shallow waters, with monopile structures that are used to support wind turbines.

For the support structure of offshore wind turbine, wave load is one of the most dangerous environmental loads during the working life of the project. Therefore, determining the most accurate wave load impacting on the structure is necessary to design the building with safety and reliability. The wave load on a structure can be determined by different methods. Normally, often use the theoretical method, typically the Morison equation, to determine the wave load on slender structures [2]. However, for the structure

D. V. K. Huynh et al. (Eds.): VSOE 2021, LNCE 208, pp. 342–349, 2022.
https://doi.org/10.1007/978-981-16-7735-9_38

near the water level, it will lead to quite a large error due to the effect on the liquid surface. Several laboratory and numerical models have investigated the impact forces caused by waves for offshore wind turbine structures. In recent years, numerical methods with the help of computers that return more accurate calculation results are increasingly used. Besides, the experimental model in wave flume or wave basin is often carried out concurrently with the numerical models to certify and calibrate the model to obtain results with the highest reliability. Some software is used for the calculation of computational fluid dynamics (CFD) such as commercial software ANSYS [3–5], ADINA, ABAQUS, COMSOL, FLOW3D or open-source software such as OpenFOAM [6] and REEF3D [7].

This paper is going to investigate wave loading on a vertical circular cylinder by experimental and numerical work. The experiment was carried out in a wave basin and the numerical model, ANSYS FLUENT [4, 5], has been applied to simulate the experiment work. The comparision between the physical and numerical data will be made.

2 Methodology

2.1 Physical Model

In the experimental, monopile support structure of offshore wind turbines modeled at 1:20 scale. The tested monopile has its diameter of 0.3 m and height of 0.9 m. The monopile structure was located at 3.45 m from the wave generator (Fig. 1). The monopile structure was equipped with eleven pressure transducers installed on the structure to measure the pressures on the structure. Pressure transducers were mounted at various levels (z/h = -0.095 to -0.76; z is the submergence of pressure sensor, h is the water depth in the basin) on the surface of the monopile as shown in Fig. 2. Three wave gauges were used to measure water elevations around the monopile (Fig. 1).

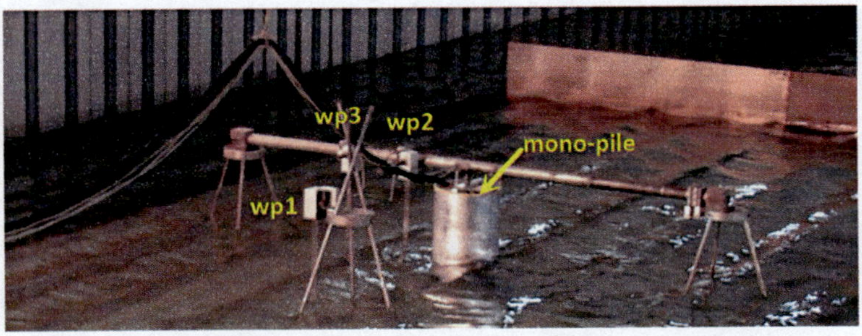

Fig. 1. Configuration of the physical model in the 3D-wave basin.

The tested wave conditions (UR7, UR8 and UR9) are the regular waves with the wave height from 0.072 m to 0.096 m and the wave period about from 1.13 s to 2.10 s. The diffraction parameter, kr_0, varies from 0.21 (corresponding to UR9) to 0.5 (corresponding to UR7) in which k is the wave number and r_0 is the radius of the mono-pile. The tests were performed within a water depth of 0.6 m.

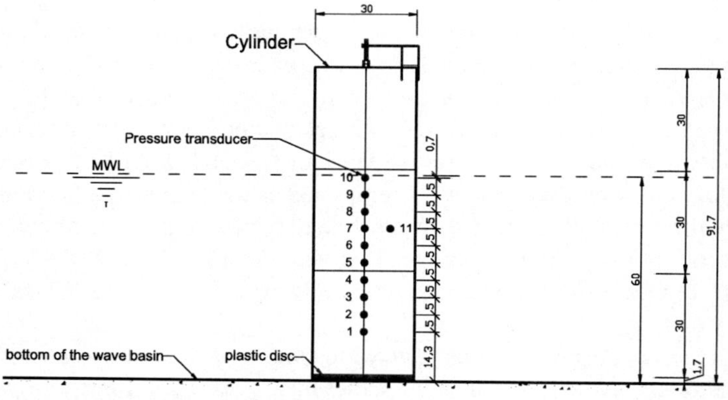

Fig. 2. Position of pressure transducers on the monopile (Note units in cm)

2.2 Numerical Model

2.2.1 Governing Equation

The governing equations of fluid flow represent mathematical statements of the conservation laws of physics: the mass of a fluid is conserved, the rate of change of momentum equals the sum of the forces on a fluid particle (Newton's second law) and the rate of change of energy is equal to the sum of the rate of heat addition to and the rate of work done on a fluid particle (first law of thermodynamics). This paper does not involve heat transfer, energy equation is not considered.

The unsteady, three-dimensional mass conservation at a point in a compressible fluid [8]:

$$\frac{\partial \rho}{\partial t} + \mathrm{div}(\rho \boldsymbol{u}) = 0 \tag{1}$$

Momentum equation in three dimensions:

$$\begin{cases} \frac{\partial(\rho u)}{\partial t} + \mathrm{div}(\rho u \mathbf{u}) = -\frac{\partial p}{\partial x} + \mathrm{div}(\mu \ \mathrm{grad}\ u) + S_{Mx} \\ \frac{\partial(\rho v)}{\partial t} + \mathrm{div}(\rho v \mathbf{u}) = -\frac{\partial p}{\partial y} + \mathrm{div}(\mu \ \mathrm{grad}\ v) + S_{My} \\ \frac{\partial(\rho w)}{\partial t} + \mathrm{div}(\rho w \mathbf{u}) = -\frac{\partial p}{\partial z} + \mathrm{div}(\mu \ \mathrm{grad}\ w) + S_{Mz} \end{cases} \tag{2}$$

where, t is the time, $\mathbf{u} = [u, v, w]^T$ is the velocity vector, $[x, y, z]^T$ is the position vector, ρ is the fluid density, p is the pressure, μ is the dynamic viscosity, and S_{Mx}, S_{My}, S_{Mz} is the momentum source.

The finite volume method (FVM) is the numerical technique applied in ANSYS Fluent software to solve the governing equations of the fluid. Fluent uses the finite volume technique to convert the general transport equation into a system of algebraic equations and it uses different iterative methods to solve the algebraic equations. In this paper, the Semi-Implicit Method for Pressure-Linked Equations (SIMPLE) is used to calculate.

The volume of fluid method (VOF) is the numerical technique used in detecting the free surface of the computational domain. The method uses a function f(x, t) whose value is 1 at any point occupied by the fluid, and 0 everywhere else. When f is averaged over the volume of a computational cell, the cell-averaged value, $\overline{f}$, represents the fractional volume of the cell occupied by fluid. Therefore, a value of $\overline{f}$ equal to 1 corresponds to a cell full of fluid while a value of $\overline{f}$ equal to 0 indicates an empty cell [9]. Furthermore, cells with values of $\overline{f}$ between 0 and 1 must contain a portion of the free surface, as shown in Fig. 3.

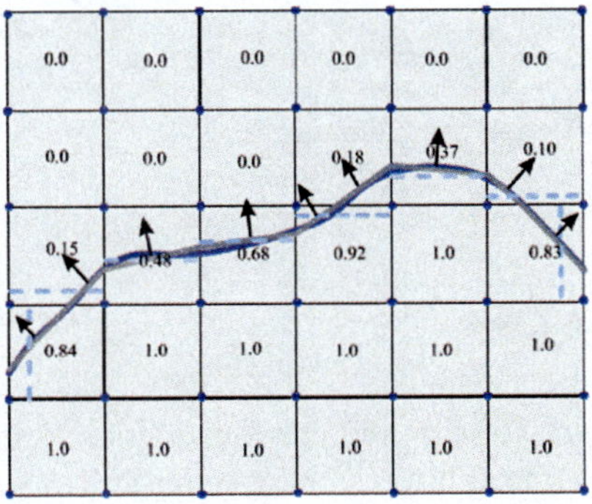

Fig. 3. Schematic of fluid volume fraction and fluid interface [9]

2.2.2 Geometry

The numerical wave tank is 22.0 m long, 2.2 m wide and 0.9 m high. The Descartes axis system in the numerical model, with the X-axis being the wave propagation direction, the Y-axis being the vertical direction.

2.2.3 Boundary Conditions

The boundary conditions applied to the boundaries of the fluid model is shown in Fig. 4. In this simulation, the open channel wave boundary condition is considered as an essential setup. The velocity-inlet and pressure-outlet are set to be the boundary condition for the inlet and the outlet of the fluid domain. The side walls are set to be symmetry, and the wall conditions are set up on the bottom and cylinder boundaries.

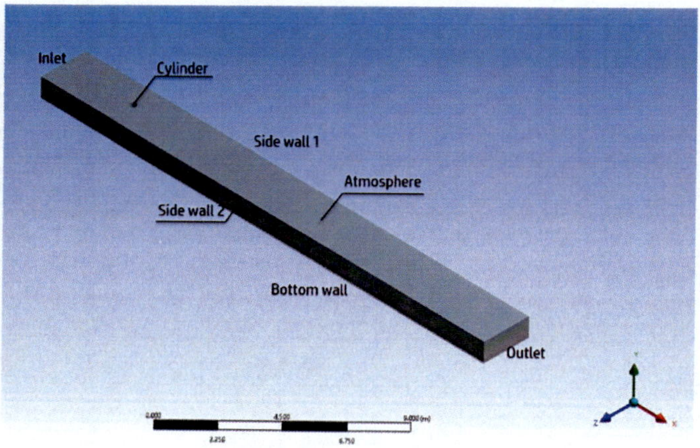

Fig. 4. Boundary setups in the fluid domain

2.2.4 Simulation Conditions

The numerical model has been applied to simulate three wave conditions tested in the ocean basin which are presented in the previous section. Summary of the simulated wave conditions are shown in Table 1. The 5th-order Stokes wave theory is used in all computational cases, this is consistent with the recommendation of the DnV standard [10].

Table 1. Summary of the simulation conditions.

Case name	H (m)	T (s)	L (m)	h (m)	H/gT^2 (-)	h/gT^2 (-)	*kr0* (-)	Wave theory
UR7	0.072	1.13	1.9171	0.60	0.0058	0.048	0.50	5th order Stokes
UR8	0.088	1.51	3.0187	0.60	0.0039	0.027	0.31	5th order Stokes
UR9	0.096	2.01	4.3886	0.60	0.0024	0.015	0.21	5th order Stokes

3 Results and Discussions

3.1 Free Surface Elevation

The free surface elevation is computed at wp1 (x = 3.30 m from the wave generator) as shown in Fig. 1. Results show that the computed free surface elevation is fitted quite well to the experimental data (see Fig. 5).

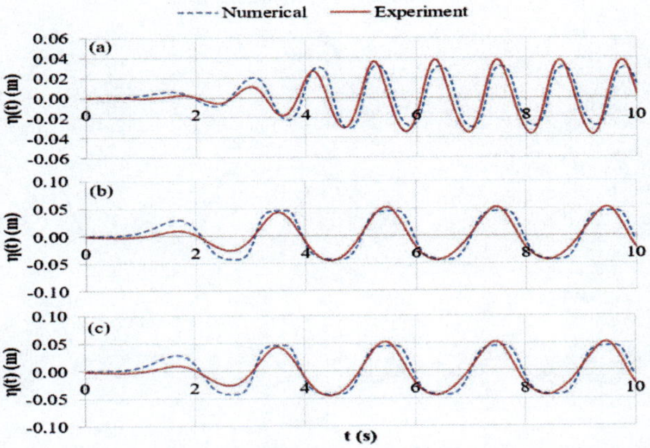

Fig. 5. Time histories of free surface elevation due to the regular wave: (a) UR7 ($kr_0 = 0.50$); (b) UR8 ($kr_0 = 0.31$); UR9 ($kr_0 = 0.21$).

3.2 Wave Loading on Monopile

The experimented and simulated pressures at different submergence levels (*z/h* = –0.095,–0.43 and –0.76) are presented in Figs. 6, 7 and 8 for the tested regular waves UR7, UR8 and UR9, respectively. The tested wave has its diffraction factor from 0.21 (UR9) to 0.50 (UR7). In general, the simulated pressures are fitted very well to the experimental pressures due to the regular waves UR7, UR8 and UR9. It is found that the submergence level (*z/h*) has a minor effect on the pressure difference of the tested and simulated results. As it is expected that the diffraction parameter, kr0, has its effects on the pressure difference of the physical and numerical investigation. The pressure difference is decreased as the diffraction parameter is decreased.

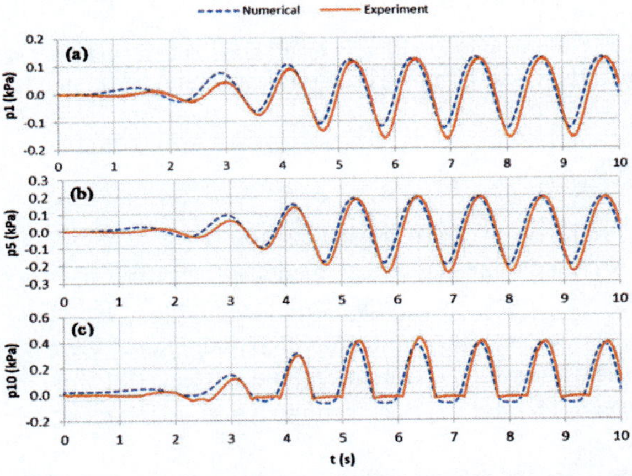

Fig. 6. Time histories of pressures at different levels due to the regular wave UR7 ($kr_0 = 0.50$): (a) *z/h* = –0.76 (p1); (b) *z/h* = -–0.43 (p5); (c) *z/h* = -–0.095 (p10).

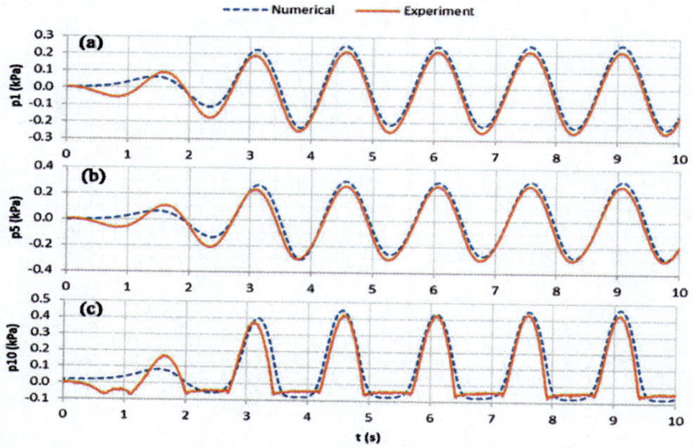

Fig. 7. Time histories of pressures at different levels due to the regular wave UR8 ($kr_0 = 0.31$): (a) $z/h = --0.76$ (p1); (b) $z/h = --0.43$ (p5); (c) $z/h = --0.095$ (p10).

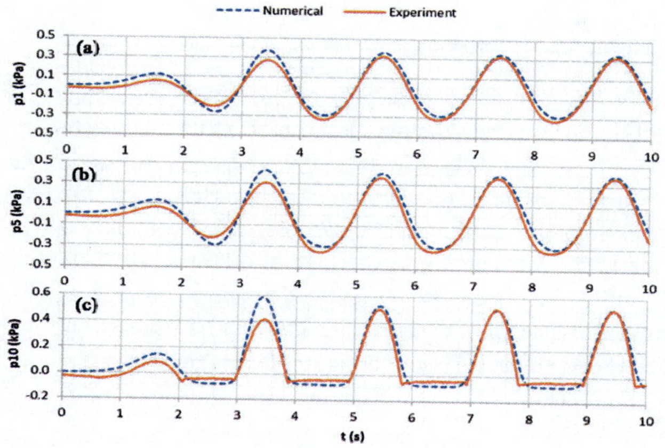

Fig. 8. Time histories of pressures at different levels due to the regular wave UR9 ($kr_0 = 0.21$): (a) $z/h = --0.76$ (p1); (b) $z/h = --0.43$ (p5); (c) $z/h = --0.095$ (p10).

4 Conclusions

The numerical simulation of wave impact pressures on monopile structure with the ANSYS Fluent software. The numerical model is based on the Reynolds-Averaged Navier-Stokes (RANS) equations together with the volume of fluid (VOF) technique for the free surface. The numerical results are compared against the experimental results for wave surface elevations and wave force. The numerically simulated wave surface elevations and wave forces are in good agreement with experimental data.

Acknowledgements. The authors gratefully acknowledge the support of the German Federal Environment Ministry (BMU) within the funded project (BMU code 0325032/A) and the support of the National University of Civil Engineering in Hanoi, Vietnam under the funded project (103–2019/KHXD).

References

1. Global Wind Energy Council: Global Wind Report 2018 (2019)
2. Sumer, B.M., Fredsøe, J.: Hydrodynamics Around Cylindrical Structures (Revised Edn.). World Scientific Publishing Co. Pte. Ltdc., Denmark (2006)
3. Lin, Y.-H., Chen, J.-F., Po-Ying, L.: A CFD model for simulating wave run-ups and wave loads in case of different wind turbine foundations influenced by nonlinear waves. Ocean Eng. **129**(2017), 428–440 (2016)
4. ANSYS: ANSYS FLUENT Theory Guide (2019a)
5. ANSYS: ANSYS FLUENT User's Guide (2019b)
6. Mohseni, M., Esperanca, P.T., Sphaier, S.H.: Numerical study of wave run-up on a fixed and vertical surface-piercing cylinder subjected to regular, non-breaking waves using OpenFOAM. Appl. Ocean Res. **79**(2018), 228–252 (2018)
7. Chella, M.A., Bihs, H., Myrhaug, D., Arntsen, Ø.A.: Numerical modeling of breaking wave kinematics and wave impact pressures on a vertical slender cylinder. J. Offshore. Mech. Arctic. Eng. **141**, 051–802 (2019)
8. Versteeg, H., Malalasekera, W.: An Introduction to Computational Fluid Dynamics: The Finite Method, 2nd edn. Prentice Hall, England (2007)
9. Katopodes, N.D.: Free-Surface Flow: Computational Methods. Elsevier, Butterworth-Heinemann, United Kingdom (2019)
10. Det Norske Veritas (DnV): OS-J101 Design of Offshore Wind Turbine Structures (2010)

Offshore Engineering: A Systems Science Perspective

Jeffrey Jackson[1] and Scott Jackson[2](✉)

[1] Bangkok, Thailand
[2] Burnham Systems, Los Angeles, USA
jackson@burnhamsystems.net

Abstract. Among the most advanced concepts in the twenty-first century is systems science. The principles of systems science can be used to describe any large scale and complex systems such as offshore platforms, their design, their management, and their operation. The premise of systems science is that the entire system and the associated organization is a single system operating as a whole and whose parts interact with each other to accomplish a purpose. The purpose of the offshore system is the recovery of petroleum and natural gas (using stringent and proven health, safety, and environmental controls) to meet the world energy needs.

Keywords: Systems science · Hierarchy · Holism

1 System Science

According to [1, p. 7], systems science "is way of thinking some part of the universe and trying to find ways to understand how things work." As we shall see below, an offshore platform is a complex entity, and a part of the universe. So, it behooves us to understand how it works, or at least, how it should work, reliably and consistently.

1.1 A System in the Systems Science Perspective

This paper presents an overview of marine and offshore structures from a systems science perspective especially from an enterprise point of view. At the seminal level it reliability assessment and risk management for assets, and of course of wider application far beyond marine and offshore structures. This paper shows that risks inherent in the maritime domain have the same roots as in the non-maritime domains. This paper is linked to the topic of Design Standards, Risk and Reliability (DSRR). However, it also treats offshore platforms at the broadest, that is, enterprise level.

System Characteristics: Before we can discuss offshore systems, we need to explain what a system is. According to [2] in the field of systems science a system has three characteristics: First it must consist of multiple parts. On an offshore platform the parts are the rig, the supply system, the management system, the management system, the pipelay vessel, and the well operations system.

D. V. K. Huynh et al. (Eds.): VSOE 2021, LNCE 208, pp. 350–357, 2022.
https://doi.org/10.1007/978-981-16-7735-9_39

Interaction of the Parts: Second, the parts have to interact with each other. The interaction is not just between the physical parts but also between the physical parts and the management parts.

Property of Emergence: Finally, the system must have a property that the individual parts do not have. This property is called emergence, to be discussed below. An emergent property cannot be determined by the examination of a single part. In the case of the case of the offshore platform the emergent property is the extraction of petroleum products.

1.2 The Relation Between Objects

The study of systems is different from other sciences according to [3] because the other sciences focus on the study of objects, while, the systems sciences study the relation between those objects. For example, the study of offshore platforms, as a system, includes the study of the relation between all the parts of the system including both the physical parts and the human parts. According to [1] "focusing on relationship and organization, see each of these [organizational] levels as distinctive emergences that have taken place within a single dynamic contextual process, that of increasing and evolving complexity." [1] tell us that "systems science undertakes the understanding of systems as such, not this kind of system or that kind of system, but the investigation of the useful attributes, dynamics, characteristics, and behaviors of systems as systems."

1.3 Systems Thinking and Systems Engineering

Two major branches of systems science are systems thinking and systems engineering.

Systems Thinking: This is the capability to think of an entire enterprise as a system. This is a fundamental for an enterprise system and its leaders discussed below. According to [1], systems thinking "entails conceptualizing the objects we encounter as systems and sub-systems that are parts of larger systems." If the leaders of the enterprise system can think of their as containing smaller systems, they qualify as systems thinkers. In the case of Deepwater Horizon, BP Group is the enterprise system; and the platform and management systems are sub-system as described below.

Systems Engineering: Another branch of systems science is systems engineering, which is the application of systems engineering to human-made systems. The basic principles of systems engineering can be found in [4]. Traditionally, systems engineering has been used in the development of aircraft and space craft. However, for the purpose of this paper, systems engineering can be used for the development of offshore platforms.

Purpose: One property of all systems for which systems engineering applies is *purpose*. According to [1, p. 114], "purpose is a word that is related to function in systems thinking that is generally reserved for those systems where we can speak of something being *aimed at*." It goes without saying that offshore platforms have a purpose, namely, the extraction of petroleum products from the sea-bed.

1.4 Risk Management

According to [4, p. 114], risk management is a major part of systems engineering. Risk is the product of the likelihood of an event, for example, a failure and the consequence of that failure. In the case of Deepwater Horizon, the well that failed was called the Macondo well. This well was indeed another element in the Deepwater Horizon system.

Roots of Risk: Risk has its roots in many sources. According to [5, p. viii], the primary source of risk on the Deepwater Horizon project was *complacency.* Hence, combating complacency is the responsibility of management at the highest levels.

Rigorous Risk Management Program: Regarding Deepwater Horizon [5, p. 266] states that "the Department of the Interior should create a rigorous, transparent, and meaningful oil spill risk analysis and planning process for the development and implementation of better oil spill response." With this recommendation in mind, it can be concluded that any offshore platform development program should have a rigorous risk management program in place.

Independent Risk Evaluation: With regard to risk management programs, it is important that the risk evaluation should be independent of the offshore platform program in question. Independence means that the risk evaluation responsibility should not be financially or organizationally part of the enterprise system for the offshore platform program. Such independent evaluations are often conducted through government organizations with cooperation by the enterprise system.

Cultural Roots: The immediate causes of the Macondo well blowout can be traced to a series of identifiable mistakes made by BP, Halliburton, and Transocean that reveal such systematic failures in risk management that they place in doubt the safety culture of the entire industry. This statement has several implications. First, the problem was enterprise wide, involving all elements of the system. Second the problem was systematic, that is resulting from the interactions among all the elements. Finally, the cause was cultural meaning that inattention to safety and complacency were features of all personnel across the enterprise.

Lack of Communication: On another project, the *Challenger* disaster, renowned scientist [6] concluded that a major problem in large-scale projects like *Challenger* and therefore like any offshore platform project, is the lack of communication between managers and engineers.

Government Oversight: In addition to enterprise failures, [5, p. 122] also cites failures of government to provide effective regulatory oversight of offshore drilling. This statement reinforces the idea that government has a key role in the success of an offshore platform. Risk evaluation is one of those roles that the government often plays.

Independent Review in Offshore Platforms: In the offshore platform domain, the MWS (Maritime Warranty Surveyors) performs the independent review function representing the insurance perspective.

An Offshore Platform as a System: All of these characteristics apply to offshore platforms as systems. One of the characteristics mentioned by Mobus and Kalton is complexity. Complexity is the one characteristic that makes the study of offshore platforms difficult. We will discuss this aspect later.

Systems Science and Deepwater Horizon: This paper uses the Deepwater Horizon disaster to illustrate the aspects of systems science as lessons learned for offshore platforms throughout the world. This case study is described by the [5] report. Aspects of that disaster that provide lessons learned for offshore platform developers include (1) the failure of the enterprise system to view the entire enterprise as a system with its subordinate parts that need to work together, (2) the complexity of the entire enterprise system, (3) the risks that emanate from that complexity, and (4) the culture of complacency that pervaded the enterprise.

2 The Offshore System

The primary systems of the offshore system are the enterprise system and the offshore management system.

2.1 The Enterprise System

The parts of the system include: The offshore enterprise, the offshore management system, the offshore rig, the supply system, the maintenance system, offshore wind farms, the pipelay vessel, and the well operations system.

The BP Group System: In the case of Deepwater Horizon, the enterprise system was the critical system. In the report [5, p. 219] it is known as the BP Group system. This report states, "BP Group and Complex Management did not detect and intervene early enough on deteriorating performance."

BP Group System Responsibilities: The [5, p. 219] gives the enterprise system an even greater role when it says, for the Deepwater Horizon system, "BP and other operators must have effective systems in place for integrating the various corporate cultures, internal procedures, and decision-making protocols of the many different contractors involved in drilling a Deepwater well."

2.2 The Offshore Management System

Within the offshore enterprise system is the offshore management system. A key role of the offshore management system is the management of the many specialty areas to make sure these areas do not conflict or overlap with each other. These specialty areas include safety, health, and environmental controls. A recognized concept for managing these areas is the integrated product development (IPD) concept. According to [7], "Integrated Product Development (IPD) is a framework to help management and active project teams reach innovation goals. Originating in government systems, IPD is a management theory that promotes simultaneous integration of multi-disciplinary teams and concurrent engineering."

3 Principles of Systems Science as Applied to Offshore Platform Systems

The principles of systems science most applicable to offshore platforms are holism, emergence, hierarchy, and complexity. These principles will be described below and how they relate to offshore platforms.

3.1 The Principles of Holism and Emergence

The most fundamental principle of systems science is holism. The main principle of holism is to view that the whole offshore system is seen in terms of interacting wholes that are more than the mere sum of its individual, elementary parts. Holism can be expanded to convey the idea of emergence, which occurs when an entity is observed to have properties that its parts do not possess on their own. These properties or behaviors emerge as a result of the parts interacting within the wider whole, i.e., the whole system demonstrates a property that a single component cannot demonstrate. As discussed above, the emergent property is the extraction pf petroleum products.

3.2 The Principle of Hierarchy

Another principle of systems science is hierarchy. This means that systems are part of other systems right from top-level off-shore system to the lowest level as shown in Fig. 1. The principle of hierarchy assures that each lower level component is an integrated part of the higher level components. This principle remains in force even if components at different levels are designed and operated by different contractors (and entities). This principle remains in force even if components at different levels are designed and operated by different contractors (and entities).

3.3 The Principle of Interaction

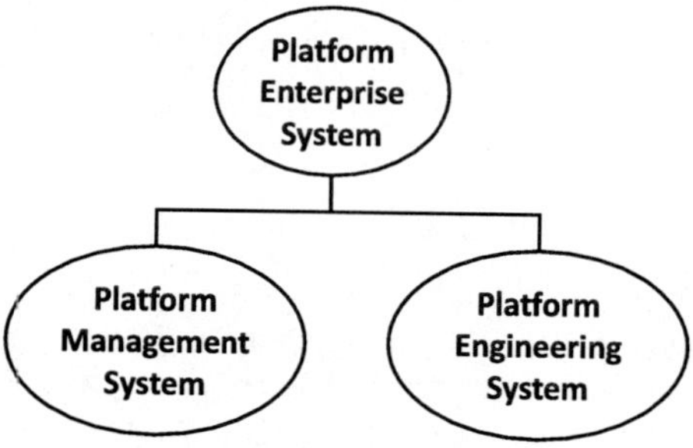

Fig. 1. Platform hierarchy of systems.

Lastly, yet another important principle of systems science is interaction. This principle assures that all elements of the entire offshore system interact with each other and that the interactions are flawless, that is, they do not allow for leakage or breakage. This is particularly important when the different subsystems are made by different contractors. Interaction management is an important part of system safety. Figure 2 illustrates interaction in an offshore platform system.

Lack of Interaction: The lack of interaction played an important role in the failure of the Deepwater Horizon system. According to [5, p. 122]:

> "the most significant failure at Macondo—and the clear root cause of the blowout—was a failure of industry management. Most, if not all, of the failures at Macondo can be traced back to underlying failures of management and communication. Better management of decision-making processes within BP and other companies, better communication within and between BP and its contractors, and effective training of key engineering and rig personnel would have prevented the Macondo incident. BP and other operators must have effective systems in place for integrating the various corporate cultures, internal procedures, and decision-making protocols of the many different contractors involved in drilling a Deepwater well."

3.4 Complexity

As mentioned above, one of the many characteristics of a platform system to be addressed by systems science is complexity. There is no doubt that platform systems are complex. This does not mean that the difficulty in understanding them is the essence of complexity. It does mean that they may consist of properties that make them difficult to understand.

Characteristics of Complexity: These properties are as follows: First, they consist of many parts. Second, they possess many interfaces between the parts. Finally, they have a high degree of stochastic uncertainty in the interfaces, the latter being the most important feature. Figure 3 illustrates the variability in the relation between two elements of a system. Since offshore platforms have many human elements, the variability will be great.

Effects of Complexity: In the context of offshore platforms [5] states that "complex systems fail in complex ways." In other words, they are saying that offshore platforms are complex systems and are likely to fail if not managed well. Furthermore, when they fail, it is difficult to analyze the failure and prescribe methods to repair it.

Complexity and Complicated: Systems science makes a distinction between complexity and complicated. A system with many parts and many interfaces will be complicated rather than complex if the relation between the parts is stable. A clock with closely fitting parts may be complicated but not complex.

Reducing Complexity: So, to minimize complexity there are three things to do: First, reduce the number of parts. Second, reduce the number of interfaces. In the context of an offshore platform system, these two things are probably the easiest to do. [8] provides the advice for the third step, namely, "maintain stability in the interfaces."

How the last is dome is up to the designer of the platform. It could be as simple as making sure the flow of oil between two elements is steady and unchanging.

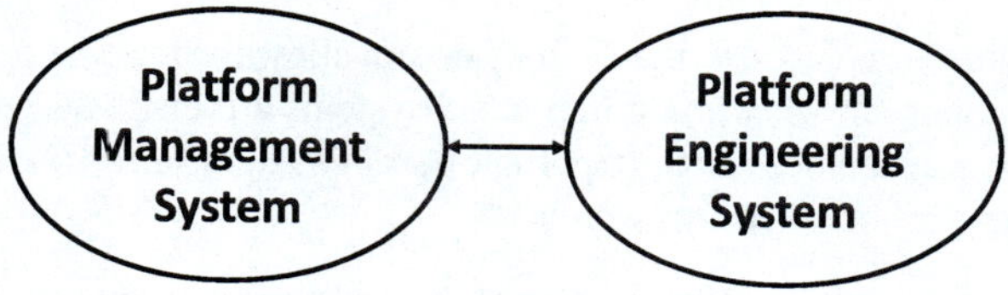

Fig. 2. Interaction of platform systems.

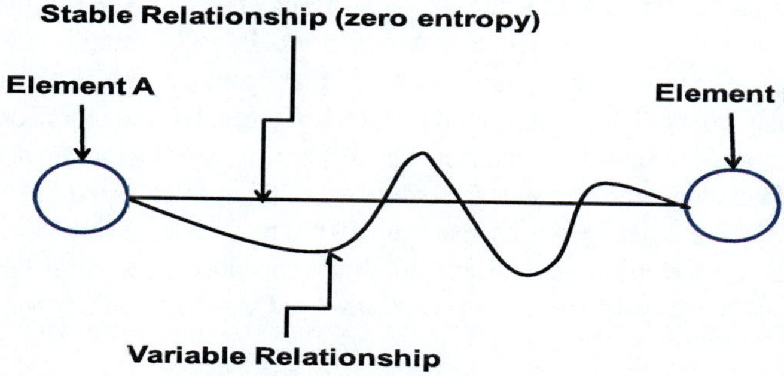

Fig. 3. Variability between elements.

4 Summary and Conclusions

In summary, the purpose of this paper is to show that if an offshore platform successfully demonstrates all the principles of systems science, and performs them rigorously, the offshore platform will be an economic benefit to the offshore enterprise while ensuring that overall safety and quality control standards are adhered to.

Transdisciplinarity: In addition, this paper has demonstrated the principle of transdisciplinarity as described by [9]. Transdisciplinarity is the execution of two or more disciplines to show both how they can be executed simultaneously but also how their principles can be synergistic to each other. In our case these disciplines are the engineering of offshore platforms and the science of systems.

References

1. Mobus, G.E., Kalton, M.C.: Principles of Systems Science (Understanding Complex Systems). Springer, New York (2015)
2. Sillitto, H., et al.: Defining "System": a comprehensive approach. INCOSE Int. Symp. Australia **27**, 170–186 (2017)
3. Checkland, P.: Systems Thinking Systems Practice. John Wiley & Sons, New York (1999)
4. INCOSE: Systems Engineering Handbook. International Council on Systems Engineering, Seattle, CA (2015)
5. Commission, N.: Deep Water: The Gulf Oil Disaster and the Future of Offshore Drilling. Environmental Protection Agency, Washington DC (2011)
6. Feynman, R.P.: An outsider's inside view of the challenger inquiry. Phys. Today **41**(2), 26–37 (1988). https://doi.org/10.1063/1.881143

7. Simple-PDH.com. Principles of Integrated Product Development. Global NP Solutions (2020). https://simple-pdh.com/principles-of-integrated-product-development/#:~:text=Integrated%20Product%20Development%20(IPD)%20is,disciplinary%20teams%20and%20concurrent%20engineering. Accessed 7 Nov 2020
8. Rechtin, E.: Systems Architecting: Creating and Building Complex Systems. CRC Press, Englewood Cliffs, NJ (1991)
9. Madni, A.M.: Transdisciplinary Systems Engineering Exploiting Convergence in a Hyper-Connected World. Springer, New York (2018)

Pile Shape Effects on the *p-y* Curves of Laterally Loaded Piles in Soft Clay

Yongqing Lai, Lizhong Wang, Yi Hong(✉), and Huan Wang

Zhejiang University, Hangzhou, China

Abstract. Pile foundations are often used to resist lateral loads to support offshore structures, transmission towers, and high-rise buildings. For the design of laterally loaded piles, the *p-y* method is widely adopted. Extensive research efforts have been paid on the development of *p-y* curves for circular piles. However, *p-y* curves for piles with other cross-section shapes have received limited attention. This paper presents a comprehensive numerical study to investigate the pile shape effects on *p-y* curves of laterally loaded piles in soft clay. Based on the numerical results, the *p-y* curves of piles with different pile shapes (i.e., circular, square, and H pile sections) are proposed. The proposed *p-y* curves are fundamentally linked to soil stress-strain response, providing engineers a simple approach to construct *p-y* curves for pile design based on soil stress-strain response measured in the laboratory, without the need for time-consuming numerical analyses.

Keywords: Pile · *p-y* curves · Lateral loading · Pile shape · Soft clay

1 Introduction

Pile foundations have been extensively used to withstand lateral loads induced by wind, wave, and earthquake. The design for a laterally loaded pile is typically based on the Winkler approach in conjunction with *p-y* models. The API [1] and DNVGL [2] standard recommend *p-y* curves, which were deduced from the field tests on slender circular piles, have been widely used for the design of circular piles supporting offshore oil and gas platforms. During the last decades, a number of improved *p-y* curves for circular piles have been also proposed based on experimental or numerical investigations [3–6]. Different from circular piles, piles with other shapes, for instance, square and H piles, have been rarely studied. Pile shape will alter the soil failure mechanism around the pile and thus influences the ultimate lateral soil resistance and initial stiffness of *p-y* curves. Suryasentana and Lehane [7] performed three-dimensional finite element analyses for piles to investigate the pile shape effects. It is found that, when the width of a square pile is the same as the diameter of a circular pile, the square pile will exhibit a higher pile capacity than the circular pile, which is more efficient at resisting lateral load. Truong and Lehane [8] conducted a series of centrifuge pile tests in clay to examine the lateral response of the circular, square, and H pile. The results also show that higher ultimate soil resistance can be developed for square and H piles compared to circular piles when the diameter of a circular pile is equal to the width of a square or H pile. Based on test

D. V. K. Huynh et al. (Eds.): VSOE 2021, LNCE 208, pp. 358–365, 2022.
https://doi.org/10.1007/978-981-16-7735-9_40

results, they developed an empirical formulation for *p-y* curves in soft clay considering pile shape effects.

This paper presents a numerical study, aiming to investigate the pile shape effects on the *p-y* curves of laterally loaded piles in soft clay. Three different pile shapes, i.e., circular, square, and H-section, have been examined. Since the *p-y* curves of circular piles have been well developed by [4], thus this study focuses on the *p-y* curves of square and H piles. Based on the numerical results, the *p-y* curves for piles with different pile shapes are proposed, which are fundamentally linked to soil stress-strain response, providing engineers a simple approach to construct *p-y* curves for pile design based on soil stress-strain response measured in the laboratory, without the need for time-consuming numerical analyses.

2 Methodology and Numerical Parametric Study

2.1 Methodology and Finite Element Model

The *p-y* response of circular, square, and H piles are analyzed. The pile cross-sections are presented in Fig. 1. Following the method of [4], a horizontal slice of a pile is analyzed in ABAQUS. The slice is 1 m in thickness, representing a 1 m horizontal segment of the pile. Vertical displacement on the top and bottom boundaries of the model is constrained, therefore enforcing a plane strain condition. Lateral displacement is applied on the top of the pile, which is modelled as a rigid body, forcing the pile to translate horizontally. The pile is displaced until the load-displacement curves become flat indicating the ultimate state. The load-displacement curve of the horizontal translating pile slice gives the *p-y* curve. It should be noted that the *p-y* curves derived from such analyses are only relevant for the part of a pile where localized flow-around mechanism upon failure is governing. The pile diameter of the circular pile and pile width of the square and H pile (B) is set as 1 m. As the results will be presented in a normalised format, the absolute dimension of the pile width does not matter. A typical value of wall thickness t of the H pile, i.e., $t = 5\%B$ [9], is considered in this study.

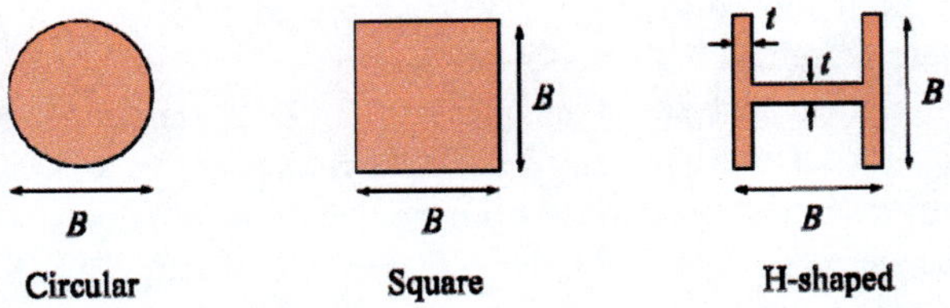

Fig. 1. Piles shapes analyzed in this study

Figure 2 shows a typical finite element mesh for piles with different shapes. The boundaries are set at $10B$ from the pile center so that the boundary effect can be ignored. The separation between the soil and pile is not allowed.

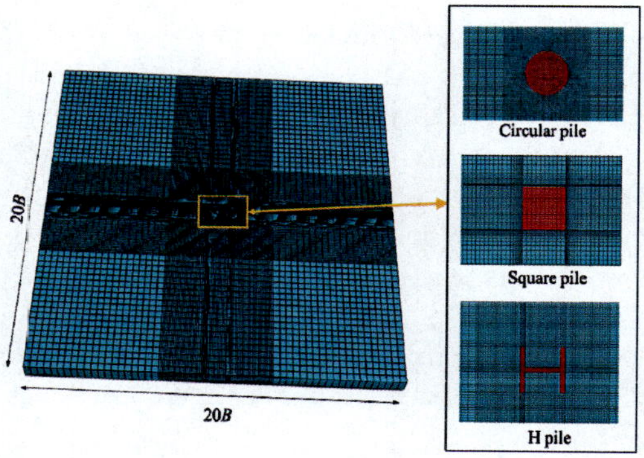

Fig. 2. Typical finite element mesh for piles with different shapes

In this study, the soil nonlinear stress-strain response is described by the NGI-ADP model [10]. The model obeys a Tresca yield criterion and features the following non-linear hardening law that links the current yield stress with the current plastic shear strain:

$$\frac{\tau}{S_u} = 2\frac{\sqrt{\gamma^P/\gamma_f^P}}{1 + \gamma^P/\gamma_f^P} \tag{1}$$

where τ = currently mobilised shear stress; s_u = the shear strength of soil (DSS strength in this case); γ^p = the current plastic shear strain; $\gamma_f{}^p$ = the plastic shear strain at failure (full mobilisation). The total shear strain γ associated with the currently mobilised shear stress τ is the summation of elastic and plastic shear strain components (γ^e and γ^p respectively):

$$\gamma = \gamma^e + \gamma^P = \frac{\tau}{G_{\max}} + \gamma^P = \frac{\tau/S_u}{G_{\max}/S_u} + \gamma^P \tag{2}$$

Three parameters, i.e., s_u, $\gamma_f{}^p$ and G_{max}/s_u are needed to define the stress-strain response of the soil. In this study, s_u = 50 kPa is adopted, while the values of the other two parameters are presented in the following section.

Parametric numerical analyses of the piles with different shapes (as shown in Fig. 2) using the pile slice model over a wide range of soil stress-strain responses are performed. The corresponding *p-y* curves are derived from the analyses. The ranges of the parameters for each pile shape are G_{max}/s_u: 500, 1000 and $\gamma_f{}^p$: 2, 5, 10, 15%. In addition, the parametric study also examines the influence of pile-soil interface roughness factor α in the resulting *p-y* responses. α = 0, 0.3, 0.5, 0.7 and 1 are considered in this study.

2.2 Validation of the Finite Element Model

To validate the finite element model, a circular pile under lateral loading is modelled. Figure 3(a) presents the evolution of the computed ultimate capacity factor $N_p = p_u/s_uB$ with pile roughness factor α, which gives an excellent agreement with the analytical solution of N_p proposed by [11]. Figure 3(b) shows the typical computed *p-y* curve for the circular pile at $G_{max}/s_u = 500$, $\gamma_f^p = 5$, and $\alpha = 1$. Zhang and Andersen [4] proposed a scaling relationship between the soil stress-strain curve and the corresponding *p-y* curve (to be detailed presented in Sect. 3.2). The predicted *p-y* curve using Zhang and Andersen [4]'s method is also plotted in Fig. 3(b). The good match demonstrated the capability of the finite element model.

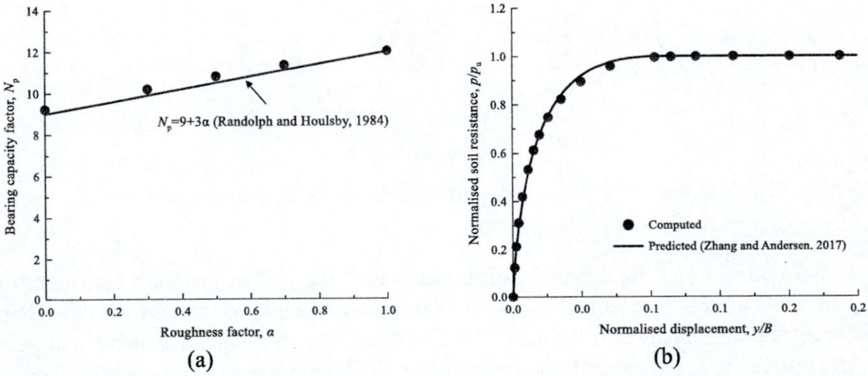

Fig. 3. Model validation of a circular pile: (a) comparison of N_p factor; (b) comparison of normalised *p-y* curve

3 Numerical Results and the Proposed *p-y* Curves Considering Pile Shape Effects

3.1 Pile Shape Effects on *p-y* Curves

Figure 4 shows the computed ultimate soil resistance p_u of the *p-y* curves (normalised by s_uB) of the circular, square, and H piles. It can be found that the ultimate soil resistance increases with the increase in roughness factor. Among the three types of pile shape, the H and square pile show the larger ultimate soil resistance than the circular pile. A similar observation was also made by Truong and Lehane [8]. This can be attributed to the larger volume of clay per unit pile width is mobilized surrounding the square and H pile. The following equations, which give excellent match with the computed results, are recommended in this study to predict the ultimate soil resistance ($p_u = N_p s_u B$) of piles with different pile shapes:

$$\text{For circular piles}: N_p = 9 + 3\alpha \tag{3}$$

$$\text{For square piles}: N_p = 12.69 + 2.3\alpha \tag{4}$$

$$\text{For H piles}: N_p = 14.69 + 0.3\alpha \tag{5}$$

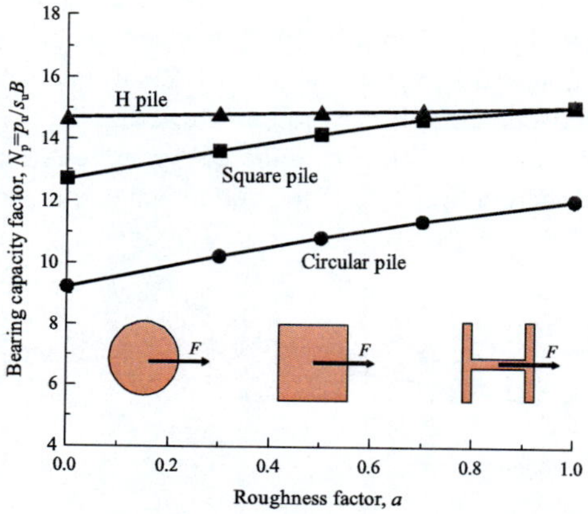

Fig. 4. Computed ultimate soil resistance of piles with different pile shapes

After determining the ultimate soil resistance of the *p-y* curves with consideration of pile shape effects (see Eqs. (3)–(5)), the remaining issue is concerned with the shape of the *p-y* curves. Zhang and Andersen [4] performed a comprehensive numerical study to investigate the *p-y* curves of circular pile in soft clay with a wide range of soil stress-strain response. The computed *p-y* curves of circular pile in this study are consistent with those presented in [4] and thus are not shown herein. In this study, more attention is paid to the *p-y* curves of square and H pile, which has been limitedly investigated.

Figure 5 shows a typical comparison of the normalised *p-y* curves of the circular, square and H pile. As illustrated, there is a tiny difference in the normalised *p-y* curves between the square and H pile, which can be ignored. While the normalised *p-y* curves of the circular pile exhibit stiffer response than those of the square and H pile. A comparison between Fig. 5(a) and Fig. 5(b) demonstrates that, when soil stress-strain response is increasingly softer, the *p-y* response also becomes increasingly softer. A similar conclusion is also made by Lai et al. [12].

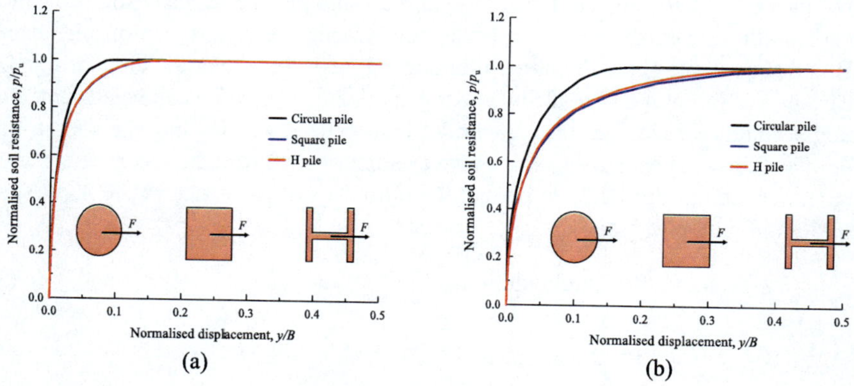

Fig. 5. Pile shape effects on the normalised *p-y* curves at (a) $G_{max}/s_u = 500$, $\gamma_f^p = 5\%$, $\alpha = 0.5$, and (b) $G_{max}/s_u = 500$, $\gamma_f^p = 15\%$, $\alpha = 0.5$

3.2 Proposed *p-y* Curves for Piles with Different Pile Shapes

Zhang and Andersen [4] proposed a framework that allows for the construction of *p-y* curves for circular piles by directly scaling the soil stress-strain curves measured in soil element tests, as schematically illustrated in Fig. 6.

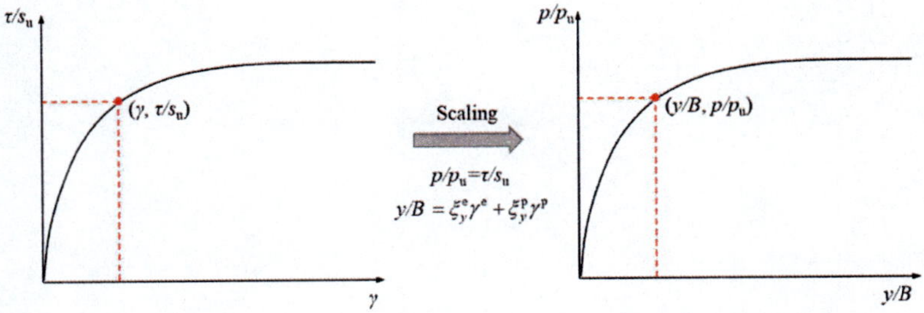

Fig. 6. Schematic illustration of scaling *p-y* curve from stress-strain curve [4]

A point on the stress-strain curve with mobilisation in shear stress (τ/s_u) corresponds to a point on the *p-y* curve with the same level of mobilisation in lateral bearing pressure (p/p_u):

$$\frac{p}{p_u} = \frac{\tau}{s_u} \tag{6}$$

The corresponding normalised displacement (*y/B*) can be scaled from the shear strain (γ) by using two scaling factors:

$$\frac{y}{B} = \xi_y^e \gamma^e + \xi_y^p \gamma^p \tag{7}$$

$$\gamma^e = \frac{\tau}{G_{max}} = \frac{\tau/s_u}{G_{max}/s_u} \tag{8}$$

$$\gamma^p = \gamma - \gamma^e \tag{9}$$

where $\xi_y{}^e$ and $\xi_y{}^p$ are the two scaling factors applied to the elastic and plastic components of the shear strain, respectively. For circular piles, the following scaling factors are proposed by Zhang and Andersen [4]:

For circular piles: $\xi_y{}^e = 2.8$, $\xi_y{}^p = 1.35 + 0.25\alpha$.

Following Zhang and Andersen [4], an attempt is made in this study to develop a scaling model for constructing *p-y* curves for square and H piles. All the computed *p-y* results for square and H piles with different pile roughness factors in varied stress-strain responses are presented in Fig. 7. It can be found from the figure that, at any given soil stress-strain response, pile shape effects and pile roughness factor show a negligible impact on the normalised *p-y* curves of square and H piles since the computed results merge into a single trend. In addition, the figure also implies that the shape of the *p-y* curves bears strong similarities with the soil stress-strain curves that were used as input to derive them. A softer soil stress-strain response leads to a softer *p-y* response. Therefore, a simple scaling model (see Fig. 6) is developed for constructing *p-y* curves for square and H piles from fundamental stress-strain response measured in laboratory tests. It is found that the following scaling factors result in a good match between computed and model-predicted results, as demonstrated in Fig. 7.

For square and H piles: $\xi_y{}^e = 2.1$, $\xi_y{}^p = 2.2$.

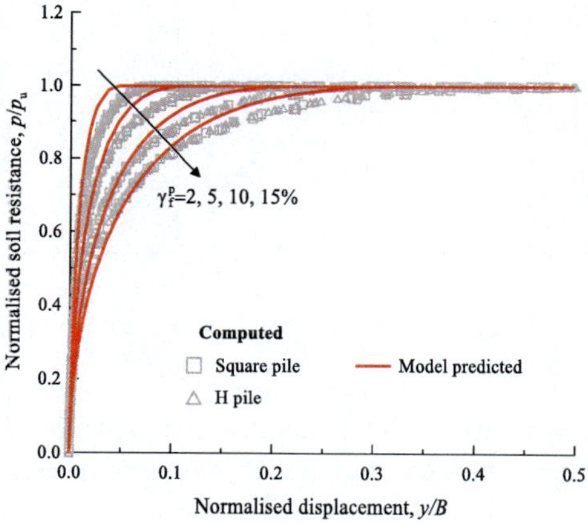

Fig. 7. Comparison of predicted *p-y* curves against numerical results

4 Limitations

There are two main limitations in this study: 1) only the pile shape effects on *p-y* curves for full-flow failure mechanism are studied. However, as revealed by Lai et al. [12] and Hong et al. [13], wedge-type flow near the ground surface and rotational flow near the lower half of a pile may also be involved, which depends on the relative pile-soil stiffness. Therefore, in addition to the *p-y* curves for full-flow mechanism, the pile shape effects on the *p-y* curves for the wedge failure and rotational failure should also be investigated. This will be our future pursuit. 2) It is worth noting that, unlike circular piles, the loading angle also plays an important role in *p-y* curves of square and H piles. However, this effect is yet to be investigated. The next step is to develop *p-y* formulations that can incorporate the combined effects of pile shape and loading angle.

5 Summary and Conclusions

This study presents a parametric numerical study to investigate the pile shape effects on the *p-y* curves of laterally loaded piles in soft clay with various soil stress-strain responses. It is found that, at any given pile width, H and square piles show the larger ultimate soil resistance than circular piles, indicating pile shape effects. At the specific loading direction investigated in this study, H piles are more efficient to resist lateral loading among these three types of piles due to the larger soil mobilization around the pile. At any given soil stress-strain response, there is a negligible difference in the normalised *p-y* curves between square and H piles. While the normalised *p-y* curves of circular piles exhibit a stiffer response than those of the square and H piles. Due to the similarities between the *p-y* curves and soil stress-strain curves that were used as input to derive them, a simple yet powerful scaling model is proposed for constructing *p-y* curves of piles with different shapes from soil element stress-strain curves that measured in the laboratory, providing engineers an efficient way to construct site-specific *p-y* curves for pile design.

References

1. API: Recommended practice planning, designing, and constructing fixed offshore platforms-working stress design. API 2A-WSD. 22nd ed. Washington, DC (2014)
2. DNVGL: DNVGL-ST-0126-Support Structure for Wind Turbines. Det Norske Veritas, Oslo (2016)
3. Jeanjean, P.: Re-assessment of *p-y* curves for soft clays from centrifuge testing and finite element modeling. In: Proceedings of Offshore Technology Conference Paper OTC20158, Houston (2009)
4. Zhang, Y.H., Andersen, K.H.: Scaling of lateral pile *p-y* response in clay from laboratory stress-strain curves. Mar. Struct. **53**, 124–135 (2017)
5. Wang, L.Z., Lai, Y.Q., Hong, Y., Masin, D.: A unified lateral soil reaction model for monopiles in soft clay considering various length-to-diameter (L/D) ratios. Ocean Eng. **212**, 107–492 (2020)
6. He, B., Lai, Y.Q., Wang, L.Z., Hong, Y., Zhu, R.H.: Scour effects on the lateral behavior of a large-diameter monopile in soft clay: role of stress history. J. Mar. Sci. Eng. **7**(170), 1–23 (2019)
7. Suryasentana, S.K., Lehane, B.M.: Updated CPT based p–y formulation for laterally loaded piles in cohesionless soil under static loading. Géotechnique **66**(6), 443–445 (2016)
8. Truong, P., Lehane, B.M.: Effects of pile shape and pile end condition on the lateral response of displacement piles in soft clay. Geotechnique **68**(9), 794–804 (2018)
9. Seo, H., Yildirim, I.Z., Prezzi, M.: Assessment of the axial load response of an h pile driven in multilayered soil. J. Geotech. Geoenviron. Eng. **135**(12), 1789–1804 (2009)
10. Grimstad, G., Andresen, L., Jostad, H.P.: NGI-ADP: Anisotropic shear strength model for clay. Int. J. Numer. Anal. Methods Geomech. **36**(4), 483–497 (2012)
11. Randolph, M.F., Houlsby, G.T.: The limiting pressure on a circular pile loaded laterally in cohesive soil. Geotechnique **34**, 613–623 (1984)
12. Lai, Y.Q., Wang, L.Z., Zhang, Y.H., Hong, Y.: Site-specific soil reaction model for monopiles in soft clay based on laboratory element stress-strain curves. Ocean Eng. (2020). https://doi.org/10.1016/j.oceaneng.2020.108437
13. Hong, Y., He, B., Wang, L.Z., Wang, Z., Ng, C.W.W., Masín, D.: Cyclic lateral response and failure mechanisms of semi-rigid pile in soft clay: centrifuge tests and numerical modelling. Can. Geotech. J. **54**(6), 86–824 (2017)

Potential Development of Floating Offshore Wind Turbine in Vietnam Offshore

Thanh-Dam Pham[1,2], Du Van Toan[3], Thi-Khang Nguyen[3], Hyunjeong Ahn[4], and Hyunkyoung Shin[4(✉)]

[1] Institute of Theoretical and Applied Research, Duy Tan University, Hanoi 100000, Vietnam
[2] Faculty of Natural Sciences, Duy Tan University, Da Nang 550000, Vietnam
[3] Vietnam Institute of Seas and Islands, Hanoi 123075, Vietnam
[4] University of Ulsan, Ulsan 44610, South Korea

Abstract. Vietnam has a vast potential for offshore wind power, which can play a key role in the country's green transition. In addition to the fixed offshore wind turbine, the floating offshore wind turbines (FOWT) open up a great prospect for making the most of domestic supply chain resources. This paper presents the potential of the floating offshore wind turbines in Viet Nam offshore area. A 10 MW offshore wind turbine was selected for this study. Two floating platforms were proposed, those are semi-submersible and spar-type. Environmental condition based on Ninh Thuan offshore area. NREL FAST code and in-house codes of the University of Ulsan (UOU) were used for designing and simulating the floating offshore wind turbine systems. Loads and performances of FOWTs were analyzed. Transportation and installation plan for each type of FOWT was proposed and discussed.

Keywords: Floating offshore wind turbine · Semi-submersible · Spar · 10 MW wind turbine

1 Introduction

The offshore wind energy in Vietnam has great potential, especially the offshore area of Ninh Thuan and Binh Thuan [1]. Most of the water depth in the Binh Thuan offshore area is shallow, so it is suitable for fixed wind turbines. In the Ninh Thuan offshore area, floating wind turbines are quite suitable because the water depths ranging from about 50 to 500 m have distance from 3 to 100 km from the land.

In this paper, a semi-submersible platform and a spar-type of the platform are designed to support a 10-MW wind turbine in the Ninh Thuan offshore area. The design and analysis methods for the semi-submersible and spar-type platforms are based on the following steps:

– Determine principal dimensions of the semi-submersible platform and the spar-type platform by upscaling a 5MW semi-submersible platform [2] and a 5 MW spar platform [3], respectively.

D. V. K. Huynh et al. (Eds.): VSOE 2021, LNCE 208, pp. 366–373, 2022.
https://doi.org/10.1007/978-981-16-7735-9_41

- Design mooring systems for two FOWT models with different water depths.
- Generate hydrodynamic coefficients for those platforms by UOU in-house codes.
- Fully coupled numerical simulation by FAST code for some design load cases (DLC) based on the IEC standard and environmental conditions of the Ninh Thuan offshore area.
- Analyze numerical simulation results to evaluate performances and responses of those FOWT models.

2 10 MW Floating Offshore Wind Turbine Concepts

2.1 10 MW Wind Turbine

The 10-MW wind turbine model was designed by extrapolating from the design of the NREL 5-MW reference wind turbine [4]. Table 1 shows the properties of the 10-MW wind turbine. The rated electric power is 10-MW at the rated wind speed of 11.2 m/s. The generator uses a supper conductor to reduce nacelle mass.

Table 1. 10-MW wind turbine specifications

Description	Value
Rated power of wind turbine	10-MW
Rotor orientation	Upwind, 3 blades
Control	Variable speed, collective pitch
Rotor diameter [m]	178.2
Hub height [m]	120
Cut-in, rated, cut-out wind speed [m/s]	11.2
Rated rotor speed [rpm]	9.48 (gearless)
Hub mass [kg]	169,440
Hub inertia about shaft [kg·m^2]	829,590
Nacelle mass [kg]	335,000
Towe mass [kg]	610,084

2.2 Floating Platforms

The semi-submersible platform consists of three offset columns and a center column supports the wind turbine. These columns are connected by braces and pontoons. The ballast material is seawater only. The spar-type platform is a steel hull slender cylinder, ballast materials are concrete and seawater. The concrete ballast is on the bottom of the platform and the seawater is above the concrete. The design draft of the semi-submersible platform and the spar-type platform are 24.6 m and 120 m, respectively. Figure 1 presents an overview of the semi-submersible and the spar-type FOWT systems and Table 2 describes main properties of two platforms.

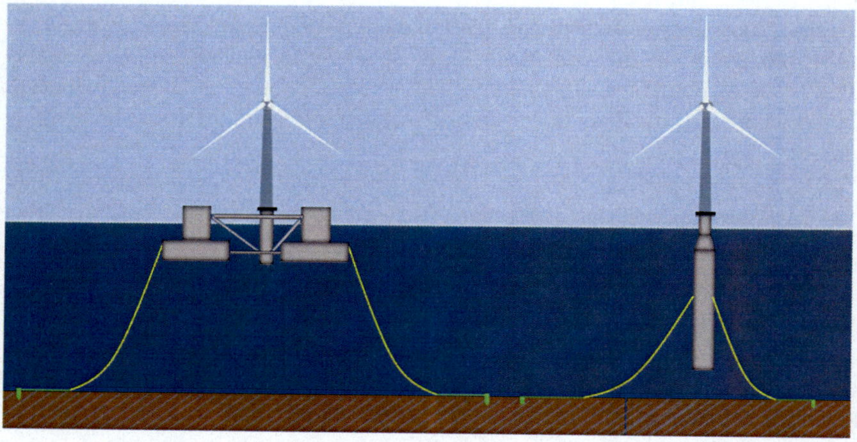

Fig. 1. Illustration of the semisubmersible and the spar-type FOWT

Table 2. Specifications of the semi-submersible and the spar

Description	Unit	Semi-Sub	Spar
Volume displacement	m^3	25,888.63	16,056
Center of Buoyancy below SWL	m	16.18	62.133
Platform mass	ton	25,297	14512
Center of Mass (CM) of platform below SWL	m	16.56	92.31
Platform roll inertia about CM	kgm^2	1.922E + 10	9.87E + 09
Platform pitch inertia about CM	kgm^2	1.922E + 10	9.87E + 09
Platform yaw inertia about CM	kgm^2	3.452E + 10	3.42E + 08

Note: SWL is still water level

2.3 Mooring Systems

Due to the small draft of the semi-submersible model, it can be installed at a water level of 50 m or more, so we chose to analyze the semi-sub model with 3 different water depths of 50, 150, and 250 m, those are three typical water depths of Ninh Thuan offshore area. The spar model is considered to install at two water depths of 150 m and 250 because of its large draft. The mooring systems were designed to sustain the position of the FOWT models in different water depths. The mooring systems need to satisfy some critical requirements such as the ultimate load of a mooring line must be smaller than the minimum breaking load, there is no vertical force at the anchor, and maximum drift smaller than 20% of water depth.

A mooring system consists of 3 studless chains spaced 120° apart. Table 3 shows the specifications of mooring systems for each model with different water depths.

Table 3. Specification of mooring systems

Description	Unit	Semi-Submersible			Spar	
Water depth	m	50	150	250	150	250
Number of mooring line		3	3	3	3	3
Mooring chain nominal diameter	mm	162	142	130	130	111
Mooring line mass density (air)	kg/m	576.8	443.2	371.5	371.5	270.8
Axial stiffness (EA)	MN	2362	1814.8	1521.0	1521.0	1108.9
Unstretched mooring line length	m	500.0	800.0	1050.0	800.0	1100.0
Depth to fairleads below SWL	m	17.2	17.2	17.2	48.0	48.0
Radius to fairlead	m	51.0	51.0	51.0	7.0	7.0
Radius to anchor	m	543.0	810.6	979.5	785.0	1057.2
Minimum Breaking Load (MBL)	MN	18.2	14.7	12.7	12.7	9.6

3 Numerical Simulation

Floating offshore wind turbines are designed and installed in water depths greater than 50 m. The wind turbines are subjected to the environmental loads arising from wind, wave, and current. The loads on these systems are dominated by hydrodynamic and aerodynamic loads. Mooring forces of a mooring system secure the floating wind turbine system in its position. The equation of motions of the FOWTs under the fully coupled analysis in the time domain can be expressed as in Eq. 1.

$$\left(M_{ij} + A_{ij}(\infty)\right)\ddot{x}_j(t) + \int_0^t K_{ij}(\tau)\dot{x}_j(t-\tau)d\tau + C_{ij}x_j(t) = F_i(t) \tag{1}$$

$$F_i(t) = F_H(t) + F_W(t) + F_V(t) + F_M(t) \tag{2}$$

where Mij is the mass and the mass inertia matrix of the FOWT structure, Aij(ω) is the added mass matrix, Cij is the hydrostatic restoring stiffness coefficient matrix, Kij is the retardation function which is expressed by the wave radiation damping matrix Bij(ω) as in Eq. 3

$$K_{ij}(t) = \frac{2}{\pi}\int_0^\infty B_{ij}(\omega)\cos(\omega t)d\omega \tag{3}$$

Fi(t) is the sum of the external forces matrix which are the hydrodynamic forces of the platform $F_H(t)$ including hydrostatic force, wave-exciting forces and current force; the restoring forces of mooring system $F_M(t)$; the viscous drag forces from Morison's equation $F_V(t)$; and the wind forces $F_W(t)$ including the aerodynamic forces of the wind turbine and tower drag forces. The total external forces are expressed in Eq. 2.

The 10 MW FOWT was simulated by the open-source code FAST v8 [5] which is the aero-hydro-servo-elastic wind turbine simulation software developed by National Renewable Energy Laboratory (NREL). It is a powerful tool to combine the simulation of wind turbines under the wind, wave, and current, and dynamic mooring system. The primary parameters such as the elastic model of the blades and tower, aerodynamic properties of the airfoil of the blades, control schedule of generator torque and blade pitch angle, hydrodynamic data of the platform are input to FAST v8 to simulate entire FOWT system under complex environmental conditions. In-house codes of the University of Ulsan (UOU) including radiation solver and diffraction solver were used for generating the hydrodynamic added mass and –damping forces and wave-exciting forces.

4 Environmental Conditions

Base on IEC 61400–3-2 standard [6], three design load cases were selected as shown in Table 4. Environmental conditions are collected and summarized based on data from Vietnam Institute of Seas and Islands. Environmental condition in the DLC1.1 include normal turbulence winds, normal sea state waves and wind-generate current speeds. DLC1.6a analyses the wind turbine systems in normal turbulence winds, a severe sea state wave and normal current speeds. Extreme condition DLC6.1a is considered 50-year-return period environmental condition with extreme wind speed, extreme wave height and extreme current speed. Detail wind, wave and current conditions are listed in Table 5. The sea states 1–11 are normal sea states and the sea state 12 and 13 is the extreme sea state over a 50-year-return period. Turbulent wind models are created by the Turbsim program [7]. Irregular waves are generated based on the JONSWAP spectrum by FAST.

5 Results and Discussions

Figure 2 compares the extreme values of the different parameters, extracted from the time history output of three design load cases for semi-submersible and spar 10 MW FOWT systems. Based on these data, it is possible to evaluate the operability as well as loads of the two FOWT systems at different water depths. For the semi-submersible model, the water depths are 50 m, 150 m, 250 m respectively, they are labeled in Fig. 2 as semi-50, semi-150, and semi-250, respectively. It can be seen that with the exception of platform surge and fairlead tension, the remaining parameters in each DLC for the three semi-submersible systems are quite similar. Since the mooring system for each water depth has to meet critical requirements there are different platform surges and fairlead tensions of the semi-submersible system operating at three different water depths. With a water depth of 50 m, the mooring system is subjected to enormous tension under extreme environmental conditions. For the spar-type FOWT systems, the water depths are 150 m, 250 m, they are labeled in Fig. 2 as spar-150 and spar-250, respectively. All parameters in each DLC for the two spar-type FOWT systems are quite similar. It can be found that the tower base bending moment, platform pitch, and nacelle acceleration of spars are much higher than those of semi-submersibles under extreme environmental conditions (DLC1.6a and DLC6.1a) but relatively similar under normal operational conditions in

DLC1.1. This difference comes from the fact that the center of mass of the spar-type FOWT system is much lower than that of the semi-submersible FOWT system, so the movement at the nacelle is higher, leading to a higher bending moment on the base tower.

Table 4. Design load cases

Item	DLC 1.1	DLC 1.6a	DLC 6.1a
Wind model	NTM	NTM	EWM
Sea state	SS1-SS11	SS12	SS13
Wind and wave directionality	0°, COD	0°, COD	MIS ± 30°
			Yaw ± 8°, 0°
Current model	NCM	NCM	ECM
Wind turbine status	Operation	Operation	Parked,
			blades:90^0

Note: COD is co-directional, ECM is extreme current model, ESS is extreme sea state, EWM is extreme wind speed model, MIS is misalignment of wind and wave, NTM is normal turbulence model

Table 5. Sea states based on data from Vietnam Institute of Seas and Islands

Sea state	Wind	Wave		Current	DLC
		Hs	Tp		
	m/s	m	s	m/s	
SS1	4	0.12	2.9	0.4	DLC1.1
SS2	6	0.47	2.9	0.4	DLC1.1
SS3	8	0.63	2.9	0.4	DLC1.1
SS4	10	1.05	3.68	0.4	DLC1.1
SS5	12	1.33	3.88	0.4	DLC1.1
SS6	14	1.81	4.51	0.4	DLC1.1
SS7	16	2.36	5.08	0.4	DLC1.1
SS8	18	3.01	5.74	0.4	DLC1.1
SS9	20	3.79	6.38	0.4	DLC1.1
SS10	22	4.71	7.18	0.4	DLC1.1
SS11	24	5.6	7.72	0.4	DLC1.1
SS12	4–24	8.8	10	0.4	DLC1.6a
SS13	42.1	8.8	10	1.47	DLC6.1a

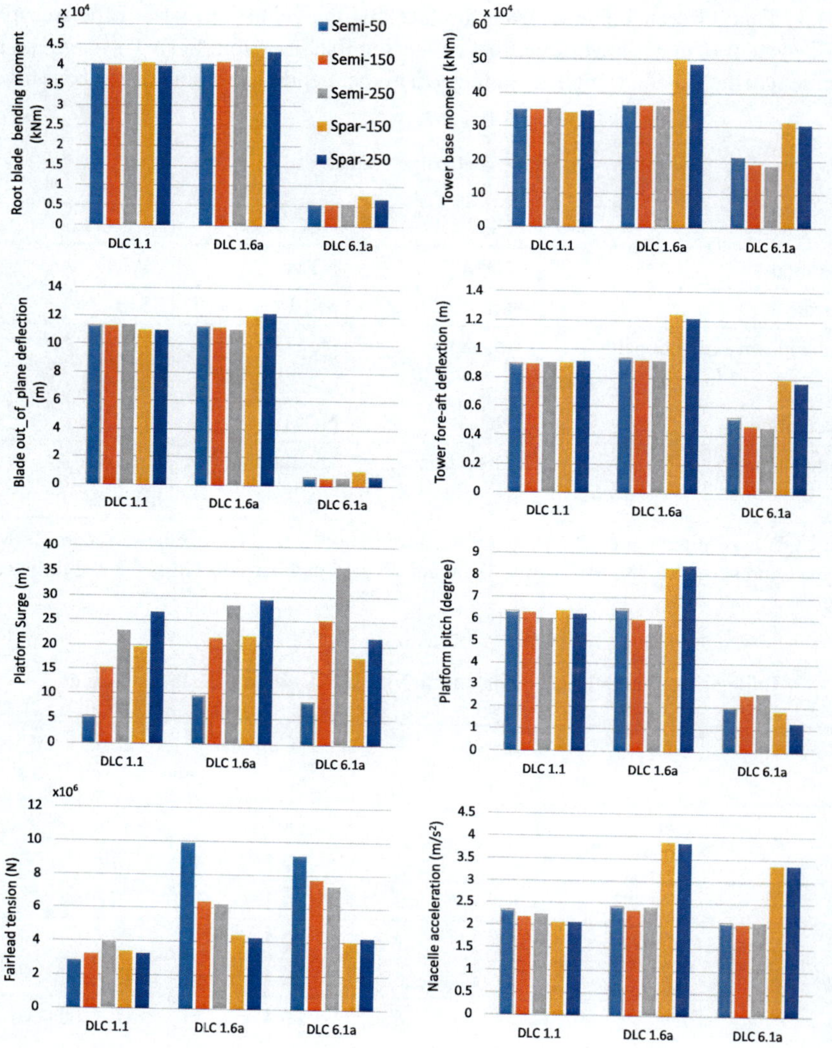

Fig. 2. Ultimate responses and loads of FOWT systems for three DLCs.

Normally floating platforms can be built at shipyards. The semi-submersible model has good stability at the small draft level so it can be completely installed with the wind turbine system at the shipyard or at the port and then pull out to the installation location. Meanwhile, the spar model can only install a whole FOWT system at the installation site. The downside of the semi-submersible model compared to the spar model is that it has a much larger platform mass than the mass of the spar model. However, the installation of the semi-submersible model is quite flexible and reliable because the installation time at sea is faster. This paper wants to emphasize that given the characteristics of the offshore area of Ninh Thuan, the 10 MW semi-submersible FOWT model is perfectly suited for large floating wind farms.

6 Conclusion

Vietnam's wind potential is huge, especially the Ninh Thuan sea area is very suitable for deploying floating wind farms. The two floating wind models studied in this paper are semi-submersible and spar-type supporting a 10 MW wind turbine and are installed at different water depth locations. These floating wind turbine models were simulated using NREL FAST code with fully coupled aero-servo-hydro-elastic. There are 3 DLCs simulated based on IEC standard and environmental conditions in the Ninh Thuan offshore area. Results from these three DLCs show that both systems are reliable at different water depths. considering the transportation and installation of these two FOWT systems, the semi-submersible model is more preferable for the deployment of floating wind farms in the Ninh Thuan offshore area. To have a more comprehensive view of the optimal platform model selection, the calculation of LCOE for each model is needed in the future.

Acknowledgment. This work was supported by the Korea Institute of Energy Technology Evaluation and Planning (KETEP), the Ministry of Trade, Industry, and Energy (MOTIE) of the Republic of Korea (No. 20184030202280), and the Korea Electric Power Corporation (No. R18XA03).

References

1. Quang, V.D., Doan, V.Q., Dinh, V.N., Duc, N.D.: Evaluation of resource spatial-temporal variation, dataset validity, infrastructures and zones for Vietnam offshore wind energy. Vietnam J. Sci. Technol. Eng. **62**(1), 03–16 (2020)
2. Robertson, A., Jonkman, J., Masciola, M.: Definition of the Semisubmersible Floating System for Phase II of OC4 (2014)
3. Jonkman, J.: Definition of the floating system for phase IV of OC3. Contract **1**, 31 (2010)
4. Jonkman, J., Butterfield, S., Musial, W., Scott, G.: Definition of a 5-MW Reference Wind Turbine for Offshore System Development. Golden (2009)
5. Jonkman, B., Jonkman, J.: FAST v8 (2018). https://nwtc.nrel.gov/FAST8
6. IEC IEC TS 61400–3–2 Design requirements for floating offshore wind turbines. Geneva, Wwiterland (2019)
7. Jonkman, B.J.: TurbSim user's guide v2. 00.00. Natl. Renew. Energy Lab. (2014)

Slope Stability and Potential Liquefaction of Tra Vinh Sea Dyke Subjected to Dynamic Loadings

Nguyen Hong Nam[1(✉)] and Vu Van Ngoc[2]

[1] ThuyLoi University, 175 Tay Son Street, Dong Da District, Hanoi, Vietnam
hongnam@tlu.edu.vn

[2] Vietnam Academy for Water Resources, No. 1, 165 Lane, Chua Boc Street, Dong Da District, Hanoi, Vietnam

Abstract. Traditionally, when studying slope stability of sea dykes, acting loads are considered statically, leading to the constant coefficient of slope stability. These assumptions, however, do not reflect the actual working conditions of the structures in practice, and not consider the potential liquefaction phenomena occurred where sea dyke structures being constructed on the loose sandy foundation. In this study, several dynamic loading scenarios that consisted of wave and/or earthquake loadings were analyzed numerically by the dynamic method and compared with the result by the static method. The calculation results showed that the stability of the sea dyke decreased significantly when the earthquake occurred, and ground liquefaction phenomenon could occur due to the wave load and/or earthquake load.

Keywords: Sea dyke · Earthquake · Liquefaction · Dynamic analysis · Stability

1 Introduction

In the context of climate change, sea level rise at present and in the near future, the research on stabilization of sea dykes is more important significance, especially the influence of earthquake and wave load. Stabilization of hydraulic structures such as sea dykes is a mandatory requirement in many design standards. During their construction and operation time, sea dykes have been frequently affected by the complex dynamic loads such as waves, tides, traffic loading or even earthquakes in separate or combining forms. Besides, many sea dyke segments have been built on soft soil and loose sandy foundation which are likely to be susceptible to liquefaction due to wave loads and/or earthquakes. For the past years, many studies have implemented on the dyke ground liquefaction caused by the wave loads (Ishihara and Yamazaki 1984; Zen and Yamazaki 1990; Ishihara and Towhata 1983, among others). However, in Vietnam, the study on the potential liquefaction of sea dykes caused by dynamic loadings is very limited. Although the territory of Vietnam is not on active earthquake-volcanic belts, it is also not a sustainable land with difficult earthquakes. There were three strong earthquakes recorded in Vietnam, namely Dien Bien (M = 6.9 in 1935), Tuan Giao (M = 6.7 in 1983) and Dien Bien Phu (M = 5.3 in 2001). The area at risk of earthquakes from 6.0

D. V. K. Huynh et al. (Eds.): VSOE 2021, LNCE 208, pp. 374–383, 2022.
https://doi.org/10.1007/978-981-16-7735-9_42

to 7.0 magnitude in Vietnam includes Central coastal zone (Xuyen 2004). The present study was implemented to a sea dyke in Tra Vinh province to evaluate the dyke stability and potential liquefaction under dynamic loading.

2 Study Project

The embankment to protect the weak section of the coast of Con Trung hamlet, Truọng Long Hoa commune, Duyen Hai district, Tra Vinh province (Tra Vinh DRD 2009) was employed in this study (Fig. 1). The elevations of the embankment top, berm, top and bottom of the cutoff are +3.2 m, +1.90 m, −0.45 m and −1.30 m, respectively. The widths of the embankment top, berm, and cutoff are 5.0 m, 4.4 m and 1.60 m, respectively. The inverse of the slopes of the upper and lower faces from the wave dissipation berm are $m_{tr} = m_d = 2$. The scour protection platform to prevent erosion of the cutoff at the seaside has elevation −0.45, width 5.07 m and the inverse of the slope $m = 2$. The inverse of the slope of the downstream face at the field side $m = 2$. The elevation of the cutoff surface at the field side is +1.00 m or +2.30 m depending on the terrain conditions. The stratigraphy of the project area consists of some layers of sandy soil and clay. Table 1 shows the soil properties of 15 samples taken along the borehole HK8 from the ground surface. In which, the fine contents of sand samples are small ($f_c \leq 6.3\%$).

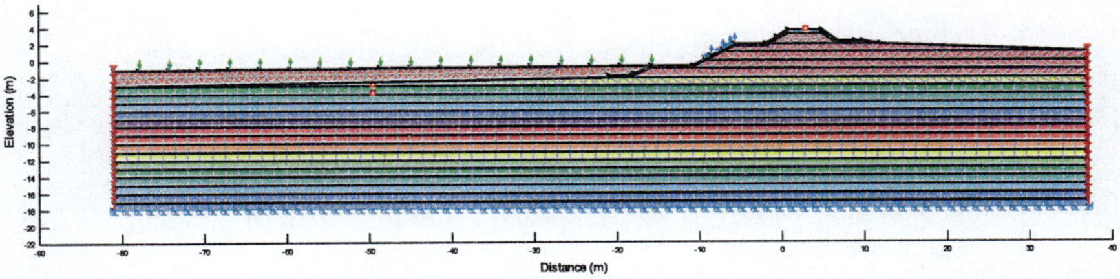

Fig. 1. A finite element mesh of the design embankment cross section.

3 Numerical Modeling

The potential liquefaction and slope stability analyses of the embankment subjected to sea wave and/or earthquake loading were analyzed based on the finite element method, using GeoStudio 2007 software (GEO-SLOPE 2008; 2010). The design cross section of the embankment was modelled as a plane strain problem (118 m long and 22 m high), which was discretized into a mesh of 1890 finite elements with the approximate size of 1.5 m (Fig. 1).

Table 1. Laboratory soil properties from borehole HK8 (Tra Vinh DRD 2009).

Soil sample	Description	PI (%)	w (%)	γ (kN/m^3)	LL (%)	ϕ (deg)	c (kPa)
HK8–1	Small sand, brownish gray, less compact	–	25.39	18.5	0	23.3	4.8
HK8–2	Fine sand, dark gray, less tight	–	29.27	18.6	0	21.3	5.0
HK8–3	Small sand, yellowish gray, less compact	–	27.92	18.8	0	22.1	4.5
HK8–4	Fine sand, dark gray, less tight	–	26.9	18.7	0	23.4	5.5
HK8–5	Small sand, dark gray, less compact	–	29.33	18.2	0	20.9	5.7
HK8–6	Small sand, yellowish gray, less compact	–	28.61	18.4	0	21.8	4.8
HK8–7	Gray-brown clay, flowing plastic	32.6	58.59	16.0	61.2	4.5	8.7
HK8–8	Fine sand, dark gray, medium tight	–	27.4	18.7	0	26.7	3.0
HK8–9	Fine sand, dark gray, less tight		28.65	18.5	0	22.8	4.1
HK8–10	Small sand, dark gray, less compact		25.51	18.2	0	24.1	3.5
HK8–11	Small sand, dark gray, less compact		26.35	18.9	0	25.2	4.5
HK8–12	Fine sand, dark gray, less tight		31.56	18.7	0	20.6	5.0
HK8–13	Gray-brown clay, flowing plastic	21.8	46.48	17.2	50.3	4.7	10.2
HK8–14	Gray-brown clay, flowing plastic	15.6	50.17	16.8	52.5	5.2	11.1
HK8–15	Clay, gray brown, soft	15.1	36.31	18.0	41.0	6.6	13.1

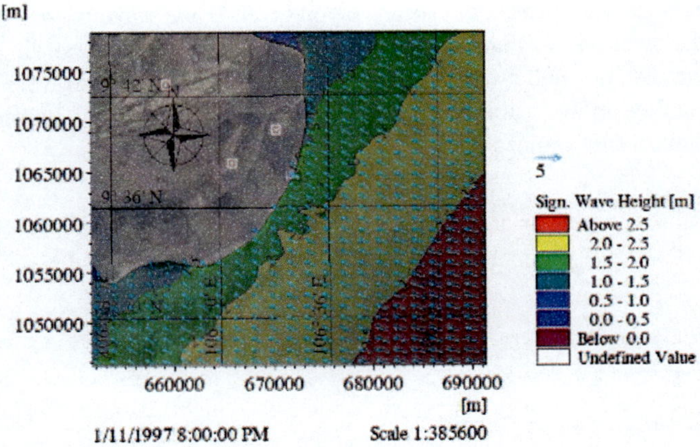

Fig. 2. Wave-height fields and directions at site.

Table 2. Simulation scenarios for slope stability and liquefaction of the embankment.

Case	Type	Load combination	Problem
1	Static	Steady state	Slope stability
2	Dynamic	Sea wave	Slope stability and liquefaction
3	Dynamic	Earthquake (T = 475 years)	Slope stability and liquefaction
4	Dynamic	Earthquake (T = 2475 years)	Slope stability and liquefaction
5	Dynamic	Sea wave and Earthquake (T = 475 years)	Slope stability and liquefaction
6	Dynamic	Sea wave and Earthquake (T = 2475 years)	Slope stability and liquefaction

Table 2 shows six simulation cases employed in this study. Case 1 analyzed dyke slope stability under static loading by limit equilibrium Morgenstern-Price method, while the remaining cases applied the finite element methods. In the finite element analysis, firstly the initial stresses were calculated; after that, the dynamic stress-strain analysis was implemented. Next, the liquefaction and slope stability analyses were carried out using the stresses calculated from previous results. Deep water waves were calculated by Bretschneider method with the conditions of still water level +1.92 m (frequency 5%), wind speed 24 m/s, fetch length 180 km. The deep-water wave parameters were used to calculate the wave propagation in the project area according to TCVN 9901: 2014 (MOST 2014) and Mike 21SW (DHI 2007). The design wave field was simulated with the wind of grade 9 (equivalent to the 1997's historic Linda storm), water depth h = 40m, wave height H = 4.68 m, wave period T = 6.37 s (Fig. 2). The results of design wave parameters at the embankment toe (H = 1.65 m, T = 6.37 s, L = 29.2 m), obtained by TCVN 9901: 2014 were used to calculate the wave loads acting on the embankment slope and on the seabed. In the process of transmitting wave load from the offshore to the dyke location, waves at consecutive points along the wave propagation direction always had slower phases than the previous point. Therefore, the coordinate origin was assumed at a distance being two times the wavelength (i.e., 58.4 m) from the dyke toe. The stress acting on the seabed could be commonly simulated by a harmonic function (Eq. 1) (Ishihara and Yamazaki 1984).

$$F = \gamma_w \frac{H}{2} \frac{1}{\cosh\left(\frac{2\pi h}{L}\right)} cos\left(\frac{2\pi}{L}x - \frac{2\pi}{L}t\right) \tag{1}$$

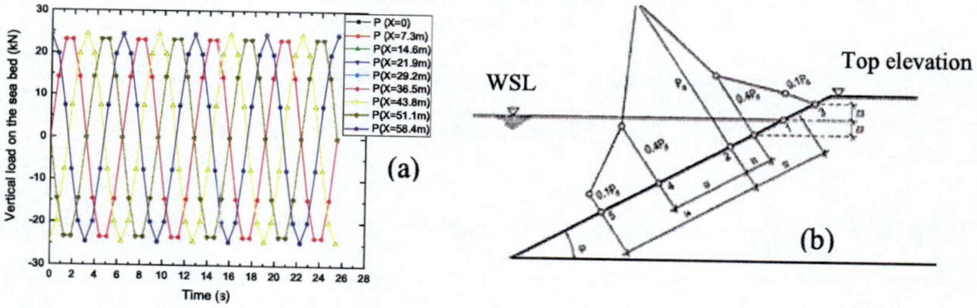

Fig. 3. a) Wave load acting on the seabed at some interval points b) Diagram of wave load pressure distribution on an inclined sea dyke (TCVN 9901: 2014).

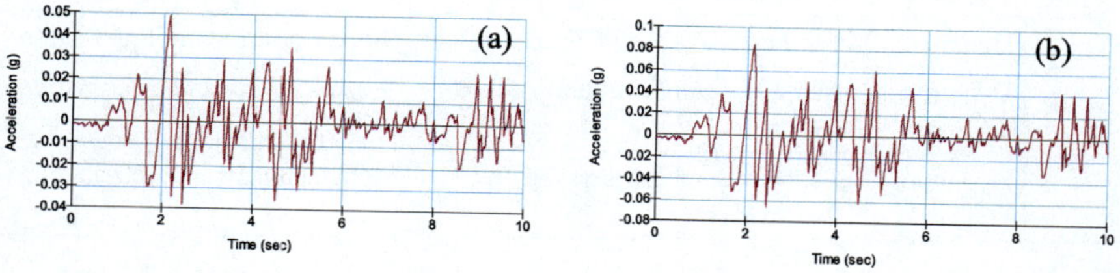

Fig. 4. The earthquake acceleration records a) T = 475 years; b) T = 2475 years.

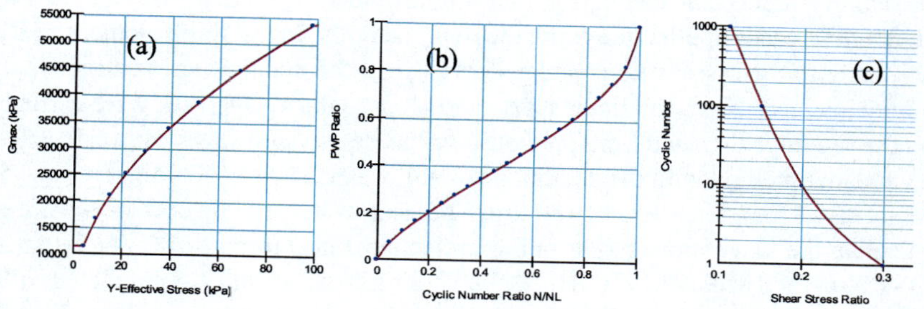

Fig. 5. Dynamic property soil functions: a) shear modulus b) PWP ratio, c) cyclic stress ratio.

To reduce the computation time, the wave pressures on seabed were transferred to the concentrated loads acting on 18 nodes with the interval distance of 6.35 m

(Fig. 1). Because the above assumption on the coordinate origin (X = 0), after a period of two times the wave period (12.74 s), the wave from the origin approaches the dyke toe. The wave pressures on dyke slope by TCVN 9901: 2014 (Fig. 3b) were converged to the point loads assigned to the FE nodes (Fig. 1). For safety and simplicity purposes, the loads acting on the upstream face were assumed time independent. The aforementioned wave function calculation was applied similarly for Cases 2, 5 and 6 with the assumptions: uniform waves, the same load function for all water from the offshore to the shore and the wave load only depending on wave parameters (such as wave height H, wavelength L, wave period T) and water depth h. The step time for case 2 was 0.0637 s, time duration was 25.48 s which was divided into 400 time steps.

Regarding earthquake loading, the project site potentially suffers from the earthquake of grade VI (MSK-64 scale) with peak ground acceleration PGA = 0. 0491 g (TCVN 9386:2012) (MOST 2012), the return period is 475 years. Due to the unavailability of seismic data at site, the time series earthquake acceleration record (Fig. 4) was employed with referring to the sample record of QUAKE/W library and the aforementioned PGA at site. In addition, we analyzed one special case with the earthquake of grade VII (PGA = 0.085 g, return period of 2475 years) (EN 1998–2) (EN 1998) using the above sample record to examine the effect of stronger earthquakes. The time step used for Cases 3 to 6 was 0.02 s, the time duration was 10 s which was divided into 500 time steps.

The FE-based slope stability analysis using SLOPE/W has been applied after getting the stress results calculated by QUAKE/W. In each time step, the minimum FOS was deduced from 968 or 1331 potential slip surfaces formed by a grid of center of slip surfaces (10 × 10) and radii (7 or 10). Thus, there were total 484,000 or 665,500 potential slip surfaces analyzed from 500 time steps during the earthquake period (10 s) (Cases 3 to 6). Similarly, 387,200 potential slip surfaces from 400 time steps during the wave loading period (28.54 s) were analyzed (Case 2); and 1331 slip surfaces for limit equilibrium analysis (Case 1).

The subsoils and embankment fill were simulated by linear elastic model for FE analyses. The model parameters were calibrated based on the laboratory test results (Table 1) and employing following assumptions due to the limitation of the dynamic tests: The damping ratio was taken 0.1 for all soil layers. The Poisson's ratio $\nu = 0.25$ for sandy layers and 0.35 for clay layers (Coduto et al. 2011). In addition, the G_{max}, the PWP ratio, the cyclic stress ratio functions were taken by the sample functions of QUAKE/W (Fig. 5). The Mohr-Coulomb soil model was applied for the stability analysis with model parameters shown in Table 1. During undrained cyclic loading, a continuous increase in the pore pressures makes the cyclic stress path intersects the collapse surface; thus, the saturated sand can liquefy (GEO-SLOPE 2010).

4 Simulation Results and Discussion

4.1 Slope Stability Results

The slope stability analysis results for the upstream and downstream slopes for different static and dynamic simulation scenarios are shown in Table 3 and Figs. 6 and 7. In general, Table 3 shows much reduction of factor of safety from the static case to the dynamic case, especially when the dynamic loading caused by earthquake only or earthquake combined with sea wave. Case 6 appears the most dangerous case where the earthquake grade VII (MSK-64) occurred during the design wave loading. Regarding earthquake loading, the higher PGA occurred, the smaller overall FOS for both dyke slopes. In case 3, the reduction in FOS was 43.5% and 61.2% for the upstream and downstream slopes, respectively. However, in case 4, the corresponding reduction values in FOS were 62% and 76%, respectively. The effect of wave loading on slope stability of dyke slope was found not large in the present study. Compared with the static case, the minimum FOS of upstream slope reduced by about 4.8%, while that of downstream slope did not change noticeably. Compared with the upstream FOS, the downstream FOS was not significantly affected by the wave loading acting on the upstream seabed boundary. The minimum FOS of downstream slope maintained a stable value of 4.78 after 2 s until the duration time of 25.48 s. This is possibly because, in the simulation, the wave load acting on the slope surface was assumed time independent by TCVN 9901:2014.

Table 3. Slope stability and liquefaction results.

FOS/Liquefaction	Case 1	Case 2	Case 3	Case 4	Case 5	Case 6
Upstream face	2.831	2.696	1.600	1.076	1.738	1.080
Downstream face	4.509	4.466	1.751	1.079	1.792	1.066
Liquefaction Potential	No	Yes	No	Yes	Yes	Yes

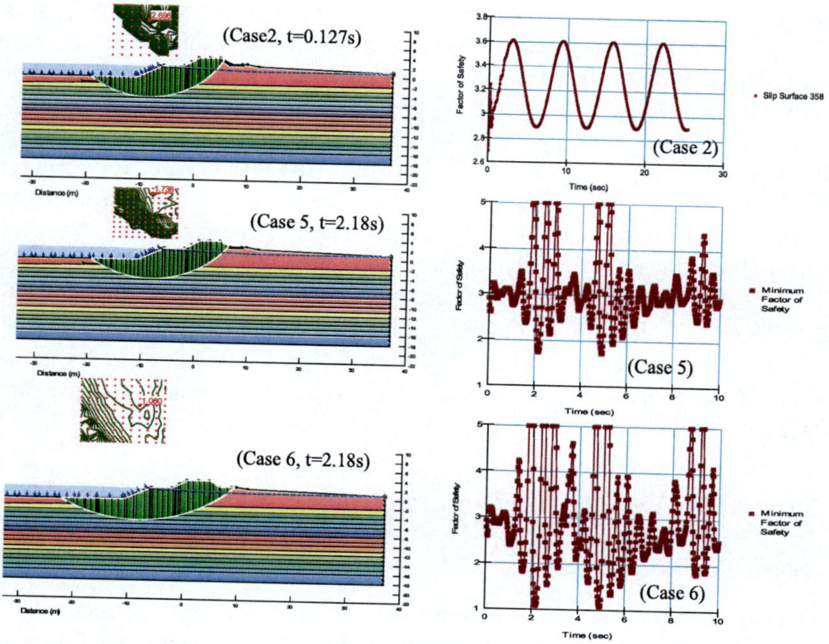

Fig. 6. Minimum FOS of upstream slopes subjected to dynamic loads.

Figure 6 shows the typical results of dynamic slope stability for the upstream slopes, while Fig. 7 showing results for the downstream slopes. Both Figs. 6 and 7 reveal that the variation in minimum FOS values with time step depended on the shape of the dynamic loading. In the analysis with the combination of wave and earthquake loads (cases 5 and 6), the minimum FOS occurred at 2.18 s for the upstream slopes and at 2.42 s for the downstream slopes. In the analysis with the wave loading only (case 2), the minimum FOS of upstream slope occurred at 0.127 s and that of downstream slope at 0.382 s.

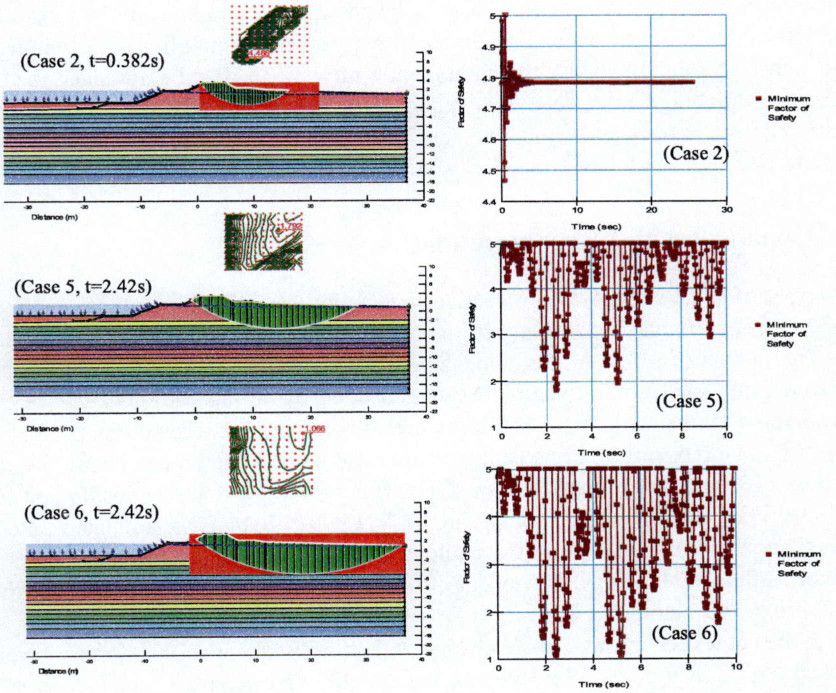

Fig. 7. Minimum FOS of downstream slopes subjected to dynamic loads.

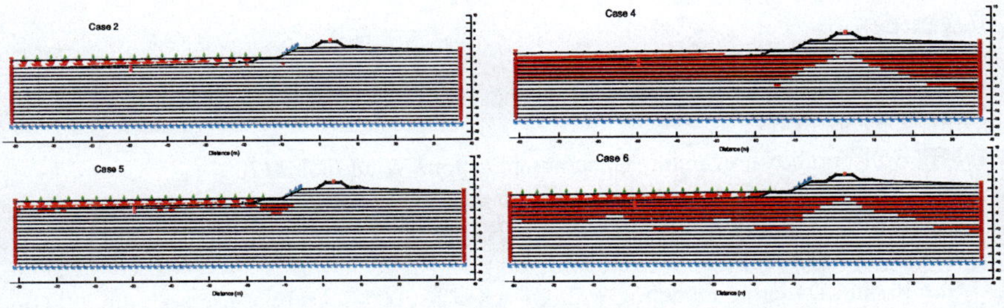

Fig. 8. Liquefied zones caused by dynamic loadings.

4.2 Liquefaction Results

Figure 8 shows the liquefied zones occurred in the seabed sand foundation, at the final time step, as a result of dynamic loads in case 2, 4, 5, 6. It can be seen from this figure that the liquefaction area is small and discontinuous in case 2 of wave loading and expands in cases 4, 5, 6. In case 2, due to the limitation in the number of concentrated loads on the seabed, the simulated zone seems not continuous. The liquefaction zone occurs at the positions of the 18 point loads. If the point loads were added more densely, the liquefied zone would become smoother. However, it should be understood that the liquefaction zone would vary according to the geological conditions of the subsoil, and the load value changes depending on the wave parameters and the water depth. The soil liquefied zone caused by the combination of wave load and earthquake load was observed with bigger area, depending much on the PGA. For cases 1 and 3, no liquefied zone occurred in the simulation. Note that effect of fine content ($f_c \leq 6.3\%$) on ground liquefaction was small (see HongNam and Van 2019).

5 Conclusion and Recommendation

Analyses of slope stability and potential liquefaction of Tra Vinh sea dykes caused by wave and/or earthquake loads have been conducted to obtain following conclusions.

The factors of safety of the embankment slopes decreased significantly when the structure subjected to the dynamic loads generated by strong earthquake/wave loads compared with the static case. The influence of peak ground acceleration on factor of safety was significantly. The seaside embankment slope suffered frequently the wave load which could reduce the factor of safety. However, effect of wave loading on slope stability FOS was not large by applying TCVN 9901: 2014. The simulated potential liquefaction could be possible in the shallow, loose sandy ground subjected strong wave and/or earthquake loadings. The extent of liquefied zone increased noticeably with the peak ground acceleration.

In the context of recent climate change with more frequent severe storms in central coastal region in Vietnam, the study on the stability and potential liquefaction of sea dykes need to be applicable for the protection of sea dykes in the coastal areas which are prone to strong earthquakes and high waves.

References

Coduto, D.P., Yeung, M.R., Kitch, W.A.: Geotech. Eng.: Principles and Practices. Pearson, USA, p. 367, 2nd edn. (2011)

DHI: MIKE 21 Toolbox user guide: Generating Cyclone Wind Fields (2007)

EN 1998–2: Eurocode 8: Design of structures for earthquake resistance – Part 2: Bridges

GEO-SLOPE Int. Stability Modeling with SLOPE/W 2007—An Eng. Method. 3rd ed. GEO-SLOPE, Alberta (2008)

GEO-SLOPE Int. Dynamic modeling with QUAKE/W 2007—An Eng. Method, 4th ed. GEO-SLOPE, Alberta (2010)

HongNam, N., Van, N.T.N.: Study on liquefaction characteristics of Red river dyke sand. In: Proceedings of the 7th International Conference on Earthquake Geotechnical Engineering, p. 8. Roma, Italy (2019)

Ishihara, K., Towhata, I.: Sand response to cyclic rotation of principal stress direction as induced by wave loads. Soils Found. **23**, 11–16 (1983)

Ishihara, K., Yamazaki, A.: Analysis of wave-induced liquefaction in seabed deposits of sand. Soils Found. **3**, 85–100 (1984)

MOST: hydraulic structures - Requirements for seadike design TCVN 9901: 2014, 1st edn. (2014). (in Vietnamese)

MOST: design of structures for earthquake resistance TCVN 9386: 2012, 1st edn. (2012). (in Vietnamese)

Tra Vinh Department of rural and development: project on the embankment to protect the weak section of the coast of Con Trung, Geotech. Eng. report (2009). (in Vietnamese)

Xuyen, N.D.: Earthquake prediction and ground motion in Vietnam, state funded project, final report (2004). (in Vietnamese)

Zen, K., Yamazaki, H.: Mechanism of wave-induced liquefaction and densification in seabed. Soils Found. **30**(4), 90–104 (1990)

Stress on the Nozzle of a Steel Vertical Storage Tank

Bau Van Nguyen(✉) and Tri Mai

Faculty of Coastal and Offshore Engineering, Hanoi University of Civil Engineering, 55 Giai Phong, Hai Ba Trung, Hanoi, Vietnam
baunv@nuce.edu.vn

Abstract. Stress on the nozzle of a steel vertical storage tank due to different types of loads along the pipe will cause different deformation on the contact between the nozzle and the tank. Therefore, this stress can be a cause of serious damage to the tank during operation of the nozzle. This paper will study stress on the nozzle by finite element method through the development of FEPipe software. Stress is processed based on the rules specified in American Society of Mechanical Engineers (ASME) Boiler and pressure vessel code, section III Rules for Construction of Nuclear facility components and section VIII Division 2 Alternative rules for Construction of pressure vessels. Some assumptions are used in this study such as: Uniform thermal expansion produces no stress in this geometry; any thermal loads will come through operating forces and moments; inner and outer element temperatures are the same throughout the model; the operating pressure is used for secondary and peak stress cases. There are 10 load case analyses mentioned and 18 stress results are given under different working conditions. The results show that there are significant effects of the bending stress in the calculation of the nozzle's stresses. In addition, the thickness of the reinforcement pad also affects the stress state of the nozzle. The results of this study will be the basis for calculating nozzle design in current and later developments.

Keywords: Stress on the nozzle · Nozzle · Vertical storage tank · FEPipe

1 Introduction

Stress on the nozzle of a steel vertical storage tank includes the primary, secondary and peak stresses. In general, the primary stress includes membrane stress and bending stress [1]. The primary stress can be normal stress developed by loading. These stresses will not decrease their values when the structure is deformed. The primary stresses are divided in primary general membrane (P_m), primary bending stress (P_b) and primary local membrane (P_L) (see Fig. 1). The secondary stress (Q) can be normal stress or shear stress, are characterized by decrease their values when the structure deforms. The peak stresses are added to primary and secondary stress present at the point of the stress concentration [2]. Peak stresses are very important in fatigue conditions or brittle materials. Peak stresses are sources of fatigue cracks and apply to membrane, bending and shear stress. In different working cases, stress on the nozzle will case different

D. V. K. Huynh et al. (Eds.): VSOE 2021, LNCE 208, pp. 384–392, 2022.
https://doi.org/10.1007/978-981-16-7735-9_43

deformation on the contact between the nozzle and the tank. Therefore, this stress can cause damage to the tank during operation of the nozzle. This paper will study stress on the nozzle by finite element method through FEPipe software which is researched and developed by Paulin Research Group [3].

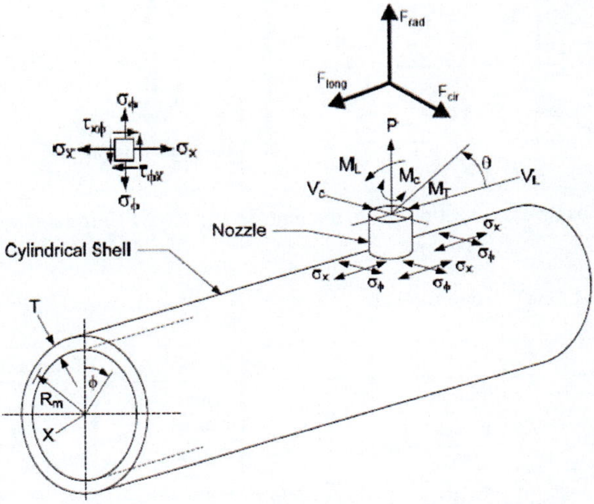

Fig. 1. Convention the force on nozzle [4]

2 Methodology

The stresses on the nozzle have been calculated by applying the finite element method through the FEPipe software. Stress is processed based on the rules specified in American Society of Mechanical Engineers (ASME) Boiler and pressure vessel code, section III Rules for Construction of Nuclear facility components and section VIII Division 2 Alternative rules for Construction of pressure vessels. The tank has an outside diameter of 54918 mm, wall thickness is 9 mm and some other parameters are given in Table 1. The diameter of nozzle is 610 mm, wall thickness is 14.27 mm and some other parameters are given in Table 2. Nozzle's properties [5]. Nozzle cross-section and FEPipe are illustrated in Fig. 2.

Uniform thermal expansion produces no stress in this geometry. Any thermal loads will come through operating forces and moments applied through the nozzle. The temperature and pressure parameters are given in Table 3. The loads on the nozzle are given in Table 4 and Fig. 3 shows the types of the loads on the nozzle. Ten load cases applied to analyse according to ASME B31 Code for Pressure Piping recommendation are given in Table 5.

3 Results and Discussions

As per ASME section VIII Division 2, the following local stresses need to be checked for clearing the nozzle: (i) Pm, Pb shall be limited to 1.0 Smh; Pl shall be limited to

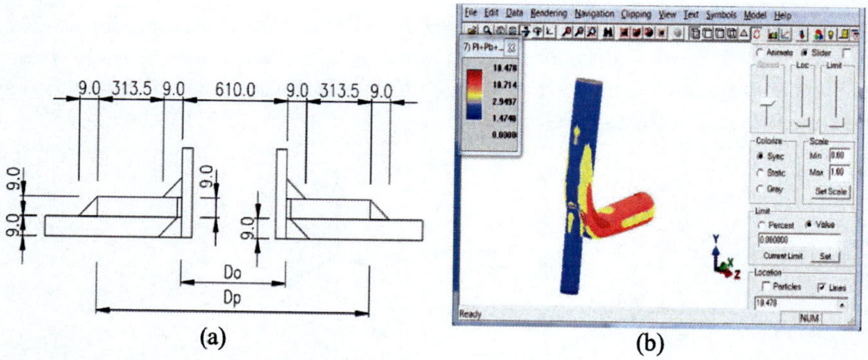

Fig. 2. (a) Nozzle cross-section (Note the units in mm); (b) FEPipe software illustration

Table 1. Tank's properties [5].

Material ID	Low Carbon Steel
Outside diameter	54918 mm
Thickness	9.0 mm
Cold allowable	137.0 MPa
Hot allowable	137.0 MPa
Ultimate tensile (Amb)	380.0 MPa
Yield strength (Amb)	205.0 MPa
Yield strength (Hot)	205.0 MPa
Elastic modulus (Amb)	199950.0 MPa
Poissons ratio	0.30
Expansion coefficient	0.1080E−04 mm/mm/deg

Table 2. Nozzle's properties [5].

Material ID	Low Carbon Steel
Outside Diameter	610 mm
Thickness	14.27 mm
Length	321 mm
RePad Width	322.5 mm
RePad Thickness	9.0 mm
Nozzle Tilt Angle	0.0°
Distance from Top	915 mm
Distance from Bottom	13085 mm
Cold Allowable	137 MPa
Hot Allowable	137 MPa
Ultimate Tensile (Amb)	380.0 MPa
Yield Strength (Amb)	205.0 MPa
Yield Strength (Hot)	205.0 MPa
Elastic Modulus (Amb)	199950.0 MPa
Poissons Ratio	0.300
Expansion Coefficient	0.1080E−04 mm/mm/deg

Table 3. Other parameters [5].

Design Operating Cycles	7000
Ambient Temperature	21.10°
Nozzle Inside Temperature	30.00°
Nozzle Outside Temperature	30.00°
Tank Inside Temperature	30.00°
Tank Outside Temperature	30.00°
Nozzle Pressure	0.001 MPa
Tank Pressure	0.001 Mpa

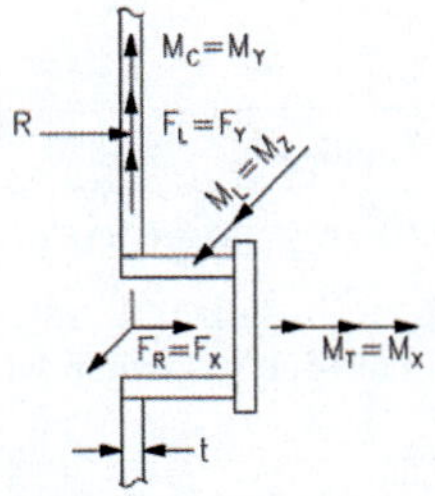

Fig. 3. Nozzle loads with values in Table 4

Table 4. Loads are given at the nozzle [5].

Load Case	F_X (N)	F_Y (N)	F_Z (N)	M_X (N.m)	M_Y (N.m)	M_Z (N.m)
WEIGHT	0.0	−2147.0	0.0	0.0	0.0	689.2
OPER	7370.0	−32990.0	−5871.0	8465.0	946.4	−15673.2
SUS	1999.0	−31397.0	24.0	−61.0	83.7	−17943.6

Table 5. Load cases are given at the nozzle [6].

Load cases	Note
1) WEIGHT ONLY (Wgt Only)	Weight ONLY case run to get the stress range between the installed and the operating states /-------- Loads in Case 1 Loads due to Weight
2) SUSTAINED (Wgt + Pr)	Sustained case run to satisfy local primary membrane and bending stress limits /-------- Loads in Case 2 Loads due to Weight Pressure Case 1
3) OPERATING	Case run to compute the operating stresses used in secondary, peak and range calculations as needed /-------- Loads in Case 3 Loads from (Operating)
4) RANGE (Fatigue Calc Performed)	Case run to get the RANGE of stresses /-------- Combinations in Range Case 4 Plus Stress Results from CASE 3 Minus Stress Results from CASE 1
5) OCCASIONAL	Occasional load case established per the requirements of the Code /-------- Loads in Case 5 Loads due to Weight Pressure Case 1 Loads from (Occasional)

(continued)

Table 5. *(continued)*

Load cases	Note
6) Program Generated -- Force Only	Case run to compute sif's and flexibilities /-------- Loads in Case 6 Loads from (Axial)
7) Program Generated -- Force Only	Case run to compute sif's and flexibilities. /-------- Loads in Case 7 Loads from (Inplane)
8) Program Generated -- Force Only	Case run to compute sif's and flexibilities. /-------- Loads in Case 8 Loads from (Outplane)
9) Program Generated -- Force Only	Case run to compute sif's and flexibilities. /-------- Loads in Case 9 Loads from (Torsion)
10) Program Generated -- Force Only	Case run to compute sif's and flexibilities. /-------- Loads in Case 10 Pressure Case 1

1.5 Smh and (ii) Pl + Pb + Q shall be limited 3.0 Sm(avg); Pl + Pb + Q + F shall be limited Sa(Sif). There are 18 stress results recommended by ASME section VIII Division 2 given under different working conditions which are shown in Table 6.

Figure 4a presents the model of the steel vertical storage tank and nozzle with the inlet parameters shown in Table 1, Table 2, Table 3 and Table 4 through FEPipe software.

Table 6. Stress case are given at the nozzle [7].

Stress cases	Stress cases
1) Pl < 1.5(k)Smh (SUS, Membrane) Case 2	10) Pl + Pb + Q + F < Sa (SIF, Outside) Case 7
2) Qb < 3(Smh) (SUS, Bending) Case 2	11) Pl + Pb + Q + F < Sa (SIF, Outside) Case 8, Case 9
3) S1 + S2 + S3 < 4S (SUS, S1 + S2 + S3) Case 2	12) Pl + Pb + Q + F < Sa (SIF, Outside) Case 10
4) Pl + Pb + Q < 3(Smavg) (OPE, Inside) Case 3	13) Pl + Pb + Q < 3(Smavg) (EXP, Inside) Case 4
5) Pl + Pb + Q < 3(Smavg) (OPE, Out) Case 3	14) Pl + Pb + Q < 3(Smavg) (EXP, Out) Case 4
6) Membrane < User (OPE, Membrane) Case 3	15) Pl + Pb + Q + F < Sa (EXP, Inside) Case 4
7) Bending < User (OPE, Bending) Case 3	16) Pl + Pb + Q + F < Sa (EXP, Outside) Case 4
8) Pl + Pb + Q + F < Sa (SIF, Outside) Case 5	17) Pl < 1.5(k)Smh (OCC, Membrane) Case 5
9) Pl + Pb + Q + F < Sa (SIF, Outside) Case 6	18) Qb < 3(Smh) (OCC, Bending) Case 5

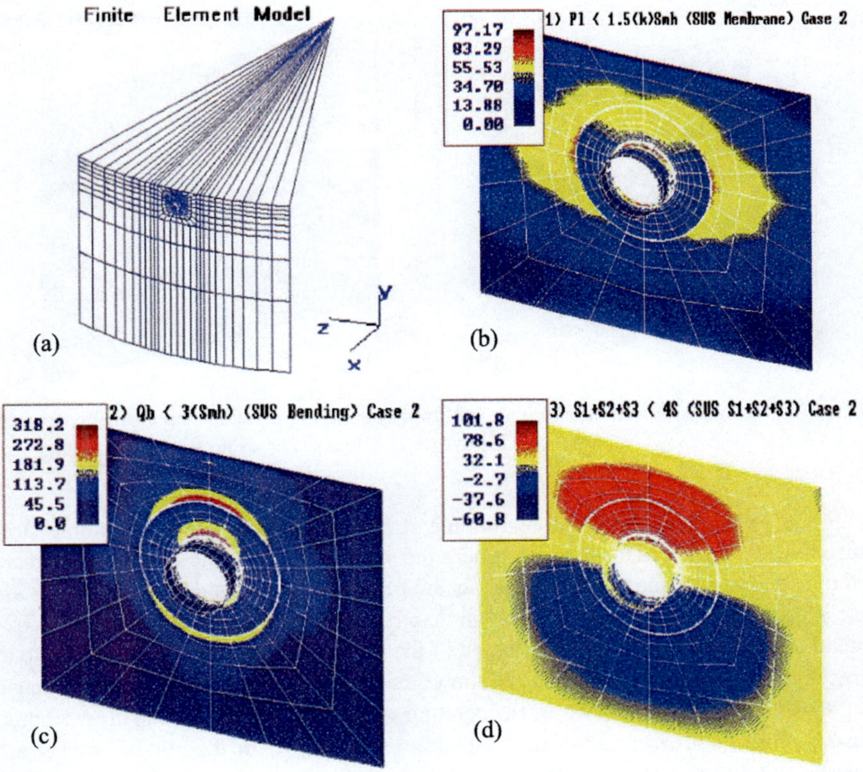

Fig. 4. (a) Finite element analysis model of the steel vertical storage tank and nozzle from FEPipe software. The calculated results for Case 2: (b) Local membrane stress due to sustained loads (Pl); (c) Secondary bending stress due to sustained loads (Qb); (d) Total stress due to sustained loads. Note the units in MPa.

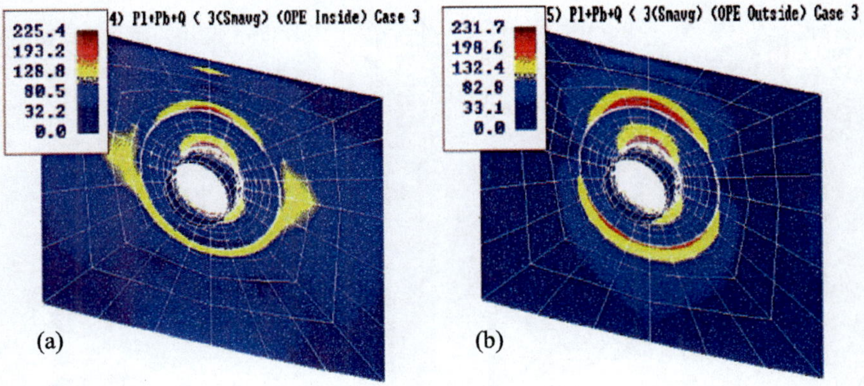

Fig. 5. (a) Inside stress and (b) Outside stress due to the membrane, bending stresses under operating loads and secondary stress caused by secondary loads for Case 3. Note the units in MPa.

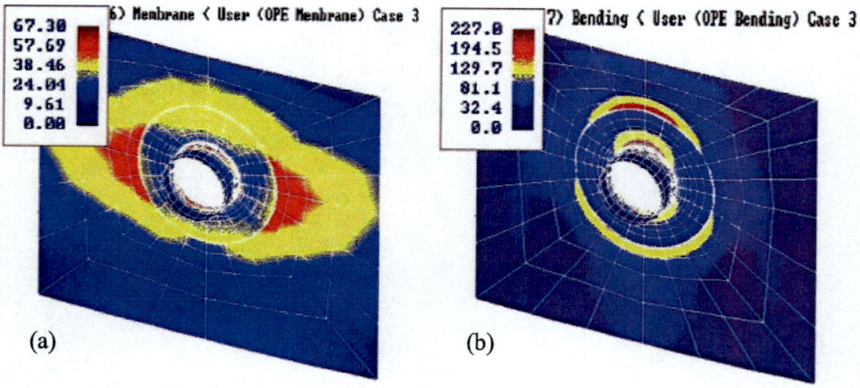

Fig. 6. The calculated results for Case 3: (a) Primary general membrane stress due to operating loads (Pm); (b) Primary bending stress due to operaing loads (Pb). Note the units in MPa.

Figure 4(b,c,d) presents the membrane stress, bending stress and combine stress caused by the sustained load case 2, where the bending stress is the greatest. The inside and outside stresses due to the membrane and bending stresses under the operating load case 3 are shown in Fig. 5 and it is shown that the outside stress is approximately 1.02% greater than inside stress due to inclusion of the reinforced pad thickness. Similar to the case 2 presented in Fig. 4(b,c,d), Fig. 6 shows the membrane and bending stresses caused by the operating load case 3, where the bending stress is much greater than the membrane stress. Figure 7 shows the inside and outside stresses due to the membrane and bending stresses under the operating load case 4, in which the outside stress is greater than the inside stress. It is found that the stress values in case 4 (Fig. 7) are much smaller than the stress values in case 3 (Fig. 5). Figure 8 presents the inside and outside stresses due to the membrane, bending and secondary stresses under the expansion loads and peak stress. It is shown that the outside stress is slightly higher than the inside stress.

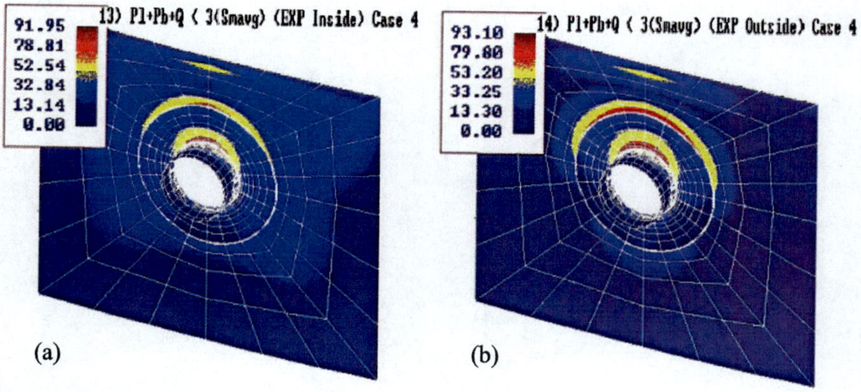

Fig. 7. (a) Inside stress and (b) Outside stress due to the membrane, bending and secondary stress under expansion loads for Case 4. Note the units in MPa.

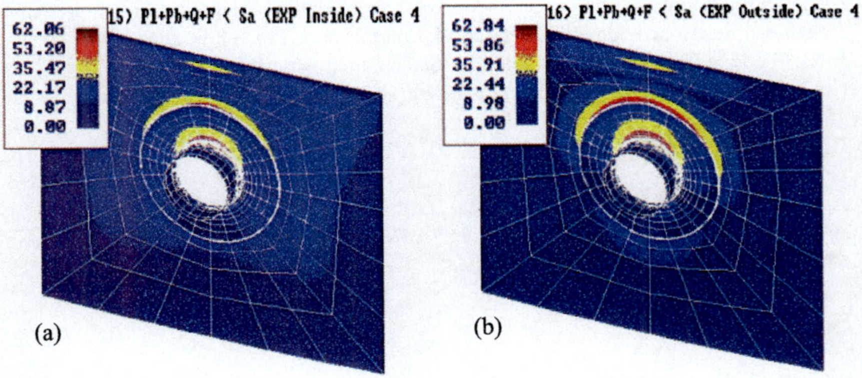

Fig. 8. (a) Inside stress and (b) Outside stress due to the membrane, bending, secondary stresses and peak stress for Case 4. Note the units in MPa.

It is found that stresses on the nozzle of the steel vertical storage tank vary for different load cases. In case 2 where there is self-weight, the secondary bending stress in both sustained and operating cases is more significant than primary local membrane stress. The bending stress is found to be the most important effect on the total stress compared to the membrane stress. In case 3, when the effect of secondary stress Q is taken into account, it is very clear that the stress on the nozzle is significantly increased and this stress is maximum under the operating case. For the fatigue calculation in case 4, it is shown that the stress on the outside and inside is not significantly different (see Fig. 8).

4 Conclusions

This paper studies stress on the nozzle of a steel vertical storage tank under different working conditions including the effect of different type of loads, based on the finite element method, using the FEPipe software. There are 10 types of loads are included in ASME B31 and 18 stress results are given as recommended in ASME section III and section VIII Division 2. From the results shown, under sustained and operating conditions, the bending stress has a significant effect on the total stress. There is a considerable difference between the inside and outside stresses of the nozzle if the thickness of reinforcement pad is taken into account.

Acknowledgements. The authors would like to express sincere thanks to the associates for their comments on this computational study. The data of this study are taken from the NRSP Complex Project.

References

1. Wichman, K.R., Hopper, A.G., Merson, J.L.: Welding research council N° 107: local stress in spherical and cylindrical shells due to external load, p. 78. Welding research Council, New York (2002)

2. Moore, S.E., Moffat, D.G.: A general theory for the construction of best-fit correlation equations for multi-dimensioned numerical data. Int. J. Press. Vessels Pip. **84**(4), 256–264 (2007)
3. PRG: "Fe/Pipe Documentation" Paulin Research Group, Houston, Tx (2007)
4. Chelu, A.: Calculation of local stress state for shells intersection. J. Eng. Stud. Res. **21**(1), 15–29 (2015)
5. PTSC: Calculation sheet for tank 101-TK-001A/B. NSRP Complex Project, Nghi Son, Thanh Hoa, Vietnam (2014)
6. American Society of Mechanical Engineers (ASME) B31 Code for Pressure Piping (2012)
7. American Society of Mechanical Engineers (ASME) section VIII Division 2 Alternative rules for Construction of pressure Vessel (2013)

Study on Degree of Influence of P-Delta Effects on Strength of Jack-Up Leg Structures in Vietnamese Conditions

Pham Hien Hau, Vu Dan Chinh(✉), and Dang Dinh Tuan

National University of Civil Engineering, Hanoi, Vietnam
{hauph,chinhvd}@nuce.edu.vn

Abstract. Jack-up is a type of offshore platforms using for oil & gas exploitation in Vietnam. In operating conditions, jack-up hull structures with large weight are lifted to high elevation and supported by slender leg structures. When subjected to environmental loading, they have large lateral displacements. In these cases, P-Delta effects are usually considered. In current standards, the effects are simulated by approximate formulas. In the previous paper, we studied on evaluating P-Delta effects on the durability of jack-up structures in static analyses according to large deformation theory and approximate methods of ABS and DnV standards. The research clarified the different influences of the models. However, in operating conditions, the hull weights and hull levels of jack-up platforms may be changed. It leads to the variability of P-Delta effects. In this paper, we propose to assess the degree of influence of P-Delta effects to strength of jack-up leg structures in Vietnamese conditions. The problem is studied by analyses of a simple model and FEM models. For the simple model, we will estimate the preliminary changing of P-Delta effects by an equivalent structural model. Then we establish global models in some operating status using SACS program to analyze jack-up leg structures in relation with different methods for consideration of P-Delta effects. The strength analysis results will be presented to evaluate real influences of the effects on jack-up structures. The research results will give useful recommendations for consultancy companies to analyze and assess jack-up platforms operating in Viet Nam offshore areas.

Keywords: Influence of P-Delta effects · Strength · Jack-up legs structures

1 Introduction

As introductions of a previous paper of authors [1], due to lateral displacements induced by environmental loading, jack-up platform weights cause increasing moments on leg structures according to P-Delta effects (Fig. 1). The effects have been researched in some studies in the world [2–4]... In the previous paper, we considered to analyze jack-up leg structures by 4 models: (1) Linear model without P-Delta effects; (2) Non-linear model based on large deformation theory; (3) Model takes into account P-Delta effects according to ABS standard [5]; (4): Model takes into account P-Delta effect according

D. V. K. Huynh et al. (Eds.): VSOE 2021, LNCE 208, pp. 393–401, 2022.
https://doi.org/10.1007/978-981-16-7735-9_44

to DnV standard [6]. The results of the paper show that there are differences between the 4 models and P-Delta has a certain effect on strength of jack-up leg structures. However, jack-up platforms operating in Vietnamese seas have own characteristics, especially in environmental data and exploiting conditions. So the degree of the effects on the jack-up leg structures is needed to estimate. It is the purpose of this paper.

Firstly, the degree of P-Delta effects is evaluated by a simple model of jack-up structures (Fig. 1). Herein, leg structures are replaced by an equivalent frame with an equivalent stiffness. Then, the effects are surveyed through global finite element models of Tam Dao 05 Jack-up structure which is operating in Vietnamese conditions. Whereas, *hull weight and hull level are various terms* in relation with different operating status of the platform. Based on the research results, authors will give comments on the efficiency of P-Delta effects when assess jack-up structures in actual projects.

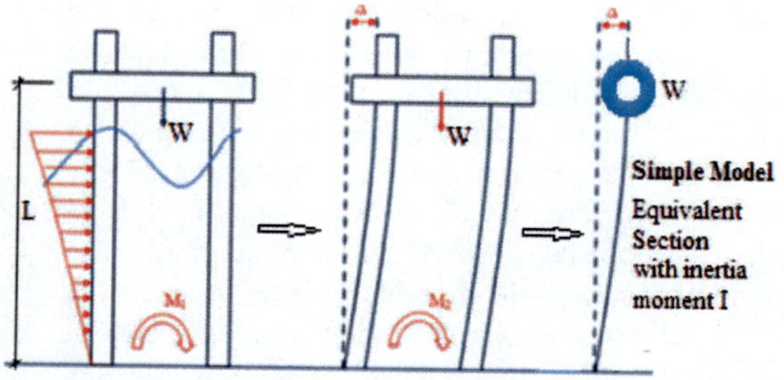

Fig. 1. P-"Δ" effect of jack-up platforms in operating status and an equivalent analysis model (Refer to [3])

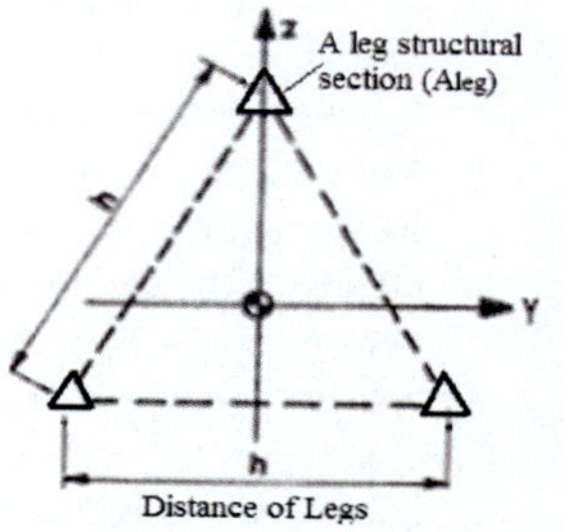

Fig. 2. Leg structural plan of a type of jack-up platform

2 Estimation of P-Delta Effects on Jack-Up Structures

2.1 Estimation of P-Delta Effects on Jack-Up Structures Using Simple Models

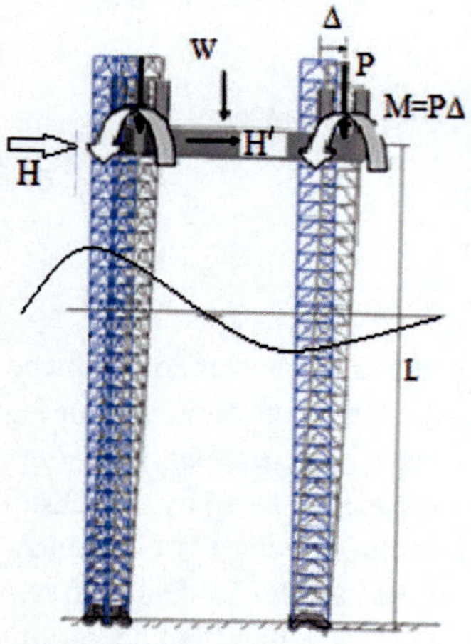

Fig. 3. P-Delta effects by horizontal loads (Refer to [6])

Considering a Jack-up structure bearing a hull load W and total lateral load H. Hull level from seabed is L. The lateral displacement of the structure is Δ. P-Delta effects can be approximated by an increased lateral load H' [6] (Fig. 3). The H' can be approximated by formula (1) for simple equivalent models.

$$H' = \frac{W\Delta}{L} \approx \frac{WHL^2}{3EI} \tag{1}$$

So the total lateral load is:

$$H_D = H + H' = H + \frac{W}{L}\frac{HL^3}{3EI} = H\left(1 + \frac{WL^2}{3EI}\right) \quad (2)$$

We consider to 2 cases as below:

Case 1: Hull weight is increased a% from W to W', so *HD* is changed to *H'D*.

$$\frac{H_D'}{H_D} = \frac{\left(1 + \frac{(1+a\%)WL^2}{3EI}\right)}{1 + \frac{WL^2}{3EI}} = 1 + a\%\frac{WL^2}{3EI + WL^2} \quad (3)$$

For a jack-up structure with 3 legs (Fig. 2), to satisfy strength requirements, the minimum area of a leg section can be estimated by the formula:

$$A_{\min(1leg)} \approx \frac{W}{3}\frac{1}{0.6F_y} \approx \frac{W}{1.8F_y} \quad (4)$$

According to [6], the inertial moment of equivalent section of 3 legs can be approximated by the formula $I_y = I_z \approx 0.5A_{leg}h^2$ with h is distance between legs (Fig. 2).

$$\text{So in this case :} \quad I_{y(\min)} = I_{z(\min)} \approx 0.5A_{\min(1leg)}h^2 \approx \frac{W}{3.6F_y}h^2 \quad (5)$$

$$\frac{H_D'}{H_D} = 1 + a\%\frac{WL^2}{3E\frac{Wh^2}{3.6F_y} + WL^2} = 1 + a\%\frac{L^2}{\frac{Eh^2}{1.2F_y} + L^2} \quad (6)$$

Normally, yield strength Fy of jack-up legs < 1000 MPa, and Young modulus E = 210000 MPa, so the maximum ratio H_D'/H_D can be evaluated by:

$$\frac{H_D'}{H_D} \approx 1 + a\%\frac{L^2}{175h^2 + L^2} \quad (7)$$

Case 2: Hull level is increased b% from L to L', so *HD* is changed to *H"D*

$$\frac{H_D''}{H_D} = \frac{\left(1 + \frac{(1+b\%)^2WL^2}{3EI}\right)}{1 + \frac{WL^2}{3EI}} = 1 + (2b\% + b\%^2)\frac{WL^2}{3EI + WL^2} \quad (8)$$

To do as similar as above with notice that b% is small, so we can obtain:

$$\frac{H_D''}{H_D} = 1 + (2b\% + b\%^2)\frac{L^2}{\frac{Eh^2}{1.2F_y} + L^2} \approx 1 + 2b\%\frac{L^2}{175h^2 + L^2} \quad (9)$$

For jack-up platforms are operating in Viet Nam, the ratio h/L is in the range of 0.3 to 0.4. For example: Tam Dao 05 platform min. of h/L ≈ 0.34, Tam Dao 03 min. of h/L ≈ 0.3, Tam Dao 02 min. of h/L ≈ 0.31, PVD III min. of h/L ≈ 0.3, PVD II min. of h/L ≈ 0.36…. So we can estimate the maximum $H_D'/H_D \approx 1 + 0.034$a% to $1 + 0.06$a% to for case 1st case, and $1 + 0.068$b% to $1 + 0.12$b% for 2nd case. According to analysis of the simple model, it can be seen that the increases of P-Delta effects are so small in comparison with the increases of jack-up hull weights or hull levels.

2.2 Global Structural Analysis Taking into Account P-Delta Effects

In reference of [1], authors presented 3 methods to analyze jack-up structures taking into account P-Delta effects. The first method is analysis of large deformed structures according to Finite Element Method. The relationship between axial strain and axial displacement follows Green formula. For more details refer to [7]. The analysis procedure according to FEM is summarized in reference of [1]. In the second method, according to ABS standard [5] P-Delta effects are modeled by springs in directions corresponding to directions of the considered waves with negative stiffness. In the third method according to DnV [6], P-Delta effects may be represented by a horizontal force at hull elevation, determined by formula (1) (Fig. 3).

2.3 Strength Conditions of Jack-Up Legs Based on AISC Standards

Performing analyses of global structural models in relation with P-Delta simulation methods as above, we obtain internal forces of the leg structures. The forces used to check strength conditions of Jack-up leg elements based on AISC [8] using UC values (Unity check). The checking formulas are summarized in reference of [1]. Details on how to determine components of above formulas are presented in reference of [8].

3 A Vietnamese Application

3.1 Input Data

Structures Data: Main structures of Tam Dao 05 jack-up platform (Fig. 4) include: spudcans, leg structures, hull structure, guide structures. Input data are summarized in Table 1 [9–11].

Environmental Data: Thien Ung field wave data [12] are used in the study. Wave data of SE (180°) and NE (270°) directions in relation with operating conditions will be chosen for the calculations. Typical wave and current data are given in Table 2.

3.2 Analysis Results

TAM DAO 05 jack-up structure is modeled and analyzed by SACS 5.8 software with 4 models: (1) Linear model without the P-Delta effects; (2) Non-linear model based on large deformation theory; (3) Model takes into account P-Delta effects according to ABS standards; (4): Model takes into account P-Delta effects according to DnV standard. In the study, authors consider to **4 cases of Hull weights: 8500T, 9000T, 9500T and 10000T and 3 levels of Hull: 139 m, 144 m and 149 m from mudline**. For each case, strength conditions (UC) of 5 typical elements of jack-up legs in 4 models as above are collected and summarized in Tables 3, 4, 5, 6, 7 and 8 as below.

These results are presented in charts as below (Fig. 5):

Table 1. Main parameters of Tam Dao 05 Jack-up structure

Main parameters of platform	Value
Water depth (m)	122
Leg length (m) included spudcan	167
Style of legs	Δ
Distance of Chords of a Leg, (m)	13
Legs distance (vertical x horizontal), (m)	47.6 × 45.7
Live loading function in operation conditions (T)	3766
Live loading function in extreme storm conditions (T)	2995
Hull loads (T) using in the study	**8500, 9000, 9500 and 10000**
Spudcan load (T)	1082

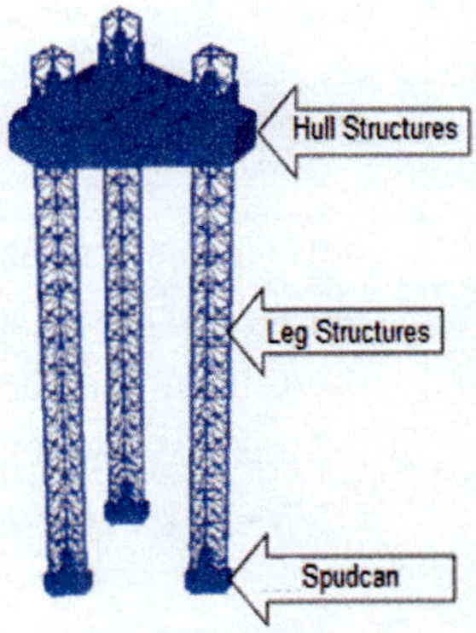

Fig. 4. Tam Dao 05 Jack-up structural configuration [1]

Table 2. Wave data for the 2 most dangerous directions

Return period	Direction 180°		Direction 270°	
	Operating	Extreme storm	Operating	Extreme storm
Wave height, Hmax (m)	6.10	9.40	10.60	15.5
Wave period, T (s)	6.20	7.40	8.90	12.40
Surface current velocity, (m/s)	0.71	0.83	1.06	2.17
Current velocity at 50 m of depth	0.36	0.47	0.76	1.53

Table 3. Strength checking results of original model

No	Members	Load conditions	UC			
			Model (1)	Model (2)	Model (3)	Model (4)
1	0BUF-0BUR	OP02	0.292	0.295	0.30	0.431
2	0BUU-0BVG	OP02	0.294	0.297	0.302	0.435
3	0BVG-0BUF	OP02	0.296	0.30	0.307	0.432
4	0BVJ-0BVV	OP02	0.292	0.294	0.303	0.452
5	0BVV-0BUU	OP02	0.293	0.295	0.303	0.433

Table 4. Strength checking results of model of hull weight 9000T and level 139 m

No	Members	Load conditions	UC			
			Model (1)	Model (2)	Model (3)	Model (4)
1	0BUF-0BUR	OP02	0.297	0.299	0.309	0.436
2	0BUU-0BVG	OP02	0.299	0.302	0.312	0.441
3	0BVG-0BUF	OP02	0.302	0.305	0.315	0.438
4	0BVJ-0BVV	OP02	0.297	0.299	0.31	0.458
5	0BVV-0BUU	OP02	0.296	0.298	0.308	0.44

Table 5. Strength checking results of model of hull weight 9500T and hull level 139 m

No	Members	Load conditions	UC			
			Model (1)	Model (2)	Model (3)	Model (4)
1	0BUF-0BUR	OP02	0.318	0.32	0.32	0.457
2	0BUU-0BVG	OP02	0.321	0.324	0.323	0.459
3	0BVG-0BUF	OP02	0.324	0.327	0.326	0.457
4	0BVJ-0BVV	OP02	0.319	0.321	0.321	0.477
5	0BVV-0BUU	OP02	0.317	0.319	0.319	0.458

Table 6. Strength checking results of model of hull weight 10000T and hull level 139 m

No	Members	Load conditions	UC			
			Model (1)	Model (2)	Model (3)	Model (4)
1	0BUF-0BUR	OP02	0.329	0.331	0.331	0.475
2	0BUU-0BVG	OP02	0.332	0.335	0.335	0.477
3	0BVG-0BUF	OP02	0.335	0.338	0.338	0.475
4	0BVJ-0BVV	OP02	0.33	0.332	0.332	0.497
5	0BVV-0BUU	OP02	0.328	0.33	0.33	0.477

Table 7. Strength checking results of model of hull weight 8500T and hull level 144 m

No	Members	Load conditions	UC			
			Model (1)	Model (2)	Model (3)	Model (4)
1	0BUF-0BUR	OP02	0.294	0.296	0.295	0.418
2	0BUU-0BVG	OP02	0.296	0.297	0.297	0.419
3	0BVG-0BUF	OP02	0.297	0.30	0.299	0.419
4	0BVJ-0BVV	OP02	0.295	0.296	0.297	0.436
5	0BVV-0BUU	OP02	0.293	0.295	0.295	0.42

Table 8. Strength checking results of model of hull weight 8500T and hull level 149 m

No	Members	Load conditions	UC			
			Model (1)	Model (2)	Model (3)	Model (4)
1	0BUF-0BUR	OP02	0.295	0.296	0.297	0.419
2	0BUU-0BVG	OP02	0.296	0.299	0.299	0.42
3	0BVG-0BUF	OP02	0.298	0.301	0.301	0.42
4	0BVJ-0BVV	OP02	0.296	0.298	0.299	0.438
5	0BVV-0BUU	OP02	0.294	0.296	0.296	0.42

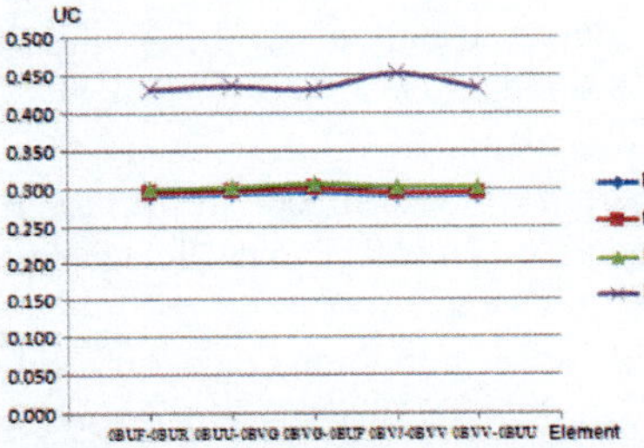

a) Strength Checking Results of Original Model

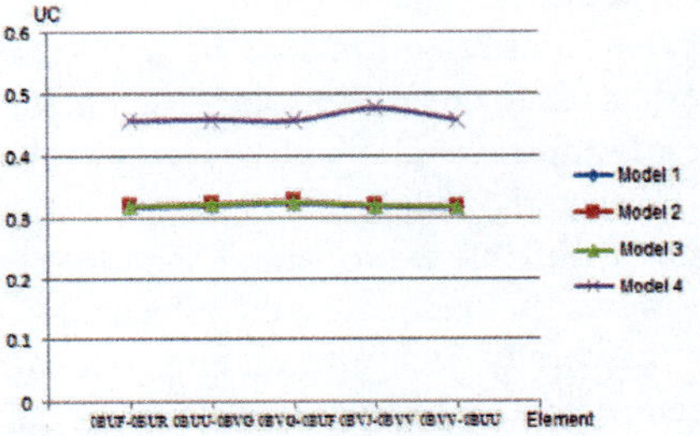

b) Strength Checking Results of Model of hull weight 9000T and hull level 139m

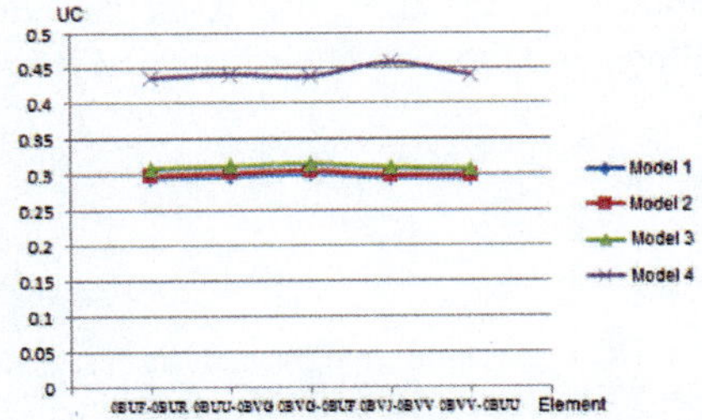

c) Strength Checking Results of Model of hull weight 9500T and hull level 139m

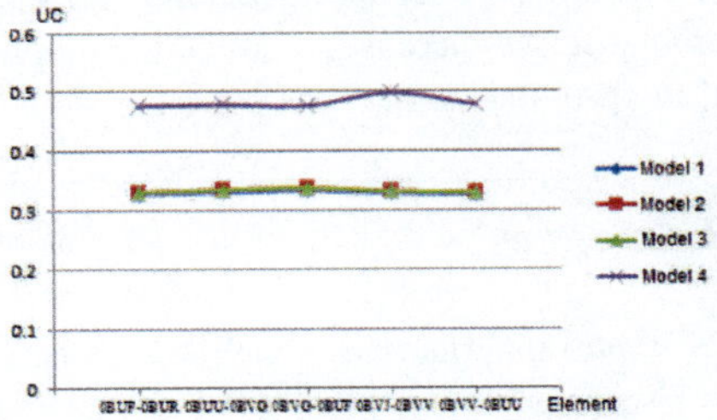

d) Strength Checking Results of Model of hull weight 10000T and hull level 139m

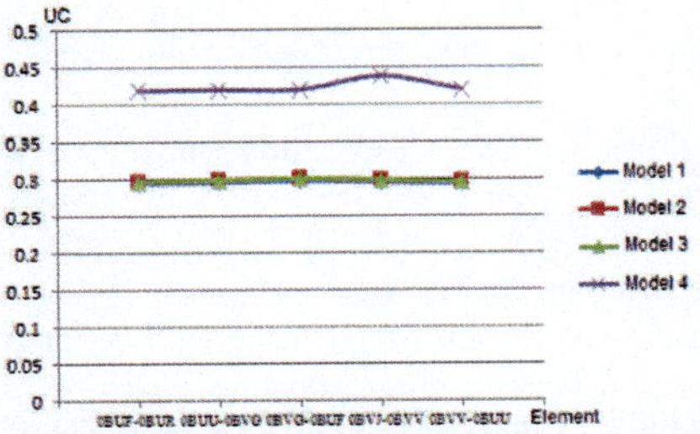

e)Strength Checking Results of Model of hull weight 8500T and hull level 144m

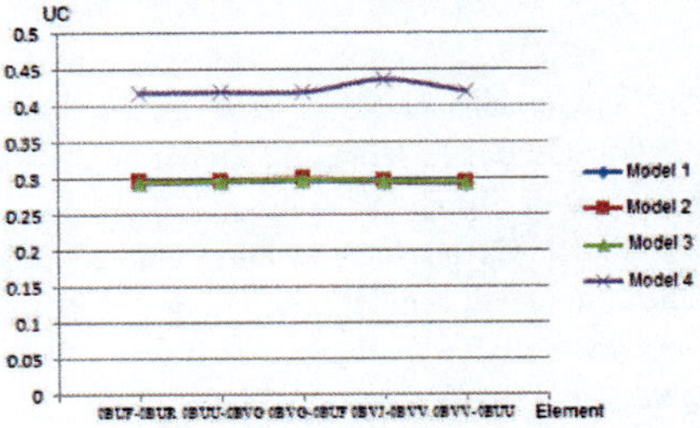

f) Strength Checking Results of Model of hull weight 8500T and hull level 149m

Fig. 5. Charts of strength checking results of Tam Dao 05 Jack-up leg structures

4 Discussion

As analyses of the simple models of jack-up structures, P-Delta effects on the structures can be simply estimated by increased horizontal forces $H' \approx H\frac{WL^2}{3EI}$. The maximum of $H' \approx \frac{HL^2}{175h^2}$. If h/L ≈ 0.3 to 0.4 then H' ≈ 0.035H to 0.063H. P-Delta effects of variable L are increased faster than the ones of variable W. However, the increased ratios are quite small (3.4% of increased hull weight ratio, 12% of increased hull level ratio).

According to analyses of FE models No 2 and No 3, P-Delta effects are quite small, too. The maximum increased percentage of UC is only about 4%. Whereas, the results of Models No 3 have significant differences, UC values are about 1.5 time the ones of Model 1. It means that DnV standard may use strict criterions in analyses and designs. When the hull weight changes about 12.5%, according to DnV the maximum effect of P-Delta to strength of Jack-up legs structures is increased only about 2.5% while the results based on large displacement models or ABS standard are almost unchanged. The results are accordant with analyses of the simple models.

When the hull level is increased, although loading is not changed but the structure is more flexible, so structural displacements are increased and UC values are decreased. However, in these cases, the effects of P-Delta are quite small, too.

Generally, jack-up structures operating in Vietnamese conditions have high safety factors. Strength checking values are quite small, so the effects of P-Delta on strength of jack-up legs are not so much. However, for existing jack-up platforms with reduced strength, P-Delta may be considered in case of assessment of life extension.

Moreover, P-Delta effects may have an influence on vibration of Jack-up platforms and cause the increases of dynamic amplifications. Authors will publish studied results of the problem in the next papers.

References

1. Tuan, D.D., Hau, P.H., Chinh, V.D., Nien, H.D.: P-Delta effects on strength analysis of Jack-up legs structures. J. Sci. Technol. Civ. Eng. NUCE 2019 **13**(1V), 46–5 (2019)
2. HSE: Interpretation of full scale monitoring data from a jack-up rig. Offshore Technology Report 2001/035. MSL Engineering Ltd., United Kingdom (2002)
3. Koole, T.: Morden jack-ups and their dynamic behaviour. Master of Science Thesis, TU Delf (2015)
4. Tamhane, V., Singh, S.K.: Behaviour of offshore structures and overturning stability analysis of jack up rigs. IJIRSET **7**(5) (2018)
5. ABS: Guidance Notes on Dynamic Analysis Procedure for Self-elevation Units. America Bureau of Shipping, Houston (2014)
6. DNV-RP-C104. Self-elevating Units. Det Norske Veritas AS (2012)
7. Soreide, H.T., Amdahl, J., Eberg, E., Holmas, T., Hellan, Φ.: USFOS: A Computer Program for Progressive Collapse Analysis of Steel Offshore Structures - Theory Manual. SINTEF, Norway (1993)
8. AISC. Steel construction Manual, Thirteenth Edition. American Institute of Steel Construction, Illinois (2005)
9. Friede and Goldman LTD. JU-2000E Construction Specifications (2012)

10. Cuong, D.Q., Chinh, V.D., Long, T.T., Dung, H.T., Manh, M.V., Hang, P.T.: Study on analysis and build of Basic design procedure for 400ft Jack-up platform. National Scientific Project No SPQG 02b.01-01 (2015–2017)
11. Petro Vietnam Shipyard. Tam Dao 05 Jack-up rig - Weight Control Report (2015)
12. Vietsovpetro. Metocean criteria and statistics 8 to 9oN, 107 to 109oE, Offshore South Vietnam. V1.0 for the Thien Ung Project Feed Design (2010)

Towards Implementation of Artificial Intelligence in Predicting Pile Driving Blow Counts

Z. L. Delimi[1](✉), A. Amavasai[1], A. Cristobal-Huerta[2], and R. Clavaud[1]

[1] Fugro, Abu Dhabi, UAE
z.delimi@fugro.com
[2] Fugro, Leidschendam, The Netherlands

Abstract. The objective of this paper is to present the potential use of Artificial Intelligence in offshore energy industry to predict pile driving blow counts at new locations based on previous installation records. To this end, a machine learning (ML) model was developed using a public dataset comprising geotechnical and pile driving records, such as cone penetration test data (CPTu), pile geometry and hammer ENTHRU energy versus depth at several locations in the North Sea. The ML model combines the principles of stacking and blending state of the art ML algorithms for better and more efficient predictions of the blow counts. The model's performance was assessed by the root mean squared error metric, in order to measure the difference between the recorded and the predicted blow counts values at each location. Results show that the adopted model allows a reasonably accurate prediction of the blow counts and faster than conventional method. Notwithstanding, the use of AI for pile drivability has some limitations and geotechnical specialists shall supervise the process and ensure the quality control of the results.

Keywords: Artificial intelligence · Machine learning · Pile installation · Blow counts prediction

1 Introduction

The modern business world is becoming increasingly technology/numerical driven. Many areas, such as healthcare/finances, have been quick to realize the benefits of Artificial Intelligence (AI)/Machine Learning (ML) in their respective industry. Nevertheless, AI and ML in oil & gas and renewable energy sectors have been slower to establish itself. Previous works in the literature have investigated the feasibility of using ML techniques to predict pile bearing capacity [1]; however, to the best of our knowledge, this paper shows the first investigation on using ML techniques to predict blow counts in pile installation problems. Fugro is investigating how to resolve offshore engineering problems using Artificial Intelligence instead of the conventional engineering workflow. Recently, a machine learning based methodology has been established by a multidisciplinary team combining geotechnical and data science specialists. The model consists of training an algorithm based on an available dataset of pile driving records in the North Sea in order to predict the blow counts for future installations.

D. V. K. Huynh et al. (Eds.): VSOE 2021, LNCE 208, pp. 402–409, 2022.
https://doi.org/10.1007/978-981-16-7735-9_45

2 Available Data

The dataset was provided as part of an open prediction event hosted in Kaggle platform [2]. The problem involved the prediction of blow counts versus depth for jacket piles installed in the North Sea. The dataset comprised 114 piles with the following data:

- CPT profiles: cone tip resistance, sleeve friction and pore pressure measurements.
- Pile outer diameter, wall thickness and tip wall thickness.
- Normalised hammer energy.
- The interdistance between each pile location pair.
- Blow counts (available for 94 locations only, remaining 20 locations to be predicted using ML).

In order to keep a certain level of data anonymous, the exact name of the field was not provided. In addition, the type of the hammer and absolute value of energy were not provided and only normalised hammer energy was made available.

3 Conventional Versus Data Science Approach

The conventional approach to predict the blow counts involves to first predict the Soil Resistance to Driving (SRD) versus depth from geotechnical data. Different methods exist such as Alm and Hamre (2001) [3] widely applied in the North Field soil conditions. The bearing graphs are then calculated from 1D wave equation analysis for selected depths using programs such as GRLWEAP [4]. The bearing graphs are then interpolated with the soil resistance to derive the blow counts.

The conventional approach is a tedious process especially when assessing a large number of piles and is certainly not time efficient. In addition, the conventional approach suffers from several uncertainties such as selecting the relevant SRD method, the continuous update of the methods depending on the new available driving records, and the dynamic soil resistance which is mainly dependent on soil damping. For those reasons, it is suggested to investigate the potential use of a data science approach to directly predict the blow count from measurable parameters. Figure 1 presents an organigram of a conventional approach versus a data science approach. The process on the left indicates the steps required to predict the blow counts using a ML model while the process on the right shows the steps required to predict the blow counts using the conventional approach.

Figure 2 provides an overview of the data science methodology adopted in this study. The workflow involved both geotechnical and data science knowledge. In Phase 1, the dataset was examined to investigate the correlations between different features comprising of geotechnical, pile and hammer information. At this stage, the data was tested with simple machine learning models to compare its performance with different models considering default parameters.

In Phase 2, the identified limitations from Phase 1 were rectified and/or refined. Furthermore, it was important to identify the dominant soil condition in this phase in order to generate the most relevant engineering features for the encountered soil

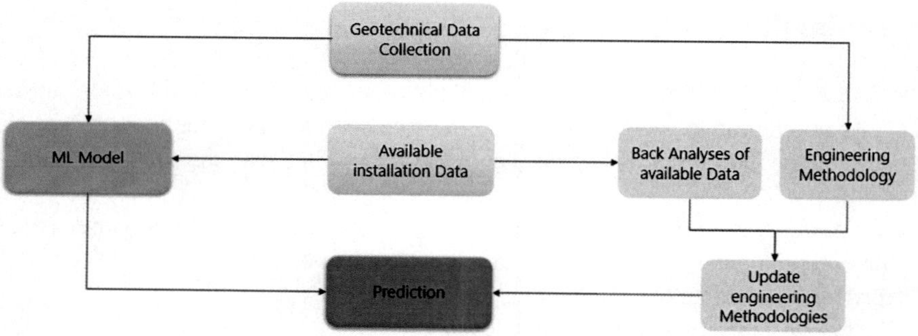

Fig. 1. Illustration of conventional vs data science approach.

conditions. Additionally, feature importance studies with previously calibrated simple ML models was conducted to identify the most significant features.

Finally, in Phase 3, the interdistance information was incorporated along with data transformation and an advanced model involving stacking and blending of several different machine learning algorithms was implemented.

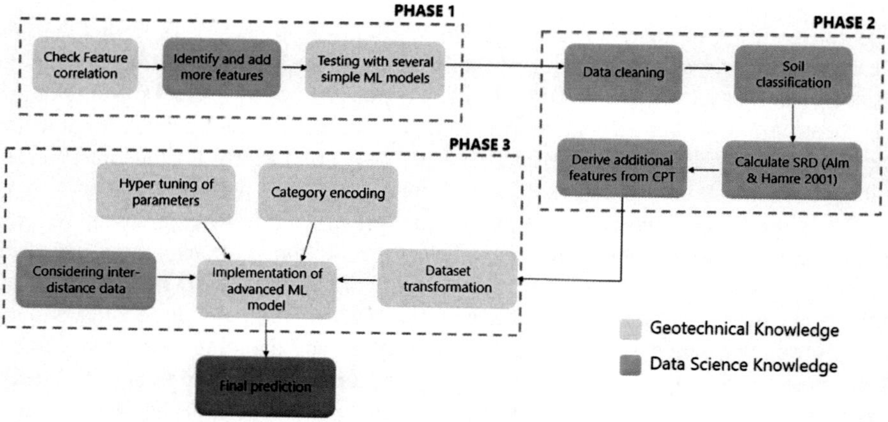

Fig. 2. Data science and geotechnical workflow.

4 Data Cleaning and Feature Engineering

The pre-processed dataset was not necessarily suffering from incorrect or incomplete data, but further cleaning was performed based on geotechnical and pile installation knowledge. This was due to some limitations related to the site investigation performed.

Features engineering consists of building additional features using domain knowledge of a particular field, in this case geotechnical engineering and pile installation, or other a priori knowledge. This was done by extracting additional data that can be directly related to blowcount, such as Soil Resistance to Driving (SRD).

The interdistance between locations was used along with CPT data to understand the lateral variability in the site. It was found that the model can be trained to predict blow counts based on only a subset of the dataset that have the most similar CPTs and energy profiles with the locations to be predicted. The features that produce the best RMSE score were found to be reasonable from a geotechnical point of view and are normally used in the conventional approach such as SRD, hammer ENTHRU energy, relative density, shaft friction and soil type index.

5 Data Transformation

Skewness or distortion of the dataset can limit the predictive efficiency and it is necessary to transform the skewed data to an approximately Gaussian distribution before progressing towards advanced data modelling implementation and parameter tuning procedures. In this study, a power transformation was used to transform the distribution of those features which did not follow a normal distribution. Figure 3 presents an example of the SRD feature distribution before and after the transformation.

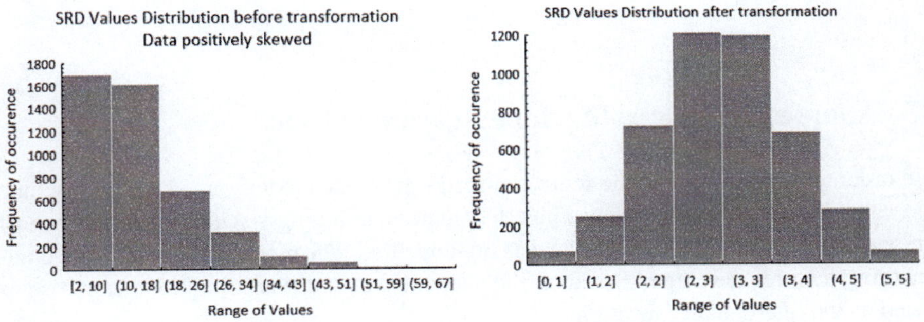

Fig. 3. Example of Data Transformation for the SRD feature.

6 Prediction Using Various Basic ML Models

A supervised regression machine learning model was used in order to relate the data set features (soil data, hammer data and pile data) to blow counts without including engineering features. The objective was to check the performance of different basic models in predicting the blowcount. Each model's performance was evaluated using the Root Mean Squared Error (RMSE) [5] metric and coefficient of determination (R^2) score [5].

In the initial stage, the dataset was split randomly into training (87 locations) and testing data (7 locations). The selected split has been optimized in subsequent phases. Figure 4 presents the performance of the different models considered. Results show that the KNN and GBR model along with its bagged version performs better than the other models while the tree models are on the contrary and suffer from overfit [6].

At this stage, there are too many factors affecting the result such as model parameters, solver, model formulation, data quality etc. and is cumbersome to quantify each of these degrees of freedom.

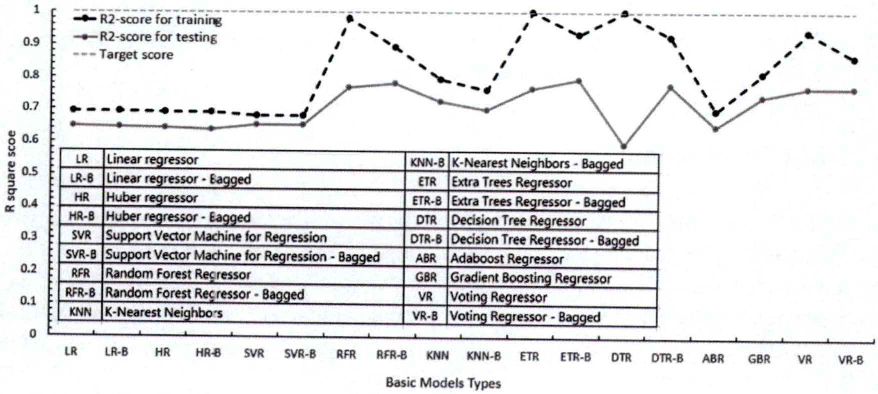

Fig. 4. Example of performance of various standard models

7 Adopted Advanced Machine Learning Model

In order to further improve the score obtained from basic models, an advanced machine learning approach based on ensemble technique combining stacking and blending of models with hyperparameter tuning was implemented. The model incorporates the engineering features developed based on the geotechnical and pile installation expertise as well as the interdistance locations.

Figure 5 shows the architecture of the models which involves two levels, the base models in the first level and Cat boost which acts as a meta learner in the second level. The individual models were blended together with specific weights to give the final prediction. This multiple layering reduces variance and bias in the models, thereby, improving its predictive efficiency. The model developed is specific to a particular region and needs further validation on other soil types and pile dimensions.

8 Results, Discussion and Limitations

The final adopted model allowed for a reasonable prediction of the blow counts compared to what is usually achieved when using a conventional approach. The model has been evaluated as part of the Kaggle competition using the RMSE score based on predicting 20 unseen locations (validation dataset) and the model yielded the best prediction [2]. Figure 6 shows an example of blow count prediction using 4 different models. Models are presented from the simplest on the left to the more complex on the right. The predictions can be obtained with a limited process of the input data which makes predictions of large number of locations time efficient.

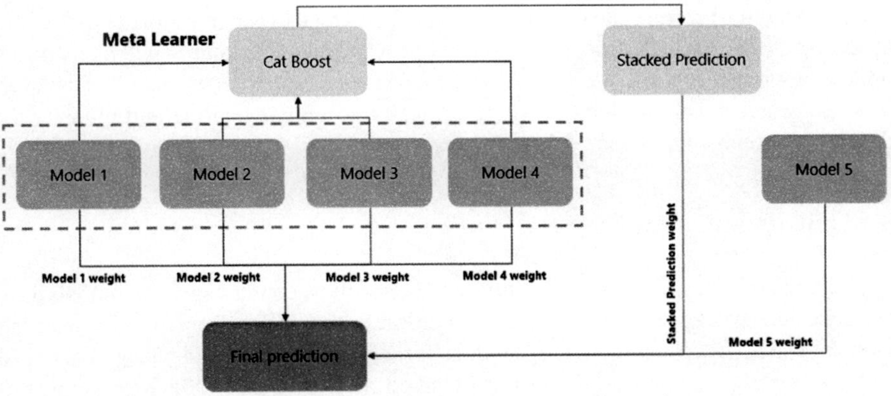

Fig. 5. Final machine learning model used for generating the predictions.

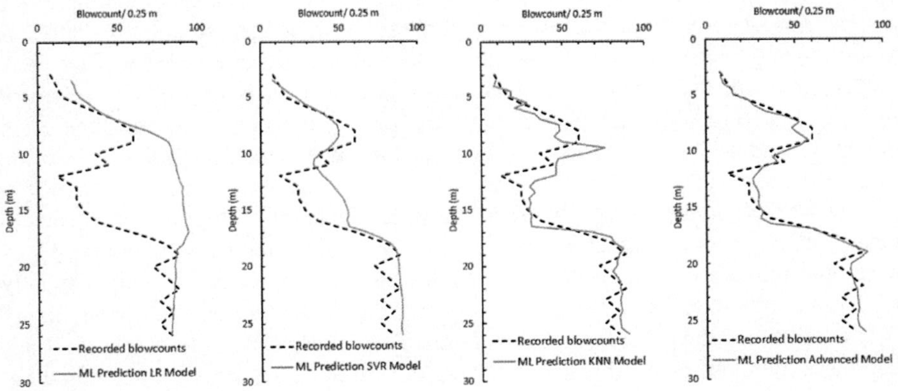

Fig. 6. Blow counts prediction using different ML models at 1 of testing locations

Although the performance of the final model (Advanced model presented in Fig. 5) is considered to be satisfactory in terms of accuracy and efficiency, attention should be made to the applicability and limitations of the model. Geotechnical specialists need to define the range of applicability which is mainly dependent on the dataset on which the model has been trained. One of the major limitations of using ML models to predict the blow counts is the availability of dataset that covers the expected soil conditions, pile dimensions and the expected hammers sizes to be used.

ML models are complex and powerful calculation tools that are sensible to numerous variables. If such tools are developed for engineering purpose, it is of utmost importance that users clearly understand the model and its limitations, in analogy to powerful geotechnical calculation tools such as finite element, small changes with inputs can lead to completely different and erroneous outcome. Therefore, with such numerical application likely to develop in the near future to solve geotechnical engineering problems, the importance of digital data storage becomes even more critical in order to efficiently use available data to train ML models.

Construction projects and geotechnical challenges tend to become more complex and Operator/Owner's schedule becoming shorter with time. ML techniques will likely be the optimal approach to resolve such growing challenges in an efficient manner. However, it is the geotechnical specialist's role to critically accompany, guide and monitor the development and application of these new practices.

9 Conclusions and Further Work

The accuracy of the predictions of the pile installation blow counts using ML models have been checked for various combination of hypotheses, assumptions, and machine learning models. The first step was to investigate the effect of data cleaning using the geotechnical and pile installation knowledge. Results obtained show a considerable improvement of predictions with a cleaned data set using geotechnical and pile installation expertise. This demonstrate the importance of combining geotechnical expertise with data science in developing ML models for geotechnical applications. After data cleaning, preliminary assessments were performed considering different basic models to allow for a better understanding of the data and the problem. Once the dataset to be used was selected, more engineering features that are known to have an impact on the drivability assessment were calculated. Finally, an advanced machine learning model was implemented using ensemble techniques such as stacking and blending with specific weights to give the final prediction.

The work performed demonstrates that the implementation of ML model for geotechnical applications involves several stages of analysis and is not straightforward. However, once implemented, it becomes a powerful and efficient predictive tool, especially for large sets of data. The efficiency of this methodology relies heavily on the quality of the dataset and the different techniques employed to inspect and clean the data. Limitations related to the available data set is considered to be the most serious limitation, since the data on which the models are predicted needs high variability to cover the range of conditions of piles to be predicted in terms of soil conditions, pile dimensions and hammer performance. Hence, the importance for the industry to digitally organize and store data as part of the acquisition process.

The model developed in this study was for a specific region and needs further validation on other soil types and pile dimensions. It is recommended to compare how the predictions compare with those from conventional techniques. This was not possible for the current work as the available hammer energy was normalised to maintain anonymity.

Acknowledgements. The authors would like to acknowledge the dataset provider (ISFOG2020) which has been used, as well as Fugro management for devoting a considerable amount of time to perform this research and development work.

References

1. Momeni, E., Nazir, R., Armaghani, D.J., Maizir, H.: Prediction of pile bearing capacity using a hybrid genetic algorithm-based ANN (2014)

2. Kaggle Competition. https://www.kaggle.com/c/isfog2020-pile-driving-predictions/overview
3. Alm, T., Hamre, L.: Soil model for pile driveability predictions based on CPT interpretations. Presented at the International Conference on Soil Mechanics and Foundation Engineering (2001)
4. GRLWEAP. Background Report, Manual for 1 dimensional wave equation analysis (2010)
5. Kotz, S., et al.: Encyclopedia of Statistical Sciences. Wiley, Hoboken (2006)
6. Ying, X.: An overview of overfitting and its solutions. J. Phys. Conf. Ser. **1168**(2), 022022 (2019)

Performance of Offshore Foundations and Structures

Bearing Capacity of Tall Monopile Offshore Wind Turbine Towers Under Environmental Loading

Y. Hu[1,2], J. Yang[1,2], and C. Baniotopoulos[2](✉)

[1] School of Naval Architecture, Ocean and Civil Engineering, Shanghai Jiao Tong University, Shanghai 200240, China

[2] School of Civil Engineering, University of Birmingham, Edgbaston, Birmingham B15 2TT, UK

c.baniotopoulos@bham.ac.uk

Abstract. Offshore wind energy is a rapidly maturing renewable energy technology that is poised to play an important role in future energy systems. The respective advances refer among others to the monopile foundation that is frequently used to support wind turbines in the marine environment. In the present research paper, the structural response of tall wind energy converters with various stiffening schemes is studied during the erection phase as the latter are manufactured in modules that are assembled in situ. Rings, vertical stiffeners, T-shaped stiffeners and orthogonal stiffeners are considered efficient stiffening schemes to strengthen the tower structures. The loading bearing capacity of offshore monopile wind turbine towers with the four types of stiffeners were modeled numerically by means of finite elements. Applying a nonlinear buckling analysis, the ultimate bearing capacity of wind turbine towers with four standard stiffening schemes were compared in order to obtain the optimum stiffening option.

Keywords: Offshore wind turbine tower · Loading bearing capacity · Stiffening scheme · Offshore wind turbine tower erection

1 Introduction

Wind energy already significantly contributes to the reduction of greenhouse negative effects while contributing to economic growth in many countries as a leading solution against climate change globally. Offshore wind energy has been developing rapidly in the world, whilst Europe as the leader in offshore wind energy has the potential to realize up to 3400 TWh of offshore wind energy within its waters in 2030 [1]. In order to keep the monopile as thin as possible, the tower can be stiffened in different ways by various stiffening schemes to resist all possible loadings e.g., the wind, the current and the wave loadings in marine environments. For wind turbine towers, stiffening schemes can be employed to strengthen the tubular towers. Several researchers studied this research topic and proposed various methods to strengthen the tower structure in order to either decrease the cost of the tower or improve its structural response. Stavridou et al. [2]

D. V. K. Huynh et al. (Eds.): VSOE 2021, LNCE 208, pp. 413–423, 2022.
https://doi.org/10.1007/978-981-16-7735-9_46

examined ring stiffeners and vertical stiffening schemes to eliminate the presence of short wavelength buckles of the tower due to bending. Gkantou et al. [3] performed the life cycle assessment of tall hybrid onshore steel wind turbine towers. Hu et al. [4–9] performed mechanical characteristics of onshore and offshore wind turbine towers under environmental loads stiffened by different stiffening solutions.

Tubular wind turbine towers are often employed as supporting structures to resist failure under external loadings; therefore, buckling analysis should be examined to explore their bearing capacity. Jay et al. [10] performed a program of flexural buckling tests of tapered circular steel wind towers with diameters between 0.7 and 1.1 m and maximum diameter-to-thickness ratios between 200 and 350 and concluded that location and orientation of the local buckling region were correlated with the spiral seam welds on the specimens. All the aforementioned studies focus on the bearing capacity and the buckling of the wind turbine system, however, they have not considered the buckling analysis and bearing capacities of the wind turbine towers during the critical erection phase. In the present study the bearing capacity of tubular wind turbine steel towers during the erection phase subjected to horizontal, vertical and bending loadings are investigated and the effect of various stiffening schemes in their ultimate bearing capacity in the marine environment is studied.

2 Bearing Capacity of the Offshore Tower with Stiffeners

2.1 Tower Models

For offshore wind turbine towers, the stiffeners are often employed to stiffen the tower wall. To compare the effect of the various stiffeners schemes on the structural performance of the offshore towers during the erection phase, towers with four stiffening schemes including a tower with rings, a tower with vertical stiffeners, a tower with T-shaped stiffeners and a tower with orthogonal stiffeners regarded, respectively, as tower I, tower II, tower III and tower IV were investigated. Figures 1 and 2 describe the geometry of the offshore wind turbine towers I, II, III and IV at hand. According to Figs. 1 and 2, the height of the tower is 73.39 m, the top and bottom diameters of the towers are, respectively, 3276 mm and 5200 mm. For tower I, its ring spacing is 9 m and it contains eight rings along the tower wall. The width and thickness of these rings are 200 mm and 100 mm, respectively. For vertical stiffeners, its width and thickness are, respectively, 200 mm and 100 mm as shown in Fig. 2 and the other edge of the vertical stiffeners is parallel to tower edge as shown in Fig. 1. The four vertical stiffeners are distributed uniformly in the tower wall. The lengths of T shaped stiffeners in two sides are, respectively, 100 mm and 200 mm as displayed in Fig. 2. The thickness of T shaped stiffeners is 50 mm. For tower IV, the ring spacing of orthogonal stiffeners is 11m and its four vertical stiffeners are also distributed uniformly in a similar way as in Tower II. The material property of FE model is Q345 steel as shown in Table 1.

For the FE models of the tower, the tie constraint was used to connect the stiffeners and the shell, i.e., the tower wall by using ABAQUS software [11]. The tower bottoms were considered to be fully fixed. On the top of the tower models, the circle center of the tower top can be setup to couple the top circle of the tower to the reference point. On the top, the tower is subjected to gravity and moment due to the eccentricity of wind turbines and the horizontal force resultant from wind loadings applied to blades of the wind turbine. The gravity, horizontal force and moment applied to the 73.39 m offshore tower are, respectively, 762,140 N, 1.0×10^6 N and 3.9479×10^8 N·mm. The tower should resist the gravity, the moment and the horizontal force on the top and therefore, its bearing capacity of the tower should be evaluated. In the present study, four stiffeners schemes are compared to explore the optimum stiffening option for the tower wall.

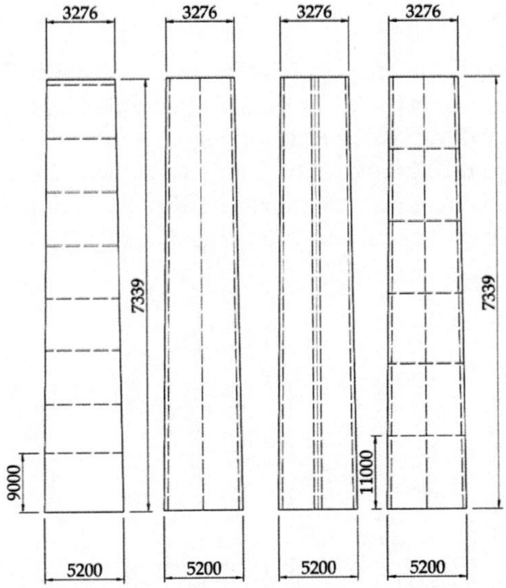

Fig. 1. Geometric data of the towers with the four stiffening schemes (in mm).

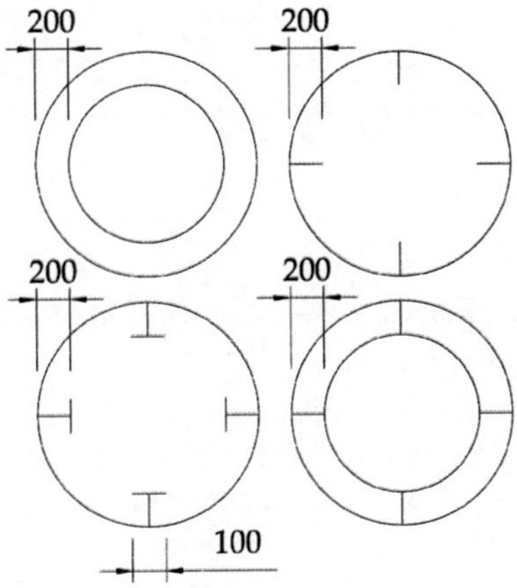

Fig. 2. Geometric data of the stiffeners in the cross-section view of the tower (in mm).

Table 1. The material properties of steel.

Material	Density	Elasticity module	Poisson's ratio	Yield stress	Plastic strain
Steel	7.85 g/cm^3	205 GPa	0.3	345 GPa	0
				428 GPa	0.1

2.2 Stiffeners Mass

To compare the different stiffening options, the rings, vertical stiffeners, T-shaped stiffeners and orthogonal stiffeners should have all the same mass in order to study their efficacy. As the T-shaped stiffeners and the orthogonal stiffeners geometrically are based on the initially designed rings and vertical stiffeners, their mass can also be obtained by using the mass formula of rings and vertical stiffeners. In Fig. 3 the geometry of the tower with rings and vertical stiffeners is presented.

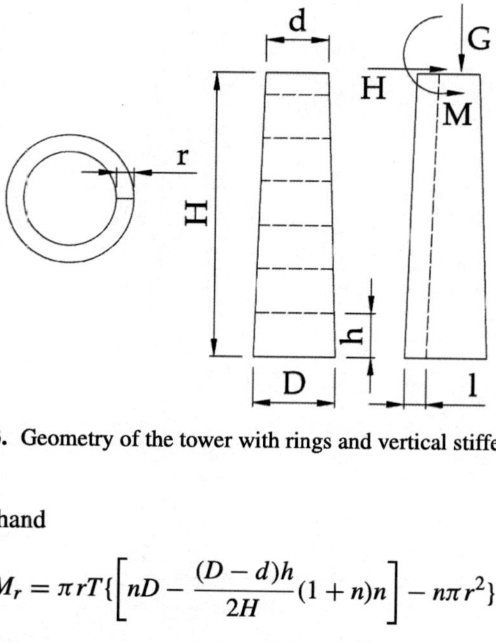

Fig. 3. Geometry of the tower with rings and vertical stiffeners.

In the model at hand

$$M_r = \pi rT\{\left[nD - \frac{(D-d)h}{2H}(1+n)n\right] - n\pi r^2\} \quad (1)$$

where M_r is the mass of rings, r the width of the rings, D the diameter of the tower base, d the diameter of the top of the tower, H the height of the tower, h the spacing of rings, T the thickness of the rings and n is the number of the rings.

$$M_v = mlHt \quad (2)$$

where M_v is the mass of the vertical stiffeners, m the number of vertical stiffeners, l the width of vertical stiffeners and t the thickness of the stiffeners.

2.3 Buckling Analysis of the Tower

When wind farms or even wind turbines finish operating at the end of their lifetime, wind farm operators often consider repowering them to save funds and enhance sustainability assets. By repowering old wind turbines into a larger size, the amount of energy is increased due to the higher efficiency of new turbines. These repowered wind turbines should however, possess enhanced ultimate bearing capacity that means that offshore wind turbine towers should be upgraded in a meticulous strengthening way.

Figure 3 shows that gravity, moment and horizontal force are applied to the top of the tower stem. The tower shell needs to be designed so that it efficiently bears the gravity load and the moment created from the eccentricity of the wind turbine placed on the top of the tower and the horizontal force being the resultant of the force of wind, the current and the wave loading. Therefore, as the loading states on the wind turbine tower are complicated in engineering practice, the gravity, the horizontal force and the moment are individually applied to each tower model in order to explore the collapse mode under each respective loading state and calculate the ultimate bearing capacity of the offshore tower under operation in marine environments. For the buckling analysis, the subspace method is applied as the eigensolver to obtain buckling modes and eigenvalues in each buckling step. For the nonlinear buckling analysis, the imperfection of the tower wall is applied to the models and certain buckling modes should be employed to acquire the nonlinear buckling modes, whilst the Riks method is used to evaluate the post buckling behavior and the critical loading of each tower. Then, the reactions on the bottom and the displacements on the top are compared to obtain the optimum stiffening scheme option.

2.3.1 On the Effect of Gravity

The gravity effect is principally due to the wind turbine placed on the top of the tower. With reference to the bearing capacity of the tower an analysis is performed to study the most appropriate stiffening option. The mass of the stiffeners in all towers I, II, III and IV is 22.07t. The buckling response of the towers under gravity is shown in Fig. 4. The nonlinear buckling analysis for gravity loading is depicted in Fig. 5. The buckling analysis shows that the modes 1, 2 and 3 of towers I, II, III, IV occur in the *y*-axis direction (Fig. 4). It can be seen that the first and second modes are symmetrical to each other and are global bending modes, whilst the third mode is a local mode. According to Fig. 5, the nonlinear behavior of towers I and II, III, IV are different. The shell/tower wall (case I) is principally compressed from the top to the vicinity of the first top ring as shown in Fig. 4. For towers II, III and IV, the shell is compressed as shown in Fig. 6. Gravity versus displacement of towers I, II, III, IV under gravity of wind turbines can be shown in Fig. 6. The displacements in the *x*-axis in Fig. 6 depict the displacement in the y direction and the reaction force is the resultant at the tower bottom. According to Fig. 6, the bearing capacity of tower I is 1.08×10^8 N, whereas the bearing capacities of towers II, III and IV are close to 1.18×10^8 N. Therefore, comparing the peak of the curves and the bearing capacities of gravity of towers II, III and IV are close each other, whilst the bearing capacity of tower I is less than those of towers II, III and IV. It can be concluded that vertical stiffeners, T-shaped stiffeners and orthogonal stiffeners are more appropriate stiffening options than rings for gravity bearing capacity.

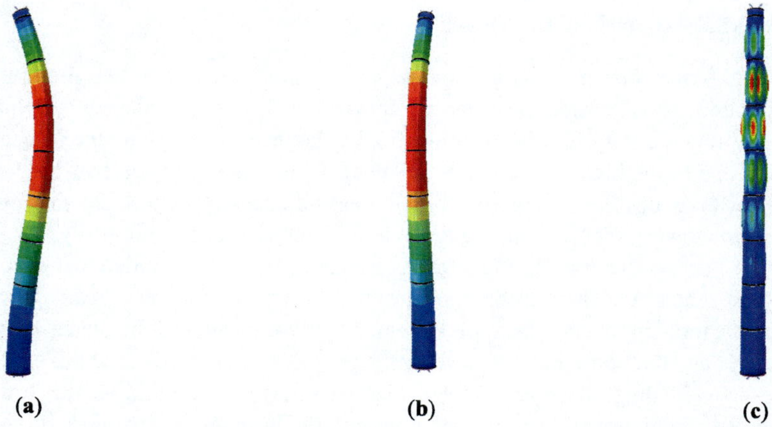

Fig. 4. Buckling analysis of Tower I under gravity load. (**a**) Mode 1, (**b**) Mode 2, (**c**) Mode 3.

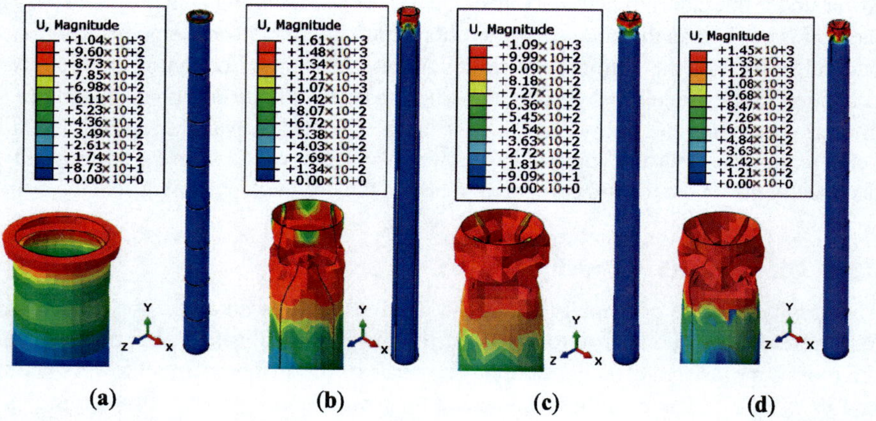

Fig. 5. Post buckling behavior of Towers I, II, III, IV under gravity loading: (**a**) Tower I, (**b**) Tower II, (**c**) Tower III, (**d**) Tower IV, (in mm).

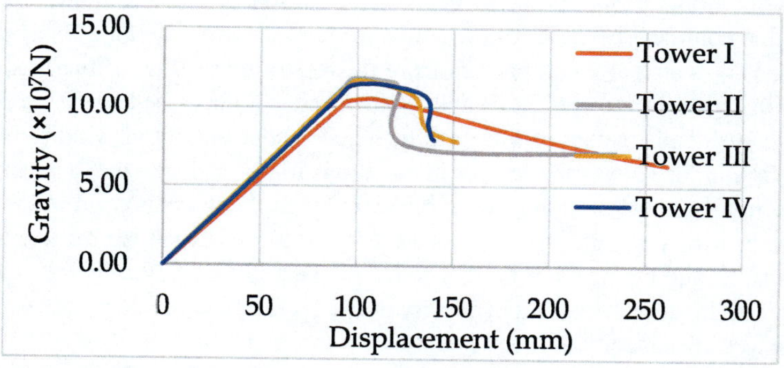

Fig. 6. Gravity versus displacement of Towers I, II, III, IV under gravity load

2.3.2 On the Effect of the Horizontal Actions

Taking into consideration the horizontal forces' action on the nacelle/top of the tower, the buckling analysis of tower I is presented in Fig. 7, whilst the nonlinear buckling responses of towers I, II, III, IV are depicted in Fig. 8, whilst in Fig. 9 the horizontal force versus displacements of towers I, II, III, IV are presented. The displacements in the x-axis in Fig. 9 represent the horizontal displacements along the x-direction that is shown in Fig. 8 (tower top) and the reaction force is the result at the tower root. It can be observed that modes 1, 2 and 3 are local modes in the vicinity of the tower base and they are similar to those of Fig. 7. For post buckling behavior, the towers can translate in the x-axis direction due to the horizontal force. The maximum post buckling displacement of tower I is less than that of towers II, III and IV (Fig. 8). According to Fig. 9, the bearing capacity of towers I and IV are, respectively, 1.1×10^8 N and 1.2×10^8 N and the bearing capacity of towers II and III are close to 1.25×10^8 N. The bearing capacity in the horizontal force of tower I is less than that of towers II, III and IV (Fig. 9), whilst the bearing capacity in horizontal force of towers II and III are greater than those of towers I and IV. Therefore, towers II and III could be selected as a more appropriate option for a tower stiffened as such.

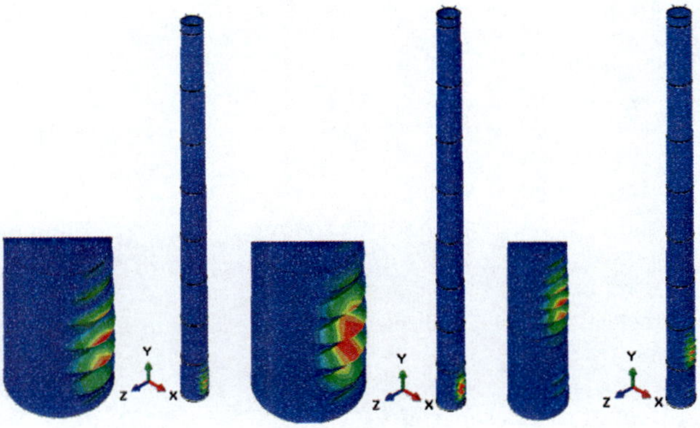

(a) (b) (c)

Fig. 7. Buckling analysis of Tower I under horizontal force. (**a**) Mode 1, (**b**) Mode 2, (**c**) Mode 3.

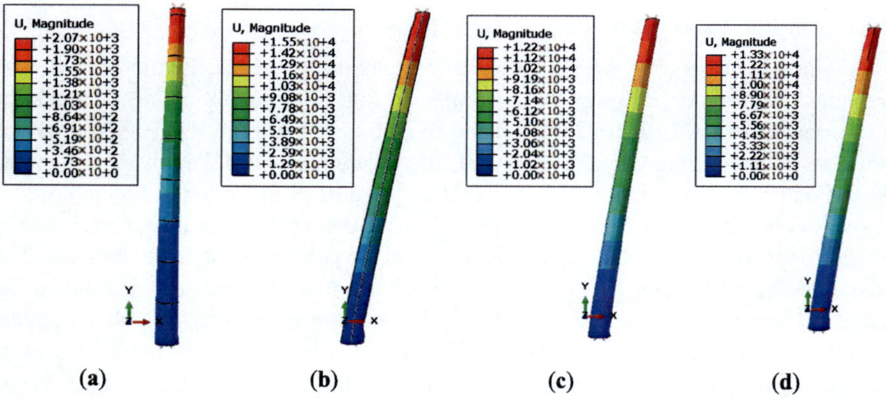

Fig. 8. Post buckling behavior of towers I, II, III, IV under horizontal force. (**a**) Tower I, (**b**) Tower II, (**c**) Tower III, (**d**) Tower IV, (in mm).

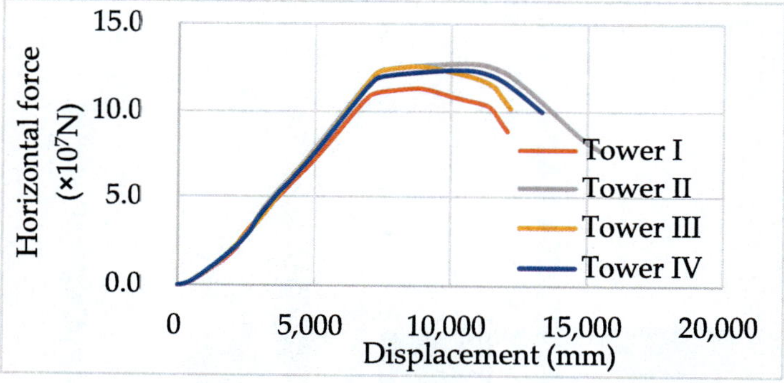

Fig. 9. Horizontal force versus displacement with regard to Towers I, II, III, IV.

2.3.3 On the Effect of the Eccentricity

The eccentricity of the wind turbine that is placed on the top of tower from the center of the cross section of the tower leads to the development of an additional bending moment on the structure. With reference to this additional bending moment due to eccentricity, the buckling behavior of tower I is shown in Fig. 10; modes 1, 2 and 3 of tower I are local modes in the vicinity of the tower top. Post buckling analysis of towers I, II, III, IV under moment is presented in Fig. 11. Nonlinear buckling behavior of towers I, II, III, IV under moment leads to local bending modes and maximum deformation occurs on tower I. The graph of moment versus rotation of towers I, II, III, IV is presented in Fig. 12. The rotation in *x*-axis in Fig. 12 represents the rotation angle around the z-direction in Fig. 11 of the tower top and the reaction force is the resultant of the tower bottom. According to Fig. 12, the bearing capacity of towers I, II, III, IV are, respectively, 9.5×10^{10} N·mm, 1.0×10^{10} N·mm, 1.05×10^{10} N·mm and 1.05×10^{10} N·mm. Therefore, the bearing capacity of tower I is the minimum and the bearing capacity of towers III and IV are greater than those of tower II. It can be concluded that T-shaped stiffeners and orthogonal stiffeners are the optimal stiffening option for the tower under investigation.

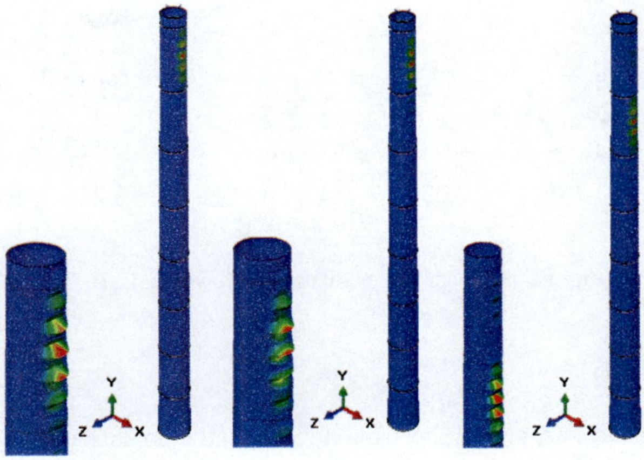

(a) (b) (c)

Fig. 10. Buckling analysis of Tower I: (**a**) Mode 1, (**b**) Mode 2, (**c**) Mode 3.

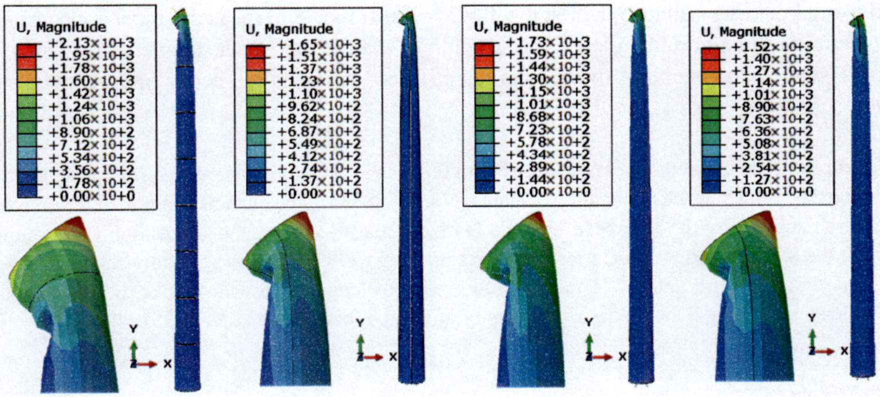

(a) (b) (c) (d)

Fig. 11. Post buckling behavior of Towers I, II, III, IV: (**a**) Tower I, (**b**) Tower II, (**c**) Tower III, (**d**) Tower IV, (in mm).

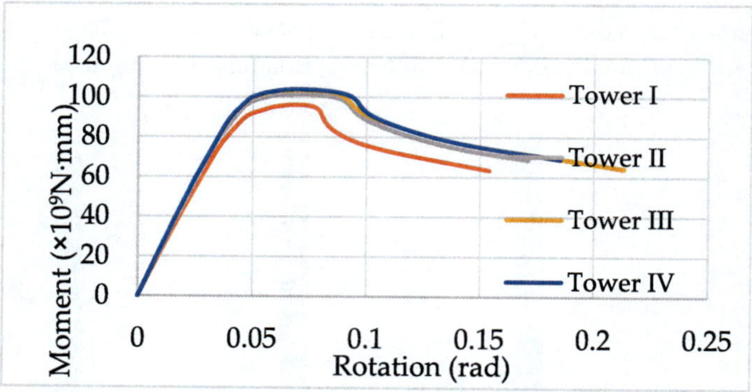

Fig. 12. Moment versus rotation of Towers I, II, III, IV.

3 Conclusions

In the buckling analysis, gravity, horizontal forces and moments are, respectively, considered as the loadings that cause the collapse of the tubular towers due to buckling. Imperfections are also imported to the tubular tower models to obtain the respective nonlinear buckling modes. The bearing capacity of the towers with four different stiffening schemes are compared each other to obtain the optimum stiffening option, and it is concluded that the stiffening ring is the weakest option as a strengthening technique to avoid buckling collapse, whilst the T-shaped stiffener scheme is the most appropriate stiffening option and to enhance the strength of the tubular tower, the vertical stiffener, the T-shaped stiffener and the orthogonal stiffener schemes are better options than the ring schemes.

Funding. The first author acknowledges with thanks the research financial support provided by National Natural Science Foundation of China (No. 51808335). The financial support of the present research activity by the European Union's Horizon 2020 Research and Innovation Programme under the Marie Sklodowska-Curie Grant Agreement No 793316 is also gratefully acknowledged by the first and third authors. The third author acknowledges also with thanks the IGI Global Challenges Funding Scheme of the University of Birmingham (Project Nr 3035).

References

1. Arapogianni, A., Moccia, J.: Wind in our sails: the coming of Europe's offshore wind energy industry. European Wind Energy Association, Brussels (2011)
2. Stavridou, N., Efthymiou, E., Gerasimidis, S., Baniotopoulos, C.C.: Investigation of stiffening scheme effectiveness towards buckling stability enhancement in tubular steel wind turbine towers. Steel Compos. Struct. **19**, 1115–1144 (2015)
3. Gkantou, M., Rebelo, C., Baniotopoulos, C.: Life cycle assessment of tall onshore hybrid steel wind turbine towers. Energies **13**, 3950 (2020)
4. Hu, Y., Yang, J., Baniotopoulos, C.: Study of the bearing capacity of stiffened tall offshore wind turbine towers during the erection phase. Energies **13**, 5102 (2020)

5. Hu, Y., Yang, J., Baniotopoulos, C., Wang, X., Deng, X.: Dynamic analysis of offshore steel wind turbine towers subjected to wind, wave and current loading during construction. Ocean Eng. **216**, 108084 (2020)
6. Hu, Y., Yang, J., Baniotopoulos, C.C.: Repowering steel tubular wind turbine towers enhancing them by internal stiffening rings. Energies **13**(7), 1538 (2020)
7. Hu, Y., Yang, J., Baniotopoulos, C.C., Wang, F.: A comparison of structural performance enhancement of horizontally and vertically stiffened tubular steel wind turbine towers. Struct. Eng. Mech. Int. J. **73**, 487–500 (2020)
8. Hu, Y., Baniotopoulos, C., Yang, J.: Effect of internal stiffening rings and wall thickness on the structural response of steel wind turbine towers. Eng. Struct. **81**, 148–161 (2014)
9. Hu, Y., Yang, J., Baniotopoulos, C., Wang, X., Deng, X.: Tall offshore steel wind turbine towers under wind, current and wave loading: an experimental and numerical study. In: XI International Conference on Structural Dynamics, Athens, Greece, 22–24 November 2020 (2020)
10. Jay, A., et al.: Large scale bending tests of slender, tapered, spirally welded steel tubes. J. Struct. Eng. **142**, 04016136 (2016)
11. ABAQUS/Standard and ABAQUS/Explicit-Version 6.8-1; Abaqus Theory Manual, Dassault System: Vélizy-Villacoublay, France (2008). https://www.doc88.com/p-439722564721.html. Accessed 26 Sept 2020

Installation Effects on Axial Performance of Monopiles in Chalk for Offshore Renewables

Matteo O. Ciantia(✉)

School of Science and Engineering, University of Dundee, Dundee, UK
m.o.ciantia@dundee.ac.uk

Abstract. Chalk is present in a large area of the East and South coast of the UK where renewables are now being installed. In most cases they are founded on displacement piles, whose design uses empirical methods or partial (wished-in-place) simulations. Installation effects such as grain crushing and pore pressure generation are conservatively estimated (if considered at all), as the change of chalk properties around the pile due to pile installation cannot be easily quantified. In this paper it is shown how advanced numerical modelling can be used to guide the design of the foundations in these complex soils. The coupled hydro-mechanical effects developing during pile installation in chalk are investigated numerically using a robust and mesh-independent implementation of an elasto-plastic constitutive model at large strains. The model, implemented into an open source Particle Finite Element (PFEM) code, is able to capture the damage of the rock until the formation of a chalk putty layer around the shaft of a small diameter pile. Installation effects are highlighted by comparing the axial performance between wished in place piles and piles which considered the full installation process. The numerical results are then used to critically address the limitation of current design specifications of displacement piles in chalk.

Keywords: Numerical modelling · Chalk · Monopiles · Installation effects · Offshore wind · PFEM · Hydro-mechanical coupling

1 Introduction

Chalk covers much of the UK and is widespread under the North and Baltic Seas where offshore wind turbine (OWT) are currently being installed and where future offshore deployment will be hosted (Fig. 1a). Large piles are often driven in chalk to support OWT and other onshore infrastructure. Dynamic installation of pile foundations in chalk causes extremely high pore-water pressure (*pwp*) build up, the intact rock surrounding the pile to crush and deteriorate into chalk putty which is characterised by a mechanical behaviour completely different from the initial intact chalk (Lord *et al.* 2002). Such complexity is the underlying reason for inadequate current design guidance for pile driveability, axial and lateral static load response (Jardine 2019). The long-term performance in service cyclic loading performance remain highly significant unknowns, making pile design highly overconservative. Considering that for OWT, foundations

D. V. K. Huynh et al. (Eds.): VSOE 2021, LNCE 208, pp. 424–432, 2022.
https://doi.org/10.1007/978-981-16-7735-9_47

account for about 20–25% of the total development cost (Ng and Ran 2016), pile design improvements in complex ground as chalk is, can be extremely beneficial from an economical and environmental perspective. Current guidelines for the design of driven piles in chalk (CIRIA C574) originate from the analysis of a limited number of pile tests (Lord *et al.* 2002). These guidelines suggest crude average ultimate unit shaft friction (τ_{sf}) design values of between 20 and 120 kPa for sites in which low-medium density and high very-high density chalk is present, respectively. These shaft friction estimates are thought to be conservative and to introduce significant increases in cost and carbon footprint (Fig. 1b). Even less guidance is available regarding cyclic or lateral loading in chalk, although this is often crucial in offshore pile design. Pile installation crushes the chalk and produces an annulus of 'putty chalk' around the pile (Alvarez Borges 2019). The pile in-service performance relies on both stiffness and strength of this putty (τ_{sf}) and the post installation stress state around the pile. Recent field tests (Buckley *et al.* 2017; Ciavaglia *et al.* 2017) have also shown that τ_{sf} can increase significantly with time post installation (set-up of up to 400% over a few months). However, such set-up was not observed when smaller piles were slowly jacked into the ground. This clearly indicates that installation effects play a major role in the in-service performance. Moreover, these time dependent phenomena appear to be site specific and, despite the large number of time consuming and expensive field tests performed, the mechanisms explaining this complex behaviour are not yet completely understood. In this paper the pile installation process is investigated through large deformation numerical analyses with a finite deformation plasticity model for structured sands to model the chalk. Installation effects are highlighted by comparing the axial performance between wished in place piles and piles which considered the full installation process. The numerical results are then used to critically address the limitation of current design specifications of displacement piles in chalk.

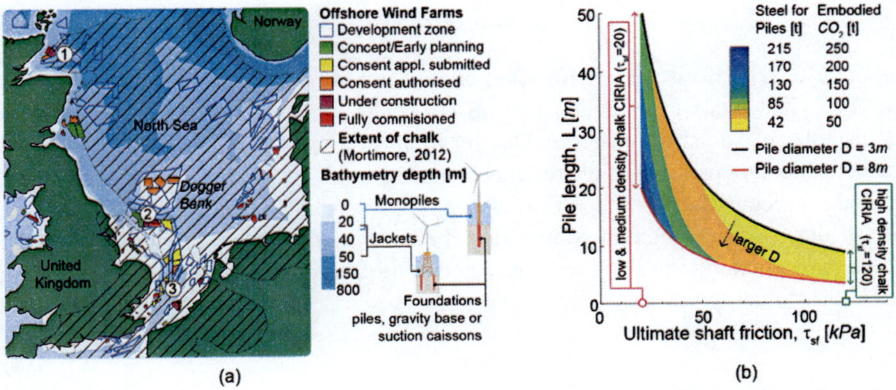

Fig. 1. a) North Sea OWT development (www.4coffshore.com) along with sea water bathymetry, extent of chalk (Mortimore 2012). (b) Embodied CO_2 and steel tonnages as a function of design interface chalk-steel strength, pile length L and diameter D. Diagram developed for a pile design axial capacity of 10MN.

2 Constitutive Model for Chalk

Soft rock behaviour can be mathematically described by extending the strain hardening plasticity framework used to model soils by increasing the size of the yield locus giving the material tensile strength (Ciantia and di Prisco 2016). In particular the intersections of the yield locus ($f = 0$) with the positive and negative p' axis are called $p_c = p_s + p_m$, and p_t; p_s acts as preconsolidation pressure, as it is for soils, and it is assumed to depend on both plastic volumetric and deviatoric strain rates. In contrast, p_t and p_m account for the effects of interparticle bonding (Fig. 2). The existence of bonds therefore implies a non-zero tensile strength ($p_t > 0$) and an increase in the yield stress along radial loading paths (Ciantia *et al.* 2015). p_m corresponds to this increase in the case of isotropic compression path.

The constitutive model has been formulated in the framework of large strains elasto-plasticity with a multiplicative decomposition of the deformation gradient into elastic and plastic parts (Simo and Hughes 1998). This framework ensures that rigid body motion of the deformable body do not produce spurious stress variations (frame indifference) and also that energy is preserved since a hyper-elastic model is employed. In this work the shape of the yield envelop is built around the Modified Cam Clay model. The yield surface is hence defined as:

$$f\left(\boldsymbol{\tau}', p_s, p_t, p_m\right) = \left(\frac{q}{M}\right)^2 + p^*(p^* - p_c^*) \tag{1}$$

where $q = \sqrt{3J_2}$ and J_2 is the second invariant of the effective Kirchoff stress tensor, $\boldsymbol{\tau}'$. M is the slope of the Critical state line in the $p' - q$ plane whereas:

$$p^* = p' + p_t \tag{2}$$

$$p_c^* = p_t + p_s + p_m \tag{3}$$

where $p' = tr(\boldsymbol{\tau}')/3$ is the first invariant of the Kirchoff stress tensor.

The yield locus in the triaxial plane is depicted in Fig. 2; that also graphically defines the plastic stress-like variables p_c^*, p_t, p_s and p_m. On the one hand, p_s is the preconsolidation pressure of the reference, unstructured soil (putty). On the other hand, p_t and p_m account for the effect of structure: p_m corresponds to the increase in the yield stress along isotropic compression paths whereas p_t is the tensile strength. Generally, these two hardening variables are considered to be proportional (Ciantia 2018), being c the proportionality factor:

$$p_t = c p_m \tag{4}$$

The evolution of the hardening variables is related to the plastic volumetric and distortional flow:

$$\dot{p_s} = \rho_s p_s \left(\mathrm{tr}\left(\boldsymbol{l}^p\right) + \chi_s \sqrt{\frac{2}{3}} \|\mathrm{dev}\left(\boldsymbol{l}^p\right)\|\right) \tag{5}$$

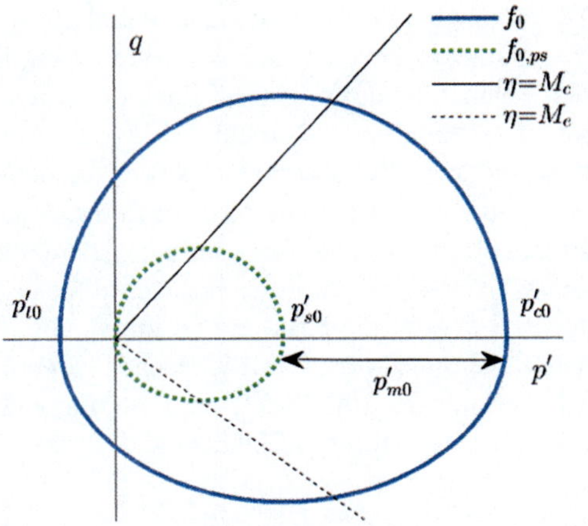

Fig. 2. Yield surface in the triaxial (left) and deviatoric (right) planes

$$\dot{p}_t = -\rho_t p_t \left(\left|\mathrm{tr}(\boldsymbol{l}^p)\right| + \chi_s \sqrt{\frac{2}{3}} \|\mathrm{dev}(\boldsymbol{l}^p)\| \right) \tag{6}$$

where ρ_s, χ_s, ρ_t and χ_s are constitutive parameters and $\boldsymbol{l}^p$ is the spatial plastic velocity gradient. The elastic response is characterized by means of an hyperelastic model incorporating a tensile range (Tamagnini and Ciantia 2016), which is formulated in terms of the Hencky strain and the Kirchhoff stress tensor. For further details of the constitutive model refer to (Monforte *et al.* 2019).

2.1 Nonlocal Integration

A nonlocal integral type regularization technique is used to mitigate the pathological mesh-dependence that exhibit numerical simulations where softening is encountered. As such, the expression of a nonlocal variable $\tilde{\beta}$ is:

$$\tilde{\beta}(\boldsymbol{x}) = \frac{\int_\Omega w(\boldsymbol{x}, \|\boldsymbol{x} - \boldsymbol{y}\|)\beta(\boldsymbol{y})d\Omega}{\int_\Omega w(\boldsymbol{x}, \|\boldsymbol{x} - \boldsymbol{y}\|)d\Omega} \tag{7}$$

where $w(\boldsymbol{x}, \|\boldsymbol{x} - \boldsymbol{y}\|)$ is the weighting function for point $\boldsymbol{x}$ controlling the influence of its neighbors in terms of their relative distance. In this work, plastic strains are considered as nonlocal variables; from these values p_s and p_t may be obtained by integrating analytically Eqs. (5) and (6). The weighting function proposed by (Galavi and Schweiger 2010) is employed here as it has been found to outperform other weighting functions in removing mesh bias.

2.2 Model Parameters

Given a set of uniaxial compressive strength (UCS) and or triaxial compression experimental data it is possible to calibrate the model following the procedure well described in Rios *et al.* (2016). The goal here is to show the potential of the numerical approach, hence the constitutive parameters have been chosen to represent a generic intact chalk with a UCS of 290 kPa, Young modulus of 100 MPa and Poisson ratio of 0.2. The values of the constitutive parameters are reported in Table 1. Figure 3 reports the results (the effective stress path and the load-displacement curve) of drained and undrained triaxial compression tests at two levels of confinement pressures. As clearly represented in the figure once the stress path reaches the yield surface, destructuration begins and for the low confinement tests deviatoric stresses decrease until the critical state is reached. For the highly confined tests, after the peak and an initial softening phase the deviatoric stress increases until the same critical state is reached. Complete destructuration is reached when p_t (hence tensile strength) goes to zero. For more details of the behavior of the chalk putty (unbonded state) refer to Ciantia and di Prisco (2016).

Table 1. Constitutive parameters adopted in this work for a generic intact chalk.

E (kPa)	ν	M	p_s (kPa)	p_t (kPa)	ρ_s	ρ_t	χ_s	χ_t	c	k (m/s)
100000	0.2	1.5	150	50	12.9	8	0	0.5	4	10^{-5}

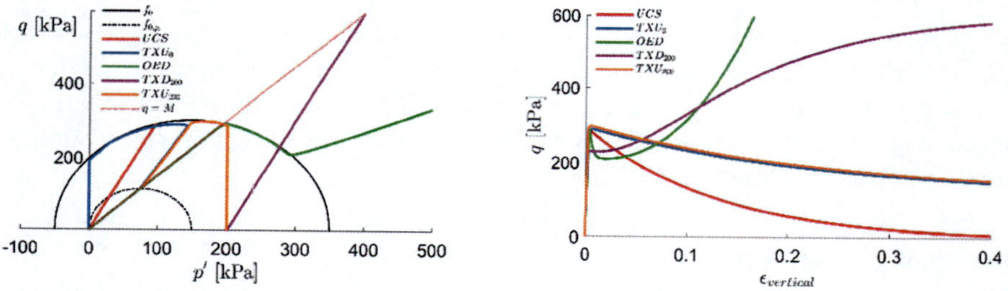

Fig. 3. UCS, oedometric compression, drained and undrained triaxial response in the p'-q plane (left) and corresponding evolution of deviatoric stresses in terms of the axial deformation (right).

3 Numerical Simulation of Pile Installation

The model, implemented in the open source Particle Finite Element (Oñate *et al.* 2004) platform KRATOS (https://www.cimne.com/kratos/default.asp), is here used to simulate the penetration of a rough cone shaped pile of 3.6 cm. To investigate the effect of installation on the axial capacity the following set of simulations are run:

- Two *wished-in-place* (WIP) piles are numerically tested in tension and compression. The response here will not consider the installation process
- Two *fully installed piles* (FIP) are tested in tension and compression. In these simulations, after the pile is installed until the same penetration depth of the WIP models, the excess *pwp* are let to dissipate before testing the pile in tension and compression respectively.

The model is now used simulate the penetration of a cone shaped pile in a calibration chamber. The model geometry initial conditions and initial mesh are represented in Fig. 4. The pile is set to be rough with a coefficient of friction $\mu = 0.2$. From the WIP initial condition the tensile capacity of the pile is tested by extracting the pile at a constant velocity of 0.02 m/s (the same employed for cone penetration tests) on the other hand the compressive capacity is determined by pushing the pile downwards. These two simulations are represented in Fig. 5 and are represented by the lines that go from point X to points Y and Z respectively. The FIP are first jacked into the ground at a velocity of 0.02 m/s until the same penetration depth of the WIP models is reached (from point A to point B). At this point the pile is unladed until the vertical force is zero (point C). At this point, after a dissipation phase of 100 days, the pile is either pushed down at 0.02 m/s until point D to determine the compressive axial capacity of the FIP model. On the other hand, the pile is extracted at a velocity of 0.02 m/s up to point E to determine the tensile capacity of the FIP. The results clearly show that installation effects have a big influence on the axial capacity. Despite the chalk is crushed into a cohesionless material (chalk-putty), the increase of confinement around the pile is responsible of an increase of both tensile and compression capacity. The tensile capacity is more than doubled while the compressive capacity increases by 35%. The model also shows a stiffer response of the FIP models.

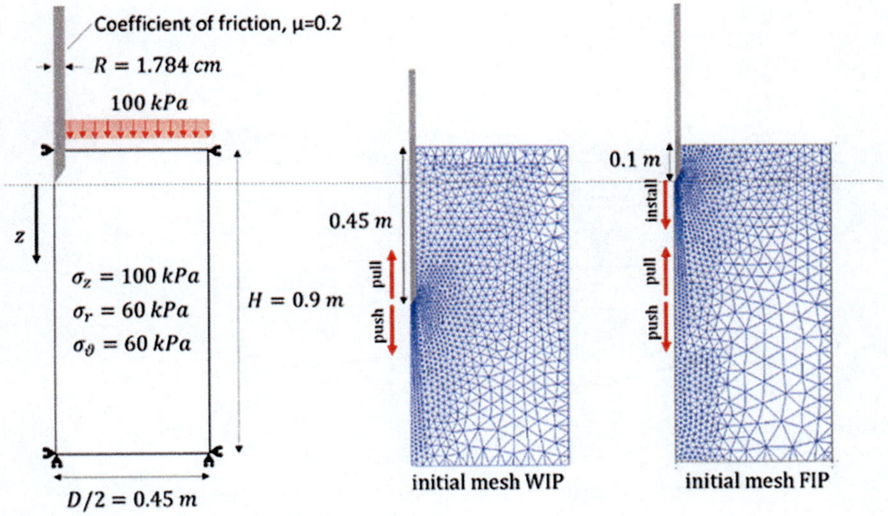

Fig. 4. Calibration chamber geometry and initial conditions

Finally Fig. 6 reports the radial stress profile of the chalk in correspondence of the shaft. The stress profile is compared with the recently developed ICP method for chalk (Buckley *et al.* 2019) developed for open ended piles driven in chalk. A numerical CPT was run in order to the determine the q_c to use in the empirical trendline. Interestingly, in order to properly fit the numerical data, the empirical relationship needs to be amplified by 6 times. Nevertheless, the stress profile evolution with h/R is captured well using the same exponential term −0.52. These differences suggest that dynamic installation and size effects play an important role in this highly coupled hydro-mechanical problem. As evinced by the radial stresses in the far field, boundary effects might have also influenced the numerical results of this preliminary study.

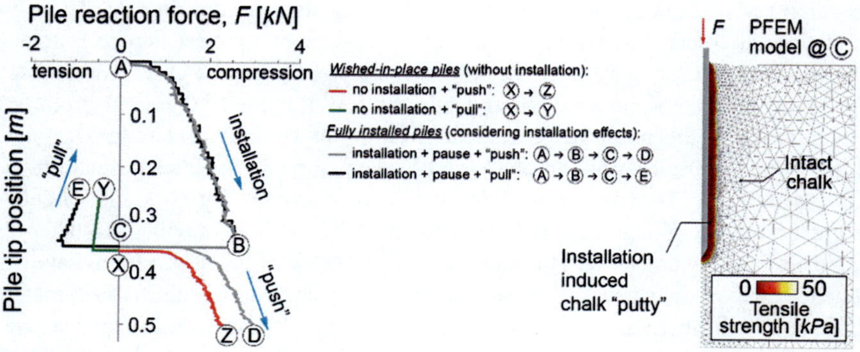

Fig. 5. Comparison of the WIP and FIP models in terms of axial response and degradation.

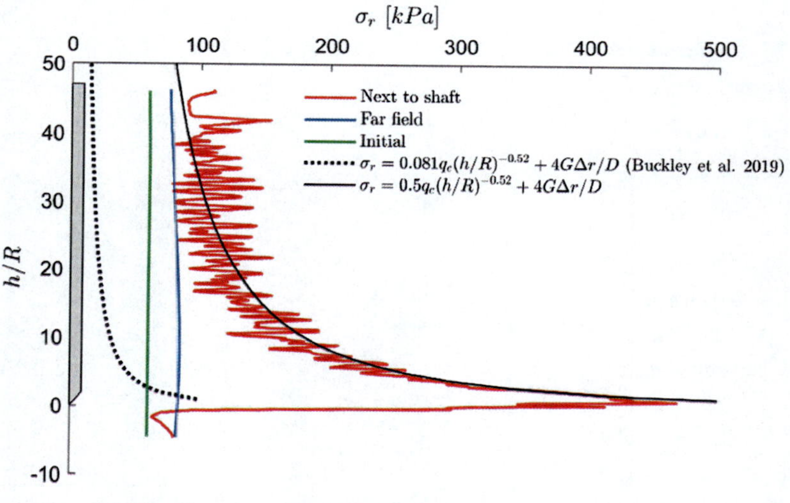

Fig. 6. Comparison of effective radial stresses on shaft

4 Conclusions

This paper shows how advanced numerical modelling can be used to guide the design of the foundations in soft rocks – cemented sands such as chalk. The coupled hydro-mechanical installation effects of pile jacking were addressed numerically using a recently developed constitutive model for soft rocks implemented in KRATOS an open source Particle Finite Element Method (PFEM) platform. Installation effects on pile performance are highlighted by comparing the axial capacity between wished in place piles (WIP) and piles which considered the full installation process (FIP). The results show that, there is a competing effect between the installation induced stress increase and the rock degradation. The former is responsible for an increased confinement of the pile whilst the latter a decrease of mechanical strength of the reacting soil. The results show that tensile and compression capacity doubled and increased by 35% respectively because of installation effects. Further research is required to better investigate such behaviour considering larger diameter piles, less permeable conditions and - or impact driving loading.

References

Alvarez Borges, F.J.: The shaft capacity of small displacement piles in chalk. University of Southampton (2019)

Buckley, R.M., Jardine, R.J., Kontoe, S., Parker, D., Schroeder, F.C.: Ageing and cyclic behaviour of axially loaded piles driven in chalk. Géotechnique **68**(2), 146–161. https://doi.org/10.1680/jgeot.17.P.012

Buckley, R.M., Jardine, R.J., Kontoe, S., Barbosa, P., Schroeder, F.C.: Full-scale observations of dynamic and static axial responses of offshore piles driven in chalk and tills. Géotechnique 1–25 (2019)

Ciantia, M.O.: A constitutive model for the hydro-chemo-mechanical behaviour of chalk. In: Engineering in Chalk, pp. 275–281. ICE Publishing (2018). https://doi.org/10.1680/eiccf.64072.275

Ciantia, M.O., Castellanza, R., di Prisco, C.: Experimental study on the water-induced weakening of calcarenites. Rock Mech. Rock Eng. **48**(2), 441–461 (2015)

Ciantia, M.O., di Prisco, C.: Extension of plasticity theory to debonding, grain dissolution, and chemical damage of calcarenites. Int. J. Numer. Anal. Methods Geomech. **40**(3), 315–343 (2016)

Ciavaglia, F., Carey, J., Diambra, A.: Monotonic and cyclic lateral tests on driven piles in Chalk. In: Proceedings of the Institution of Civil Engineers - Geotechnical Engineering, vol. 170, no. 4, pp. 353–366. Thomas Telford Ltd. (2017)

Galavi, V., Schweiger, H.F.: Nonlocal multilaminate model for strain softening analysis. Int. J. Geomech. **10**(1), 30–44 (2010)

Jardine, R.J.: Geotechnics, energy and climate change: the 56th Rankine lecture. Géotechnique **70**(1), 3–59. https://doi.org/10.1680/jgeot.18.RL.001

Lord, J.A., Clayton, C.R.I., Mortimore, R.N.: Construction Industry Research and Information Association. Engineering in chalk. CIRIA (2002)

Monforte, L., Ciantia, M.O., Carbonell, J.M., Arroyo, M., Gens, A.: A stable mesh-independent approach for numerical modelling of structured soils at large strains. Comput. Geotech. **116**, 103215 (2019)

Mortimore, R.N.: Making sense of Chalk: a total-rock approach to its Engineering Geology. Q. J. Eng. Geol. Hydrogeol. **45**(3), 252–334 (2012)

Ng, C., Ran, L.: Offshore Wind Farms: Technologies, Design and Operation, Offshore Wind Farms: Technologies, Design and Operation. Elsevier Inc. (2016)

Oñate, E., Idelsohn, S.R., del Pin, F., Aubry, R.: The particle finite element method – an overview. Int. J. Comput. Methods **01**(02), 267–307 (2004)

Rios, S., Ciantia, M., Gonzalez, N., Arroyo, M., da Fonseca, A.V.: Simplifying calibration of bonded elasto-plastic models. Comput. Geotech. **73**, 100–108 (2016)

Simo, J.C., Hughes, T.J.R.: Computational Inelasticity. Springer, New York (1998). https://doi.org/10.1007/b98904

Tamagnini, C., Ciantia, M.O.: Plasticity with generalized hardening: constitutive modeling and computational aspects. Acta Geotech. **11**(3), 595–623 (2016)

Optimisation of Predictions for Driven Piles Performance in Carbonate Silts for Offshore Structures in the Arabian Gulf

Romain Clavaud, Zine Labidine Delimi(✉), Luke Martin, and Muhammad Bilal Mumtaz

Fugro, Abu Dhabi, UAE
z.delimi@fugro.com

Abstract. The objective of this paper is to bring focus on behavior of carbonate silt layers encountered offshore in the Arabian Gulf interacting with driven pile foundations.

The shallow geology of the Arabian Gulf was subject to complex processes that originated significant heterogeneities in the soil conditions with many different facies that directly affect the foundations of offshore structures. For instance, carbonate soils are quite widely spread in the Arabian Gulf, with a behaviour which is well known to affect the performance of driven pile foundations. The particularity of carbonate silts is not well documented and the industry would provide recommendation for either drained or undrained approaches which are not always representative of the actual behaviour.

As part of this study the authors have reviewed geotechnical and monitoring data from four offshore platforms located in the same field in the Arabian Gulf. Measurements from pile monitoring allow to assess set up of the soil with time which is a very important parameter when assessing pile capacity for driven piles of future platforms or for re-qualification of ageing assets. A closer analysis of the pile monitoring data allowed to better understand the shaft resistance after short setup in the silt layer.

A framework is therefore proposed to support operators and installation contractors. This will aim to better optimise the design of their driven pile foundations for offshore structures according to the soil conditions at both design and life extension stages. Moreover, recommendations are outlined to further reduce uncertainties with prediction of driven pile performances beyond the use of pile monitoring data and geotechnical data.

Keywords: Carbonate silt · Driven piles · Pile design optimisation · Pile monitoring · Arabian Gulf

1 Introduction

Soils with high carbonate contents are commonly found throughout the Arabian Gulf. The peculiar behaviour of carbonate soils has been reported extensively in geotechnical

D. V. K. Huynh et al. (Eds.): VSOE 2021, LNCE 208, pp. 433–440, 2022.
https://doi.org/10.1007/978-981-16-7735-9_48

literature for several decades, especially for the detrimental influence it has on the performance of deep foundations. Nevertheless, carbonate soils are still an area of active research and many aspects remain unexplored. One such aspect is the behavior of carbonate silts, which is the subject of this paper. The behavior of carbonate soils is not always well captured in the codes and standards on which offshore operators heavily rely on, most particularly in the Arabian Gulf. The gap from industry guidelines leaves room for a variety interpretations which affect the design of pile foundations. In other regions where carbonate soils are observed such as the Australian north-western shelf, the regional industry has invested in research and development to compensate the gaps in industry guidelines and developed framework specific to their regional needs. The purpose of this paper is to present a framework that will support operators in developing their own processes specific to their needs and that will address their challenges. By following this framework operators would also be able to reduce uncertainties, and therefore risk on their assets resulting in cost savings.

Carbonate sediments typically exhibit high intergranular and intragranular porosities. This is responsible for many of the engineering properties exhibited by soils with high carbonate contents including the fragile soil structure, crushability, high levels of degradation upon loading, high lateral and vertical variability etc. The extremely high sensitivity of carbonate sediments also leads to the mobilization of very low pile shaft friction during driving [1].

This study aims at highlighting the challenges associated with axial pile capacity prediction methods that have been used on fine grained carbonate silts in the Arabian Gulf. Pile capacities measured from four platforms in the same offshore field in the Arabian Gulf, either at end of driving or restrike, were compared to the predictions made by two different axial pile capacity methods. Differences in the measured and predicted capacities as well as opportunities for pile design optimization are discussed.

2 Site Conditions

The study presented in this paper focuses on four platform locations in the same offshore field in the Arabian Gulf. Although the available information for each platform varied, the results of the geotechnical site investigation and pile installation reports were generally available to the authors.

The soil conditions at the four locations are somewhat similar to each other in terms of soil types encountered and general engineering properties. However, lateral variability exists as far as degree of cementation, layer thickness and strength are concerned. All four locations consist predominantly of fine grained carbonate silt, with other soil types such as carbonate sands, carbonate clays and soft carbonate rocks (calcarenite, calcisiltite, calcilutite) with varying thicknesses. A generalized summary of soil stratigraphy between the four locations is presented in Table 1. Cone penetration test (CPT) data and undrained shear strength measurements are presented on Fig. 1 for all four locations.

The carbonate silt layer (Unit 5) is usually encountered 25–45 m below mudline and is quite complex in nature. The silt layer is usually slightly to well cemented and contains medium to widely spaced, variably thick (ranging from 20 cm to several meters)

Table 1. Generalized overview of soil stratigraphy at the four locations

Unit	Soil type	Thickness (m)	Cone tip resistance, qc (MPa)	Carbonate content	Plasticity index	Fines content
1	Silt/Sand	1–8	0.5–30	>80%	2–12	15–99%
2	Calcarenite/Calcisiltite *	5–11	20 – >60 (Refusal)	>80%	N/A	N/A
3	Silt/Sand	0–15	4 – >60 (Refusal)	>80%	4–21	55–90%
4	Clay	14–22	2–5	10–60%	16–32	95–100%
5	Silt	>55 **	15 – >60 (Refusal)	>85%	2–18	55–98%

* Rock in Unit 2 was usually recovered as sandy/silty material due to very low strength of the rock.
** The silt layer (Unit 5) was observed till the maximum investigation depth of 100 m at all four locations.

layers of gypsum, with occasional layers of calcarenite/calcisiltite. This complex nature of the carbonate silt layers makes it difficult to recover high quality intact samples and therefore carry out advanced laboratory testing. Similarly, CPT data exhibit very heterogenous traces with erratic pore pressure responses. CPT cones often face refusals ($q_c > 60$ MPa) after a penetration of 5–20 cm. The use of CPT classification chart or soil behavior index can show misleading soil behavior classification due to the very high cone tip resistance and erratic pore pressure behavior. Hence, there can be a lot of room for interpretation during soil characterization, and different geotechnical engineers may classify the silt layer as fully drained or undrained based on their interpretation, leading to inaccurate foundation design and potential foundation performance issues that will be further discussed in the paper.

3 Methodologies Used in Arabian Gulf for Carbonate Sediments

Pile design in the Arabian Gulf has historically been performed using API standards (API RP 2A-WSD and API RP 2GEO) [2, 3]. API main text guidelines have been predominantly based on data from tests performed in silica soils (commonly high plasticity clays) from the Gulf of Mexico and Europe decades ago. In recent years API has introduced commentary and recommendations for design in carbonate sands and Kolk (2000) [6] has been accepted and considered by the industry in the Arabian Gulf. Nevertheless, recommendations for soils such as carbonate clays and silt remain absent from API standards.

An alternative method based on pile static test data and derived from [4] and [5] was recently proposed to derive axial pile capacities in specific offshore fields in the Arabian Gulf. The method was adjusted by geotechnical specialists to account for the effect of carbonate soils and uses the cone tip resistance for frictional soils, and undrained shear strength and plasticity index for cohesive soils. To derive shaft friction resistance, frictional soils with high carbonate content considered a relationship based on submerged

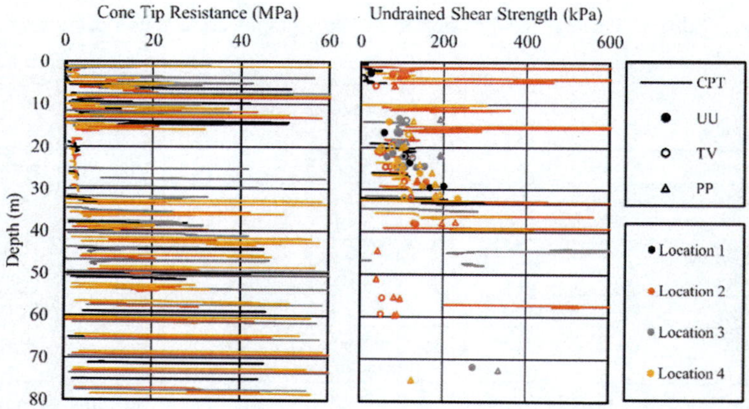

Fig. 1. Cone tip resistance and undrained shear strength measurements across the four locations (with s_u derived from CPT assuming $N_k = 25$)

unit weight with a friction fatigue component, whereas for soils with intermediate carbonate content, logarithmic interpolation between silica soils and high carbonate soils was considered as per [6]. It shall be noted that no distinction was made in the method to derive shaft resistance in either silt or sand, a common frictional approach was used for both. This method is referred to as "Method B" in the rest of the paper, whereas the API RP 2GEO method following Kolk's approach [6] is referred to as "Method A".

4 Pile Monitoring Measurements

Pile monitoring data is a valuable source of information as it can help in estimating the axial pile capacity during installation.

CAPWAP (Case Pile Wave Analysis Program) [7], which is a signal matching software that uses measurements taken on a pile upon hammer impact, is used to estimate the achieved pile capacity at the time of measurement, generally at end of driving (EOD) and upon restrike. CAPWAP results enable to estimate the ratio between shaft and toe resistance.

CAPWAP analyses at EOD and for restrike were available at the studied locations, therefore information on set-up build up were available for time period ranging between 1 h and 24 h. Most of axial pile capacity prediction methods, including the ones considered in this study, provide an estimate of the axial pile capacity at about 100 days after end of driving. Therefore the set-up measured from CAPWAP are likely to under-estimate final set up at 100 days. As part of this study we compared predicted pile capacities from method presented in previous section and measured ones from CAPWAP results.

5 Carbonate Silt Response to Driven Pile Foundations

The complex nature of the carbonate silt layers was discussed in Sect. 2. Hence, understanding the behaviour of such material may be challenging, if solely based on geotechnical soil information. This is illustrated by the differences of interpretation between geotechnical engineers that have been working on this peculiar type of soil. The two methods discussed in Sect. 3 treat soil either as fully drained or undrained. This simplification may not be suitable when considering silts since it is typically expected to behave in a partially drained manner during axial pile loading [8].

In the case of the carbonate silts present in this offshore field, it was observed that the shaft resistance would be underestimated if considered as fully drained material and overestimated if considered as fully undrained. Moreover, with the presence of cementations and gypsum layers, hard driving with risk of refusal and high end bearing may be expected, which is quite different than what is generally experienced with frictional carbonate sands. In order to increase accuracy in predictions for both installation and end bearing, it is recommended to be cautious during geotechnical data acquisition in order to clearly identify where high end bearing and hard driving may be expected.

It was therefore concluded that the carbonate silt was to be modelled with a specific partially drained approach in order to predict more reliable static pile capacity, and special care should be brought into consideration for end bearing. The authors of this paper also recommend that a more advanced laboratory testing protocol be developed to further assess and measure the response of such soils when interacting with driven pile foundations in the future.

6 Results and Discussions

A comparison of the predicted pile capacity using both methods A and B with the measured pile capacity (using CAPWAP) at the four locations is presented in Fig. 2. For the purpose of this study, the silt materials are treated as fully drained, except for when thin cohesive silt layers were occasionally observed.

The accuracy of method A predictions vary with depth, with the relatively shallower depth (~50 m) predictions are generally higher than the measured capacity at EOD (end of driving), whereas the deeper (>70 m) predictions are lower than measured capacity at EOD, which doesn't account for set-up effect. Measured capacities at restrike are similar to, or more often, greater than the predicted capacities. It is very likely that the long term capacity (at 100 days) is underestimated. Very similar conclusions can be drawn from the method B predictions.

A common trend observed is that both methods struggle to accurately predict pile capacities in the deeper carbonate silt layer. The measured capacities at EOD are similar to or greater than the capacities predicted by both methods, whereas the measured capacities upon restrike are always greater than predicted capacities. Hence, for the platform locations studied in this paper, it can be concluded that the actual pile capacity after a period of 100 days (full set-up) will be higher than the predicted capacities. This can partly be attributed to the partially drained behavior due to the high percentage of fines in the silt layer. Modelling this silt layer as a fully drained material leads to underprediction of capacity for both methods. Conversely, modelling it as a fully undrained material would probably result in overpredicting the contribution of the shaft resistance and under estimate the toe resistance leading to much more uncertain results. Accurate prediction of shaft friction resistance can be critical in cases of conductor piles or composite piles where only the shaft resistance contribute to the axial pile capacity.

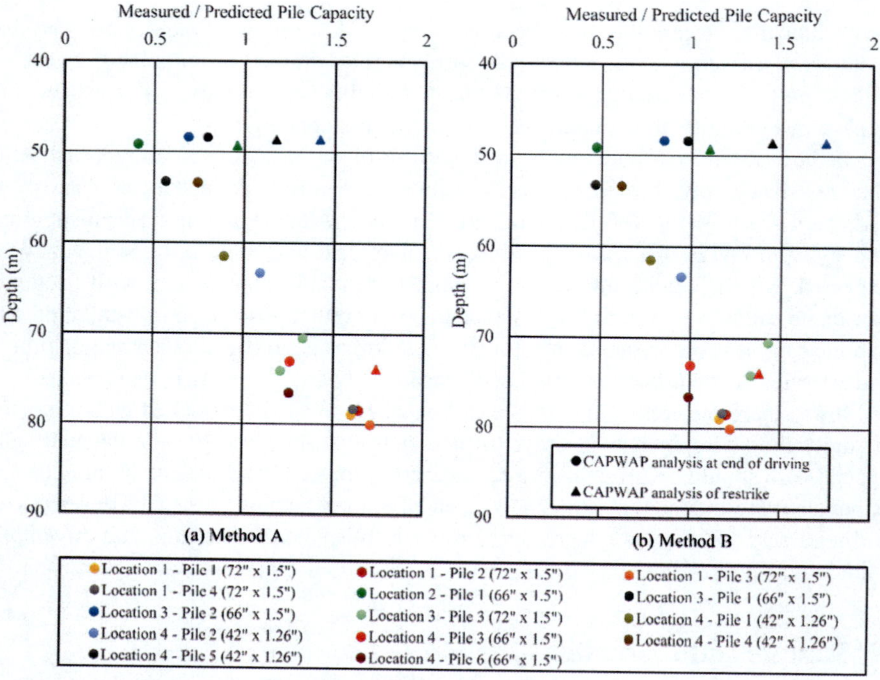

Fig. 2. Comparison of measured pile capacities at end of driving and restrike (at 1 to 24 h after end of driving) with pile capacities predicted by two different methods

7 Proposed Framework

In order to limit the exposure level of operators to design risks associated with uncertainties on soil behaviour when interacting with driven pile foundations, especially for soil types that are not well addressed by industry guidelines in matured fields, it is recommended to consider the following elements:

1. Carry out pile driving monitoring and CAPWAP analyses at EOD and after restrike test to build a database of short-term set-up measurements. It is recommended to perform restrike tests at least 24 h after EOD.
2. Review geotechnical data, selected design parameters and pile monitoring data from the area/field of interest and develop geotechnical ground models considering experience from previous pile installations data/CAPWAP analyses.
3. Back-analyse the SRD profiles using the available pile installations records in the area in order to improve the drivability predictions for the future installations.
4. Assess the performance of different axial pile capacity methods versus measurements based pile monitoring.

5. If a large database of pile monitoring data is available in the same field, then a site specific method may be developed based on an extensive review of geotechnical and pile monitoring data in the field. The site-specific methods shall be developed by an experienced specialist with domain expertise in both geotechnical and pile monitoring fields. This generally allow for a considerable optimisation of pile axial capacity which can be used for platform repurposing in the future.

The optimisation of the static pile capacity can result in a significant increase of the ratio of maximum cyclic loading to static pile capacity. In this case, the geotechnical specialist shall explicitly assess the pile foundation response to axial cyclic loading. This can be achieved by performing advanced cyclic tests on selected high-quality soil samples to understand soil response to cyclic loading. The pile response can be then assessed using specialized software such as CYCLOPS [9].

We also encourage operators and contractors to explore alternatives to conventional pile monitoring by using technologies such as Osterberg cell or use of fiber optic with monitoring [10]. Both technologies have potential to evaluate shaft resistance distribution along the pile with greater level of accuracy. With an additional advantage with Osterberg cell that testing can measure directly the static capacity and can be carried out after much longer set-up period without requirement to keep installation spread on standby.

Uncertainties related to the performance of the jacket structure and pile foundations under different loading conditions can be refined by considering a combination of pile monitoring, structural monitoring and metocean measurements simultaneously. The proposed comprehensive monitoring approach would allow to increase the reliability of design and avoid over-conservatism which can help in re-qualification of existing platforms.

8 Conclusions

A comparison of pile capacities predicted by two different methods with measured pile capacities in the field (using pile monitoring data) was performed for four platform locations and for different pile sizes in an offshore field in the Arabian Gulf. It is observed that typical axial pile capacity methods developed based on pile testing data from other regions present limitations to accurately predict the pile performance in fine grained carbonate silts in the Arabian Gulf. The shaft friction and end bearing resistance are either overestimated or underestimated, leading to an unoptimized design and in extreme cases, pose severe risks to asset integrity. A simple framework was presented for utilizing pile monitoring data of previous installation in the same field in order to make more realistic pile capacity predictions. It is recommended to calibrate the ground model, selected soil parameters or the pile capacity methods to deliver site specific predictions. It is also recommended that site specific pile capacity methods are developed, if a large enough database of site conditions and pile monitoring data is available, to reduce uncertainties and risks.

References

1. Watson, P., et al.: Foundation design in offshore carbonate sediments – building on knowledge to address future challenges. In: From Research to Applied Geotechnics: Invited Lectures of the XVI Pan-American Conference on Soil Mechanics and Geotechnical Engineering, 17–20 November 2019, Cancun, Mexico, vol. 7, p. 240. IOS Press (2019)
2. American Petroleum Institute: Recommended Practice for Geotechnical and Foundation Design Considerations. API Recommended Practice 2GEO (API RP 2GEO), 1st edn. (With Addendum 1 October 2014) (2014)
3. American Petroleum Institute: Recommended Practice for Planning, Designing and Constructing Fixed Offshore Platforms - Working Stress Design. API Recommended Practice 2A-WSD (RP 2A-WSD), 22nd edn. (2014)
4. Karlsrud, K., Clausen, C.J.F., Aas, P.M.: Bearing capacity of driven piles in clay, the NGI approach. In: Proceedings of International Symposium on Frontiers in Offshore Geotechnics, vol. 1, pp. 775–782. A.A. Balkema Publishers, Perth (2005)
5. Clausen, C.J.F., Aas, P.M., Karlsrud, K.: Bearing capacity of driven piles in sand, the NGI approach. In: Proceedings of International Symposium on Frontiers in Offshore Geotechnics, vol. 1, pp. 677–682. A.A. Balkema Publishers, Perth (2005)
6. Kolk, H.J.: Deep Foundations in calcareous sediments. In: Al-Shafei, K.A. (ed.) Engineering for Calcareous Sediments: Proceedings of the Second International Conference on Engineering for Calcareous Sediments, vol. 2, pp. 313–344. A.A. Balkema, Rotterdam (2000)
7. CAPWAP. https://www.pile.com/products/capwap. Accessed 15 Oct 2020
8. Robertson, P.K., Woeller, D.J., Gillespie, D.: Evaluation of excess pore pressures and drainage conditions around driven piles using the cone penetration test with pore pressure measurements. Can. Geotech. J. **27**(2), 249–254 (1990)
9. Erbrich, C.T., O'Neill, M.P., Clancy, P., Randolph, M.F: Axial and lateral pipe design in carbonate soils. In: Proceedings of 2nd International Symposium on Frontiers in Offshore Geotechnics, pp. 125–154. Taylor and Francis, Perth (2010)
10. Buckley, R.M., et al.: Optimization of impact pile driving using optical fiber Bragg-grating measurements. ASCE J. Geotech. Geoenviron. Eng. **146**(9), 04020082 (2020)

Revised Soil Damping Using Realistic Pile-Soil Interaction Models

Bruno Stuyts(✉), Wout Weijtjens, and Christof Devriendt

OWI-Lab, Vrije Universiteit Brussel, Pleinlaan 2, 1050 Brussels, Belgium
bruno.stuyts@vub.be

Abstract. The contribution of soil damping to the overall damping of offshore wind turbine structures has proven to be elusive. The strain-dependent soil stiffness and material damping can be quantified from small-scale element tests but applying the results from these tests in an integrated structural model requires a realistic assessment of the shear strains in the subsoil (Kagawa and Kraft 1980). Recent developments in pile-soil interaction modelling for large-diameter monopiles (Byrne et al. 2017) have led to more accurate models of pile response under operational loads. Moreover, field measurements of offshore wind turbine foundations before installation of the superstructure are available which allow improved identification of the effect of soil damping. Applying the conventional pile-soil interaction modelling did not provide a satisfactory explanation for the differences in soil damping.

This paper applies improved pile-soil interaction modelling with site-specific soil parameters to three different offshore monopile foundations for a Belgian offshore wind project to identify whether the improved models can explain the differences in measured soil damping. The results show the influence of the pile-soil interaction modelling on the derived soil damping.

Keywords: Damping · Monopile · Response · Monitoring

1 Introduction

Damping of offshore wind turbine structures plays a crucial role in the assessment of the foundation lifetime as the amplitudes of vibrations at the resonance frequency is inversely proportional to the damping. The overall damping of an offshore wind turbine structure consists of multiple components as shown in Eq. 1.

$$D_{tot} = D_{structural} + D_{soil} + D_{aero} + D_{hydro} + D_{mass} \tag{1}$$

The overall damping includes contributions from the material damping of the structure and the energy dissipation due to internal shearing in the soil. In addition to this, aerodynamic damping from the interaction of blades and tower with the surrounding air and hydrodynamic damping from the interaction of the submerged section of the monopile and transition piece with the water need to be considered. Finally, offshore wind turbine structures often have tuned mass dampers which also contribute to damping.

D. V. K. Huynh et al. (Eds.): VSOE 2021, LNCE 208, pp. 441–448, 2022.
https://doi.org/10.1007/978-981-16-7735-9_49

1.1 Damping Measurements in the Field

The overall damping of the structure can be derived from acceleration monitoring data at different levels on the offshore wind turbine structure during parked condition (Shirzadeh et al. 2015) or by examining the response during emergency stops (Devriendt et al. 2013).

Damping estimates for the 5 dominant modes are shown in Fig. 1 for a 5 m diameter monopile during low and high wind speed. The average overall damping of the first two modes ranges between, 1.5% and 3.0% depending on the mode and loading condition considered. It is observed that damping increases with increasing wind speed and therefore increase mobilization of soil resistance.

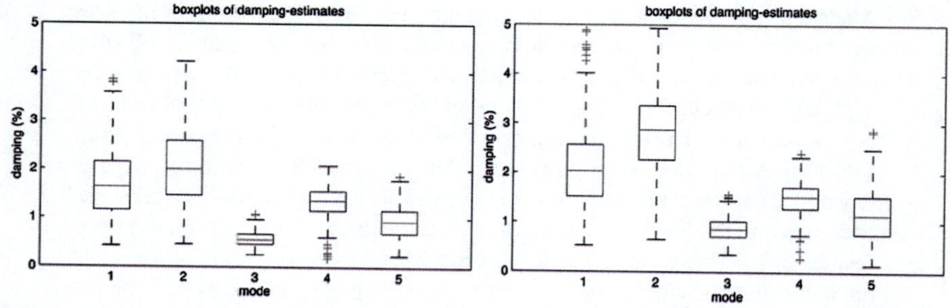

Fig. 1. Damping estimates for a 3.3 MW offshore wind turbine on a 5 m diameter monopile (Devriendt et al. 2014).

While damping measurements can be inferred from monitoring data, there is no widely accepted methodology to calculate the contribution of soil damping to the overall damping of the structure. Moreover, it is unclear how damping measurements from soil element tests can be used to determine the soil-related damping for offshore monopile foundations. This paper describes a critical review of the methodology by Kagawa and Kraft (1980) which was developed for pile design in seismic areas.

1.2 Methodology for Monopile Damping Calculations

The methodology presented by Kagawa and Kraft (1980) starts off from a lateral pile response calculation under a given excitation. The pile deflection y under this load is then converted in a profile of shear strain γ vs depth according to Eq. 2.

$$\gamma(z) = \frac{1+\nu}{2.5D} \cdot y(z) \tag{2}$$

Where υ is the Poisson's ratio of the material and D is the pile diameter. This relation was developed for lateral pile analysis of slender piles in soft clay and might not be applicable to stiff monopiles in layered soils.

A profile of small-strain shear modulus G_{max} needs to be estimate based on the results of field or laboratory tests and the degradation of shear modulus from the calculated strain level in Eq. 2 is then taken into account in the elastic energy (E_s) profile vs depth (Eq. 3).

$$E_s(z) = 0.5 \cdot G(z, \gamma) \cdot \gamma(z)^2 \tag{3}$$

Similarly, the profile of dissipated energy (E_d) vs depth can be obtained from the depth- and strain-dependent damping ratio derived from soil element tests (Eq. 4).

$$E_d(z) = 4 \cdot \pi \cdot E_s(z) \cdot \xi(z, \gamma) \tag{4}$$

The elastic energy and dissipated energy can then be integrated over the penetrated pile length to obtain the soil damping ξ_{soil} as shown in Eq. 5.

$$\xi_{soil} = \frac{\int E_d(z)dz}{4 \cdot \pi \cdot \int E_s(z)dz} \tag{5}$$

The methodology is a three-step approach requiring the following calculations:

- Pile lateral response analysis under load levels representative for the loading episode under consideration;
- Shear modulus degradation calculation using results from field and laboratory tests;
- Soil damping ratio determination from laboratory tests.

Moreover, this methodology requires knowledge of the dynamic soil properties and their dependency on soil type, stress level and strain magnitude.

1.3 Soil Conditions and Load Levels at Monitored Locations

In this paper, the different components of the soil damping calculation are applied at three example monopile locations for which measured total damping is available. The soil profiles at the monitored locations consist of stiff overconsolidated clay, interbedded sand-clay and dense to very dense sand. Location 1 has clay at the surface while locations 2 and 3 predominately have sand at the surface.

The pile penetrations at the site range between 32 m and 40 m for pile diameters between 7 m and 8 m.

Design soil profiles for each monitored location were determined based on the results from in-situ PCPT tests, S-PCPT tests and onshore laboratory testing. Soil profiles for lateral response analysis were defined for both the API RP2 GEO method (API RP 2GEO 2011) and the more recent PISA methods (Byrne et al. 2017). The latter method is deemed to be more accurate for operational loading. This should also have an impact on the damping assessment, which should in turn become more accurate.

Correlations between cone tip resistance and small-strain shear modulus were calibrated to available S-PCPT measurements (Stuyts et al. 2020) to obtain design profiles of the small-strain shear modulus.

The load levels considered for the lateral response analysis were based on estimated operational loads during the monitored episodes. The load levels considered were far below the ULS load (Fig. 2).

The lateral response analysis for API RP2 GEO and PISA methods were carried out using a non-linear 1D beam column finite element analysis to obtain the deflected shape of the pile. The pile head deflections for the considered operational load levels were very small, ranging between 0.05 to 0.5 mm depending on the soil profile and method considered. The deflected pile shape is presented in subsequent sections.

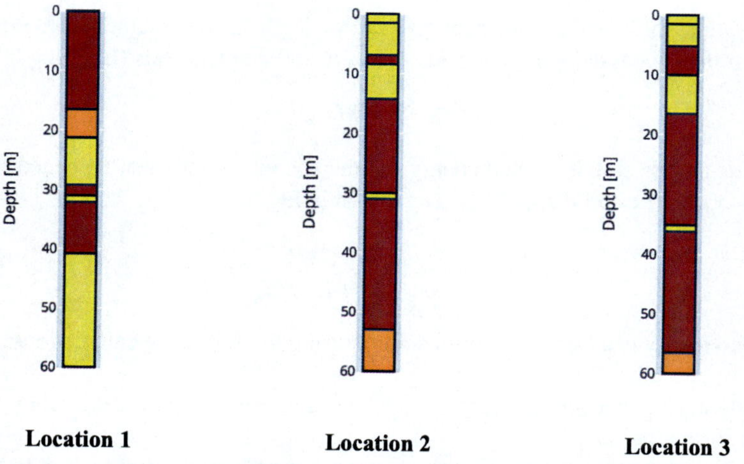

Fig. 2. Soil profiles at monitored locations (brown: stiff overconsolidated clay, orange: interbedded sand-clay, yellow: dense to very dense sand)

2 Soil Damping Measurements for North Sea Soils

2.1 Test Description

At the selected project site, resonant column tests are available on three different cohesive samples. Two cohesionless samples were also tested at three different stress levels, resulting in a total of 6 tests on sandy soil.

2.2 Empirical Correlations for Shear Modulus Degradation and Damping Ratio

Ishibashi and Zhang (1993) present empirical formulae to calculate shear modulus and degradation for sands and clays. The shear modulus degradation depends on the plasticity index of the soil. For cohesive soils, greater plasticity means less rapid degradation of the shear modulus. For cohesionless soils, the equations can be used by setting the plasticity index of the soil equal to zero.

The dependence of shear modulus degradation on the mean effective stress at the considered depth is equally taken into account. At lower effective stresses, more degradation is expected.

Ishibashi and Zhang (1993) also propose an empirical formula for the damping ratio of cohesionless and cohesive soil.

The formulae by Ishibashi and Zhang were applied to the resonant column tests for the project site to check their applicability for overconsolidated North Sea soils.

2.3 Damping Measurements for Cohesionless Soils

The tests on cohesionless soils were carried out at consolidation pressures ranging between 50 and 400 kPa. Comparison with the empirical formulae by Ishibashi and Zhang showed good agreement between the empirical shear modulus degradation curves and the test data. For the damping ratio however, agreement was poor with damping measurements indicating lower damping than the proposed trends. For use in further calculations, alternative formulae which capture the dependency on stress level were suggested as shown in Fig. 3.

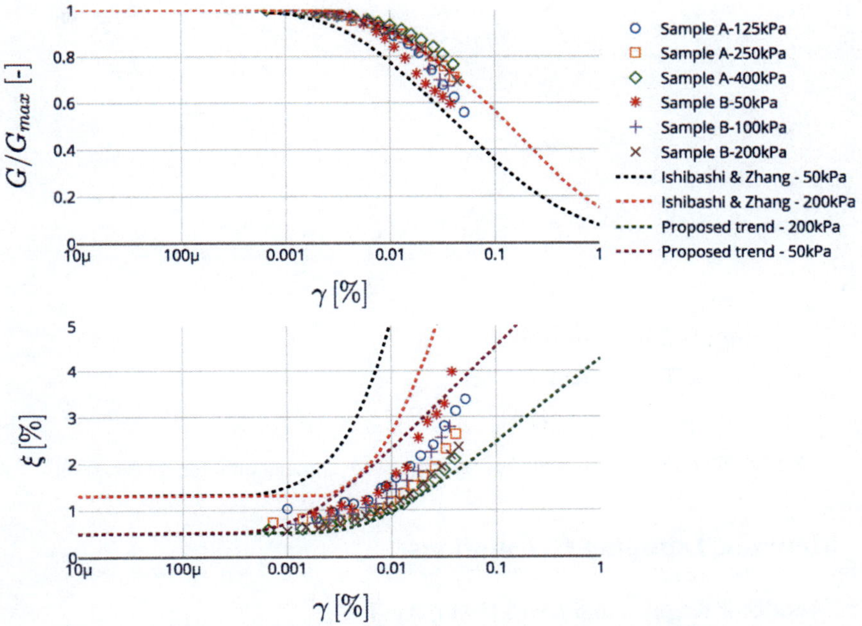

Fig. 3. Strain-dependence of shear modulus and damping ratio for cohesionless soil

2.4 Damping Measurements for Cohesive Soils

For cohesive soils, testing was carried out at two different stress levels and three initial void ratios. The plasticity index of the material was not determined directly on the samples used for testing but Atterberg limit tests on other samples from the boreholes indicated a relatively wide range of plasticity index between 50 and 100%.

Similar to the cohesionless material, the shear modulus degradation trend showed good agreement with the test data. For cohesive soils, the empirical formulae overestimated the damping ratio and alternative formulae were again proposed as shown in Fig. 4. As the exact value of plasticity index was not known, low, best and high estimate trends were proposed.

The soil profiles determined at the foundation locations were combined with the modulus and damping ratio degradation laws and the calculated shear strains along the pile depth to obtain profiles of shear modulus and damping vs depth.

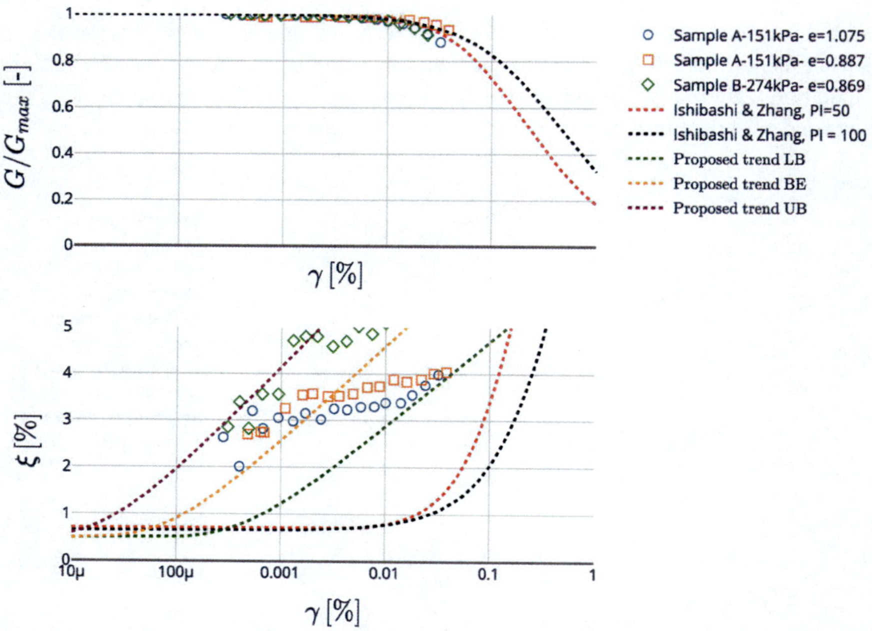

Fig. 4. Strain-dependence of shear modulus and damping ratio for cohesive soil

3 Monopile Damping Calculations

3.1 Results of Kagawa and Kraft (1980) Analysis

The damping assessment described in Eq. 2–5 was performed for the three selected locations. The results of the profiles of shear modulus degradation, damping ratio and elastic and dissipated energy vs depth are shown for Location 1 in Fig. 5; both for the API RP2 GEO and PISA response analyses. The results show that deflections calculated with the PISA methods are smaller, resulting in lower shear modulus degradation and damping. This observation is in line with the greater accuracy of the PISA methods for operational loading.

The elastic and dissipated energy were integrated over the depth according to Eq. 5 and an estimate of the soil-related damping was obtained. This estimate was compared to the average measured total damping of the structure (Table 1).

If the methodology would provide reliable estimates of damping, the ratio of calculated soil damping would be uniform across the locations. This is clearly not the case here, with the calculated soil damping according to the API response analysis at Location 1 even being higher than the total measured damping. The PISA method already improves the estimate at Location 1 although there is an inconsistency with Location 2 and 3 where soil damping forms a much lower proportion of the measured overall damping. With Location 1 being the only location with clay at the surface, a higher value of damping is expected based on the geotechnical properties of the soil (higher damping observed for cohesive soil and lower shear modulus). However, this is not reflected in

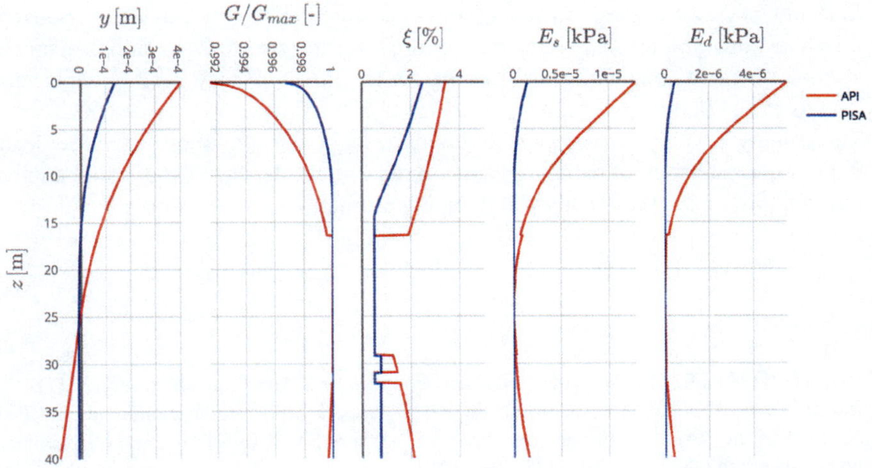

Fig. 5. Results of damping assessment at location 1 according to Kagawa and Kraft (1980)

the measurement. Further multi-disciplinary analysis is required to determine the causes of this discrepancy.

Table 1. Comparison of calculated soil damping and measured overall damping.

Location	Calculated soil damping/measured average total damping	
	API	PISA
Location 1	2.407	0.866
Location 2	0.413	0.222
Location 3	0.784	0.286

4 Conclusions and Recommendations

This paper presents a critical review of analysis method by Kagawa and Kraft (1980) to determine soil-related damping for monopiles based on damping measurements from resonant column tests.

The method was applied to three North Sea monopile locations where overall damping measurements and resonant column test results were available. The analysis revealed three areas where the analysis could be improved:

- Using the more recent PISA guidance for monopile lateral deflection calculations resulted in more realistic pile deflections at low load levels. The inaccuracy of the API method for these low load levels compromises the application of the Kagawa and Kraft (1980) method from the onset;

- Selection of the small-strain stiffness at monopile foundation locations should be based on a data integration from in-situ and onshore laboratory tests. Correlations of small-strain stiffness with cone tip resistance from the literature may need re-calibration (Stuyts et al. 2020);
- The paucity or absence of resonant column tests at offshore wind farm sites leads to poor quantification of damping values and the strain-dependence of soil damping, thus compromising the accuracy of damping estimates.

References

API RP 2GEO: API RP 2GEO Geotechnical and Foundation Design Considerations (2011)

Byrne, B.W., et al.: PISA: new design methods for offshore wind turbine monopoles. In: Proceedings of the 8th International Conference on Offshore Site Investigation and Geotechnics. Presented at the OSIG 2017, pp. 142–161 (2017)

Devriendt, C., Jordaens, P.J., De Sitter, G., Guillaume, P.: Damping estimation of an offshore wind turbine on a monopile foundation. IET Renew. Power Gener. **7**, 401–412 (2013). https://doi.org/10.1049/iet-rpg.2012.0276

Devriendt, C., Weijtjens, W., El-Kafafy, M., De Sitter, G.: Monitoring resonant frequencies and damping values of an offshore wind turbine in parked conditions. IET Renew. Power Gener. **8**, 433–441 (2014). https://doi.org/10.1049/iet-rpg.2013.0229

Ishibashi, I., Zhang, X.: Unified dynamic shear moduli and damping ratios of sand and clay. Soils Found. **33**, 182–191 (1993)

Kagawa, T., Kraft, L.M.: Lateral load-deflection relationships of piles subjected to dynamic loadings. Soils Found. **20**, 19–36 (1980)

Shirzadeh, R., Weijtjens, W., Guillaume, P., Devriendt, C.: The dynamics of an offshore wind turbine in parked conditions: a comparison between simulations and measurements. Wind Energy **18**, 1685–1702 (2015). https://doi.org/10.1002/we.1781

Stuyts, B., Weijtjens, W., Devriendt, C., Versteele, H., Vanden Haute, C.: Monopile lateral response calibration from in-situ monitoring data. In: Proceedings of the Fourth International Symposium Frontiers in Offshore Geotechnics. Presented at the ISFOG2020, Austin, Texas (2020)

Fossil Energy: Technology Opportunities and Challenges

Applied Machine Learning and Deep Learning to Predict Oil and Gas Production

Nguyen Van Hung[1], Luong Khanh Loc[1(✉)], Nguyen Thi Thanh Huong[2], Nguyen Van Dung[2], and Nguyen Minh Quy[3]

[1] PetroVietnam University, 762 Cach Mang Thang 8, Long Toan Ward, Ba Ria, Vietnam
hungnv@pvu.edu.vn

[2] Campus in Ho Chi Minh City, University of Transport and Communications, No. 450-451 Le Van Viet Street, Tang Nhon Phu A, District 9, Ho Chi Minh City, Vietnam
{huongntt_ph,ngvdung}@utc.edu.vn

[3] Vietnam Petroleum Institute, 167 Trung Kinh, Yen Hoa, Cau Giay, Hanoi, Vietnam

Abstract. This study proposed an approach using Regression algorithms and Ensemble methods to leverage the results. A typical Recurrent Neural Networks include Long Short-Term Memory (LSTM), Gated Recurrent Unit (GRU) and Simple RNN has been developed to predict oil and gas production. The results have shown that machine learning gained good results in the early stage of the production phase and the recurrent neural networks have improved the prediction in the final stage of the production phase. Compare with conventional decline curve analysis (Arps), LSTM has outperformed due to the ability in time-series prediction. In Volve field studies, LSTM is better than reservoir simulation at well F14 and F15 and close to the accuracy of actual values in all five wells. It demonstrates the ability to use Artificial Intelligence in production forecasting. While reservoir modeling needs a great amount of time and effort to construct a reservoir model with extensive domain knowledge of geology, geophysics, petrophysics, reservoir engineering, and reservoir simulation, LSTM only based the historical data to complete the tasks. So, in this circumstance LSTM is working on cutting-edge methods, and need to be studied carefully in the future in order to give better and accurate results.

Keywords: Machine learning · Deep learning · Production forecasting · Artificial intelligent · Volve dataset

1 Introduction

Machine Learning (ML) is basically the prediction of the future with or without, prior knowledge of the past (supervised learning and unsupervised learning. A study from McKinsey in 2016 mentioned that digital organization, which covers AI and Machine Learning is reshaping the future of the industry. Digitizing and automating work will result in better safety and productivity as fewer people will be at risk and the risk of human error is reduced. Moreover, in the low oil price environment, oil companies often have no choice but to cut their operational budget, reduce the headcount, and/or fully optimize

D. V. K. Huynh et al. (Eds.): VSOE 2021, LNCE 208, pp. 451–458, 2022.
https://doi.org/10.1007/978-981-16-7735-9_50

their remaining assets. This further encourages the adoption of artificial intelligence in the industry, with the hope to simplify and/or automate inefficient processes [1].

There have been some implementations of AI in the field of oil and gas production. Boomer (1995) [2] used the Neural Network to forecast oil production in Vacuum Field, Texas and the model outperformed the professionals 93% of the time. Cao et al. (2016) found that Neural Network performed better than conventional decline curve methods [3]. Suhag et al. (2017) [4] used Neural Network to predict the oil rate in Bakken shale wells and yielded better results than Duong's method. J. Sun, X. Ma, and M. Kazi (2018) [5] have recently proposed a novel method of using Recurrent Neural Networks for production forecast, and it gives outperformed results compared to conventional methods like DCA, SEPD and Duong.

In this paper, a workflow of using ML and DL technique will be used to predict future production rate with limited parameter data: (1) oil production data; (2) gas production data; and (3) water production data. In first part, the traditional Decline Curve Analysis (DCA) method is used for forecasting, whereas in second step, a software is utilized to predict the oil and gas production. Then some AI methods were used to compare the results between different stages.

2 Methodology

2.1 Input Data

The production dataset used in this paper is from Volve field on Norwegian continental shelf with around 8 years of history (Table 1). Daily oil, gas, and water production data are all recorded consist with pressure, temperature and choke size as operation constraint. These production data are the main input variables for the machine learning and deep learning model, and it gives multi-phase production predictions in the future as model output.

In this research, bottom hole pressure (BHP), tubing head pressure (THP), bottom hole temperature (BHT), well head temperature (WHT), different pressure in the casing (DP), choke size (CS) in percentage is the input for training machine learning and deep learning algorithms.

Table 1. Description of data set (from historical data of oil production)

	BHP	BHT	DP	CS	WHP	WHT	OIL	GAS	WATER
Count	2365	2365	2365	2365	2365	2365	2365	2365	2365
Std	14.47	3.35	22.44	24.82	12.61	2.23	1052	149356	934.56
Min	135.63	54.64	103.14	20.30	27.19	72.14	100.33	5928.54	100.16
Max	281.3	106.7	239.8	100.0	115.0	93.51	5644.3	786328.3	5691.7

2.2 Workflow

Almost every machine learning and deep learning algorithm have followed the workflow in Fig. 1. Machine learning models are implemented with Scikit-learn library, and deep learning models are also implemented with Keras library, two of these very useful and popular libraries for working with ML/DL algorithms. During the training process, we have used a grid search for finding the best combination of hyperparameters of algorithms.

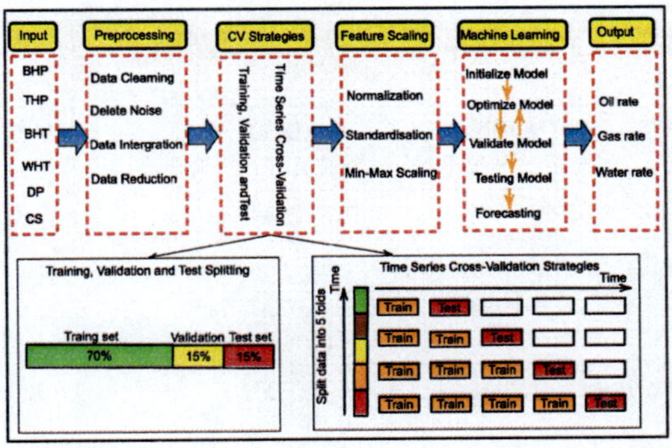

Fig. 1. Detailed work flow for using machine leaning and deep learning algorithms.

3 Result and Discussion

3.1 Machine Learning

The Table 2 shows the result of every algorithm we have used. The coefficient of determination (R2) and mean squared error (MSE) has been used to evaluation algorithm.

After comparing the result on validation and test set, we just picked up five algorithms that yield the best score on the test set. In order to evaluate how possible, the algorithms can be applied to forecast a new set of data, we use the Pearson correlation for evaluation. Table 3 show the Pearson correlation for the best five algorithms.

While all machine learning algorithms we have used, data need to be shuffled before training, the dependence of production rate followed time have vanished. Therefore, we continue to use the time series cross-validation strategy on the same data set. Instead of using gradient boosting machine regressor, authors used LGBM and XGB algorithms because these algorithms have better result compare with GBM. It is observed that the results of using the time series cross-validation method on oil and gas production still depend on the phase of cross-validation data. Specifically, at the early stage, these methods cannot catch the oil and gas trend, and it gives better results near the end of the production phase. There is a gap between actual values and prediction that demonstrate that these algorithms are not sufficient for time series prediction. Hence, we continue to utilize Neural networks and Recurrent neural networks.

Table 2. Score and error of machine learning algorithms on oil

No.	Model	Validation score	Validation error	Test score	Test error
1	Linear regression	0.82132	0.00598	0.80275	0.00609
2	Ridge	0.8257	0.00583	0.78125	0.00676
3	Kneighbors regression	0.9218	0.00262	0.95795	0.00130
4	Support vector regression	0.8556	0.00483	0.84564	0.00477
5	Decision tree regression	0.94392	0.001876	0.95973	0.00124
6	Random forest regression	0.9687	0.00138	0.9585	0.00128
7	Extra tree regression	0.9569	0.00144	0.96031	0.00123
8	AdaBoost regression	0.90031	0.00334	0.90247	0.00301
9	Gradient boosting regression	0.9554	0.00149	0.9597	0.00124
10	XgBoost regression	0.92087	0.00265	0.93438	0.00203
11	LightGBM regression	0.89392	0.00355	0.93392	0.00204

Table 3. Pearson correlation of best five algorithms

No.	Model	Pearson correlation (oil)	Pearson correlation (gas)	Pearson correlation (water)
1	Exreame gradient boosting	0.96667	0.96995	0.8988
2	Decision tree regression	0.9798	0.9596	0.8970
3	Randon forest regression	0.97927	0.9757	0.9161
4	Gradient boosting machine	96689	0.9745	0.92
5	Extra tree regression	0.98065	0.98205	0.9351

3.2 Neural Networks and Recurrent Neural Networks

a. Neural Networks

Neural networks have been used widely in the oil and gas industry from the time it is created. Compare with other machine learning algorithms, NNs have given better or worse results depend on the process of training. While shallow neural networks with one hidden layer (25 nodes) give worse score and error is bigger, deep neural networks with three hidden layers give the increase the score and reduce error. In this research, we have tuned hyperparameters of neural networks to find the best combination of them for giving good results on the validation test (Table 4, 5, 6 and 7).

Table 4. Score and error of oil prediction results

No.	Model	Validation score	Validation error	Test score	Test error
1	0.8563	0.0048	0.0048	0.8426	0.00486
2	0.9492	0.001568	0.001568	0.94	0.00199

Table 5. Score and error of gas prediction results

No.	Model	Validation	Validation error	Test score	Test error
1	Shallow neural networks	0.6826	0.0101	0.6068	0.00948
2	Deep neural networks	0.7824	0.00551	0.7767	0.00574

Table 6. Score and error of water prediction results

No.	Model	Validation	Validation error	Test score	Test error
1	Shallow neural networks	0.6826	0.0101	0.6068	0.00948
2	Deep neural networks	0.7824	0.00551	0.7767	0.00574

Moreover, we have used some very novel techniques in AI, dropout for example, and ReLU activation function, to train our neural networks. Thus, some problems can be solved by using recurrent neural networks.

It is obvious that neural networks have given better results when increasing the number of hidden layers from 1 to 3. However, like other machine learning algorithms, data also need to be shuffled before training. That lead to the loss of temporal characteristic. It is the main problem of using machine learning algorithms include neural networks if

Table 7. Pearson correlation of neural networks

No.	Model	Pearson correlation (oil)	Pearson correlation (gas)	Pearson correlation (water)
1	Shallow neural networks	0.9225	0.9372	0.8314
2	Deep neural networks	0.9798	0.9711	0.9014

data is shuffled. This problem can be solved by using recurrent neural networks which been proven to be one of the most successful algorithms of time-series problems.

b. Recurrent Neural Networks
Recurrent neural network algorithms are very powerful with time series-data. We split data set in three parts: train, validation, and final year. Compare with machine learning models, RNNs have improved significantly when predicted in the last year. Since GRU and LSTM give the best result on oil prediction, and the SimpleRNN result is better with water forecasting. Applied Long Short-Term Memory as the best algorithms.

After the comparison of different algorithm results, the recurrent neural network seems to be the most appropriate remedy for time series production prediction. Hence, Long - short term memory results have been selected to compare with convention and reservoir modeling methods, namely Decline Curve Analysis (DCA) and reservoir simulation. The history production data is also divided into a training set and test set.

c. Comparison with Decline Analysis Curve (DCA)
There are eight wells in Volve field including five produced wells and three injection wells. Due to the limitation of decline curve analysis methods that are only used with the boundary-dominated flow condition of the well. This means that the DCA is possible with F14 and F12 wells prediction and the F14 well is chosen for using decline curve analysis. A production profile from 01/01/2015 to 07/12/2016 has been used for comparison between LSTM and DCA.

Ideally, the optimized error should be as small as possible and could not be excessed 10%. From the table below, LSTM algorithms (6.04% of error) have outperformed decline curve analysis methods in predicting cumulative oil production. Our result shows that there is the fluctuation of LSTM results at the end of production, and it has been the closest curve with the actual curve compare with the other decline analysis curve.

Different from a simulator where an average of 10 days of production has been matched with actual values, using LSTM memory can predict each day during the production period. According to our results, the LSTM algorithm has given better results compared with model in software for well F1, F14, and F15. In addition, LSTM prediction results are more likely to catch the trend of production evaluation. However, in the well F11 and F12, the Software model has better matching. This might be the limit of

LSTM algorithms because of the data noise. Next, cumulative production will be compared between Software output and LSTM. Oil rate will be chosen for this comparison in this study.

Table 8. Cumulative oil production by well.

	Well F1	Well F11	Well F12	Well F14	Well F15
Actual cumulative production (m3)	177709.3	1227203	4579610	3942233	148518.6
Cumulative oil of LSTM (m3)	181144.8	1224349	4912712	3960222	150325.6

On well F11 and F14, LSTM gives the prediction of cumulative oil production with 0.23% and 0.45% error respectively. On well F1 and F15, errors are 1.93% and 1.21 and F12 well is 7.27%. These errors are supposed to be acceptable for further use of the LSTM algorithm (Table 8).

d. Total Cumulative Oil Production Comparison
The Volve reservoir model produced 9,980,819 m3 of oil for 3,197 days, and the cumulative oil from historical production was 10,037,081 m3. Long short-term memory gives the output of 1,043,256 m3, slightly greater than actual values due to the mismatch of well F12 (7.97%). Although the error given by LSTM is slightly bigger than the reservoir model, it still a very good match of production rate.

In addition, compared with building a complex reservoir model, LSTM algorithms just cost less time and effort to construct. Since the objective is to study the potential of artificial intelligence in production prediction, it is practical to use AI in brownfield which has a great amount of historical data. But in case of a lack of data, we can use it as a tool for production optimization in a short period of time.

From the case studies, the performance evaluation criteria indicate that better oil production forecasting can be achieved using some AI techniques with only one type of input parameter i.e. oil/water/gas production data.

4 Conclusion

In this paper, the author has applied some AI techniques include machine and deep learning algorithms to predict the production rate of an oil well in different scenarios. Different algorithms have been used and comparisons between them have shown that while machine learning gains good results in the production phases with the shuffle, the recurrent neural networks have improved the prediction with.

The following conclusions from this study are:

- The machine learning and deep learning algorithms used in this study significantly rely on the amount of data to develop the capacity of production prediction.
- The quality of prediction highly depends on the data input. If the input is messy and inconsistent, then the production prediction can be unreliable.

- Since machine learning has gained good result when the data is shuffled, hence in order to predict time series production rate, recurrent neural networks are performed. Machine learning algorithms can also be used for production optimization. The constraint conditions of producing well can be modified and regulated based on a complicated relationship that was acquired from using machine learning algorithms.
- Compare with conventional decline curve analysis (Arps), LSTM has outperformed due to the ability in time-series prediction. In Volve field studies, LSTM is better than reservoir simulation at well F14 and F15 and close to the accuracy of actual values in all five wells. It demonstrates the ability to use Artificial Intelligence in production forecasting.
- While reservoir modeling needs a great amount of time and efforts to construct a reservoir model with extensive domain knowledge of geology, geophysics, petrophysics, reservoir engineering, and reservoir simulation, LSTM only based the historical data to complete the tasks. So, in this circumstance LSTM is working on cutting-edge methods, and need to be studied carefully in the future in order to give better and accurate results.

Also, the workflow of using machine learning and deep learning algorithms has been built with detailed explanations. This study has demonstrated that without domain knowledge in the oil and gas industry and complicated physics-based modeling, Artificial Intelligent can help engineer in production forecasting in effective ways. AI has been proven their potential and could be useful tools in the oil and gas industry.

References

1. Tita, R.: Machine learning applied to multiphase production problems. MS thesis, Standford University (2018)
2. Boomer, R.J.: Predicting production using a neural network (artificial intelligence beats human intelligence). In: Society of Petroleum Engineers, SPE, p. 30202 (1995)
3. Cao, Q., Banerjee, R., Gupta, S., Li, J., Zhou, W., Jeyachandra, B.: Data driven production forecasting using machine learning. In: SPE-180984-MS, Argentina, 1–3 June 2016 (2016)
4. Suhag, A., Ranjith, R., Aminzadeh, F.: Comparison of shale oil production fore- casting using empirical methods and artificial neural networks. In: SPE-187112- MS. University of Southern California (2017)
5. Sun, J., Ma, X., Kazi, M.: Comparison of decline curve analysis DCA with recursive neural networks RNN for production forecast of multiple wells. In: SPE-190104-MS (2018). https://doi.org/10.2118/190104-MS

Challenges in Cementing Deep Offshore Wells

S. Gruber and J. Plank(✉)

Chair for Construction Chemistry, Technical University Munich (TUM), Munich, Germany
sekretariat@bauchemie.ch.tum.de

Abstract. Common cement retarders effective at high temperatures (>150 °C) typically also disperse cement. Consequently, they provide a rheology which allows to easily mix the slurry at surface. However, when such slurry is circulated downhole and becomes exposed to high temperature, then thermal thinning occurs and the slurry becomes unstable, resulting in cement settling and development of free water which hampers zonal isolation. In view of this, the aim of our research was to develop a viscosifying HT stabilizing agent which is released only at high temperature and which can offset the negative effects associated with thermal thinning. For this purpose we developed an encapsulation technology to coat a viscosifying CaATBS-NNDMA copolymer such as that at surface, the polymer particles do not dissolve and impact rheology. It is however gradually released at T > 125 °C and then imparts sufficient viscosity to mitigate thermal thinning of the slurry. The properties and performance of this novel additive is presented by means of rheological data and HTHP thickening time curves including on/off cycles. The new additive allows to cement deep offshore wells more safely and to achieve zonal isolation more successfully.

Keywords: Oil well cementing · Zonal isolation · HT stabilizing additive

1 Introduction

It is known that at high temperatures (≥140 °C) the effect of thermal thinning of the cement slurry endangers slurry stability under HT/HP conditions [1]. Under these circumstances, the slurry loses its carrying capacity [2]. This is evidenced by the settling of cement particles, resulting in a density gradient [3]. Consequently, the cement which had settled at the bottom cannot hydrate completely, while in the upper part a layer of free water will develop. Especially when cementing horizontal sections it must be ensured that the cement slurry possesses sufficient carrying capacity to avoid free water formation at the top of the slurry. In order to achieve complete zonal isolation, such sedimentation of cement and the formation of free water at the top of the cement column must be avoided by any means [4–6].

Anti-settling agents (also referred to as stabilizers) are commonly applied to overcome this problem. Their working mechanism is based on an increase of the yield point of the cement slurry. Typical examples of anti-settling agents include guar derivatives, cellulose ethers and microbial biopolymers such as welan gum [7]. Unfortunately, all of them are effective only at lower to medium temperatures (<140 °C). The only promising

D. V. K. Huynh et al. (Eds.): VSOE 2021, LNCE 208, pp. 459–466, 2022.
https://doi.org/10.1007/978-981-16-7735-9_51

concept of a high temperature stabilizing agent was presented by Funkhouser et al. [8, 9]. They developed a crosslinked polymer which contains degradable ester and amide linkages. At increasing temperatures (>115 °C), the linkages gradually break, the de-crosslinked polymer then dissolves and provides sufficient viscosity to prevent cement settling. Unfortunately, this highly innovative polymer proved to be difficult to produce and therefore did not make it to regular field use. Thus, still a need exists for a technically sustainable solution for the settling problem of cement at temperatures exceeding 150 °C.

Our approach to solve this problem was to coat a viscosifying polymer in such way that the coating released the polymer at elevated temperatures where thermal thinning started to commence. This way, we hoped to achieve an additive which did not impact slurry viscosity and pumpability at the surface, yet provided sufficient viscosity to compensate the negative effect of thermal thinning.

2 Materials and Methods

2.1 Materials

In our work, a high molecular weight copolymer with strong viscosifying property was encapsulated using rubber latex. The copolymer was synthesized from calcium acrylamide tert. butyl sulfonate (CaATBS) and N,N-dimethyl acrylamide (NNDMA) monomers. Its chemical structure is provided in Fig. 1. The rubber coating shields the polymer in the core and prevents its premature dissolution. A commercial sample of natural rubber latex was utilized as coating material.

An API (American Petroleum Institute) Class G well cement "black label" (Dyckerhoff AG) corresponding to API Specification 10 A [10] was employed as base cement.

Furthermore, a commercial sample of silica flour (SSA-1 from Halliburton, Celle/Germany) with a specific density of 2.65 kg/L was used. Commercial samples of retarders (SCR-100® and HR-25® from Halliburton) were admixed to adjust commonly required pumping times. All dosages of the additives applied in cement slurries are given in % by weight of cement (% bwoc).

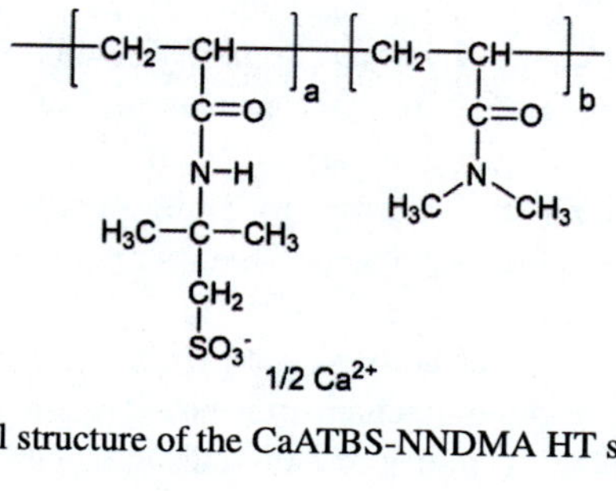

Fig. 1. Chemical structure of the CaATBS-NNDMA HT stabilizing agent.

2.2 Methods

Cement slurry preparation followed "Recommended Practice for Testing Well Cements", API Recommended Practice 10B-2, issued by the American Petroleum Institute [11]. First, the cement was dry-blended with silica flour (35% bwoc), then all additives were added to the cement - silica flour blend as powders and were homogenized. The slurries were mixed at a water-to-cement (w/c) ratio of 0.44 using a blade-type laboratory blender (Waring blender from OFI, Houston/TX). Within 30 s, the dry mixture was added to the mixing water placed in the cup of the Waring blender while mixing at 4,000 rpm and was then mixed for 60 s at 12,000 rpm. The cement slurry formulations applied in the tests performed at different temperatures are presented in Table 1.

Table 1. Composition of the cement slurries (w/c = 0.44) containing the HT stabilizing agent tested at different temperatures.

Components	150 °C		180 °C		200 °C	
	Amount (g)	Dosage (% bwoc)	Amount (g)	Dosage (% bwoc)	Amount (g)	Dosage (% bwoc)
Cement API Class G	620.0	–	620.0	–	620.0	–
Silica flour SSA-1	217.0	35	217.0	35	217.0	35
DI water	272.8		272.8		272.8	
SCR-100® retarder	4.34	0.7	5.58	0.9	11.16	1.8
HR-25® retarder	4.34	0.7	5.58	0.9	11.16	1.8
CaATBS-co-NNDMA	3.1	0.5	3.1	0.5	3.1	0.5
Anti-settling agent	9.92	1.6	15.5	2.5	18.6	3.0
Defoamer	1.5 mL	–	1.5 mL	–	1.5 mL	–

The thermal thinning effect was assessed on a Couette-type coaxial cylinder rotational viscometer (Model PVS Rheometer, Brookfield, Middleboro/USA). Measurements were carried out at room temperature and 150 °C, at an N_2 pressure of 70 bar, using a B5H bob and 25 mL of cement slurry. The slurry was heated to the selected temperature at 30 rpm mixing speed and this temperature was held for 30 min before measurements were taken by shearing the slurry at a sequence of 3-6-100-200-300-200-100-6-3-600 rpm.

Pumping times for the cement slurry and its consistency were determined on a high-temperature/high-pressure (HT/HP) consistometer (Model 8240, Chandler engineering) according to API Spec 10A [10]. Slurry viscosity was measured in Bearden Units of Consistency (Bc). In this test, setting of cement was indicated by an increase of slurry viscosity from an initial 10–40 Bc to a value of >70 Bc. In order to simulate different wellbore conditions, the cement slurry was heated to 150 °C, 180 °C and 200 °C respectively, with a heating rate of 2.5 °C/min at a final pressure of 700 bar by applying varying pressurization rates. Slurry stability was tested by keeping the temperature and pressure in the HT/HP consistometer constant for 30 min before the motor was shut down for

15 min (off-cycle). In case of particle settling, a spike in consistency would occur upon restart of the motor (on-cycle). In total, three off/on-cycles were conducted at intervals of 15 min or 1 h.

3 Results and Discussion

3.1 Temperature-thinning Effect Demonstrated via Rheological Measurements

Figure 2 illustrates the problem of thermal thinning of cement slurries with increasing temperature when no stabilizing agent is present. It shows the rheometer readings for a cement slurry at room temperature and 150 °C. At higher temperature, the cement slurry almost completely loses its viscosity and thus its carrying capacity. The carrying capacity depends on the yield point of the slurry. In this case the yield point of the cement slurry is reduced from 117 lbf/100 ft^2 at room temperature to 10 lbf/100 ft^2 at 150 °C, thus demonstrating severe temperature thinning.

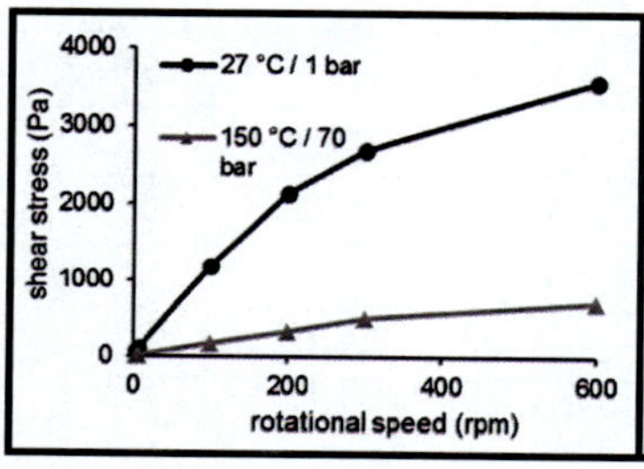

T (°C)	P (bar)	viscosity (Pa) at rotational speed (rpm) 3	6	100	200	300	600
27	atm	57	98	1193	2137	2692	3559
150	70	5	8	177	332	494	706

Fig. 2. Demonstration of thermal thinning of a conventional cement slurry without stabilizing agent by comparison of slurry viscosity at room temperature and 150 °C.

3.2 Temperature-Thinning Effect Demonstrated in HTHP Consistometer

Here, a cement slurry without anti-settling agent was tested. According to Fig. 3, this slurry shows a spike in consistency after an off/on-cycle period of 15 min at 150 °C. This spike indicates that massive particle settling had occurred. As a result, the consistometer paddle got stuck in sedimented cement and the test had to be discontinued at this point.

This result confirms that at such temperature it is mandatory to add a powerful anti-settling agent.

3.3 Performance of Anti-settling Agent at 150 °C

In the following, the rheology of cement slurries containing the anti-settling agent at 150 °C, 180 °C and 200 °C, respectively, and at 700 bar pressure was assessed in the HTHP consistometer. The off/on-cycle interval was varied between 15 min and 1 h.

At first, a slurry holding 1.6% of the stabilizing agent was tested in the HT/HP consistometer at 150 °C and 700 bar. As can be seen in Fig. 4 a sufficiently high viscosity

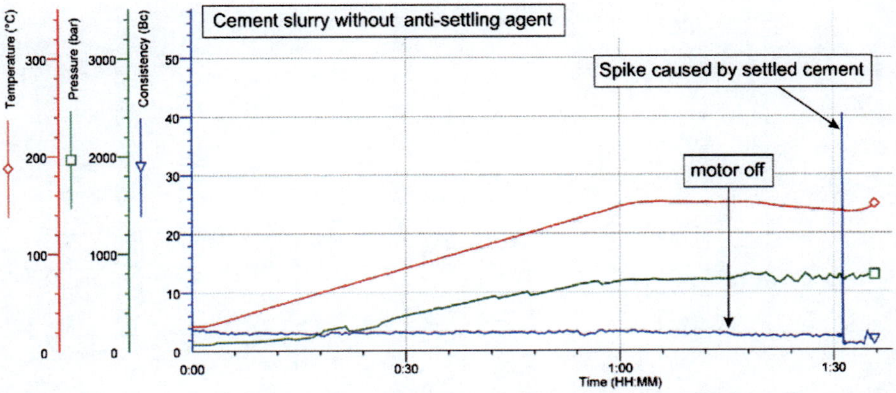

Fig. 3. Cement slurry aged at 150 °C and 700 bar in the HT/HP consistometer; this slurry without anti-settling agent shows sedimentation after a 15 min off-on-cycle.

was maintained throughout the test which is supported by the behavior during the off/on-cycles where no spikes as in Fig. 3 are observed, indicating cement slurry stability. Moreover, the rheology of the cement slurry at surface mixing was relatively low, thus facilitating proper surface mixing and pumping of the slurry.

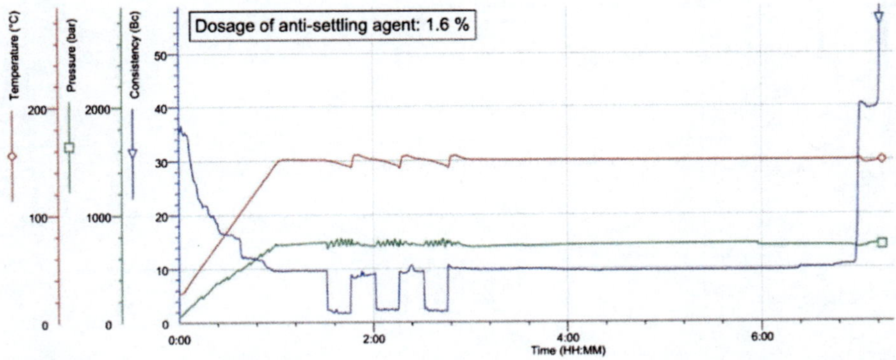

Fig. 4. Cement slurry holding 1.6% of anti-settling agent aged at 150 °C and 700 bar in the HT/HP consistometer, tested at 15 min off/on-cycle intervals.

Furthermore, a slurry was formulated holding an increased dosage of 2% of the anti-settling agent to study its influence on the cement slurry. As can be seen from Fig. 5, even at such high stabilizer dosage low surface viscosity is accomplished and no indication of particle settling from the cement was observed.

3.4 Performance of Anti-settling Agent at 180 °C

When temperature was further increased to 180 °C, dosages of the retarders and the anti-settling agent had to be increased. Also at this temperature, sufficient slurry stability

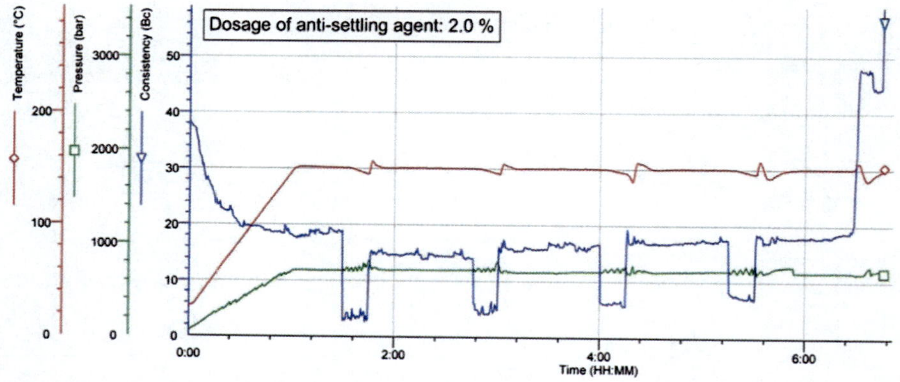

Fig. 5. Cement slurry holding 2% of anti-settling agent aged at 150 °C and 700 bar in the HT/HP consistometer.

was accomplished, as is demonstrated in Fig. 6. Off/on-cycles produced no evidence for particle settling. Initial consistency attained a field-acceptable value of about 38 Bc. At increasing temperature, the cement slurry initially starts to thin which however comes to a halt as soon as the stabilizing copolymer is gradually released.

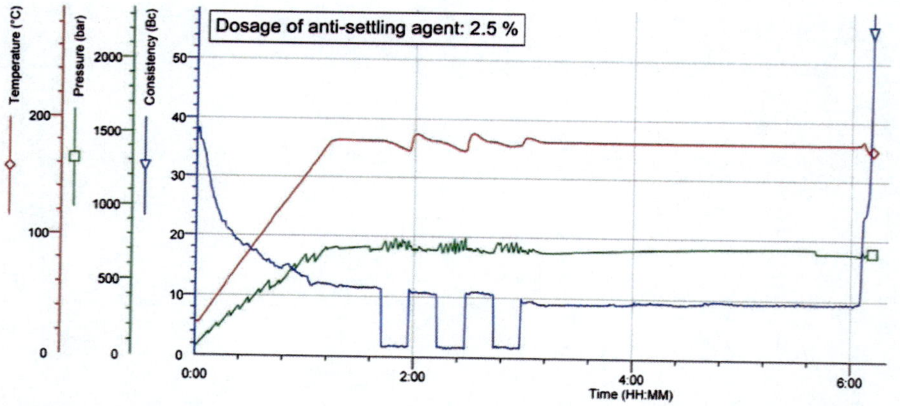

Fig. 6. Cement slurry holding 2.5% of anti-settling agent at 180 °C and 700 bar in the HT/HP consistometer; 15 min off/on-cycle intervals.

3.5 Performance of Anti-settling Agent at 200 °C

Finally, measurements were taken at 200 °C and at 700 bar to explore the temperature limitation of the stabilizing additive which now was dosed in at 3%.

The results displayed in Fig. 7 indicate that at 200 °C minor slurry instability is detected. During off/on-cycles, a very small spike is recorded upon restarting the motor. Also, the slurry consistency slightly decreases with ageing time, thus suggesting that the stabilizing copolymer begins to lose its viscosifying effect at 200 °C.

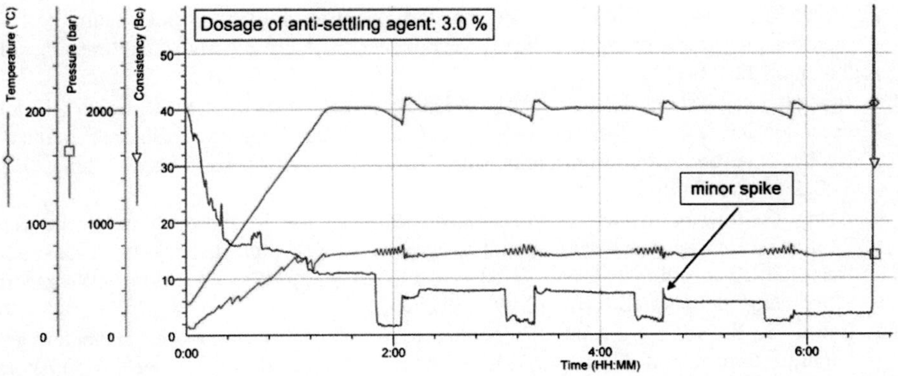

Fig. 7. Cement slurry holding 3% of the anti-settling agent at 200 °C and 700 bar in the HT/HP consistometer; 1 h off/on-cycle intervals.

A potential explanation for the decreased performance of the anti-settling agent at 200 °C is the limited temperature stability of the CaATBS-NNDMA copolymer. Apparently, this copolymer starts to gradually become less effective at temperatures >180 °C owed to chemical degradation. For this reason, the copolymer exhibits a decreased effectiveness at 200 °C.

4 Conclusion

I this work a conventional viscosifying fluid loss polymer made from AMPS and NNDMA monomers was encapsulated into a matrix of natural rubber latex. This way, the polymer did not affect viscosity and pumpability of the cement slurry at the surface. However, when heated up in the borehole the rubber coating softened and gradually started to release the viscosifying polymer. Through this mechanism, the negative effect of thermal thinning was counteracted, and no slurry sedimentation occurred, as was demonstrated by off-on cycles during the HTHP consistometer tests. Moreover, this composite additive was found to effectively stabilize the cement slurry up to at least 180 °C.

Our work emphasizes the importance of controlling cement slurry rheology at high temperature in deep offshore well cementing to prevent particle settling which can lead to unsuccessful zonal isolation and presents a severe safety risk.

Acknowledgement. S. Gruber wishes to thank Jürgen Manchot Foundation for financial support of her study.

References

1. Reddy, B.R., Patil, R., Patil, S.: Chemical modification of biopolymers to design cement slurries with temperature-activated viscosification–a laboratory study. SPE Drilling Completion **27**(01), 94–102, SPE-141005-PA (2012)

2. Abbas, G., Irawan, S., Kumar, S., Elrayah, A.A.: Improving oil well cement slurry performance using hydroxypropylmethylcellulose polymer. In: Advanced Materials Research, vol. 787, pp. 222–227 (2013)
3. Bassett, J.A., Watters, J.T., Sabins, F.L., Febbraro, A.: Focus on cement design and job execution increases success for shale cementing operations. In: Proceedings SPE Americas Unconventional Resources Conference, Pittsburgh, Pennsylvania, USA, 5–7 June 2012, SPE-155757-MS (2012)
4. Abbas, G., Irawan, S., Memon, M.K., Kalwar, S.A., Kumar, S.: Hydroxypropylmethylcellulose as a free water and settling control agent in oil well cement slurry. In: Awang, M., Negash, B., Md Akhir, N., Lubis, L. (eds.) ICIPEG 2014, pp. 121–128. Springer, Singapore (2015). https://doi.org/10.1007/978-981-287-368-2_11
5. Irawan, S., Kalwar, S.A., Abbas, G., Kumar, S.: An innovative solution for preventing gas migration through cement slurry in the lower Indus basin of Pakistan. In: Proceedings Offshore Technology Conference Asia, Kuala Lumpur, Malaysia, 20–23 March 2018, OTC-28219-MS (2018)
6. Keller, S., Crook, R., Haut, R., Kulakofaky, D.: Deviated-wellbore cementing: part 1-problems. J. Petrol. Technol. **39**(08), 955–960 (1987)
7. Ganguli, K.: Biopolymers as free water and settling control agent. In: Proceedings SPE Production Operations Symposium, Oklahoma City, Oklahoma, 21–23 March 1993 (1993)
8. Funkhouser, G.P.: Thermally-activated, high temperature particulate suspending agents and methods relating thereto, US Patent 8,739,871 B2 (2014)
9. Funkhouser, G.P., Leotaud, L., Bratcher, J.: Delayed-release suspending aid provides cement-slurry stability in high-temperature, horizontal wells. In: Proceedings SPE International Symposium on Oilfield Chemistry, The Woodlands, Texas, USA, 13–15 April 2015, SPE-173725-MS (2015)
10. API RP 10A: API Recommended Practice 10A, Specification for Cements and Materials for Well Cementing, 25th edn. Institute, A. P., Washington D.C. (2019)
11. API RP 10B-2: API Recommended Practice 10B-2, Recommended Practice for Testing Well Cements, 2nd edn. Institute, A. P., Washington D.C. (2013)

Interfacial Tension of Hydrocarbon/Water System with Mixed Anionic Surfactants at High Pressures and Temperatures: Optimization of Alkane Number and Salinity

Abdulrazag Y. Zekri[1](✉), Mamdouh T. Ghannam[1], and Hamdi Ben Zekri[2]

[1] United Arab Emirates University, POBox 15551, Al-Ain, UAE
a.zekri@uaeu.ac.ae
[2] Schlumberger - Libya, That El Imad, Tower 1, 7th Floor, Tripoli 91931, Libya

Abstract. The effect of temperature on three commercial surfactants (TRS 1080, Petrostep 450, and Petrostep 465) was studied. The surfactants were mixed in different proportions to vary the equivalent alkane number. The solutions contained 2g/L of surfactant mixture. Interfacial tensions were measured against mineral oil and n-dodecane. Sodium Chloride concentration was varied from 0.00 to 20 g/L. The experiments were conducted in the temperature range of 25 °C to 175 °C, using the modified spinning drop tension meter. The interfacial tension for surfactant mixtures decreased above about 150 °C. A minimum interfacial tension of about 10^{-3} dyne/cm was observed with a salt concentration of 10 g/L at a temperature of 175 °C. The effect of various electrolytes (sodium–chloride, sodium sulphate, sodium carbonate, and sodium phosphate) was investigated as well. Of these, sodium chloride gave the lowest interfacial tension at all employed temperatures.

Mixtures of anionic surfactant solutions have been aged for periods up to a year at room temperature. Interfacial tensions of these aged mixtures were higher than the fresh solutions. Since these solutions were not protected from air, the results are probably a result of slow decomposition of surfactants in the presence of oxygen.

Keywords: Interfacial tension · Surfactants · Alkane number · Salinity

1 Introduction

The Interfacial tension between a given crude oil and a surfactant solution of the sulfonate type may be modelled by replacing the crude oil with one particular alkane. The number of carbon atoms in the equivalent alkane is referred to as the equivalent alkane carbon number (EACN) of rule [1]:

$$(EACN)\_m = (EACN)\_ixi \quad (1)$$

Where Xi is the mole fraction of the ith component extended over all the crude oil components. We have to keep in mind this cannot be accomplished in practice since all

D. V. K. Huynh et al. (Eds.): VSOE 2021, LNCE 208, pp. 467–474, 2022.
https://doi.org/10.1007/978-981-16-7735-9_52

the components of a crude oil have never been identified [2]. Rossini [2] has reported that a particular crude oil contains no more than 2% of any one chemical species and at last count had isolated and identified 175 hydrocarbon molecules. However, we could accomplish the applicability of Eq. (1) by applying it to complex but known composition hydrocarbon mixtures [3]. Cayias et al. [3] have developed a procedure wherein the EACN of a crude oil can be determined based on the interfacial tension behavior of dilute surfactant system against the crude in comparison to the same surfactant against pure hydrocarbons. Glinsman [4] used another method to measure the EACN of a crude Oil.

Tichelkamp et al. [5] employed four different alkylbenzene sulfonate surfactants in low salinity water environment to assess their potential of improved oil recovery. The results indicate a considerable EOR potential of surfactant free alkylbenzenesulfonate surfactants at low salinity conditions. Posocco et al. [6] investigated oil/water emulsions using two common non-ionic surfactants, Tween 80 and Span 20, in the concentration range C (0.3–1 wt.%) well above their respective CMCs. They found a remarkable collegial effect between Tween 80 and Span 20. Campanelli and Xiao [7] measured the dynamic interfacial tension of mixtures of two surfactants: Triton X-100, palmitic acid, and Span 80. They reported that the adsorption kinetics of the mixtures were influenced by the Span 80 surfactant at lower concentration than the critical micelles concentration. Li et al. [8] have studies the possibility of using a mixture of anionic/cationic surfactants flooding in a sandstone reservoir. These mixtures produced very low interfacial tension (IFT) of 10–4 mN/m and low adsorption of 0.27 mg/g in sand stone environment at high temperature of 81 °C. Li et al. [9] reported an increase in oil recovery by 8.0% due to the injection of a mixture anionic-cationic surfactants in a sandstone reservoir having permeability of 41.5 mD, at a temperature of 83 °C, salinity of 15,000 mg/L, and clay of 12. Xu et al. [10] investigated the interaction between anionic polyacrylamides and cationic surfactants in different temperature and salinity. Based on the literature review done in this project, it is clear that more work is required in finding mixtures of surfactants that can produce ultra-low interfacial tension at very high temperature environment. In this work the interfacial tension of different mixtures of three anionic surfactants at high pressures and temperatures against mineral oil and n-Dodecane were measured. Optimization of Salinity and surfactant mixture also conducted.

2 Methods and Material

2.1 Surfactants

Three commercial anionic surfactants were employed in this study. The first of these surfactants was Petronate TRS 10-80. It is manufactured by Witco Chemical Corporation. TRS 10–80 contained 79.7 weight percent (797 g/Kg) active sulfonate, 11 weight percent (110 g/Kg) unsulfonated oil, 9.2% wt. (92 g/Kg) water and about 0.1% weight (1 g/Kg) sodium sulphate (Na2S04). Purification to a 100% active basis was done by liquid chromatographic technique. It has a wide range of sodium alkyl benzene sulfonates with an average molecular weight of 415 and critical micelle concentration at 25 °C, of 48 mol/L. TRS 10-80 (2 g/L and 10 g/L NaCl) has an EACN of 7.

The second commercial surfactant employed in this study was Petrostep 465 manufactured by Stepan Chemical Corporation. It has an average molar mass of 465, and a 64 EACN of 10.64. The third commercial sulfonate used in this study was Petrostep 450 (control No, 538) from Stepan Chemical Company. It contained 61.3% active sulfonate, 18.0% free oil, 18.4% water and 2.3% inorganic salts. It was purified by the technique described in detail in ASTM 2548. Petrostep 450 has EACN 4.5 (at 2 g/L surfactant concentration and 10 g/L NaC1).

2.2 Mineral Oil

The mineral oil used was a Sonneborn white oil obtained from Chevron U.S.A. It is a homogeneous mixture of saturated aliphatic and alicyclic hydrocarbon and is entry-free from aromatic. Table 1 gives complete analysis of this oil. The hydrocarbons used in this study were 99- mole percent pure.

Table 1. Physical and chemical properties of mineral oil # 9.

Property	Value
Gravity, °API	30.7
Specific gravity at 25 °C	0.87
Surface Tesion, Mn/m	29.4
Viscosity, cp	6.0

2.3 Interfacial Tension Measurements

A constant speed (3600 RPM) spinning drop apparatus was modified for high temperatures application. A schematic diagram of the modified Amoco tensiometer is shown in Fig. 1.

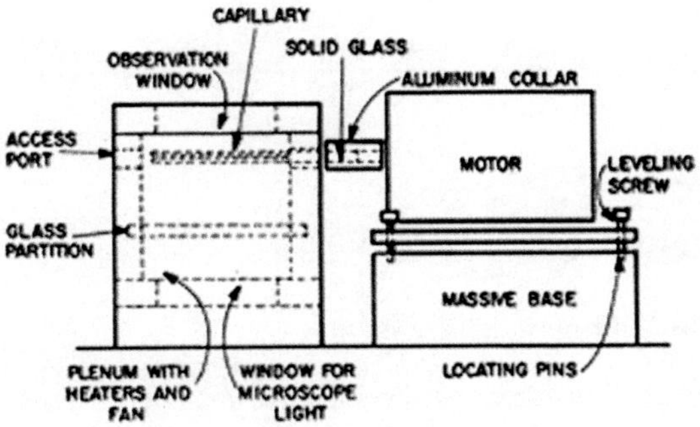

Fig. 1. Typical oil drop in spinning drop interfacial tension meter.

2.4 Sample Preparation

All of the three commercial surfactants (TRS 10-80, Petrostep 465 and 450) were deoiled as described in ASTM D3712-78. The three commercial surfactants were mixed at different proportions to vary the EACN between 4–10. The solution contained 2 g/L of surfactant mixture. Jacobson et al. [11] showed the ECAN of TRS 10-80, Petrostep 465 and Petrostep 450 to be equal to 7, 10 and 4.5, respectively. This greatly extend the number of surfactants that may be considered as candidates for use in low interfacial tension flooding. The range of the salt concentration was varied from 0.0 g/L to 20 g/L. The values of ECAN of each mixture and the mole percent of each component in the mixture are presented in Table 2 and Table 3. Various electrolytes solutions (sodium chloride, sodium sulphate and sodium phosphate) were prepared using de-aired distilled water. The ionic Strength of these solutions was 0.2.

Table 2. Averaging of multicomponent surfactant mixtures 1–7.

Component/Misture	1	2	3	4	5	6	7
TRS 10-80	0	0.492	0.466	0.594	0.688	0.355	0
Petrostep 465	0	0.055	0.104	0.132	0.153	0.317	0.492
Petrostep 450	1	0.453	0.43	0.274	0.159	0.328	0.508
EACN	4.5	6.033	7.277	6.7	7.062	7.1	7.206
MW of mixture	450	433.605	435.25	431.19	428.22	442.34	457.38

Table 3. Averaging of multicomponent surfactant mixtures 8–12.

Component/Misture	8	9	10	11	12
TRS 10-80	0.399	0.308	0.263	0.089	0
Petrostep 465	0.396	0.515	0.586	0.789	1
Petrostep 450	0.205	0.177	0.151	0.122	0
EACN	7.206	8.1	8.38	9	10
MW of mixture	441.98	446.95	449.95	458.73	465

3 Experimental Results

3.1 Mixture of Surfactants vs Mineral Oil

The tension behaviour of different surfactant mixtures against mineral oil #9 showed similar trends. The concentration of the mixtures were fixed at 2 g/L and the EACN was varied between 4 to 10 and the NaCl concentration was varied between 0 to 20 g/L. The interfacial tension varied from 1.0 to 0.002 mN/m as shown in Figs. 2 to 3.

Effect of Temperature and NaCl on IFT of Mineral Oil. At zero NaCl concentration the IFT passed through a minimum. The temperature at which the minimum occurred was about 140 °C. In the presence of NaCl (2, 5, 10, 20 g/L) IFT was not affected by temperatures up to 100 °C above which started to decrease. Figures 2 and 3 displays the effect of temperature and Na Cl concentration on IFT respectively. Increasing NaCl concentrations decreases the interfacial tension for low temperature ranges 30 °C. For 10 and 20 g/L NaCl concentrations, we observed large reduction in the interfacial tension as the temperature approached 175 °C, which is about the temperature one would expect in a high temperature oil reservoirs. Mixtures #2 and #4 produced IFT as low as 0.02 mN/m for NaCl concentrations of 10 and 20 g/L. For all mixtures for various NaCl concentrations, IFT didn't pass through a minimum value for the range of temperatures checked 25 to 175 °C. There is indication, however, that a minimum IFT might occur at higher temperatures. The one component system (Petrostep 450 at 2 g/L) passed through a minimum for all NaCl concentrations.

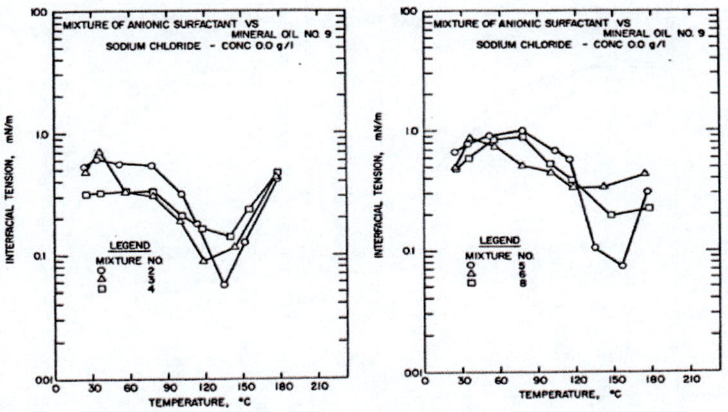

Fig. 2. Effect of temperature on IFT of mixtures of anionic surfactants vs mineral oil.

Effect of Surfactant Composition. Various mixtures of anionic surfactants were observed to exhibit similar IFT behaviour. The interfacial tension for surfactant mixtures decreased above 120 °C. As shown in Figs. 3 these mixtures show an abrupt decrease in interfacial tension at temperatures in excess of 120 °C. Overall there is a slight difference in IFT values for different mixtures. At high temperatures low EACN mixtures showed better interfacial activities at elevated temperatures 176 °C (350 °F).

3.2 Mixture of Surfactants vs n-Dodecane

The same mixtures employed in the previous section were tested against n-dodecane. The measured interfacial tensions of these mixtures ranged from a low of 0.002 mN/m at 176 °C for the mixture containing 0.114 wt.% TRS 10-80 and 10 g/L NaCl to a high of 0.4 mN/m at 125 °C for a mixture containing equal weight of Petrostep 465 and Petrostep 450 with 2 g/L NaCl.

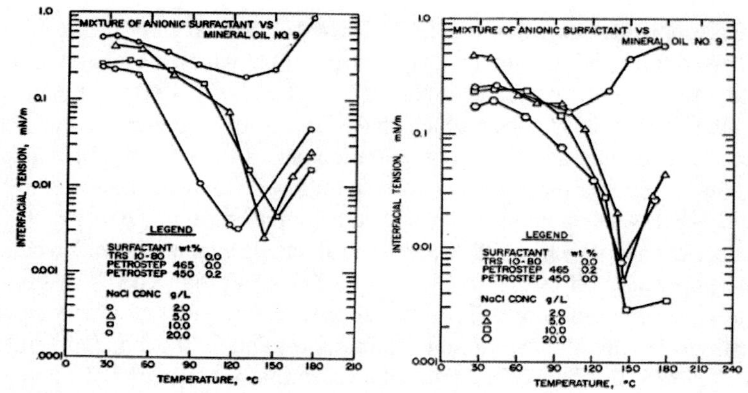

Fig. 3. IFT of mixtures of surfactants as function of temperature and salinity vs mineral oil.

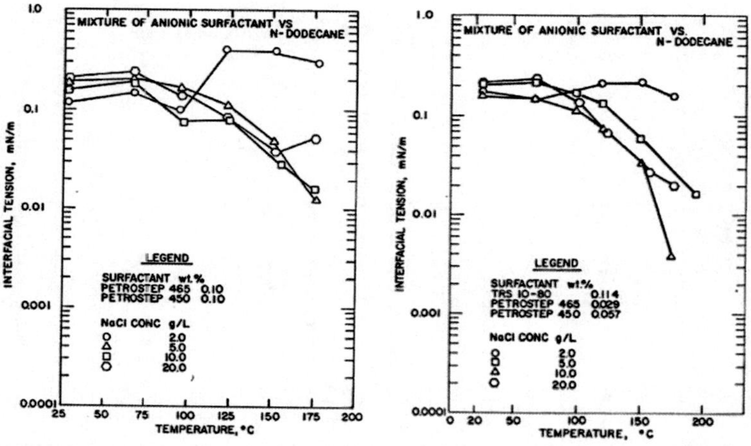

Fig. 4. IFT of mixtures of surfactants as function of temperature and salinity vs n-Dodecane.

Effect of Temperature and NaCl on IFT of n-Dodecane. In the low range of temperatures between 25–140 °C, no noticeable effect of temperature on the measured interfacial tension was observed. At temperatures above 140 °C, we observed a rapid decline in IFT with increasing temperature for all mixtures in the presence of sodium chloride. Temperature effects on mixtures of two surfactants (0.1 wt.% Petrostep 465, 0.1 wt.% TRS 10-80) were small compared to mixtures of three surfactants (TRS 10-80, Petrostep 456 and Petrostep 450) as presented in Fig. 4. All data indicate that the interfacial tensions of these mixtures of anionic surfactants versus n-Dodecane move toward minima. No minimum was observed for the range of temperatures tested -However, an extrapolation of the data would suggest a minimum at higher temperatures.

Effect of Electrolytes on IFT of n-Dodecane. The effect of different electrolytes on the interfacial tension of mixture of surfactants (0.066 wt.% TRS 10-80, 0.066 wt.% Petrostep 465 and 0.066 wt.% Petrostep 450) versus n-dodecane is shown in Fig. 5.

The electrolytes were all compared at the same ionic strength of 0.2, Electrolytes studied were sodium chloride, sodium sulphate, sodium carbonate and sodium phosphate. Of these sodium chlorides gave the lowest interfacial tensions at all temperatures. At 176 °C (350 °F) sodium chloride produced ultra-low IFT around 0.0004 mN/m. The IFT measured using sodium carbonate solution was about 0.0009 mN/m.

Effect of Aging on the IFT of Mixture of Surfactants versus n-Dodecane. Mixtures of anionic surfactant solutions have been aged up to a year at room temperature. Figure 6 present samples of the measured data of the effect of aging on the IFT. A substantial difference in IFT was discovered (particularly at the temperature of the minimum interfacial tension) when the data for two surfactant solutions of different ages against n-dodecane were compared. NaCl concentration had no effect on the aging process, Fresh aqueous solution produces IFT as low as 0.0004 mN/m at 20 g/. Aged solution of the same mixture produces IFT as high as 0.04 mN/m.

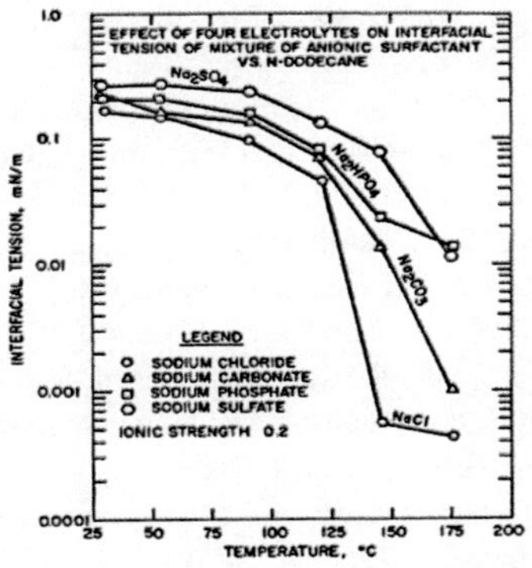

Fig. 5. IFT of a mixture of surfactant for four electrolytes with ionic strength of 2.

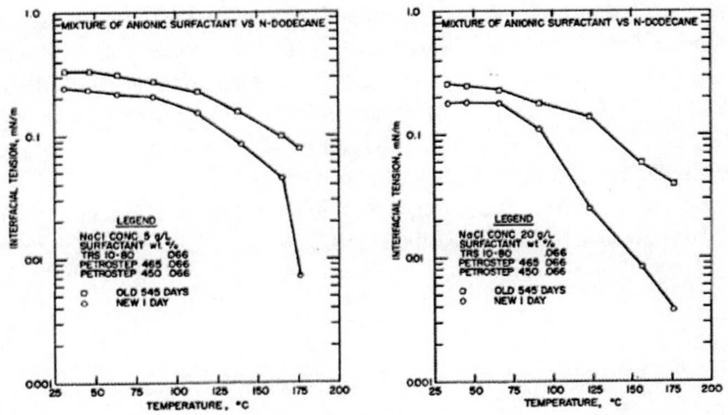

Fig. 6. IFT of New and Aged Mixtures of Surfactants versus n-Dodecane.

4 Conclusions

The following conclusion may be inferred from the data presented here.

Temperature was observed to have negligible effects on the interfacial tensions of different mixtures of anionic surfactant solutions versus crude oil at low NaCl concentrations. Interfacial tensions of aged mixtures were higher than fresh solutions, particularly at the temperature of the minimum interfacial tension. Ultra-low IFT of 0.0004 mN/m was achieved at a temperature of 176 °C and 0.2 NaCl ionic strength for a mixture of surfactants consist of equal 0.066 wt.% of TRS 10-80, Petro step 465, and Petrostep 450.

References

1. Cash, R.L., et al.: The application of low interfacial tension scaling rules to binary hydrocarbon mixtures. J. Coll. Int. Sci. **57**, 39–44 (1976)
2. Rossini, F.R.: J. Chem. Educ. **37**, 554 (1960)
3. Cayias, J.L., Schecher, R.S., Wade, W.H.: Modelling crude oil for low interfacial tension. Soc. Pet. Eng. J. 351–357 (1976)
4. Glinsmann, G.R.: Surfactant flooding with microemulsions formed in in situ-effect of oil characteristics. Paper SPE 8326, Presented at the 54th Annual Fall Technical Conference of SPE-AIME, Las Vegas, Nevada (1979)
5. Tichelkamp, T., Khanamiri, H.H., Nourani, M., Åge Stensen, J., Torsæter, O., Øye, G.: EOR potential of mixed alkylbenzenesulfonate surfactant at low salinity and the effect of calcium on optimal ionic strength. Energy Fuels **30**, 2919–2924 (2016)
6. Posocco, P., Perazzo, A., Preziosi, V., Laurini, E., Pricl, S., Guido, S.: Interfacial tension of oil/water emulsions with mixed non-ionic surfactants: comparison between experiments and molecular simulations. RSC Advances (6) (2016)
7. Campanelli, J., Xiao, H.: Dynamic interfacial tension of surfactant mixtures at liquid–liquid interfaces. J. Colloid Interface Sci. **213**(2), 15, 340–351 (1999)
8. Li, Y., et al.: Mixtures of anionic-cationic surfactants: a new approach for enhanced oil recovery in low-salinity, high-temperature sandstone reservoir. SPE-169051-MS, Proceeding of SPE Improved Oil Recovery Symposium, Tulsa, Oklahoma, USA, 12–16 April (2014)
9. Li, Y., et al.: Pilot test of surfactant-polymer flood with mixtures of anionic-cationic surfactants for high temperature low permeability sandstone reservoir. SPE-195056-MS, Proceeding of SPE Middle East Oil and Gas Show and Conference, Manama, Bahrain (2019)
10. Xu, L., Han, M., Zhang, X., Fuseni, A.: Study on oppositely charged polymer and surfactant mixture for enhancing oil production. SPE-198094-MS, Preceding of SPE Kuwait Oil & Gas Show and Conference, Mishref, Kuwait, 13–16 October (2019)
11. Jacobson, J.K., Morgan, J.C., Schechter, R.S., Wade, W.H.: Low interfacial tensions involving mixtures of surfactants. Soc. of Pet. Eng. J. (1977)

Temperature Effect on Forecast Gas Production

Nguyen Thanh Phu[1,2] and Ta Quoc Dung[1,2](✉)

[1] Faculty of Geology and Petroleum Engineering, Ho Chi Minh City University of Technology (HCMUT), 268 Ly Thuong Kiet Street, District 10, Ho Chi Minh City, Vietnam
tqdung@hcmut.edu.vn

[2] Vietnam National University Ho Chi Minh City, Linh Trung Ward, Thu Duc District, Ho Chi Minh City, Vietnam

Abstract. High gas reservoir pressure and temperature usually occur pressure loss and the appearance of condensation that blocks the gas flow. To optimize the production, choosing the most effective production tubing size to reduce pressure gradient and prevent condensation is essential. The paper demonstrates the use of Gray correlation to predict pressure gradient, design operated pressure and applied Enthalpy Balance Model to evaluate wellhead temperature to avoid condensation blocking inside tubing. By combining the temperature and pressure profile with gas phase diagram, the result proposes large production tubing allows low pressure loss and less potential of the condensation while producing gas.

Keywords: Gray correlation · Enthalpy balance model · Condensation · Production tubing

1 Introduction

Temperature model is utilized to examine the fluid condition. On the basis of thermodynamic energy balance and the usual gas flow equations, it is possible to establish a simple temperature gradient calculation by prediction of pressures and temperatures at the wellbore region [1]. One of the key parameters required for accurate temperature modeling is the geothermal gradient [2]. A primary value of the natural gas - geothermal gradient used black – oil model to get a quick approximate solution for temperature drop of natural gas flow in production well [3]. According to that research, a study about temperature effect can be performed by using the steady-state model in order to directly measure the geothermal gradient and wellhead temperature [3, 4]. To enhance production profile of gas well, the steady-state model is evaluated with different lithology as coal, sand and shale. It shows different geothermal gradients that the operator can enhance their reservoir characterization, completion strategies and production performance of future offset well [5].

In practical, a temperature effect can be evaluated using an enthalpy energy conservation model, taking into consideration: kinetic energy, frictional heating, Joule-Thomson heating or cooling as well as conduction and convection heat transfer between the flow fluids and formation rock [6]. Another method could be based on an integrated workflow with the focus on thermal analysis of completion string, wellhead and surface network.

D. V. K. Huynh et al. (Eds.): VSOE 2021, LNCE 208, pp. 475–482, 2022.
https://doi.org/10.1007/978-981-16-7735-9_53

The research outputs are the temperature profiles under various production scenarios. The higher is the heat capacity of the produced fluid, the lower is the heat loss along the production conduit, considering the pressure constant. Decrease in pressure with increase liquid production will counter the effect of higher temperature conductivity from downhole to surface [8].

It is essential to prevent two phase flow inside the tubing to enhance gas production. When the pressure and the temperature reduce lower than dewpoint conditions, the liquid phase appears in gas flow. As reported by Bahadori [3], all the tests express the near-surface region commonly releasing condensation. To prevent condensation, high well head pressure does not ensure complete possibility of no liquid phase in production tubing. Gas condensation detection enables operations to be timely for notification by displaying phase diagram to find the phase change [9]. In this paper, based on the idea of enthalpy balance temperature model, pressure loss during flow and initial phase diagram, an integrated workflow has been developed to identify the flow condition.

2 Methodology

2.1 Enthalpy Balance Temperature Model

Depending on the bottom hole temperature, there are generally three basic mechanisms to find the wellhead temperature as conduction, convection and radiation.

Conduction in formation, cement, and casing illustrates the transfer of energy between neighboring regions of production tubing. In principal, the more energetic material will transfer energy or the heat to the less ones. The rate at which conduction occurs, $\Delta Q1$, is dependent on the geometry of the medium, thermal conductivity of the material, and the temperature thermal gradient.

$$\Delta Q_1 = 2\pi.k.\Delta L \frac{T_{outside} - T_{inside}}{\ln\left(\frac{r_2}{r_1}\right)} \tag{1}$$

$$k = k_{gas}.(1 - H_L) + k.H_L \tag{2}$$

ΔQ_1: Heat transfer by conduction (British thermal unit/hr - BTU/hr) 1 BTU/hr ~ 1 kJ/hr
k: Average conductivity
H_L: Holdup liquid (if there is no liquid phase let $H_L = 0$).
r_1: Tubing radius (in)
r_2: Tubing diameter (in)
$T_{outside}$: Surrounding temperature (°F)
T_{inside}: Tubing temperature (°F)

Convection is the transfer of heat between a solid surface and an adjacent, moving liquid or gas. Convection occurs through the combination of conduction and fluid motion. There are two typical convections: forced convection in tubing and free convection in annulus.

Natural or free convection exists when there is change in temperature from bottom to the wellhead. Forced convection appears by artificially forced to flow over the surface

Table 1. Conductivity (k) and Specific Heat of Fluid (Cp) [7]

Fluid type	Cp (BTU/lb/°F)	Conductivity (BTU/hr/ft/°F)
Water (low salinity)	1	0.35
Water (high salinity)	1.02	0.345
Heavy oil	1.04	0.34
Medium oil	0.49	0.089
Light oil	0.5	0.0815
Gas	0.26	0.0215

subjected to any external operation units. The rate of convection, ΔQ_2, increases at an increasing rate in case the fluid-motion exists. Table 1 summaries some typical values of conductivity and specific heat of fluid for different fluid types.

$$\Delta Q_2 = 2\pi.r_2.h_f.\Delta L.(T_2 - T_1) \tag{3}$$

$$h_f = \frac{0.023.k.R_{en}^{0.8}.P_r^{1/3}}{r_2} \tag{4}$$

ΔQ_2: heat transfer by convection (BTU/hr)
R_{en}: Reynolds number
P_r: Prandtl number

$$P_r = \frac{C_p\mu}{k} \text{where } C_{p_{avg}} = C_{p_{gas}}(1 - H_L) + C_{p_{fluid}}(H_L)$$

Cp: Specific heat of fluid (BTU/lb/°F)
T_1: Wellhead temperature (°F) (measured from surface)
T_2: Bottom hole temperature (°F)
μ: Gas viscosity (cp)
ΔL: Different in tubing length (ft)

Radiation takes effect when there is at least one gas component significantly evaporating under high temperature. Each gas component has its own boiling temperature, if the temperature is higher than that boiling temperature, the component will evaporate leading to the reduction in heat of fluid. That mechanism called radiation and it co-occurs with either conduction or convection. In most cases, radiation appears in pipe wall area and can its rate, ΔQ_3, can be calculated as follows:

$$\Delta Q_3 = 2\pi.r_1.(h_c + h_r).(T_2 - T_1) \tag{5}$$

$$h_c = \frac{0.049.(G_y.P_r)^{1/3}.P_r^{0.0074}.k}{r_2.\ln\left(\frac{r_1}{r_2}\right)} \quad (6)$$

$$h_r = \frac{\sigma.\left(T_{outside}^2 - T_{inside}^2\right).(T_{outside} + T_{inside})}{\frac{1}{\varepsilon + \frac{r_1}{r_2}.\left(\frac{1}{\varepsilon} - 1\right)}} \quad (7)$$

ΔQ_3: Heat transfer by radiation (BTU/hr)
ε: Tubing emissivity
σ: Stefan-Boltzmann constant, approximately $5.67 \times 10^{-8}(W.m^{-2}.K^{-4})$
G_y: Grashof number where $G_y = \frac{\beta.g.(T_{outside} - T_{inside}).r_2^3}{\mu^2}$.
β: The coefficient of thermal expansion

Table 2 provides some values of the conduction heat transfer coefficient and the emissivity for different types of tubing.

Table 2. Tubing Emissivity [7]

	Conduction Heat Transfer Coefficient (BTU/ft/hr/°F)	Emissivity
Mild steel tubing	26	0.65
Plastic coated tubing	20	0.65
Stainless steel (13%)	18	0.4
Stainless steel (15%)	15	0.3
Line pipe	27	0.9

Total heat loss by depth:

$$\Delta T = \frac{\sum \Delta Q}{U.\Delta D^2} = \frac{\Delta Q_1 + \Delta Q_2 + \Delta Q_3}{U.\Delta D^2} \quad (8)$$

ΔD: Difference in depth (ft)
ΔT: Temperature decrease when flowing up (°F)

For measuring the well head temperature, the U – over heat transfer coefficient is used to calculate the heat loss by depth, however by the experience, U value is:

- Dry Gas: 1–3 BTU/(hr.ft2.degF)
- Retrograde condensate fluid: 5–7 BTU/(hr.ft2.degF)
- Oil: 8–9 BTU/(hr.ft2.degF)

2.2 Gray Correlation

The total pressure loss versus depth is demonstrated in Eq. (9). There are three factors affecting the pressure change: friction force, potential and kinetic energy. If the tubing is divided into small segments, then the pressure loss by kinetic energy is not considerable. In gray correlation, it can be applied for high rate condensate gas ratio (CGR), more than 50 bbl/MMscf and large tubing inside diameter, 3.5 or 4.5 in.

$$\frac{dP}{dZ} = \frac{f.p_n.v_m^2}{2.(r_1 + r_1)} + \frac{g_c}{g}\rho_s.\sin\theta \tag{9}$$

f: Friction factor number
v_m: Mixture velocity, ft/s
ρ_n: Mixture average density of liquid and gas phase, lbm/ft^3
ρ_s: Slip mixture density of liquid and gas phase lbm/ft^3
θ: Well deviation angle, degree

3 Result and Discussion

The main objective in measuring the change of pressure and temperature is to analyze gas flowing to surface. Bottom hole temperature and well head pressure must be kept constant as input into the calculation procedure.

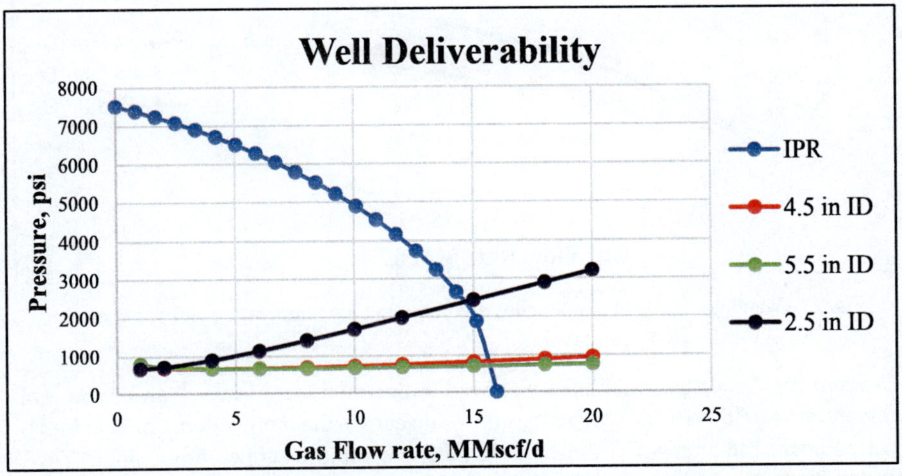

Fig. 1. Apply Gray correlation to measure a rate of gas well will produce

By generating three different sizes of inner diameter tubing: 2.5, 4.5 and 5.5 in, the VLP curve shows a clear performance in optimal maximize production if the wellhead pressure is 500 psi. The value is shown in Table 3.

Table 3. Well deliverability value

Tubing ID (in)	Well deliverability	
	Bottom hole pressure (psi)	Gas flow rate (MMscf/d)
2.5	2422	15
4.5	710	16
5.5	716	16

At constant wellhead pressure, the bottom hole pressure of 2.5 in tubing inner diameter is extremely higher than the two others if applied the Gray correlation. Three tubing sizes allow nearly same gas flow rate, around 16 MMscf/d, however, both the 4.5 and 5.5 in ID give better performance in high flow rate and low bottom hole pressure.

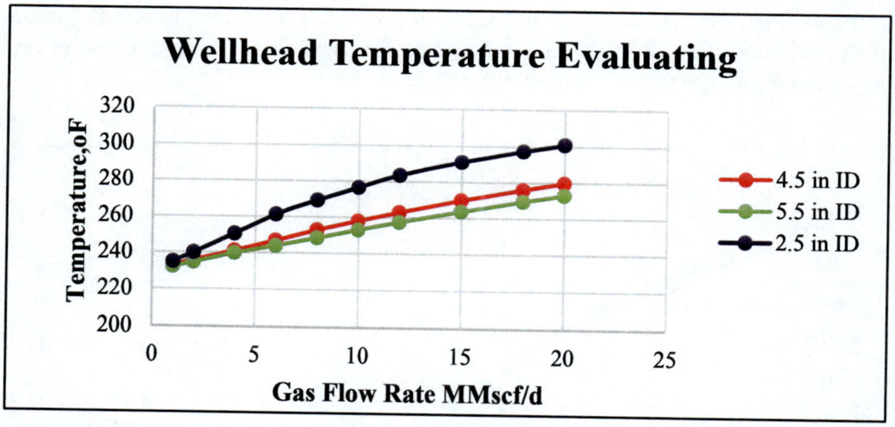

Fig. 2. Calculate wellhead temperature used bottom hole temperature and gas flow rate

From Fig. 2, all the calculated wellhead temperatures are lower than bottom hole temperature (at 350 °F). The higher gas flow rate causes the more volume of heat loss by friction inside the tubing. For the 4.5 in ID, the temperature at operating point is 276 °F display the temperature reduce 74 °F/8000 ft, it means 0.01 °F/ft. For the 5.5 in ID, the temperature is 268 °F at well head, it means 82 °F/8000ft, 0.01 °F/ft as same as 4.5 in ID. From this result, we can conclude the tubing ID sizes is not the parameter that affect the change of pressure and temperature.

It can be observed the operating point in Fig. 1, that the bottom hole pressure of the 2.5 in ID is higher compared to 4.5 and 5.5 in ID. Consequently, the 2.5 in ID - wellhead temperature will be higher than two previous diameters. As shown in Fig. 3. If

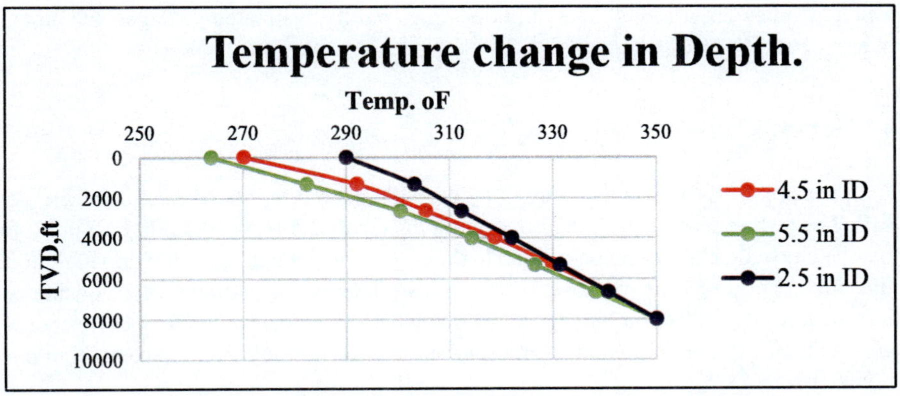

Fig. 3. Temperature change during flow to surface

the well produces at high rate, due to the impact of conduction, the heat loss mainly will mostly occurs at the bottom of the well. When flowing near the surface, a low effect of convection and radiation mechanism, the heat loss rate of small production tubing ID is smaller than the larger one.

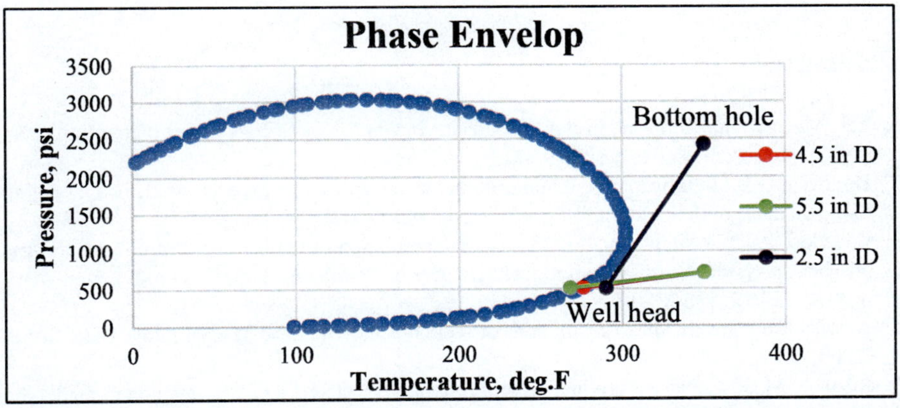

Fig. 4. Combine the temperature change and pressure with fluid phase diagram to identify phase change

Figure 4 illustrates the phase diagram describes regions of fluid condition inside the tubing. It is agreed that one phase flow is more secure than multi – flow. It is observed from the figure that if the estimation of temperature and pressure is inside the diagram, it tends to be divided in two phases: the outside is in single phase: the left side is liquid and the other is in gas stage. In fact, pressure and temperature at wellhead are lower than the base bottom hole condition. At initial stage, the gas exists at the bottom.

The 2.5 in ID shows a good performance while it produces in just gas phase. Additionally, 4.5 in ID and 5.5 in ID have a similar production rate with the stream at bottom is gas. Nonetheless, at near surface area, the stream is lightly divided into two phase which is appeared in Fig. 4. The fluid deliveries hardly, estimated 95% gas, 5% liquid stage.

4 Conclusion

The pressure gradient calculated by Gray correlation as it is the most used for the gas well. The testing results for three different tubing sizes (2.5, 4.5 and 5.5 in ID), indicate that the Gray correlation can be applied to the large tubing inner size, allowing high flow rate and low bottom hole pressure. For the gas well, Gray correlation is also considered the effective roughness of tubing in calculation which is critical parameter used to choose suitable tubing material in order to prevent erosion and friction force caused by high rate of gas flow.

There are three typical mechanisms causing temperature loss: conduction, convection and radiation. Heat transfer from casing to formation named conduction, convection exist inside the production tubing. Radiation mostly appears in pipe wall. The temperature model combined with the pressure model is needed to check the efficiency of production tubing size. At high gas flow rate, preventing heat loss inside production tubing is highly essential. The results presented in this paper demonstrates that large tubing size allows high gas flow rate at low operating pressure. However, it can cause higher heat loss per depth.

References

1. Tek, M.R.: Temperature and Pressure Gradients in Gas Wells. Society of Petroleum Engineers (1978). https://doi.org/10.2118/7495-MS
2. Donovan, G.D.: Improved production profiling using thermal balance and statistical modeling in the pinedale anticline of the US rocky mountains (2008)
3. Bahadori, A:L.: A simple method for accurate prediction of temperature drops in natural gas production systems for black-oil models. Society of Petroleum Engineers (2009). https://doi.org/10.2118/125510-MS
4. Aarvik, A.D.: Steady State Calculations of Wellbore and Wellhead Temperatures. BHR Group (2009)
5. Johnson, M.D.: Enhanced production profiling of gas wells using a new method to model the geothermal temperature gradient. Society of Petroleum Engineers (2010). https://doi.org/10.2118/135321-MS
6. Gysen: Analyzing temperature data. production logging analysis tool: PLATA Reference Guide, pp. 159–162 (2008)
7. E Petroleum: PROSPER Single Well Systems Analysis, 8th ed. Petroleum Experts, Scotland (2003)
8. Ahmed, H.A.: Temperature simulation: an integrated approach. Society of Petroleum Engineers (2017)
9. Pathak, U.: Gas condensation and phase diagram in multiphase flow systems. Pipeline Simulation Interest Group (2019)

Tubing Performance with Different Multiphase Flow Correlations

Tran Duy Ngoc Giao[1], Nguyen Han Thinh[1], and Ta Quoc Dung[2](✉)

[1] Faculty of Geology and Petroleum Engineering, Ho Chi Minh City University of Technology (HCMUT), 268 Ly Thuong Kiet Street, District 10, Ho Chi Minh City, Vietnam
[2] Vietnam National University Ho Chi Minh City, Linh Trung Ward, Thu Duc District, Ho Chi Minh City, Vietnam
tqdung@hcmut.edu.vn

Abstract. Multiphase flow in tubing string has been receiving much attention in upstream petroleum industry as it still remains some challenging issues over past few decades. Ones of those is to select the most suitable tubing size and finding the best correlation of pressure loss prediction under certain operating conditions. This research work involved the use of data from oil well X to quantify the pressure loss by Hagedorn-Brown and Beggs-Brill correlations (multiphase flow correlations) and subsequently suggesting the ways for production optimization. Following that, a sensitivity analysis has been done to observe the parameters that affected the tubing performance. The results show that Hagedorn and Brown correlation appeared to be the best-fit correlation for oil well X with the error approximately 15% comparing to the observed values.

Keywords: Multiphase flow correlations · Sensitivity analysis · Tubing performance relationship (TPR)

1 Introduction

Reliable prediction of two or more phase flow behaviour often requires a correlation which can be developed from several experiments. Every multiphase flow correlation has its limitations and only works well when subjected to a certain range of well conditions. Assessing the complexity of the multiphase flow behavior by correlations have been developed since 1914. Several studies have been performed to evaluate various multiphase-flow correlations. In 1967, Orkiszewski [1] was the first to develop his own correlation and applying to 148 well cases, his method is an extension of the work done by Griffth and Wallis [2]. A later study by Espanol [3] on 44 wells confirmed the conclusion of Orkiszewski. A study by Messulam [4] using data from 434 wells concluded that the Hagedorn-Brown correlation was the most accurate. But afterward, Rossland [5] showed that Hagedorn-Brown tended to under-predict pressure drops due to the under-prediction of liquid holdup. Thompson [6] added that, the modified Hagedorn-Brown correlation tended to over-predict pressure loss in bubble flow, while it tended to under-predict for slug flow.

D. V. K. Huynh et al. (Eds.): VSOE 2021, LNCE 208, pp. 483–490, 2022.
https://doi.org/10.1007/978-981-16-7735-9_54

Although many correlations calculating the pressure drop in pipes are available, the evaluations among these correlations with the effect of tubing size and tubing roughness have not been clearly studied. Therefore, in the present study, an attempt was to make an in-depth analysis of Hagedorn-Brown and Beggs-Brill correlations with sensitivity analysis of tubing parameters, and find the most suitable correlation for the oil well X.

2 Methodology

2.1 Fundamental of Nodal Analysis

Nodal Analysis is one of the most well-known approaches when studying a system consisting of interacting components. This technique was first introduced to the petroleum production system in 1954 by Gilbert [7] and later discussed by Nind [8] in 1964 and Brown [9] in 1978.

The basic concept behind this method is to set a division point or a node at specific locations in the system for a thorough examination of every associating components. In regard to the petroleum-producing system, a common selection of the node is at the bottom of a well. In this arrangement, all of the components upstream of the node comprise the inflow section, while the outflow section consists of all of the components downstream of the node. The relationship between flow rate and pressure must be available for each component in the system. The operating flow rate of the system can be determined once the following requirements are satisfied:

- The flow into the node equals the flow out of the node.
- Only one pressure can exist at a node.

At a specific time during the life of the well, there are always two fixed pressures are not function of the flowrate. One of these pressures is the average reservoir pressure, P_r, and the other is the outlet pressure of the petroleum production system, usually the separator pressure, P_{sep} or the wellhead pressure, P_{wh}. Then, the node pressure, P_{node}, is calculated from both directions starting at the fixed pressure points.

Inflow to the node:

$$P_{node} = P_r - \Delta P_{upstream} \tag{1}$$

Outflow from the node:

$$P_{node} = P_r + \Delta P_{downstream} \tag{2}$$

Where the pressure drop, ΔP, in any components depend on the flowrate. Therefore, a plot of node pressure versus flowrate will create two curves. Finally, the intersection between two curves represents the operating point of the petroleum production system. However, if any upstream components change, the outflow curves remain constants. Also, the curves shift when the fixed pressures change.

To apply this excellent procedure to a well's tubing performance, it is essential to have a good estimation of pressure drop throughout the piping system. These pressure drops depend not only on the flow rate but on the size and other characteristics of the assigned components. If some incorrect method can be recognized in the process of calculating the pressure drops, the systems analysis can provide erroneous results.

2.2 Tubing Performance Relationship (TPR)

When the fluid flows inside the tubing from the bottom hole to the surface, it is hard to keep the pressure above the bubble point pressure (P_B). Hence, the flowing fluid will be in two or more phases. As this circumstance happens, the flow behavior is much more complex comparing to single-phase flow. The increased complexity of multiphase flow in pipes logically resulted in a higher degree of empirism for predicting flow behavior. Many empirical correlations have been developed to predict flow pattern, slippage between phases, friction factors, and other such parameters.

Hagedorn and Brown Method. Hagedorn and Brown is a correlation for multi-phase flow, which developed from the experiment on 1500 ft experimental well using 1-inch to 4-inch tubing [10]. Experiments included three-phase flow (water, oil, and gas). This correlation is independent of flow patterns. The pressure drop calculation for this correlation can be expressed in oilfield unit as:

$$144\frac{dp}{dz} = \overline{\rho} + \frac{f\dot{m}^2}{(7.413 \times 10^{10}D^5)\overline{\rho}} + \overline{\rho}\frac{\Delta(u_m^2/2g_c)}{\Delta z} \tag{3}$$

Where:

$\frac{dp}{dz}$: pressure gradient (psi/ft)
$\overline{\rho}$: mixture density (lb_m/ft^3)
$\dot{m}$: the total mass flowrate (lb_m/d)
D: the inner diameter (ft)
u_m : the mixture of superficial velocity (ft/sec)
f : the friction factor
g_c: gravitational constant (ft/sec^2)
Δz: depth of segment (ft)

Beggs and Brill Method. The Beggs and Brill correlation was developed from experimental data in a small scale test facility. The equation presented Beggs and Brill apply to flow in a pipe at any angle of inclination, including downward flow [11]. Although the method has been found to slightly over predict pressure gradient in vertical wells in some cases, it gives good results for pipeline calculation. Here is the equation are made to account for the holdup behavior with inclination.

$$\frac{dp}{dz} = \frac{g}{g_c}\overline{\rho}\sin\theta + \frac{2f_{2f}\overline{\rho}{u_m}^2}{g_cD} \tag{4}$$

Where:

$\overline{\rho}$: mixture density (lb_m/ft^3).
θ: inclination angle.
u_m: mixture velocity (ft/sec).
f_{2f}: 2 phase friction factor.
D: inner diameter (in).

2.3 Calculation Procedure

Initially, by correlation formulations for PVT calculations, we can determine bubble point pressure of oil well X with corresponding reservoir pressure and temperature. Bubble point pressure is a key element for subsequent computations.

These calculations are complicated and multistep to complete, therefore the basic working principle shown in Fig. 1.

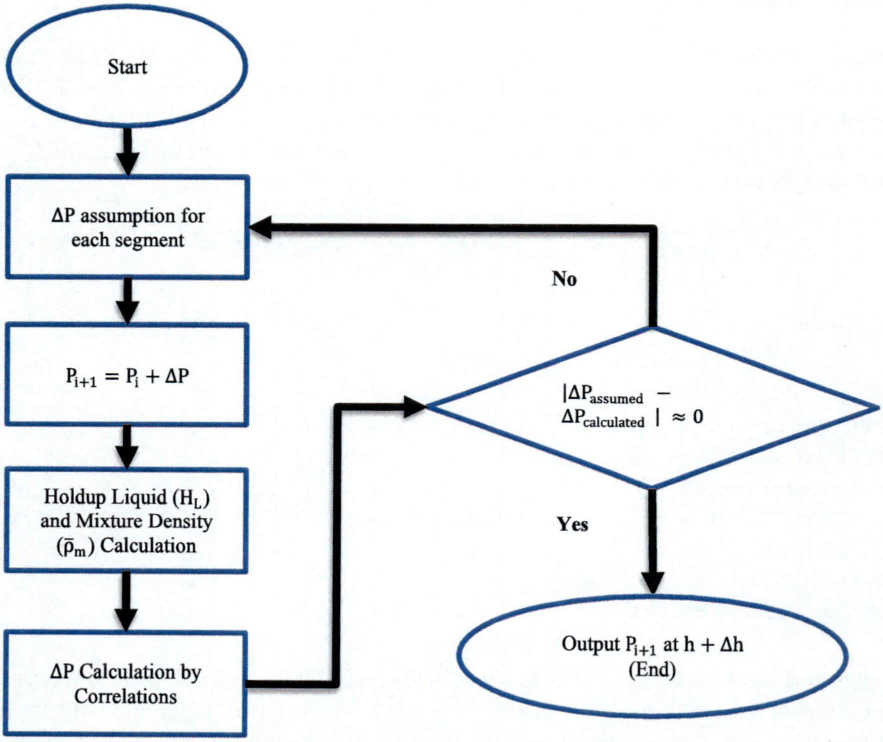

Fig. 1. TPR procedures for pressure drop calculation

2.4 Erorr Estimation

$$\text{Error} = \frac{|\text{Predicted Value} - \text{Measured Value}|}{\text{Measured Value}} \tag{5}$$

Erorr was calculated by Eq. (5) to find the difference between predicted value from the multiphase flow correlations and measured value collected from the specific wells.

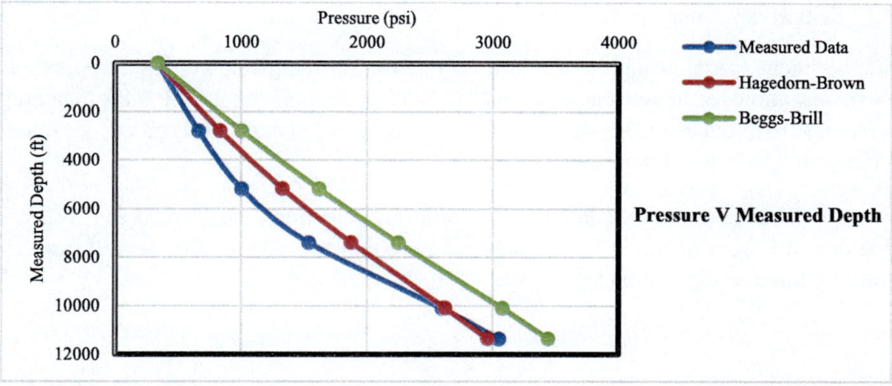

Fig. 2. Gradient matching using Hagedorn-Brown and Beggs-Brill correlations

3 Results and Discussion

3.1 Gradient Matching

Matching multiphase flow correlation is the first imperative step for quality control of a production model. Figure 2 shows that the pressure gradient predicted by Hagedorn-Brown correlation provides a good match over 6 measured data points although there are some over predictions from depth of 2000 ft to 9000 ft with the average error about 15%. Besides, Beggs-Brill tends to over-predict throughout the depth of 11349 ft.

Table 1. Difference between measured pressure and predicted pressure

Data from oil well X		Error of pressure	
Measured depth (ft)	Measured pressure (psi)	Hagedorn-Brown (%)	Beggs-Brill (%)
0	339	0	0
2792.13	662.35	25.61334642	51.4305126
5180.7	1003.18	32.37903467	61.4366315
7378.97	1538	21.71651495	46.0988296
10086.8	2597.44	0.89857706	18.3857952
11349	3047.44	2.977253039	12.8816318

The variation of measured pressure with predicted pressure by Hagedorn-Brown and Beggs-Brill is expressed in Table 1. In general, at the range of 10000 ft to 11349 ft, the error of the estimation is small. The average error of Hagedorn-Brown gives about 15% while the Beggs-Brill prediction is less accurate with mean variation of nearly 40%.

3.2 Sensitivity Analysis

The pertinent model of multiphase flow correlation for the given well data in previous work was employed to assist in the sensitivity studies. In this section, IPR/TPR matching is required to initialize the sensitivity analysis. The calculation of TPR is proceeded under a range of flow rates and pressure values.

Effect of Tubing Size. By adding more tubing size selections, the graph in Fig. 3 shows that pressure were obtained for the flowrate ranging from 100 to 8900 barrels per day with 2.2 in. to 3.5 in. tubing size.

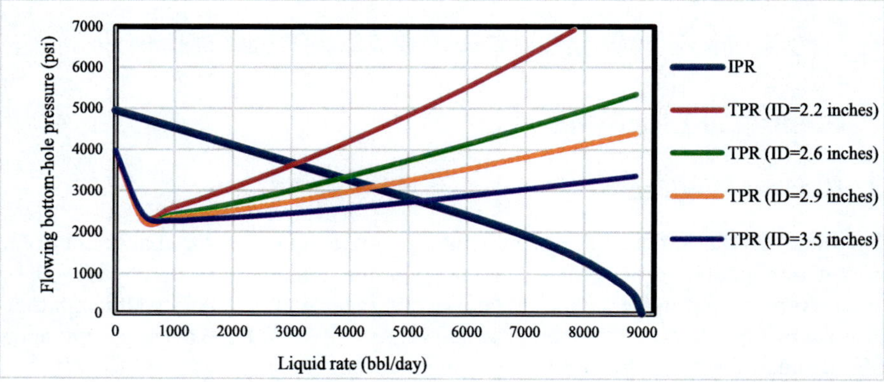

Fig. 3. Hagedorn & Brown TPR curve for multiple tubing sizes

Table 2 shows that increasing tubing diameter at constant wellhead pressure (339 psi) will expand the optimum production rate. At the tubing size of 2.2 in., the optimum production rate is about 3070 bbl/day. The following tubing diameter of 2.6 in, 2.9 in, and 3.5 in yield optimum liquid rate of 3877 bbl/day, 4393.7 bbl/day, and 5195 bbl/day respectively. Aside from that, the variance between the observed value of the operating production rate with the calculated ones is shown in the "Change in production rate". It is clear that if decreasing the tubing size below 2.9 in., the changes of production rate at the actual operating condition decline significantly. On the contrary, at the range of 2.9 in. to 3.5 in. tubing diameter, the rate of variation is relatively small.

It was claimed that the use of smaller tubing size has reduced the flow area and consequently it increases the resistance to flow. This will resist the production rate and subsequently reduces the amount of fluid that can be produced. Conversely, larger tubing size will cause excessive downhole liquid loading during lifting. It leads to the impact on economic aspect of the operating equipments [12].

Effect of Tubing Roughness

Figure 4 shows that tubing roughness ranging from 0.0006 in. to 0.0045 in. gives a downward trend of the production rate. Specifically, the production rates decline from 4900 bbl/day to approximately 4580 bbl/day. It indicates that variation with regard

Table 2. Effect of various tubing sizes on production rate

Inner tubing diameter (inches)	Operating production rate (bbl/day)	Change in production rate (%)
2.2	3070	35.02646
2.6	3877	17.94709
2.9	4393.7	7.01164
3.5	5195	9.94709

to actual production rate is ranging from 1% to 3%. Overall, the decrease of tubing roughness will provide the increased amount of produced liquid rate.

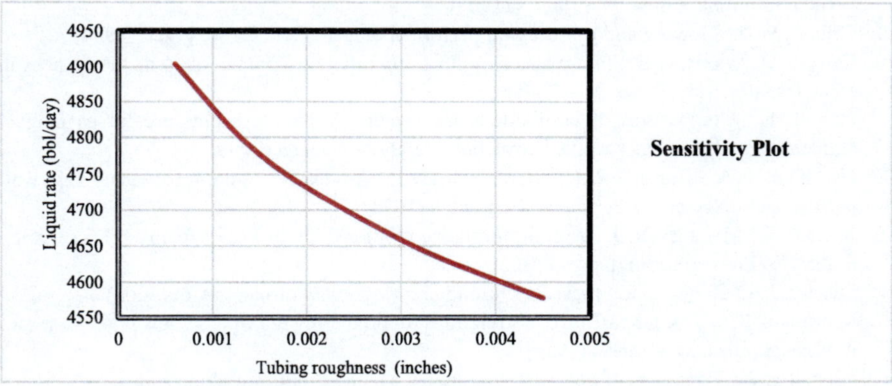

Fig. 4. Effect of tubing roughness on liquid production rate of oil well X

4 Conclusion

Hagedorn-Brown correlation is the better choice for applying to oil well X as it gives a small variance of pressure prediction which is approximately 15% comparing to actual values in the field. In contrast, the Beggs-Brill model provides inconsistent pressure drop estimations with a larger variance of nearly 40% when matching 6 measured pressure survey data points.

Based on the sensitivity analysis, the following conclusions are listed below:

1. The effect of tubing size adjustment is significantly important to production optimization. Scaling down the inner tubing diameter can reduce 35% of the actual value.
2. Increment of tubing roughness also affects the tubing performance in general. From 0.0006 in. to 0.0045, the variance of produced rate ranges from 1% to 3% regarding to actual production rate at the given operating conditions.

References

1. Abu, T.: Optimization of Gas Lift System in Varg Field. M.S. Thesis. Stavanger: Univ. of Stavanger (2007)
2. Ahmed, T.: Reservoir Engineering Handbook. Gulf Professional, Massachusetts (2010)
3. Beggs, H.D.: Production Optimization Using Nodal Analysis. OGCI and Petroskills, Tulsa (2003)
4. Guo, B., Lyons, W.C., Ghalambor, A.: Petroleum Production Engineering: A Computer-assisted Approach. Elsevier Science & Technology Books, Massachusetts (2007)
5. Bradley, H.B.: Petroleum Engineering Handbook. Society of Petroleum Engineers, Texas (1987)
6. Brill, J.: Multiphase flow in wells. J. Petrol. Technol. **39**(01), 15–21 (1987)
7. Brown, K.E., Beggs, H.D.: The Technology of Artificial Lift Methods, vol. 1. Penn Well, Tulsa (1978)
8. Espanol Herrera, J.H.: Comparison of three methods for calculating a pressure tranverse in vertical multiphase flow. M.S. thesis (1968)
9. Gilbert, W.E.: Flowing and gas-lift well performance. API Drill. Prod. Pract. (1954)
10. Griffith, P., Wallis, G.B.: Two-phase slug flow. In: Taitel, Y., Barnea, D. (eds.) Advances in Heat Transfer, vol. 20, pp. 84–128 (1961)
11. Hamid, K.: Comparison of correlations for predicting wellbore pressure losses in gas-condensate and gas-water wells. Petroleum Society of Canada (2003)
12. Hutchins, L.A., Burton, R.K., MacIntosh, D.J.: An expert system for analyzing well performance. Society of Petroleum Engineers (1996)
13. Kumar, K., Mahadevan, J.: Well performance relationships in heavy-foamy oil reservoirs. Society of Petroleum Engineers (2012)
14. Lasater, J.A.: Bubble point pressure correlation. Magnolia Petroleum Co. (2009)
15. Messulam, S.A.G.: Comparison of correlations for predicting multiphase flow pressure losses in vertical pipes. M.S. thesis (1970)
16. Nind, T.E.W.: Principles of Oil Well Production. McGraw-Hill, London (1964)
17. Orkiszewski, J.: Predicting two-phase pressure drops in vertical pipe. J. Petrol. Technol. **19**(06), 829–838 (1967)
18. Rossland, L.: Investigation of the performance of pressure loss correlations for high capacity wells. M.S. thesis (1981)
19. Thompson, L.G.: Determination of optimum combination of pressure losses and PVT property correlations for predicting pressure gradients in upward two-phase flow. M.S. thesis (1982)
20. Whitson, C.H.: Reservoir well performance and predicting deliverability. Society of Petroleum Engineers (1983)

Offshore Renewable Energy Systems

Fabrication of Concrete Gravity Foundations as Support Structures for Offshore Wind Turbine Facilities

Ove Tobias Gudmestad(✉)

University of Stavanger, Ullandhaug, 4036 Stavanger, Norway
ove.t.gudmestad@uis.no

Abstract. Offshore wind turbines in shallow waters are most often supported by monopiles driven into the ground. As a means to reduce the installation costs, alternative concepts should be investigated, as for example tower structures with gravity foundation that potentially could bring the tower, the nacelle and the blades to the offshore location for easy installation. In this paper it is suggested that concrete gravity foundations be considered for support of wind turbines offshore Vietnam. As an additional effect, the structures can be prepared in a dry dock using only local goods and services and towed to site by smaller tugs available in the country. The fabrication site should be sufficiently large to ensure an assembly line for fabrication of a series of same design wind turbines. The most optimal site for a construction yard is identified with sufficient water depth to tow fully commissioned wind turbines to the actual offshore locations.

Keywords: Wind power potential · Selection of foundation concept · Concrete gravity support structures · Fabrication site considerations · Installation

1 Background

1.1 The Scope of the Study

The offshore wind power potential in water depths up to 30 m is large in some of the Vietnamese provinces, Fig. 1, [1]. The technical potential is approximately 100 GW in up to 30 m depth, 160 GW in 30–60 m, and for floating wind the potential is 210 GW, in total 470 GW [2]. The most promising region is the coastal area of the Binh Thuan and Ca Mau provinces with a power capacity density of nearly 1000 W/m^2 [3, 4].

Support structures for offshore wind turbine facilities in shallow water (in the range of up to 30 m) are most often large mono-columns driven into the ground by pile hammers. In this case, competent sand is needed to support the foundation in order to provide a stable foundation and avoid large vibrations. For water depths exceeding 30 m, the most common support structures are jacket type piled structures. In even deeper waters, there is a range of depths where jackets and floating support structures are competing, however; for depths of more than, say 120 m, floating support structures take over.

D. V. K. Huynh et al. (Eds.): VSOE 2021, LNCE 208, pp. 493–500, 2022.
https://doi.org/10.1007/978-981-16-7735-9_55

The installation process requires use of heavy equipment like jackups with crane arrangements or large floating lift barges. This equipment is sensitive to the wave environment and even low height swell conditions with long periodic waves could lead to resonant motions whereby the installation work has to be set on hold, waiting for suitable weather conditions [5].

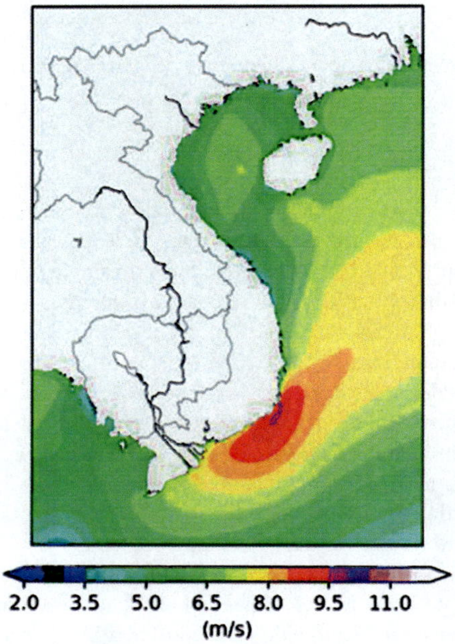

Fig. 1. Wind Energy resources offshore Vietnam, wind average velocity measured at 100 m height [1].

A more elegant foundation solution is to use concrete gravity foundations with large base areas. These have been employed in European waters, Fig. 2, in case of large loading (ice loading in the Baltic or breaking waves on shoals). The turbines fabricated for Thornton Bank (Fig. 3) were installed with use of crane vessels, however, in case of deeper tow out channels from the construction site, these structures should be in position to carry the nacelle and the rotor/the blades to the offshore site during a tow operation, see discussion below.

The requirements to site condition are important as a concrete gravity foundation structure has to be placed on sandy soils or hard clay [6]. The foundation plate is to be outfitted with a steel ring (a "skirt") that will penetrate into the soil to avoid erosion. The length of the steel "skirt" will be determined by the erosion potential and the depth to competent soils that can withstand the vertical weight and the moments due to the wave actions. The bottom should be plain without stones, however, to adjust for unevenness, grout could be injected under the foundation plate. Checking of these requirements will demonstrate which sites in Vietnam are suitable for that type of foundation.

Fig. 2. Assembly line for fabrication of offshore wind turbine foundations for Thornton Bank, Belgium [7].

Fig. 3. Installation of wind turbines on Thornton Bank, Belgium [7].

1.2 The Experience with Offshore Concrete Gravity Structures

The main region for the application of platforms with concrete gravity foundations has been the North Sea. The search for oil and gas reached the North Sea area in the late 1960's. Almost at the same time the increasing price of hydrocarbons substantiated economic development of the findings in this area. As a result, a rapid development of structures fit for the North Sea was seen. It should be kept in mind that the average weather conditions of the North Sea are considerably more hostile than in the Gulf of Mexico, on which early North Sea developments were modelled. Technology developed for the Gulf of Mexico failed in the North Sea, in particular due to larger fatigue loading.

In the late 1960's and the early 1970's platforms with concrete gravity foundations were selected in preference to steel jackets by several oil companies for production platforms in water depths from 70 m to 150 m. The lack of deep "fjords" in the UK, where deck transfer can take place, was, however, a drawback for the selection of concrete gravity foundations in the UK. In Norwegian waters, where deep water sites are available and where deep tow out channels are present with water depths up to 241 m, concrete gravity foundations were chosen for several major offshore developments up to the mid 1990's. The availability of suitable local locations for construction of concrete gravity foundations was important. Since the 1990's, design fabrication and installation techniques of both steel and concrete platforms improved. Concrete gravity foundations as support structures for the drilling and process equipment were still selected for several offshore field developments offshore Sakhalin (Russian Federation) and Newfoundland (Canada).

1.3 Wind Turbine Concepts Made of Concrete

As was discussed above, wind turbine foundations made of concrete have been installed on the Thornton Banks. Furthermore, such foundations have been installed in the Baltic due to their resistance against potential ice drift. The idea of using a self-installing concrete mono-tower was also presented by Bouygues Offshore in 2002 [8].

The process of designing an efficient wind tower foundation fabricated in concrete must include the following aspects:

- Size the foundation structure to bring the tower, the nacelle and the rotor (the turbine blades) to the offshore location.
- The foundation slab will be dimensioned to resist the wave and wind loads using the actual physical environmental data and the soil conditions at the offshore location [9].
- Sufficient initial stability during the tow to the offshore location must be secured by ensuring deep draft along the tow route, adequate amount of ballast and a sufficient water plane.
- By designing the *tower* as a telescope [10], the nacelle and the blades could be lowered into the gravity foundation to lower the center of gravity during the tow. Furthermore, the design of the concrete tower must ensure a *water plane area* sufficient to secure stability when the structure is lowered to the seafloor at the offshore location.

  - Should there, in the case a telescopic tower is selected, be any conflict between the blades and the required height above surface during the tow, it is proposed that the tower of a two-bladed wind turbine can be lowered down closer to the sea surface as compared to a three bladed rotor.

2 Identification of Site for Fabrication

2.1 Potential Fabrication Site in Vietnam

During visits to Vietnam in the early 1990's, Vietnamese authorities associated with Petro Vietnam drew our attention to a site north of Nha Trang that would be a strong candidate for fabrication of concrete support structures for offshore oil and gas production platforms [11], see Figs. 4 and 5.

In the "fjord" system north of Nha Trang, in Dam Mon Bay, a dry dock with a depth of 14 m can be established for fabrication of large offshore concrete structures. Tow out drafts of 25 m (or possibly near to 30 m, with some dredging) can be achieved and the towing route is straight into the open ocean. Therefore, concrete foundation structures

Fig. 4. Location of possible fabrication site

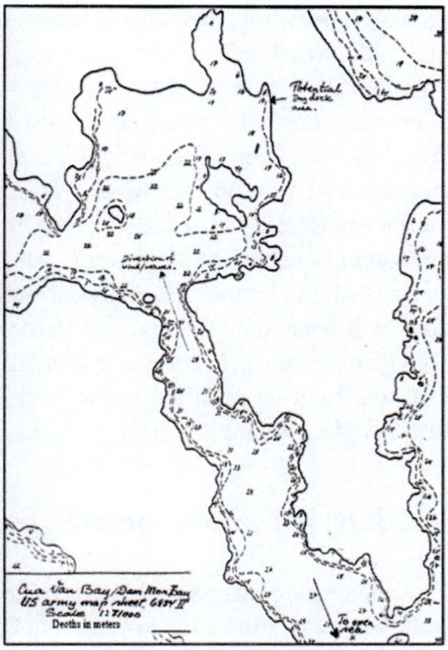

Fig. 5. Water depth in the area where a fabrication site could be established

fabricated in a dry dock established here, can be considered for offshore wind turbines. This site would also represent an excellent site for construction of floating wind turbines for deeper waters.

2.2 Fabrication Dock Requirements

A dry dock where wind turbine foundation fabrication is necessary. The size of the dock depends on how many structures should be fabricated simultaneously in an assembly line production scenario. In addition, there needs to be an area for rigging, storage, prefabrication, workshops and offices. The size of this area depends on the amount of work which is expected to be performed in the yard. It is further assumed that all structural steel and pipework prefabrication is performed at a different location. For the suitable concrete platform yard, it is necessary to have sufficient deep water for tow out of the completed structure. Furthermore, dry dock excavation can be expensive so it is important to find an area which offers minimum rock blasting. The dock must be stable and not leak. Also, the dock floor has to support the completed structures Construction of concrete platforms is labor intensive and it is important that there is an adequate supply of skilled as well as unskilled labor in the vicinity of the yard. The main utilities required for concrete platform construction is water and electrical power. The consumption of water and the electricity demand depend on the activity in the dock.

A 14 m deep dry dock (below mean sea level) have large potentials. This depth, however, means that water ingress and dock foundation stability must be considered carefully. For dry dock general arrangement, see [11]. Dry dock development based on experience from Norway requires steel sheet pile walls supported by granular fill materials. Steel sheet piles, piled to rock, will give a water tight dock. Further soil investigations and review of available resources may, however, result in other methods being more cost effective. A dock gate that can be opened when the concrete foundations are ready must be installed. It is also assumed that a quay is required to offload equipment and material transported by sea. The size of this quay will depend on the expected marine traffic. It is anticipated that bulk materials such as sand, cement aggregates and reinforcing steel to be transported by ship. The dock floor may need a concrete foundation slab in order to support the heavy structures. The size and type of foundation in the dry dock will depend on the distance to rock below the dock floor.

Out of the dock, the concrete structure will be lowered to tow draft and the equipment that could not be installed in the dock, for example the nacelle and the rotor, will be lifted in place in the Bay prior to tow to offshore location.

3 Materials and Concrete Quality

The concrete structure should be designed in accordance with the International Standardization Organization's Fixed Concrete Offshore Structures Standard [12]. Highlights are that:

- Durable offshore concrete should have high strength, be dense and have at least 40 mm cover over the reinforcing bars to avoid rebar corrosion and spalling.
- To obtain a concrete quality of C55 or better, it is important to have the right mixture of cement, aggregates, and water, as well as additives to control the concrete hardening and to obtain good workability.
- The cement normally used for offshore concrete is Portland cement with C_3A content less than 8% and with no CaC1. Test certificates from the cement manufacturer should be required for every cement delivery. Special cement with smaller particle size, like Portland P30 Rapid cement, will harden faster than normal cement but will have about the same quality after 28 days. Special care must be taken to avoid excessive heat development during curing.
- The aggregates consisting of natural or manufactured hard sand and gravel are an important part of the concrete mix. The gravel usually is up to 32 mm. A good aggregate mix has a certain sieve grade distribution. The minerals that react with alkalis in the cement must be limited so that no deterioration will occur. Only a small amount of organic impurities is allowed. In many cases, the impurities will be removed by washing the sand and gravel with fresh water.
- The water should be fresh, clean, and free of injurious amounts of oil, acids, alcohols, salts, or organic materials. Seawater should not be used.
- The concrete quality and strength depend on the water/cement ratio (<0.45), cement content (>400 kg/m^3), sieve grade of aggregates and cement quality. Furthermore, compacting of the concrete with a vibrator is necessary. During concrete curing, the temperature should not exceed 70 °C and the temperature gradient should not be higher than 30 °C/300 mm.

There are several types of additives: retarders to slow the hardening process if necessary, accelerators for casting by slip forming, and water-reducing additives to obtain good workability of the concrete mix when the water/cement ratio is low. Silica fume may be used when high-strength concrete is required. The silica also improves fresh concrete properties like workability, stability and pumpability.

- Reinforcing steels normally are classified on the basis of their size (bar diameter); characteristic yield stress, good bond properties and weldability. The most common nominal diameters for bars in offshore structures are 16, 20, 25 and 32 mm; 12 mm bars are used for stirrups.
- Characteristic yield stresses of 400 MPa (Grade 400) and 50 MPa (Grade 500) are used. The value of yield corresponds to 0.2% permanent elongation. The bond properties depend on rib size and form. Bars that do not satisfy the requirements should be treated as plain bars with respect to bond. Carbon, Silicon, Mangan, phosphor, sulphur and nitrogen content in the steel is limited.
- The prestressing steels usually are classified on the basis of their characteristic tensile strength, f_{pt}. The prestressing steel should withstand without failure 2×10^6 cycles at 200-MPa stress, with a maximum stress of 0.8 f_{pt}.

4 Discussion of Potential Benefits

The potential benefits of fabricating wind turbine foundations in concrete are considerable:

- The fabrication can be undertaken with a limited workforce of skilled laborers and materials (cement and aggregate) produced in the country can be utilized to a maximum degree.
- The fabrication of concrete gravity foundations will enhance the competence of those who are involved in the work at the fabrication site and in the engineering related tasks.
- The completed turbines can be assembled and be fully commissioned in a dry dock/in a Bay and sailed to the offshore location with the use of easily accessible harbor tugs, thereby saving the cost of hiring expensive international installation vessels.
- A suitable location for dock facilities in Vietnam is identified and export to other countries where potential deep-water fabrication sites do not exist seems possible.
- Furthermore, dense concrete has demonstrated its use in harsh North Sea conditions over 50 years of utilization and requires less maintenance as compared to steel structures.

The author will suggest that a thorough detailed study of the benefits and the cost savings be undertaken to the benefit of Vietnamese industry and for the provision of electric energy at affordable costs.

References

1. Toan, D.V., Doan, Q.V., Le Duy Anh, P., Dinh, V.N.: The zoning of offshore wind energy resources in the Vietnam Sea. In: Randolph, M., Doan, D., Tang, A., Bui, M., Dinh, V. (eds.) Proceedings of the 1st Vietnam Symposium on Advances in Offshore Engineering. VSOE 2018. Lecture Notes in Civil Engineering, vol. 18. Springer, Singapore (2019). https://doi.org/10.1007/978-981-13-2306-5_34
2. ESMAP: Going Global: Expanding Offshore Wind to Emerging Markets. Washington, DC: World Bank (2019)
3. Doan, V.Q., et al.: Usability and challenges of offshore wind energy in Vietnam revealed by the regional climate model simulation. SOLA **15**, 113–118 (2019). https://doi.org/10.2151/sola.2019-021
4. Ha-Duonga, M., Teskeb, S., Pesciac, D., Pujantoro, M.: Options for wind power in Vietnam by 2030. HAL ouvertes (2020). https://hal-enpc.archives-ouvertes.fr/hal-02329698v2/document
5. Gudmestad, O.T.: Waiting on suitable weather to perform marine operations. In: Murali, K., Sriram, V., Samad, A., Saha, N. (eds.) Proceedings of the Fourth International Conference in Ocean Engineering (ICOE2018). Lecture Notes in Civil Engineering, vol. 22. Springer, Singapore (2019). https://doi.org/10.1007/978-981-13-3119-0_1, 978-981-13-3118-3, 456464_1_En, (1)
6. Gudmestad, O.T., Warland, T.A., Stead, B.L.: Concrete structures for development of offshore fields. J. Pet. Technol. **45**(08), 762–770 (1993). https://doi.org/10.2118/22376-PA. Paper Number: SPE-22376-PA
7. Esteban, M.D., Couñago, B., Santos Lopez-Gutierrez, J., Negro, V., Vellisco, F.: Gravity based support structures for offshore wind turbine generators: review of the installation process. Ocean Eng. **110**, 281–291 (2015). https://www.researchgate.net/publication/284078201_Gravity_based_support_structures_for_offshore_wind_turbine_generators_Review_of_the_installation_process
8. Ruer, J., Coche, E., Jego, G.: Bouygues offshore, Self-installing wind turbine limits costs, provides improved stability, Offshore Magazine, 1st August 2002. https://www.offshore-mag.com/business-briefs/equipment-engineering/article/16759948/france-selfinstalling-wind-turbine-limits-costs-provides-improved-stability
9. Gudmestad, O.T., Stead, B., Sparbye, B.: Reusable production systems. Presented by Statoil at SPE South East Asia, Singapore, February 1993
10. Campione, G., Cannella, F., Papia, M.: Structural behavior of telescopic steel pipe for a full-scale 60 kW wind turbine tower. Int. J. Steel Struct. (2020). https://doi.org/10.1007/s13296-020-00313-9
11. Gudmestad, O.T., Pollard, N.: Requirements for construction of offshore concrete platforms. In: Proceedings of the ISOPE-94, vol. 4, pp. 381–388. Osaka, April 1994
12. International Standardization Organization ISO: ISO 19903:2019. Petroleum and natural gas industries — Fixed concrete offshore structures. ISO Geneva, Switzerland (2019)

Microstructure and Mechanical Properties of Methane Hydrate-Bearing Sand

Thi Xiu Le[1(✉)], Anh Minh Tang[1], Patrick Aimedieu[1], Michel Bornert[1], and Andrew King[2]

[1] Laboratoire Navier, Ecole des Ponts ParisTech, CNRS UMR 8205, Univ. Gustave Eiffel, Marne-la-Vallée, France
thi-xiu.le@enpc.fr
[2] Psiché beamline, Synchrotron SOLEIL, Paris, France

Abstract. Methane hydrates (MHs), solid ice-like compounds of methane gas and water, form naturally at high pressure and low temperature in marine or permafrost settings. They represent a great alternative energy resource but also a source of geo-hazards and climate change. Knowledge of physical/mechanical properties of sediments containing MHs, depending considerably on methane hydrate morphologies and distribution at the pore scale, is of major importance to be able to minimize the environmental impacts of future exploitation of methane gas from methane hydrate-bearing sands (MHBS). Much of the reported experimental work consists in laboratory tests on synthetic samples due to the extreme difficulty to get intact cored natural methane hydrate-bearing sediment samples. Various methods have been proposed to form methane hydrates in sandy sediments to mimic natural MHBS, but with limited success. This paper discusses the morphologies and pore habits of MHs formed in synthetic MHBS, considered as model materials for real MHBS and their effects on the mechanical properties of MHBS, at various scales and via different methods mainly as synchrotron X-ray computed tomography, magnetic resonance imaging and triaxial tests. An alternative MH formation method is proposed to better mimic natural MHBS and the validity of existing idealized models used to describe MH pore habits is discussed.

Keywords: Methane hydrate · Microstructure · Mechanical behaviors

1 Introduction

Natural gas hydrates (mainly methane hydrates), most often located in marine and permafrost settings, represent a great potential energy resource (estimates range over several orders of magnitude: ~3100 Tm^3 to ~7,650,000 Tm^3) but also a source of greenhouse gas (methane is an active "green-house" gas that has a global warming potential twenty times greater than an equivalent weight of carbon dioxide when integrated over 100 years) and geo-hazards (slope instability and wide-scale gas venting due to the dissociation of MHs replacing rigid components with free gas and excess water, which may significantly reduce the geo-mechanical stability of affected sediments) [1]. Detailed knowledge about

D. V. K. Huynh et al. (Eds.): VSOE 2021, LNCE 208, pp. 501–508, 2022.
https://doi.org/10.1007/978-981-16-7735-9_56

how released methane gas from MHs reaches atmosphere needs to be clarified as that is a key issue in understanding the connections between MH as an energy resource and also a potential player in future climate scenarios [2].

The formation of MH in sandy sediments modifies the microstructure and as consequence the physical/mechanical properties of MHBS. The study of the microstructure of MHBS, MH morphologies and distribution at the pore scale (pore habits), is of major importance in order to minimize the environmental impacts of future exploitation of methane gas from MHBS [2].

The majority of the existing experimental works focus on synthetic samples in laboratory due to challenges to get undisturbed cored MHBS samples. Various methods have been proposed to form MH in sandy sediments in laboratory and conceptual models proposed by Dvorkin et al. [3] have usually been used to indirectly assess the MH pore habits in sandy sediments [4–8]. Laboratory and synchrotron X-ray computed tomography (XRCT & SXRCT) have been recently used to investigate the microstructure of gas hydrate-bearing sediments [9–12]. It should be noted that due to the poor contrast between water and methane hydrate in a XRCT image, other types of gas or saline water solutions have been usually used.

In this paper, results obtained from the project HYDRE [13] on MH formation and its effects on the microstructure and the mechanical properties of synthetic MHBS under both excess-gas and excess-water conditions are discussed. The MH formation following the excess-gas method is first analyzed. Afterward, MH morphology and pore habit changes after the water saturation with or without an additional temperature cycle and their corresponding effects on the microstructure and the mechanical properties of MHBS are debated.

2 MH Formation Under Excess-Gas Conditions

2.1 MH Formation Procedure

In the project HYDRE [13], MHBS was formed following the partial water saturation method by injecting methane gas at 7 MPa and maintaining temperature at 3–4°, see [8, 12–14] for more details.

The initial water saturation was 25–50%. The stabilization of the compressional wave velocity (V_P) at the end of the MH formation suggested that almost the entire quantity of available water has been transformed into methane hydrates [8]. The final MH saturation increased with an increase in initial water in this range. Furthermore, two experiments at 25% of initial water saturation, observed by magnetic resonance imaging (MRI) [14], showed that almost 100% of water became MH in the end. Ten percent of volume increase was supposed for the water-methane hydrate conversion. However, in the work of Zhang et al. [15], 5 to 15% of pore water was found to remain unconverted to MH at the end of the experiments with 25–50% of initial water saturation. Actually, in these works, the signal of methane gas was not considered and that would induce errors in water content estimation in the specimen. In addition, the experiment duration was only few hours for these works compared to several days for triaxial and MRI tests in the present work [8, 13, 14].

2.2 MH Morphologies and Pore Habits Under Excess-Gas Conditions

The results of eight samples shown in Le et al. [8] show that MH formation in gas-saturated media increased V_p, and this increase reached the stabilization state at the end of the MH formation (before shearing). It was supposed that almost the distribution of MHs was similar to that of water prior to their creation. Comparison between the experimental data and Dvorkin's model [3] shown in Le et al. [8] suggests that at a low initial water saturation (MH saturation), methane hydrates were mainly distributed at grain surfaces (cementing - mineral coating), while at a high initial water saturation, the role of MHs at grain contacts (cementing - grain contacts) dominated. These macroscopic results are in agreement with that of Waite et al. [4] by using similar approaches (velocity measurements and Dvorkin's model). Note that the cementing - grain contacts model was only considered at low gas hydrate saturation.

At the pore scale, Zhao et al. [16] observed MHs (formed following the excess-gas method) at gas-water interfaces, floated between sand grains without coating on grain surfaces. A thick water layer was found to envelop grain surfaces by using image segmentation. Besides, polyhedral crystals of Xe hydrates around grains and loosely connected aggregates of Xe hydrate crystals in the pore space (formed following the excess-gas method), were observed by SXRCT [10]. Furthermore, thin layers of water were observed to exist between sand grains and Xe hydrates. It should be noted that the Xe hydrate formation time was just some ten minutes, remaining water could still be transformed into MHs. Furthermore, the morphology of gas hydrates depends on the type of used gas.

Within spatial resolution of sub-micron voxel size of SXRCT images in the work of Le et al. [12], MH morphologies and pore habits were observed directly without the need of segmentation (see Fig. 1a–d). That allows to discuss the four types of MH pore habits, proposed by Dvorkin et al. [3]: cementings, load-bearing and pore-filling, usually used for predictions of physical/mechanical behaviors of MHBS. It is usually supposed that MH formed under excess-gas conditions in sandy sediments (following the excess-gas or ice-seeding method) have the shape of pendular water menisci at grain-grain contacts or thin uniform MH layers coating sand surfaces (cementing - mineral coating) [4, 5]. The present study and that of Lei et al. [11] showed irregular shapes of MH at sand surfaces often accompanied by finger-like spreading patterns of MHs at low MH saturation. It appears that methane hydrates were at grain-grain contacts and/or grains surfaces (cementing) at low hydrate saturation while they tended to fill the pore space at high hydrate saturation. Besides, methane hydrates could be porous if they were not totally clogged over time. It is clear that MH morphologies and pore habits in the sample are not only heterogeneous at the pore scale but also along the sample height (the sample scale) due to water migrations [12]. Different types of MH morphologies and pore habits could exist in the sample. Not only MH pore habits but also MH morphologies in sandy sediments (which can only be observed at high image spatial resolution) are important for studies on physical/mechanical behaviors of methane hydrate-bearing sediments. At similar MH saturation, different MH morphologies (filaments, "crystals" or layers) at sand grain surfaces (cementing - mineral coating) or at contacts of grains (cementing - grain contacts) can play different roles in mechanical behaviors of methane hydrate-bearing sediments. We suppose here an intermediary model of MHs at the grain scale (see Fig. 1e) with complex geometry of MHs that could eventually be porous. MHs could play more role at grain contacts compared to the conceptual cementing-mineral coating model by forming bulk MHs. Furthermore, effects of thin water film supposed existing between sand grain and MH on physical/mechanical behaviors of sediments have not been well studied yet. It seems vital that numerical studies on mechanical behaviors of MHBS, which have been based on four idealized MH pore habits, should take into account more realistic and less caricatural MH morphologies and pore habits.

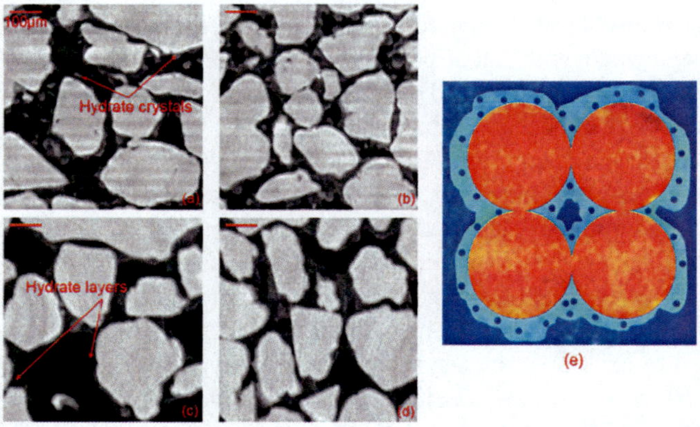

Fig. 1. Some examples of real MH microstructure in MHBS under excess-gas conditions (a–d); (e): An alternative model of MH pore habits under excess-gas conditions (Blue: methane gas; Cyan: MHs; and Red: sand grains).

3 MH Under Excess-Water Conditions

3.1 MH Formation Procedure

Following the MH formation under excess-gas conditions aforementioned, two different procedures were used: a water saturation by replacing the excess gas in the sample by water at 7 MPa (Procedure A) or a water saturation accompanied by a temperature cycle (Procedure B). Please see [8] for more details.

3.2 MH Morphologies and Pore Habits Under Excess-Water Conditions

Water Saturation

The subsequent water saturation phase significantly decreased V_p for triaxial tests [8]. Kneafsey et al. [17] observed a similar decrease in sonic velocities when saturating the gas-saturated MHBS with water. These results suggest that water saturation modified the MH distribution at the grain scale. MHs located at grain contacts and/or at grain surfaces would be progressively converted or/and redistributed into the pore space [7].

For triaxial tests shown in Le et al. [8] the effects of water injection in gas-saturated MHBS were investigated at different MH saturations. The results indicate that this process took longer time for higher MH saturation. For some specimens (B2, B4, A3, and A4), this transformation was not complete when the subsequent step (triaxial compression for the specimens A and heating/cooling cycle for the specimens B) was applied. This may explain the higher measured values of V_p after the water saturation than the predicted values of the load-bearing model. Besides, for the two tests following the procedure A at high MH saturation (A3 and A4), even after waiting long time for the water saturation to make sure that MHBS was well saturated with water, V_p was still higher than the value predicted for the load-bearing model.

The results of MRI tests in the work of Le et al. [14] show heterogeneous water distribution along the sample height after the water saturation. Note that similar water saturation procedure was applied for triaxial tests and MRI tests [8, 13, 14]. However, for triaxial tests, water was injected from both top and bottom inlets while water was only injected from the bottom inlet for MRI tests as the sample cell was put into the MRI system. That explains why MH saturation was higher at the top of the sample compared to that at the bottom for MRI tests. Furthermore, the water injection time of MRI tests was longer than that of the triaxial tests. That could induce more MH dissociation during the water injection. MH distribution after the water saturation of triaxial tests would be more homogeneous as water was injected from both top and bottom inlets.

Temperature Cycle

Choi et al. [7] formed MHs in unsaturated sand following first the excess gas method, then injected saline water at conditions just outside of the MH stability zone for saline water and conducted a temperature cycle. It was concluded that slow saline water injection was a key to initiate the formation of non-cementing hydrates and the temperature cycle ensured this formation. *Vp* after warming (the MH dissociation) was quite high while the sample was not saturated. The MH dissociation was perhaps not completed before the MH reformation. This could explain why the difference of *Vp* after the saline water injection and that after the temperature cycle was small.

In the present work [8, 13, 14], normal water was injected to saturate the sample and MHs were completely dissociated before being reformed. The measurement of *Vp* at the end of the temperature cycle was smaller than that obtained after the water saturation. In addition, *Vp* at the end of the temperature cycle fitted with the pore-filling or the load-bearing model. It can then be expected that in the range of 0–50% of MH saturation, the heating/cooling cycle allows for the complete conversion of MHs with cementing pore habit into the non-cementing type either when the water saturation is not finished before the temperature cycle or when MH saturation is in a range of 40–50% (*Vp* after the complete water saturation is still higher than the value predicted for the load-bearing model).

Furthermore, MRI results [14] show more homogeneous water distribution (MH distribution) along the sample height after the temperature cycle, which suggests that the temperature cycle can redistribute MHs at the grain scale as well as at the sample scale.

To mimic natural methane hydrate-bearing marine sediments, the dissolved gas method is considered as the best method but it was time-consuming especially at high

MH saturation due to the low solubility of methane gas in water [18]. The water-excess method, proposed by Priest et al. [6], was suggested to create load-bearing MHs in sandy sediments at MH saturation lower than 40% according to sonic wave velocity measurements. However, MH was observed to be formed heterogeneously inside their sample via XRCT [16]. Kerkar et al. [9] confirmed patchy MH distribution and heterogeneous MH accumulation with XRCT at higher image spatial resolution. In the work of Lei et al. [11], the effect of saline water injection on the conversion/redistribution of MHs was confirmed by XRCT scans at high image spatial resolution. Furthermore, the obtained images showed round MH particles under excess-water conditions. Via SXRCT images [12] of the present work, MHs were initially formed following the excess-gas method. However, after multiple water migrations, MHs in both local excess-gas and excess-water media existed in the sample. MHs in local excess-gas media were in cementing forms (mineral coating and/or grain contacts, see Fig. 1) while round MH particles were found mixed with saline water in the pore space under excess-water conditions (Fig. 2). That confirmed the pore-filling/load-bearing distribution of MHs in excess-water media (round MH particles mixed in saline water in the pore space of porous media). It is supposed that MH distribution at the grain scale of MHBS after the water saturation at low MH saturation and after the temperature cycle of triaxial and MRI tests [8, 14] looks alike.

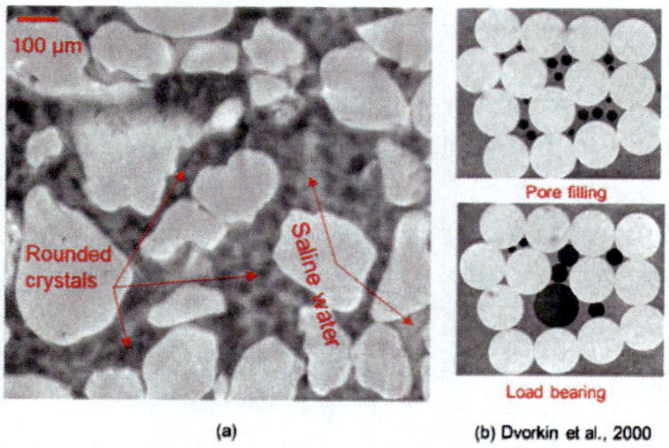

Fig. 2. (a) An example of real MH morphologies and pore habits under excess-water conditions; (b) two conceptual models proposed by Dvorkin et al. [3].

4 Mechanical Properties of MHBS

Hyodo et al. [19] shows that water-saturated sand exhibited a strain hardening and shear contraction behavior while methane hydrate-bearing sediments showed a strain softening and a shear dilation behavior. The higher the MH saturation was, the higher the strength and the more apparent the shear dilation behavior were. Furthermore, by comparing the mechanical properties of MHBS under excess-gas conditions with those under excess-water conditions (using the same sand, under the same stress conditions and at similar MH saturation) higher stiffness and higher failure strength were found for specimens under excess-gas conditions.

The experimental results obtained from triaxial tests in the present work of Le et al. [8] showed higher values for the maximum deviator stress, secant Young's modulus, residual deviator stress, and dilation angle at a higher MH saturation of MHBS under excess-water conditions [8]. Hyodo et al. [19] found similar effects of MH saturation on the maximum deviator stress and the secant Young's modulus while experimentally testing MHBS prepared following a procedure similar to the procedure A. The effects of MH saturation on the stiffness and the failure strength of MHBS were explained by particle bonding. However, the measurements of V_p in the work of Le et al. [8] suggest that cementing hydrates have been significantly converted into non-cementing types at the end of the procedure A, as explained above. In addition, the results obtained by the procedure B (where the conversion has already been completed) showed similar effects of MH saturation on the stiffness and the failure strength of MHBS. This indifference could be explained by the fact that these results were at high deformation when available MH grain bonds were broken.

5 Conclusions

In this paper, the MH formation under both excess-gas and excess-water conditions and the effects of the MH microstructure on the mechanical properties of MHBS are discussed based on the literature and the some results of the project HYDRE [13]. The following conclusions can be drawn:

MH morphologies and distribution at the grain scale of MHs under excess-gas conditions are much more complex than that described by the both cementing models (mineral coating and grain contacts) while for the media with MHs under excess-water conditions, the two idealized models (pore-filling and load-bearing) could be used to describe the MH distribution at the grain scale;

Saturating MHBS by replacing excess methane gas by water is a delicate process that can dissociate a part of MHs and redistribute MHs at the sample scale. An additional temperature cycle (under undrained conditions) makes MHs distributed more homogeneously in the sample compared to that after the water saturation. Furthermore, it is supposed that the water saturation converts (and/or redistributes) MHs in cementing form (grain contacts and/or mineral coating) to the pore space and the temperature cycle allowed for the completion of this conversion for high methane hydrate saturation samples. The MH morphologies and distribution at the grain scale of formed MHBS are supposed to be similar to that of natural MHBS (pore-filling/load-bearing habits);

The higher the MH saturation was, the higher the mechanical properties of MHBS were. However, the mechanical properties of MHBS were higher under excess-gas conditions compared to that under excess-water conditions. The effects of the temperature cycle on the physical/mechanical properties of MHBS under excess-water conditions could only be detected at high methane hydrate saturation. That is why to mimic natural methane hydrate-bearing sediments for the studies of their mechanical behaviors, with a range in MH saturation of 0–50%, the procedure A (without a temperature cycle) could be used when methane hydrate saturation was smaller than 40% to reduce the time of the MH formation, while at higher methane hydrate saturation (40–50%), a temperature cycle should be added (the procedure B should be used) to complete the MH redistribution in the pore space.

References

1. Boswell, R., Collett, T.S.: Current perspectives on gas hydrate resources. Energy Environ. Sci. **4**(4), 1206–1215 (2011)
2. Le T.X., Tang, A.M., Aimedieu, P., Bornert, M., Chabot, B., Rodts, S.: Methane hydrate-bearing sand - an energy resource? In: Randolph, M., Doan, D., Tang, A., Bui, M., Dinh, V. (eds.) Proceedings of the 1st Vietnam Symposium on Advances in Offshore Engineering. VSOE 2018. Lecture Notes in Civil Engineering, vol. 18, pp. 158–163. Springer, Singapore (2019). https://doi.org/10.1007/978-981-13-2306-5_20
3. Dvorkin, J., Helgerud, M.B., Waite, W.F., Kirby, S.H., Nur, A.: Introduction to Physical Properties and Elasticity Models. Natural Gas Hydrate. Springer, Dordrecht (2000)
4. Waite, W.F., Winters, W.J., Mason, D.H.: Methane hydrate formation in partially water-saturated Ottawa sand. Am. Miner. **89**(8–9), 1202–1207 (2004)
5. Priest, J.A., Best, A.I., Clayton, C.R.: A laboratory investigation into the seismic velocities of methane gas hydrate-bearing sand. J. Geophys. Res. Solid Earth **110**(B4) (2005)
6. Priest, J.A., Rees, E.V., Clayton, C.R.: Influence of gas hydrate morphology on the seismic velocities of sands. J. Geophys. Res. Solid Earth **114**(B11) (2009)
7. Choi, J.H., Dai, S., Cha, J.H., Seol, Y.: Laboratory formation of noncementing hydrates in sandy sediments. Geochem. Geophys. Geosyst. **15**(4), 1648–1656 (2014)
8. Le, T.X., Aimedieu, P., Bornert, M., Chabot, B., Rodts, S., Tang, A.M.: Effect of temperature cycle on mechanical properties of methane hydrate-bearing sediment. Soils Found. **59**(4), 814–827 (2019)
9. Kerkar, P.B., Horvat, K., Jones, K.W., Mahajan, D.: Imaging methane hydrates growth dynamics in porous media using synchrotron X-ray computed microtomography. Geochem. Geophys. Geosyst. **15**(12), 4759–4768 (2014)
10. Chaouachi, M., et al.: Microstructural evolution of gas hydrates in sedimentary matrices observed with synchrotron X-ray computed tomographic microscopy. Geochem. Geophys. Geosyst. **16**(6), 1711–1722 (2015)
11. Lei, L., Seol, Y., Choi, J.H., Kneafsey, T.J.: Pore habit of methane hydrate and its evolution in sediment matrix–Laboratory visualization with phase-contrast micro-CT. Mar. Pet. Geol. **104**, 451–467 (2019)
12. Le, T.X., Bornert, M., Aimedieu, P., Chabot, B., King, A., Tang, A.M.: An experimental investigation on methane hydrate morphologies and pore habits in sandy sediment using synchrotron X-ray computed tomography. Mar. Pet. Geol. **122**, 104646 (2020)
13. Le, T.X.: Experimental study on the mechanical properties and the microstructure of methane hydrate-bearing sandy sediment. PhD Thesis, Paris-Est University (2019)
14. Le, T.X., et al.: Kinetics of methane hydrate formation and dissociation in sand sediment. Geomech. Energy Environ. **23**, 100103 (2020)
15. Zhang, L., et al.: Analyzing spatially and temporally visualized formation behavior of methane hydrate in unconsolidated porous media. Magn. Reson. Imaging **61**, 224–230 (2019)
16. Zhao, J., Yang, L., Liu, Y., Song, Y.: Microstructural characteristics of natural gas hydrates hosted in various sand sediments. Phys. Chem. Chem. Phys. **17**(35), 22632–22641 (2015)
17. Kneafsey, T.J.: Examination of hydrate formation methods: trying to create representative samples (2011)
18. Spangenberg, E., Kulenkampff, J., Naumann, R., Erzinger, J.: Pore space hydrate formation in a glass bead sample from methane dissolved in water. Geophys. Res. Lett. **32**(24) (2005)
19. Hyodo, M., et al.: Mechanical behavior of gas-saturated methane hydrate-bearing sediments. J. Geophys. Res. Solid Earth **118**(10), 5185–5194 (2013)

Options for Zonation and Grid Integration of Offshore Wind in Vietnam

Phuong H. Nguyen[1,2], Van Nguyen Dinh[1,3], Hoang Anh Nguyen Trinh[1], To Nhien Ngo[1], An Ha Truong[1](✉), and Minh Ha-Duong[1,4]

[1] Vietnam Initiative for Energy Transition, Hanoi, Vietnam
anha.truong@vietse.vn
[2] TU/e University, Eindhoven, The Netherlands
[3] University College Cork, Cork, Ireland
[4] Centre International de Recherche sur l'Environnement et le Développement, Paris, France

Abstract. To unlock more than 400 GW potential of offshore wind in Vietnam, this study presents possible options for zonation and grid integration from now till 2030. We identified four potential zones along the national coast line for offshore wind development by assessing the technical resource areas based on wind speed and variation, ports, synchronous power plants, oil and gas infrastructures, shipping routes, and sensitive sea areas. We then investigate the capability to integrate wind capacity regarding the transmission expansion plan in the period 2020–2030 by calculating generation adequacy and transmission adequacy. The results specify a possible path for the development of offshore wind which balances out the onshore grid capacity with the quantified capacity from four different wind resource zones. By calculating the loss of load probability with various penetration levels of offshore wind, we find out that 10.5 GW of offshore wind by 2030, spreading across four zones, makes an adequate reliable power supply.

Keywords: Offshore wind · Zoning · Grid integration · Vietnam · Wind power · Renewable energy

1 Introduction

Offshore wind integration plays a crucial role to determine technical boundaries and preliminary economic assessment of integration options [1]. Due to the variability and intermittency of this renewable source, several aspects including the possibility to ensure generation capacity, and stable system operation need to be analyzed thoroughly.

This study aims to explore the potential of offshore wind generation in Vietnam and to provide a preliminary assessment on system adequacy with the given power development plan to help policymakers sharping a roadmap for offshore wind development. In particular, this study investigates the penetration levels of offshore wind capacity while addressing wind integration aspects including generation adequacy and transmission adequacy. A focus will be on the period from now till 2030 to provide an insight into how much offshore wind capacity the grid can accommodate based on the current development of the infrastructure.

D. V. K. Huynh et al. (Eds.): VSOE 2021, LNCE 208, pp. 509–517, 2022.
https://doi.org/10.1007/978-981-16-7735-9_57

2 Methods and Models

2.1 Offshore Wind Potential

Elements for preliminary assessment of strategic zones for offshore wind development are based on the UK methodology [2] and recent research about potential offshore wind resources in Vietnam [3], which assessed the technical resource based on wind speed and variation, ports, and synchronous power plants. This assessment considers oil and gas infrastructures, shipping routes, and Particularly Sensitive Sea Areas to be excluded from the technical resources areas. The spatial distribution of the annual mean wind power density at 100 m height and capacity factor was obtained by using numerical simulations with Weather Research and Forecasting (WRF) model at a 10-km resolution for 10 years (2006–2015) and validated with two observational data sets [4]: (i) wind data observed for 10 years 2006–2015 at six ground-based weather stations located in islands off the coast of Vietnam and run by the Vietnam Center of Hydro-Meteorological Data; and (ii) the QuikSCAT (Quick Scatterometer) data. QuikSCAT is NASA's Earth observation satellite carrying the sea winds scatterometer [5, 6]. The reference turbine is Vestas V164-8.0 with a rated power of 8 MW, a blade length of 80 m, and an approximate hub height of 105 m [4].

2.2 Grid Integration Assessment

The method used in this study refers to the recommended practices from the International Energy Agency (IEA) – Wind expert group [1] with an adaptation suit the scope of the study. The ***load demand*** is adopted from the demand scenarios developed in the Power Development Plan 7 Revised (PDPR7), recently updated directives (58/BC-BCT in 2019, and 724/BCT-DL in 2020), as well as other studies like the one published by the World Bank [7]. The ***transmission expansion plan module*** includes mainly 500 and 220 kV transmission lines and substations. The ***offshore wind generation module*** includes offshore wind capacity fitting into the national energy mix scenarios and associated transmission possibilities to connect to the (nearest) onshore substations.

To analyze ***generation adequacy*** with various penetration levels of offshore wind, this study developed energy mix models for Vietnam's power system in 2025 and 2030 accordingly, including information reported in PDPR7 and document No. 2491/BCT-DL. DIgSILENT PowerFactory is used as the simulation platform that can integrate the energy mix model along with stochastic wind models to analyze the impact of offshore wind uncertainty in terms of ***Loss of Load Probability (LOLP)***. With the same simulation platform, this study developed further a network model simulating current Vietnam's 500 kV and 220 kV grids and its expansion planned in 2025 and 2030 according to PDPR7 and critical update from Document No. 1968/TTr-BCT (2020) and 2491/BCT-DL (2020) to accommodate the high renewable installed capacity recently, i.e. ***transmission adequacy***.

To support future integration of offshore wind, some assumptions have been made:

- Network models: Integrating a complete 500 kV grid model and some important 220 kV grids.

- Generation models are based on the high generation mix scenario from the PDPR7 with the recent update regarding the commissioning progresses of some important power plants.
  - Integrating detailed models of important conventional generators (>600 MW) and large-scale solar PV and onshore wind installations (>100 MW);
  - Aggregating smaller renewable resources in provinces to the equivalent models;
  - Simulating offshore wind farms (OWFs) with the standard platform of 600 MW for the simulation with various transmission options in different voltage levels, i.e. 220 kV or 500 kV, and transmission technologies, i.e. High-Voltage Alternative Current (HVAC) or High-Voltage Direct Current (HVDC).
- Load models: Scaling load demand in the North, Central, and South based on given scenarios developed from PDPR7 with the recent update from Ministry of Industry and Trade (MOIT) as in Table 1.
- Simulation cases:
  - Simulating the "worst-case" scenarios only (P_{max} scenario at the end of the dry season when the demand of the South is at peak and electricity generation from main hydropower plants in the North is limited).
  - Not taking voltage violations as the limit of calculation.
  - Not taking contingency study (n − 1) into account.

Table 1. Load demand scenarios in regions

	2020	2025	2030
Total peak load demand [MW]	44,244	68,367	100,215
Load demand from the North [MW]	17,698	27,347	40,086
Load demand from the Central [MW]	4,424	6,837	10,022
Load demand from the South [MW]	22,122	34,184	50,108

The residual capacity of the onshore grid is estimated to determine which resource zones might be exploited first (Table 2). The residual capacity is based on the maximum loading of the existing near 500 kV substations to the resource zones and the grid expansion capacity reflected by new planned 500 kV substations in the region. The estimation is based on the thermal rating of the grid assets, i.e. transformer rating, according to the PDPR7. The contingency calculation ($n - 1$) and stability analysis, which are not taken into account in this study, might lead to lower levels of the residual capacity.

Table 2. Estimation of residual capacity for the onshore transmission to OW resource zones

	Current		2025		2030	
	Onshore existing 500 kV substations	Average max. loading [%]	New onshore 500 kV substations	Residual cap. [MW]	New onshore 500 kV substations	Residual cap. [MW]
Zone 1	Quang Ninh, Pho Noi, Dong Anh, Hiep Hoa, Thuong Tin	78	Bac Giang, Bac Ninh, Hai Phong	4,032	Gia Loc, Long Bien	5,382
Zone 2	Da Nang, Doc Soi, Thach My	33		3,003	Binh Dinh	3,228
Zone 3	Song May, Tan Uyen, Chon Thanh	51	Long Thanh	3,335	Dong Nai 2, Binh Duong 1	5,135
Zone 4	Cau Bong, Phu Lam, Nha Be	62	Duc Hoa, Long An	3,743	Thot Not, Cu Chi	5,318

3 Result and Discussions

3.1 Offshore Wind Zonation

The Energy Sector Management Assistance Program (ESMAP) [8] estimates the country's total technical potential within 200 km from the shoreline is 261 GW for fixed wind and 214 GW for floating wind. The offshore wind potential and distribution in the ESMAP report agree well with those in [4] that verify the potential of offshore wind in Vietnam.

Based on the potential and variation of offshore wind energy in [4] and [8], and consultations with experts, the offshore wind resource in Vietnam is proposed to be divided into four zones as in Fig. 1 [3].

Zone 1 is the region with the coldest winter and consisting of eight provincial sea areas from Quang Ninh to Ha Tinh. In the northern area of this Zone (in the Gulf of Tonkin), there is a sizeable area with a technical potential for fixed foundation wind of 88 GW and an area with a technical potential of 39 GW for floating foundation [8]. Zone 2, where the winter is moderately cold, has a sea area of seven coastal provinces from Quang Binh to Binh Dinh and marginal potential for offshore wind. Zone 3, comprising of five provincial seas from Phu Yen to Ba Ria – Vung Tau, is less affected by the winter monsoon. Zone 4 is the sea region from Ho Chi Minh city to Kien Giang, least affected by the winter [3]. Off the coast of Binh Thuan and Ninh Thuan, the mean wind speeds exceed 10 m/s. This area then extends south and, with average wind speeds falling to 7 m/s the area extends up to 125 km from shore and covers both Zones 3 and 4. In water depths, less than 50 m, this area alone has a technical potential for 165 GW of fixed offshore wind [8]. Outside there is a very large area below 1,000 m water depth and running from the South to the Central region (south of Hue), covering a part of Zone 2, Zone 3, and Zone 4, with a technical potential of 175 GW for floating foundation [8].

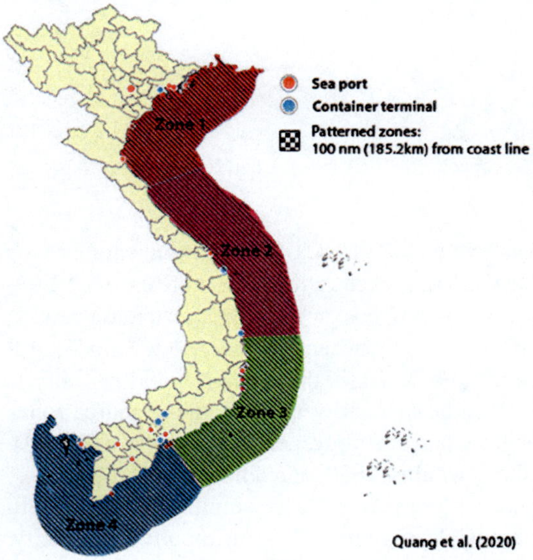

Fig. 1. Location of major ports and container terminals and proposed four zones for evaluation of offshore wind technical resources in Vietnam.

The major seaports and container terminals with channel depths greater than 10 m and maximum vessel size acceptable of 30,000 Dead Weight Tonnage (DWT) all located in Zone 1, the southern of Zone 2, 3 and 4 [3]. Three container ports in Vietnam: Hai Phong and Dinh Vu in Zone 1, and Tan Cang Sai Gon in Zone 4 are listed in the top 20 of Southeast Asia [9]. Especially, the Van Phong International Transshipment Terminal is being developed in Van Phong Bay, Khanh Hoa of Zone 3, which has a depth range of 15–20 m, a large area, and expects to accept the maximum vessel size of 9,000 TEUs (Twenty-foot Equivalent Units) or approximately 120,000 DWT. Considering the important information including draft/channel depth, vessel accepted and the available area, the port facilities in Zone 3 are most favorable to offshore wind farm development. Those in Zone 1 and 4 are also in good capacity.

3.2 Generation Adequacy – Loss of Load Probability

LOLP Calculation Within the scope of our study, we took certain assumptions from the PDPR7 and calculate the LOLP values without having complete data (even the historical data for dispatchable generation units is needed to develop the stochastic models). It is assumed that the additional offshore wind capacity aims to cover 10% of the thermal plant capacity to be added in the corresponding period 2020–2025 and 2025–2030. The generation adequacy analysis considers the fixed demand level according to the load scenarios presented in Table 2 for the stochastic assessment. To calculate the above LOLP results, a Monte Carlo simulation is conducted, taking the stochastic wind models into account. The wind penetration levels were added to Vietnam's generation mix in 2025 and 2030 scenarios.

Figure 2 shows the results of LOLPs according to the increase in offshore wind penetration levels in 2025 and 2030. This study takes the import capacity for the generation

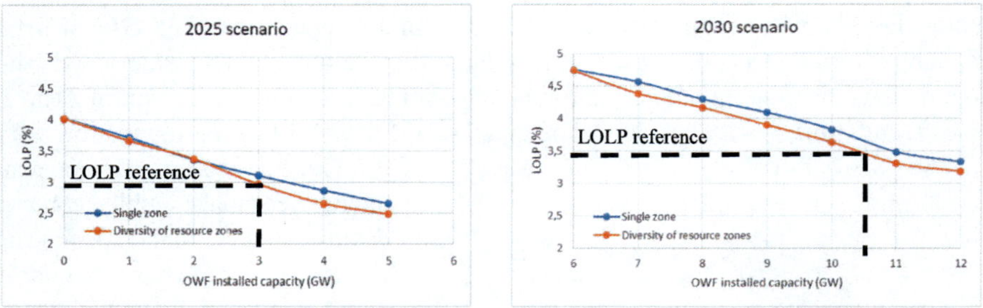

Fig. 2. Impact of offshore wind capacity on the Loss of Load Probability.

mix high case scenario reported in document No. 2491/BCT-DL as the indication of the expected LOLP value. It was planned that the import capacity is set at the levels of 2.9% in 2025 and 3.4% in 2030 accordingly. As it can be seen from the calculation results, a penetration level of above 3 GW offshore wind can secure the 2.9% level of LOLP in 2025. To retain LOLP at the level of 3.4% in 2030, the offshore wind capacity is expected to be above 10,5 GW. In addition, the spatial diversity of wind resource zones would likely have higher effects on securing power supply, i.e. decreasing LOLP levels.

We find out that 10.5 GW of offshore wind in 2030, spreading across four zones, makes an adequate reliable power source. More precisely, we simulated the LOLP to be 3.4%, which compares to the reliability level expected from imported electricity. To have more insight into the wind contribution to power demand adequacy, extensive measurement-based data crossing the wide geographic areas coupling with available flexibility resources is required [10].

3.3 Transmission Adequacy

This section takes a step further from the estimation of the onshore transmission residual capacity summarized in Table 2 to calculate possibilities for connecting around 10 GW offshore wind in 2030 as specified in the previous section. Based on the wind resource potential and transmission expansion plan, three main stages for offshore wind integration to reach 10 GW in 2030 are proposed, as follows:

- Integrating the first 100 × 6 MW OWF from the southern part of Zone 3 (Binh Thuan sea) by 2023 (see Fig. 3 left frame);
- Expanding six additional OWFs in the South (mainly Zone 3 and partially Zone 4) by 2025 with the first VSC-HVDC transmission line (see Fig. 3 middle frame);
- Adding ten additional OWFs over the period 2025–2030, half of them in the North (Zone 1) connecting via the second VSC-HVDC transmission line (Fig. 3, right).

Based on the simulated results, the maximum offshore wind capacity that the grid can absorb is about 4.2 GW until 2025 (Table 3), coming mainly from Zone 3 and partially Zone 4. Besides the high potential of wind resources in these zones, the high load demand in the South urges for having higher regional wind penetration.

Figure 3 left and middle frames show snapshots of electricity flows between main regions in 2023 and 2025. The simulations are calculated with the worst-case scenario during peak periods in April-May when the demand from the South is max while having high production from renewable sources, including offshore wind generation. In 2023, the major part from 3 GW solar PV in Quy Nhon, Ninh Thuan, Binh Thuan together with the first OWF platform (100 × 6 MW) will be delivered to the central load in the South via 500 kV Song May and Tan Uyen substations. Another 1.5 GW flow comes from the Highland areas which have a high amount of small hydropower plants together with solar PV and onshore wind. By 2025, five additional OWF platforms can be added to ensure sufficient power supply for the South and Central areas. A significant surplus of electricity (especially from onshore wind collected from Quang Tri, Ha Tinh, and Nghe An) will be delivered back to the North that approaches the transmission limit.

The calculation shows that about 6 GW capacity of offshore wind can be integrated from mainly Zone 1 (3 GW) and partially Zone 3 (1.2 GW) and Zone 4 (1.8 GW) (Table 3) in the 2026–2030 period. This possible development plan considers the residual transmission capacity of the onshore grids and the urgency to meet the peak load demand in the regions concerned. After 2025, the peak load demand in the North will increase and the regional generation capacity will not be sufficient, especially if commissioning new thermal plants is delayed. As can be seen from Fig. 3 right frame, exploiting offshore wind resources from Zone 1 earlier in this period would contribute to reduce the risk of not having generation adequacy for the North and relieve the transmission stress of the Central – North interconnection. A total of 10.2 GW offshore wind capacity can be integrated from 2020–2030.

Fig. 3. Load flow snapshot in 2023, 2025, and 2030 (from left to right) - Cases of high renewable sources including max capacity from OWFs.

Table 3. A possible development plan of offshore wind in the period 2023–2030

Zone	Platform	Trans. Tech	2023	2024	2025	2026	2027	2028	2029	2030	Total
3	1	Point-to-point HVAC	600								3,000
	2	VSC-HVDC		600							
	3			600							
	4				600						
	5				600						
4	6	Point-to-point HVAC		600							1,200
	7				600						
1	8	VSC-HVDC				600					3,000
	9						600				
	10							600			
	11								600		
	12									600	
3	13	Upgrading VSC-HVDC					600				1,200
	14	Point-to-point HVAC							600		
4	15	VSC-HVDC				600					1,800
	16							600			
	17									600	
Sum	**[MW]**		**600**	**1,800**	**1,800**	**1,200**	**1,200**	**1,200**	**1,200**	**1,200**	**10,200**

4 Conclusions

This study explores the options for zonation and grid integration to 2030. Four offshore wind resource zones have been identified with the highest potential in the sea from Phu Yen to Ba Ria – Vung Tau.

The grid integration simulations shows that in the period 2023–2025, some first fixed-bottom OWFs could be developed in Binh Thuan, Ba Ria – Vung Tau, and Bac Lieu sea with a total capacity of 4.2 GW to meet the peak load demand in the South. In the period 2025–2030, the development of OWFs near the coastline in the North with a total capacity of 3 GW can play an important role to relieve inter-region transmission stress. This together with further development of OWFs along with the coastlines of the South Central and the South can add up to 10.2 GW. We find that 10.5 GW of offshore wind in 2030, spreading across four zones, makes an adequate reliable power source. We simulated the Loss of Load Probability to be 3.4%, which compares to the reliability level expected from imported electricity.

More analyses are needed to explore the robustness of alternative infrastructure plan, because this study's result appears sensitive to yet-to-be-decided parameters, such as feed-in-tariffs (FITs) for solar PV and wind. Still, this study confirms that offshore wind can contribute to the energy transition in Vietnam. It is time to start including this technology in Vietnam's power development plans.

References

1. Holttinen, H., et al.: Expert Group Report on Recommended Practices: 16. Wind Integration Studies (2013)
2. The Crown Estate: Resource and Constraints Assessment for Offshore Wind: Methodology Report (No. 38255-TCE-REP-024) (2019)
3. Quang, V.D., et al.: Evaluation of resource spatial-temporal variation, dataset validity, infrastructures and zones for Vietnam offshore wind energy. Vietnam J. Sci. Technol. **62**(1), 3–16 (2020). https://doi.org/10.31276/VJSTE.62(1).03-16
4. Doan, V.Q., et al.: Usability and challenges of offshore wind energy in vietnam revealed by the regional climate model simulation. SOLA **15**, 113–118 (2019). https://doi.org/10.2151/sola.2019-021
5. Draper, D.W., Long, D.G.: Evaluating the effect of rain on SeaWinds scatterometer measurements. J. Geophys. Res. Oceans **109**(C12), (2004). https://doi.org/10.1029/2002JC001741
6. Said, F., Long, D.G.: Determining selected tropical cyclone characteristics using QuikSCAT's ultra-high resolution images. IEEE J-STARS **4**(4), 857–869 (2011). https://doi.org/10.1109/JSTARS.2011.2138119
7. IES: Integration of Renewable Energy Targets into Vietnam's Power System - Report for Power System Planning Study to the World Bank Group and MOIT (2018)
8. ESMAP: Going Global: Expanding Offshore Wind to Emerging Markets (2019)
9. Dang, V.L., Yeo, G.T.: A competitive strategic position analysis of major container ports in Southeast Asia. Asian J. Shipp. Logist. **33**(1), 19–25 (2017). https://doi.org/10.1016/j.ajsl.2017.03.003
10. Ackermann, T., Kuwahata, R.: Lessons learned from international wind integration studies (2011)

Preliminary Assessment Potential of Underground Energy Storage for Renewable Energy in Cuu Long Basin, Vietnam

Duy Thong Kieu[1](✉), Pham Huy Giao[2], Bui Viet Dung[2], Doan Huy Hien[2], Pham Quy Ngoc[2], and Vu Dinh Hieu[3]

[1] Hanoi University of Mining and Geology, Hanoi, Vietnam
kieuduythong@humg.edu.vn
[2] Vietnam Petroleum Institute, Hanoi, Vietnam
[3] Vietnam Institute of Seas and Islands, Vietnam Administration of Seas and Islands, Hanoi, Vietnam

Abstract. Vietnam is in the monsoon climate zone and has more than 3,000 km of coastline that has extensive resources to develop offshore wind energy. Moreover, offshore wind has the advantage of not taking up land. The rapid development of technology in recent years has resulted in a drastic decline in the cost of wind power. However, the biggest drawback of wind power is its fluctuating electricity supply, which depends on daily and seasonal variation of wind conditions. Therefore, energy storage will play an important role in facilitating the development of wind energy. In this study, an overview on how to select a georeservoir for energy storage and a preliminary evaluation of the underground energy storage potential in the Cuu Long basin, which is situated closest to the area of highest wind energy potential and is also the largest petroleum-producing basin in the country, making it a promising target for underground energy storage. In this study some geological and petrophysical information were collected and used for a preliminary evaluation of energy storage capability for a number of geological formations, based on porosity, permeability and depth to the top. As a result, two formations of Dong Nai and Bien Dong show to be possible good candidates, however more studies should be conducted in future.

Keywords: Underground energy storage · Cuu Long Basin · Renewable energy

1 Introduction

To reduce the carbon footprint, renewable energy resources including wind, biomass, solar and geothermal energy are increasingly utilized [1, 2]. However, using solar and wind energy is that it heavily depends on the weather and geography. Therefore, electricity production fluctuations require cheap, reliable and accessible storage of energy on a scale of seasonal usage to stabilize the power system [1–3]. Energy storage systems comprise of mechanical, chemical, biological, magnetic, thermal and thermochemical types [4]. Each technique can be used for different purposes depending on the storage capacity and discharging time (Fig. 1).

D. V. K. Huynh et al. (Eds.): VSOE 2021, LNCE 208, pp. 518–526, 2022.
https://doi.org/10.1007/978-981-16-7735-9_58

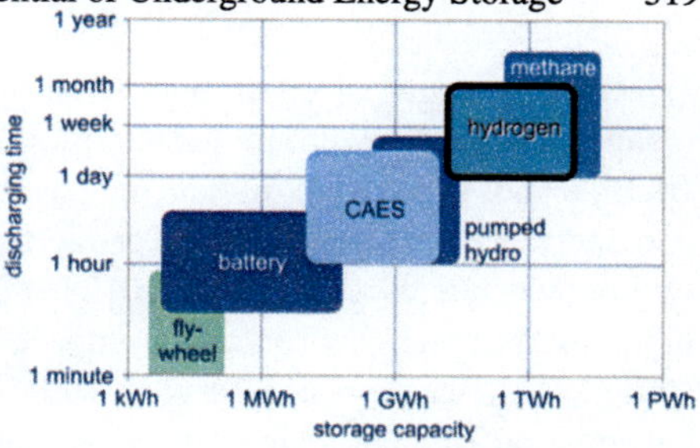

Fig. 1. Energy storage options (After Chen, Cong [4])

The solar and wind energy storage will compete directly with fossil-based electricity options [5]. In the system, storage can play a significant role in reducing the cost. An example of hydrogen storage in France [6], shows that investing in storage is more economical than building more renewable capacity to meet demand. The energy storage facilities are man-made [3] or georeserviors [1, 2, 7]. A comparison among storage techniques [4] shows that for long duration and large scale storage, pumped hydro storage (PHS), compressed air energy storage (CAES) and underground hydrogen storage (UHS) are dominants. The large storage demand for renewable energy (ranging from GWh to TWh) can be primarily satisfied by underground geological formation [5]. PHS also have large capacity storage; however, this storage system requires a large surface footprint and high cost [2]. In contrast, the underground storage system has a small surface footprint, is long term and of low cost [8]. This is the cheapest method for storing a large amount of gaseous hydrogen [9]. But, it is entirely dependent on the existence of suitable geological conditions and rock types. Therefore, research on the geological formation for underground storage is vital. For the Cuu Long Basin geological conditions, there are two main suitable reservoirs, which are depleted gas or oil fields and porous rock geologies.

2 Wind Energy Status in Vietnam

Vietnam has a great potential for offshore wind energy (Fig. 2). A recent analysis by the Danish Energy Agency (DEA) and the World Bank projected that Vietnam offshore wind, if fully harnessed, could generate up to 160 GW of power, and Vietnam is capable of generating 10 GW of electricity by offshore wind farms by 2030 [25]. Wind power in Vietnam is projected to develop rapidly in the future. Before 2018, there were only a few wind plants with a small amount of capacity. The capacity has increased significantly in 2019, obtaining 0.45 GW of wind at the end of June 2019 (EVN, 2019). The expectation of the investment costs of wind will decrease drastically in the coming 30 years. It is projected that wind powerplants at optimal locations will be more cost-effective than coal plants in 2030 for the first 20 GW [10].

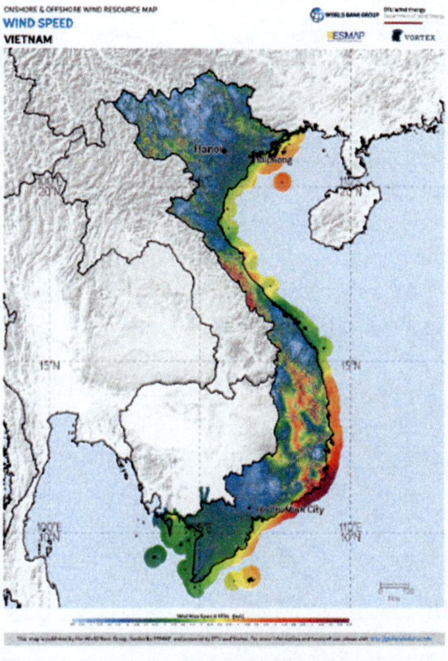

Fig. 2. Map of wind potential energy in Vietnam. Source: Ministry of Industry and trade/GIZ Energy support program.

The challenges of developing wind power sources are the ability to integrate and balance the wind power into the energy system. Power systems must balance in periods from seconds to seasons. In Vietnam, base-load generation has used coal plants. Medium demand range and peak demand range have been stabilized by natural gas-based and hydropower plants, respectively [10]. However, these balance-based are of low flexibility, high cost and high environmental impact. The solution for these issues may be energy storage in the geological reservoirs with large scale, long term, and low cost. Vietnam has great potential storage in the sedimentary basin [11]. Thus, research on the basins for energy storage is vital. In this work, we focus on the Cuu Long basin as it not only has the potential of storing 1,757 Mt of CO2 [11]. It also has great potential of wind energy (Fig. 2). Moreover, it is also the largest petroleum-producing basin in the country. The data in the petroleum exploration and production can be transferred into the research of underground energy storage. The petroleum reservoirs may prove promising for underground energy storage. Oil and platform rigs in defunct fields can be used as offshore structure base of wind turbines [12].

3 Assessment of Underground Storage Potential

3.1 Suitable Geologies for Energy Storage

CAES is currently the commercially mature technology for energy storage in salt caverns. The CAES could be applied to porous rocks in sedimentary basins, where legacy data from hydrocarbon exploration are available, and if geographically close to renewable energy sources [2]. Generally, CAES operation requires an anticline that is comprised of permeable, porous media [7]. Porosity needed for CAES is shown in Table 1 and Table 2. There have been several studies carried out in Israel using fractured rock aquifers [14]. Underground gas storage options include depleted gas/oil reservoirs, aquifers, and salt caverns. The geologic conditions of hydrogen storage have not been researched in detail like CAES as this technique was developed later. The fundamental geologic aspects of hydrogen storage are stated in the work of Liebscher, Wackerl [15]. To select the potential underground hydrogen storage sites, geological criteria such as the volume of the reservoir, lithology and tectonic activity, the depth of occurrence, are taken into account [16].

Table 1. Score-based system for identifying the candidate sites for aquifer CAES (after [13]).

Score	Permeability (md)	Porosity (%)	Type of reservoir
Unusable	<100	<7	Highly discontinuous
Marginal	100–200	7–10	Moderately vulgar limestone & dolomite
OK	200–300	10–13	Reefs, highly vulgar limestone & dolomite
Good	300–500	13–16	Channel sandstones
Excellent	>500	>16	Blanket sands

Table 2. Rank the suitability of potential sites for porous reservoir, modified from [17]

Score	1	2	3	4	5
Score interpretation	Unusable	Marginal	OK	Good	Excellent
Permeability (mD)	<100	100–200	200–300	300–500	>500
Porosity (%)	<7	7–10	10–13	13–16	>16
Total reservoir volume (VR/VS)	<0.5		0.5–0.8 or >3.0	0.8–1.0 or 1.2–3.0	1.0–1.2
Total closure rating (h/H)	<0.5		0.5–0.75	0.75–0.95	0.95–1.0
Depth to top of reservoir (m)	<137 or >760	140–170	170–260 or 670–760	260–430 or 550–670	430–550
Reservoir pressure (MPa)	<1.3 or >6.9	1.3–1.5	1.5–2.3 or 6.1–6.9	2.3–3.9 or 5.0–6.1	3.9–5.0
Type of reservoir	Highly discontinuous	Moderately vugular limestone & dolomite	Reefs, highly vugular limestone & dolomite	Channel sandstones	Blanket sands
Cap rock permeability (mD)			$>10^{-5}$	$<10^{-5}$	
Cap rock threshold pressure (MPa)			2.1–5.5	>5.5	
Cap rock thickness (m)			<6		>6

Table 3: Worldwide hydrogen storage potential sites (after [1])

Field/project name	Storage type	Properties	Note
Diadema, Argentina	Depleted gas reservoir	Pressure = 10 bar Temperature = 50 °C Depth = 600 m	Volume and status of operating are not reported
Underground Sun Storage, Austria	Depleted gas reservoir	Pressure = 78 bar Temperature = 40 °C Depth = 1000 m	Volume is not reported, status is operating
Rough gas storage facility, UK	Depleted natural gas reservoir	Capacity = 48 million m^3 $\phi = 20\%$ K = 75 mD Pressure = 50–100 bar	- Chemical stability - Biological consumption - Leakage - Operational condition
Midland Valley, UK	Oil reservoir	K = 60–80 mD Thickness = 100–1000	- Geological uncertainty - Storage capacity
Rhaetian, Schleswing-Holstein, Germany	Gas reservoir	$\phi = 13–33\%$ K = 2.1 575.2 mD Pressure = 65 bar Depth = 460–490 Thickness = 5–30 m	- Feasibility of hydrogen storage into the proven possible natural gas reservoir - Storage performance (deliverability)

The depleted hydrocarbon reservoirs, which are usually well-understood geological conditions, have good tightness, integrity of caprock and pre-existing surface and subsurface installations, are the most appropriate options for underground gas storage (Table 2 and Table 3). This enables the conversion of the depleted hydrocarbon reservoirs into the storage in a shorter time and at a lower cost. However, the conversion requires a comprehensive study because the possibility of chemical reactions and conversion of hydrogen into, e.g., methane will increase.

3.2 Methodology for Assessment of Storage Potential in Porous Formations

In general, CAES and UHS are in the stage of research and lack industrial experience. There has not been a specific norm of geological criteria. Several studies have illustrated that geological criteria for CAES and UHS [17, 18] have many similarities to natural gas storage. Therefore, we infer the key requirements of the natural gas storage industry to evaluate the potential CAES and UHS sites (Table 2 and Table 3).

4 Preliminary Assessment of Underground Storage Potential in Cuu Long Basin

4.1 Geological Setting

The Cuu Long Basin is an Early Tertiary age pulled apart rift basin. It intersects the most northern extent of the Cuu Long Basin. It is part of the Con Son Swell and the Nam Con Son Basin (Fig. 3). The Cuu Long Basin was formed as an extensional or a pull-apart basin. It was initiated during the Early Tertiary period as a result of the movement and collision of various tectonic plates, especially the collision between the Indian and Asian plates [20]. The relative movements of these tectonic plates created differential movement along pre-existing zones of weakness between Sundaland and Indochina subplates. This resulted in the creation of rift basins such as the Cuu Long rift basin [21]. The tectonic elements and evolution of Cuu Long basin consist of three main elements, which are rifting, compression and thermal sag. There are described very clearly in the research of Schmidt et al. (2019) [21]. We adopted their definition of sequence stratigraphic scheme (Fig. 4).

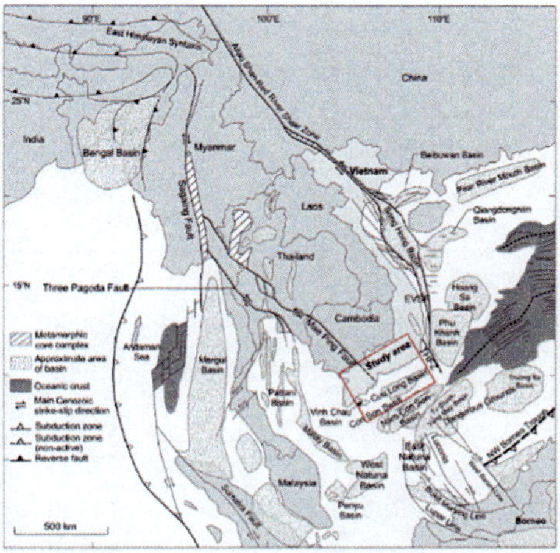

Fig. 3. Location of major structural features and Cenozoic basins for the greater IndoChina area, after Fyhn et al. (2009) [19]. The study area is marked by the red rectangle.

4.2 Potential for Energy Storage in Cuu Long Basin

Storage Potential in Porous Sediments

Based on the stratigraphy of the Cuu Long basin (Fig. 4) and some of the criteria of the storage reservoir (Table 2), the most favorable sedimentary units are Late Miocene to Quaternary (Dong Nai and Bien Dong Formation). The depth of top porous reservoirs should be from 137 to 760 m (Table 2). Thus, below Con Son Formation (Fig. 4) the depth will exceed that range. Moreover, these formations contain sandstone layers that are favorable lithology for reservoir storage [2]. The seismic data [19] show that the sediments of Late Miocene to Recent are characterized by relatively continuous, intermediate-amplitude reflections interrupted by patchy channel features. This suggests continued alluvial to shallow marine deposition in the area in accordance with well data. It is worth noting that these formations are not the objects of the oil and gas industry. Thus, the data are limited. We have not had well information to estimate directly the porosity and permeability of these formations. However, the porosity and permeability of reservoir of these formations should larger than the below reservoir in Con Son formation where the values of porosity varies from 10 to 31% and the permeability from 100 to 3066 mD [22].

Storage Potential in the Depleted Oil and Gas Reservoirs
The main oil and gas reservoirs in Cuu Long basin (Table 4) are fractured basement reservoirs. The porosity and permeability are very good, in comparison with the conditions shown in (Table 2 and Table 3) for the underground storage. However, further research on the conditions of the reservoirs is needed [23].

Table 4. Summarize basic information of the main reservoirs in Cuu Long basin

Reservoir names			Reservoir properties			
			Depth (m)	Temperature (oF)	Permeability (mD)	Porosity (%)
Miocene	White Tiger	North	3100	110	50	16.5
		Central	3050	100	30	15
		South	3000	95	160	18
	Dragon		2800	90	253	19
	Rang Dong		2700	91	185	23.2
	Ruby		1800	84	300	19
	STD		2100	87	1292	26
	TGT		3100	105	293	18.8
Oligocene	White Tiger	Upper Oligocene	3200	120	31	12.04
		Lower Oligocene	3300	125	20	10.25
	Dragon		3200	115	10	15
	STD		3200	105	247	22
	TGT		3200	110	24.8	15

5 Discussion

Georeservoirs can store large amounts of hydrogen or compressed that can help to balance production fluctuation of offshore wind energy. The Cuu Long Basin that is closest to the area of highest wind energy potential in Vietnam and also one the largest petroleum-producing basins in the country was selected for a preliminary screening regarding its potential for underground energy storage. In this study, an overview on the criteria to screen a gereservoir for hydrogen or compressed air storage capacity was conducted, based on which a preliminary investigation was done to show that the depleted oil and gas reservoirs and the porous sedimentary rock in Dong Nai and Bien Dong formations can be the possible target candidates for underground energy storage in the Cuu Long basin in future.

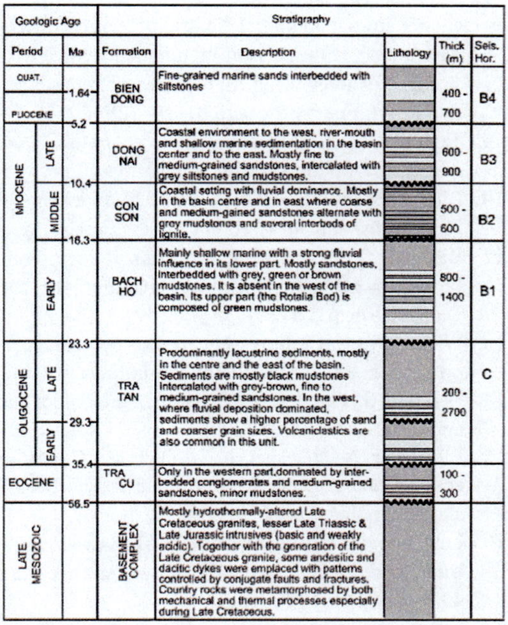

Fig. 4. Stratigraphic units of the Cuu Long basin (After Cuong and Warren [24])

It is also recommended more systematic studies on this issue should be done immediately to keep pace with the rapid development of offshore wind energy development activities in the southern Vietnam these days.

Acknowledgments. This research is funded by the Vietnam National Foundation for Science and Technology Development (NAFOSTED) under the grant number 105.99-2019.324.

References

1. Zivar, D., Kumar, S., Foroozesh, J.: Underground hydrogen storage: a comprehensive review. Int. J. Hydrogen Energy **46**, 23436–23462 (2020)
2. Mouli-Castillo, J., et al.: Inter-seasonal compressed-air energy storage using saline aquifers. Nat. Energy **4**(2), 131–139 (2019)
3. Wang, Z., et al.: A review of marine renewable energy storage. Int. J. Energy Res. **43**(12), 6108–6150 (2019)
4. Chen, H., et al.: Progress in electrical energy storage system: a critical review. Prog. Nat. Sci. **19**(3), 291–312 (2009)
5. Crotogino, F., Schneider, G.-S., Evans, D.J.: Renewable energy storage in geological formations. Proc. Inst. Mech. Eng. Part A J. Power Energy **232**(1), 100–114 (2017)

6. Le Duigou, A., et al.: Relevance and costs of large scale underground hydrogen storage in France. Int. J. Hydrogen Energy **42**(36), 22987–23003 (2017)
7. Aghahosseini, A., Breyer, C.: Assessment of geological resource potential for compressed air energy storage in global electricity supply. Energy Convers. Manage. **169**, 161–173 (2018)
8. Klumpp, F.: Comparison of pumped hydro, hydrogen storage and compressed air energy storage for integrating high shares of renewable energies—potential, cost-comparison and ranking. J. Energy Storage **8**, 119–128 (2016)
9. Taylor, J.B., et al.: Technical and economic assessment of methods for the storage of large quantities of hydrogen. Int. J. Hydrogen Energy **11**(1), 5–22 (1986)
10. Linh, N.H., et al.: EREA & DEA: Vietnam Energy Outlook Report 2019 (2019)
11. ADB: Prospects for Carbon Capture and Storage in Southeast Asia. ADB (2013)
12. Barros, J.C., et al.: Fixed platforms at ageing oil fields - feasibility study for reuse to wind farms. In: Offshore Technology Conference, Offshore Technology Conference, Houston, Texas, USA, p. 18 (2017)
13. Walke, W.C.: Compressed air energy storage: preliminary design and site development program in an aquifer. Plainfield, Indiana (1980)
14. Eckroad, S., Gyuk, I.: EPRI-DOE handbook of energy storage for transmission & distribution applications, pp. 3–35 (2003)
15. Liebscher, A., Wackerl, J., Streibel, M.: Geologic storage of hydrogen – fundamentals, processing, and projects. In: Hydrogen Science and Engineering: Materials, Processes, Systems and Technology, pp. 629–658 (2016)
16. Lewandowska-Śmierzchalska, J., Tarkowski, R., Uliasz-Misiak, B.: Screening and ranking framework for underground hydrogen storage site selection in Poland. Int. J. Hydrogen Energy **43**(9), 4401–4414 (2018)
17. Succar, S., Williams, R.H.: Compressed air energy storage: theory, resources, and applications for wind power **8**, 81 (2008)
18. Kabuth, A., et al.: Energy storage in the geological subsurface: dimensioning, risk analysis and spatial planning: the ANGUS+ project. Environ. Earth Sci. **76**(1), 1–17 (2016). https://doi.org/10.1007/s12665-016-6319-5
19. Fyhn, M.B.W., Boldreel, L.O., Nielsen, L.H.: Geological development of the Central and South Vietnamese margin: implications for the establishment of the South China Sea, Indochinese escape tectonics and Cenozoic volcanism. Tectonophysics **478**(3), 184–214 (2009)
20. Hutchison, C.S.: Geological Evolution of South-east Asia, vol. 13. Clarendon Press, Oxford (1989)
21. Schmidt, W.J., et al.: Tectonic evolution and regional setting of the Cuu Long Basin, Vietnam. Tectonophysics **757**, 36–57 (2019)
22. Bui, L.T.: Evaluation of oil shows in Middle-Miocene at block 02 of Cuu Long Basin based on some of well data. Sci. Technol. Dev. J. **16**(2), 48–63 (2013)
23. Flesch, S., et al.: Hydrogen underground storage—petrographic and petrophysical variations in reservoir sandstones from laboratory experiments under simulated reservoir conditions. Int. J. Hydrogen Energy **43**(45), 20822–20835 (2018)
24. Cuong, T.X., Warren, J.K.: Bach Hofield, a fractured granitic basement reservoir, Cuu Long Basin, Offshore SE Vietnam: a "buried-hill" play. J. Petrol. Geol **32**(2), 129–156 (2009)
25. The Danish Energy Agency and the World Bank Group: Roadmap to Offshore Wind Power Development and Policy Recommendations for Vietnam (2020)

Marine Spatial Planning

GIS-Based Methods for the Analysis of the Main Ports in Europe

Aleksandar Valjarević(✉)

Faculty of Geography, University of Belgrade, Studenstki Trg 3/III, 11000 Belgrade, Serbia
aleksandar.valjarevic@gef.bg.ac.rs

Abstract. In recent years, the interest for safe marine traffic in Europe has been growing. This type of traffic has bad connection with other traffic systems. Europe has central geographical position in the world for many vessel routes. The location of European ports gives the possibility of improving shorelines and traffic near the ports' area. Maritime transport could be one of the most important type of traffic in the years to come, due to low cost of freight and high ecological potential. The assessment of shorelines' changes, the capacity of railway, road, air traffic, and comparison with population density distribution will be very important for spatial analysis. Geographical Information System (GIS) and digital mapping of the shorelines in Europe can be visualized to improve the traffic potential of the ports. For the purpose of this research, geographical coordinates of the main ports in Europe were used for calculations and analyses. GIS and numerical methods showed the advantages and disadvantages of marine traffic in Europe. This research also compared the maritime transport with other traffic systems in Europe. The calculations showed whether there is a potential for new ports or not and how these new ports would be connected with other traffic systems. The final estimation presented marine traffic with population density in an area 100 km away from European shorelines.

Keywords: Marine traffic · Ports · Digital cartography · Geographical Information Systems · Numerical analysis

1 Introduction

Europe has a central geographical position in the Atlantic Ocean. Throughout history one of the first maritime traffic flows was in Europe. In this research, 581 ports were analyzed. GIS (Geographical Information System) analysis and numerical methods may give satisfying results in the investigation of geographical position of the ports and connection with traffic. The analysis of ports' position in Europe is of utmost significance for marine traffic. The connection of ports with another traffic systems is always interesting and in the process of constant change. This change depends on new traffic capacity, the necessity for a new strategy of the European Union, and the countries that declared that they want to be in the European Union. The analysis in this research included the geographical position of ports and distance from railways and roads' access points as

D. V. K. Huynh et al. (Eds.): VSOE 2021, LNCE 208, pp. 529–536, 2022.
https://doi.org/10.1007/978-981-16-7735-9_59

well as airports. Electronic navigation charts (maps) and Geographical information systems may describe a lot of routes and positions of nodes (ports) in the geospace [1]. The specific geographical information and nautical systems are always an essential part of safe travelling on the sea. GIS experts, professional cartographers usually use GIS methods and results [2]. GIS technics and algorithms like buffer and interpolation methods, numerical analysis of networks could enhance the knowledge of marine traffic. A systematic approach and the visualization of layers of all types of traffic, conducted in this research, present the way of making better connection between ports and access points. Apart from maps, digital maps, GIS, the system for safety navigation, many other spatial and mathematical models must be included in the investigation of ports' positions. The Gulf of Finland has an excellent geographical position. The increase of marine traffic in this dense area could be a big problem in the future. This rapid increase in marine transportation of various cargoes, for example the transport of oil from Russia may additionally burden the ports in Finland. The highest risks are caused by the passenger ships between Helsinki and Tallinn together with the high traffic intensity from Russians ports. With the help of digital mapping and statistical analysis, an estimate of transport safety and better connection to the ports was made [3]. The port of Istanbul presents one of the most dangerous places because of the strait and its position. With the support of Marine Traffic Fast Time Simulation (MTFTS) an attempt is made to improve navigation safety by proposing Local Traffic Separation Schemes (LTSS). These systems are supported by digital mapping and with cartographic analysis [4]. The coastal belts in China are under great pressure of marine traffic. Rapid development of maritime industry, maritime traffic and transport systems influenced coastal areas and ports across the world. The water area of China was analyzed in terms of hazards and accidents, with the support of applied Formal Safety Assessment (FSA) approach, endorsed by the International Maritime Organization. The whole research was supported by cartographic analysis and GIS [5]. Many ports across Europe are positioned in inaccessible areas. Specific operational models involving analysis of the traffic in the channels may be useful for sea traffic analysis. These methods included marine traffic control, speed regulation, GIS analysis, mapping [6]. The Automatic Identification System (AIS) is applied to many seas. In the Portuguese coastal Vessel Traffic Service (VTS) system along with digital mapping was used on Portuguese coasts in order to provide safe transport to the ports. By means of this method it is possible to calculate and reduce collision risk between vessels [7]. Collisions in the ports, excluding human factors, depend also on the seasonality of sea physical properties, thus affecting the safety of ports. In South Korean ports, traffic safety assessments showed extreme danger when weather was unstable, increasing by 82% [8]. Apart from the dispersion of ports and their geographical positions, the physical properties of the world sea would have strong effect on the safety of marine traffic. The sea currents, tidal periods, seasonal waves, strong winds, change of the sea level and climate change effects have had a strong influence on the ports in recent periods [9].

2 Materials and Methods

In this research free, open-source data were used for the calculations of the properties of ports across Europe. These data were in vector and raster formats. All data were collected in two ways. First way was collecting the data from the web page of Natural Earth [10]. These vector data are in the resolution of 10 m. The data of ports and populated places were downloaded from the same page. The raster data for the land and coastal areas were downloaded from the Land Sat 8 Satellite and the official web page of USGS (United States Geological Survey). These data were in 30 m of resolution [11]. Other raster data were downloaded from the Earth data web page; these data were in the resolution of 10 m [12]. The data were in DEM (Digital Elevation Model), which enables manipulation. In the second stage of GIS, numerical data were downloaded and transferred to shp extension to represent the coastal belts of Europe. With the support of spatial information tools in QGIS open-source software, models of ports' data were obtained by means of algorithms. Another open-source software Gephi 0.9.2 calculated and analyzed the properties of marine traffic in Europe. This software also compared graph properties (marine traffic) and ports with other types of traffic such as air, road, railway. The main elaborated traffic properties are mobility, connectivity, accessibility, modularity, centrality, and clustering coefficient. Marine network with ports included was compared with a specially created Likert scale [13]. The results are presented on a scale between 0.0 and 1.0, where 0.0–0.2 presents low properties of traffic, 0.3–0.7 medium, and 0.7–1.0 high. The data of road and railway networks and nodes were downloaded from the official page of USGS. The network and nodes of air traffic were downloaded from the web page of open-source data Open flights [14].

2.1 GIS Analysis

For the purpose of this research GIS, numerical methods, and algorithms were used. When data of all European countries were downloaded in the shape file, the data of the location of ports were downloaded too. GIS and modeling geospatial data is a very powerful tool for estimating and evaluating the geographical distribution of ports in Europe. Geospatial analysis of coastal areas, including the data of other types of traffic, gives possibilities to compare the obtained results with marine traffic. To estimate the density of the population Interpolation, Kriging and buffer methods were used within QGIS and SAGA software. Although there are a few other methods, the priority is given to both of them. The reason for that is in the possibility to include semi-variogram and autocorrelation or numerical and statistical relationship among the measured points between ports in Kriging methods [15, 16]. In this research, geographical position or data of longitude and latitude, were provided for all the points of ports. The advantage of this method lies in the fact that weights are not only based on the distance between the measured points, but that this distance will be very important for future calculations. The prediction and the analysis of ports distribution across Europe could be performed by means of semi Kriging methods which minimize the errors in spatial analysis and define semi-variogram [17]. The analysis of raster data was performed with the processing button and acceleration in QGIS (Quantum Geographical Information System). Other algorithms used in the estimation were zonal statistics and segmental interpolation. For the analysis of ports with their geographical position, buffer analysis was used to estimate the properties. Buffer analysis was conducted by means of MMGIS in QGIS software, within which advanced GIS tools are operating. For the analysis of road, railway, and air traffic in comparison with the position of ports, proximity and nearest neighbor algorithms were used. The proximity and nearest neighbor algorithms belong to advanced GIS methods. In the open-source software, Gephi 0.9.2 the properties of networks and nodes of the traffics were estimated and analyzed. This software proved as efficient in the numerical properties of the networks. In this software centrality, connectivity, availability, modularity, and clustering coefficient were estimated for all kinds of traffic. The ports (marine traffic) and their geographical position were thus compared with other types of traffic. The data on populated places across the world were downloaded from the official web page of the United Nations, with special insight of Europe population dispersion and density [18, 19].

3 Results and Discussion

GIS and numerical analysis estimated the results for (Marine Traffic) and the position of ports in Europe. The mobility is 0.4, the connectivity 0.2, and the availability 0.6; density of total lines is 0.0009; the modularity is 0.4; the average clustering is 0.300; the centrality is 0.008. The number of possible connected lines per year is 0.4 X 10^6. The results of air traffic are; the mobility is 1.0; the connectivity is 0.8; the availability is 0.9; the density of total lines is 0.0003, the modularity is 0.1; the average clustering coefficient is 0.370 and the centrality is 0.004. The number of possible connected lines per year is 1.1×10^7. The road network has the following results; the mobility is 0.9; the connectivity is 1.0, the availability is 0.8; the density of total lines is 0.006; the modularity is 0.2; the average clustering is 0.440; the centrality is 0.008. The number of possible connected lines per year is 0.2×10^{11}. The railway traffic network has the following results: the mobility is 0.9; the connectivity is 0.5; the availability is 0.9; the density of total lines is 0.0005; the modularity is 0.5; the average clustering is 0.840; the centrality is 0.008. the number of possible connected lines per year is 0.7×10^8.

The biggest nodes of air traffic are at Heathrow Airport in the United Kingdom, Charles de Gaulle Airport in France, Amsterdam Airport Schiphol in Netherland, Frankfurt am Main Airport in Germany, Adolfo Suárez Madrid–Barajas Airport in Spain, Barcelona–El Prat Josep Tarradellas Airport in Spain, Istanbul Airport in Turkey, Sheremetyevo International Airport in the Russian Federation, Munich Airport in Germany, Gatwick Airport in the United Kingdom. The highest nodes of all types of traffic are in the United Kingdom, Germany, France. The biggest nodes of railway in Europe are Gare du Nord in France, Hamburg Hauptbahnhof in Germany, Châtelet–Les Halles in France, Frankfurt (Main) Hauptbahnhof in Germany, Zürich Hauptbahnhof in Swiss, Roma Termini in Italy, München Hauptbahnhof in Germany, Milano Centrale in Italy, Madrid Atocha in Spain, Berlin Hauptbahnhof in Germany. The biggest bus stations are in Germany, Italy, the United Kingdom, the Russian Federation, Spain, Poland, Turkey, etc. The biggest nodes of marine traffic are at Rotterdam in the Netherlands, Antwerp in Belgium, Hamburg in Germany, Bremerhaven in Germany, Valencia in Spain, Athens in Greece, Algeciras in Spain, Gioia Tauro in Italy, Felixstowe in United Kingdom, Barcelona in Spain. The densest road networks according to the situation from 2019 belong to Germany, Italy, Switzerland, France, the Netherlands, Spain, the United Kingdom, Hungary, Czech Republic, Poland (see Fig. 1).

Fig. 1. Road network in Europe according to the situation from 2019.

The densest railway networks in Europe belong to Germany, France, Italy, Spain, the United Kingdom, Austria, Belgium, the Netherlands, Hungary, Poland, the Russian Federation (see Fig. 2).

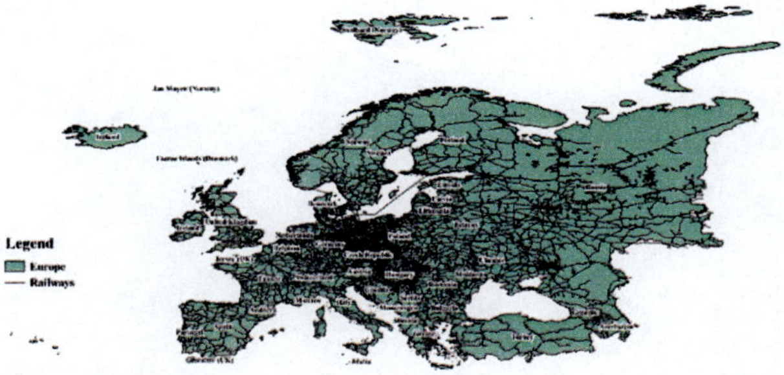

Fig. 2. Railway networks in Europe according to the situation from 2019.

In comparison with populated places in Europe, the ports are not equally distributed. The agglomerations in Europe and areas near big cities are very far from the main ports, excluding some countries. The main agglomerations are dispersed in north Italy, south Switzerland, Belgium, the Netherlands, the south of the United Kingdom, south-east Spain, central Germany and France, south France, central Russian federation, western Turkey, east Ukraine. The average density in Europe may be divided into five classes. The first class presents the areas with average density between 0 and 10 inhabitants. This class covers 57% of the territory. The second class is average density between 10 and 51 inhabitants, covering 14% of the territory. The third class has average density between 51 and 100 inhabitants and it covers 18% of the territory. The fourth class has average density between 101 and 500 inhabitants, covering 5% of the territory. The fifth class which covers 6% has more than 500 inhabitants. This result is connected with very low density of population in the Russian Federation. These areas are, on average, between 150 km and 250 km away from the European ports. Denmark, the Netherlands, Belgium, Germany, north France and south England have myriad of ports. The ports in Denmark are 20 km away from the main traffic access points. In the Netherlands this distance is 17.5 km in Belgium 22.5 km, in Germany 18.5 km, in France 23.5 in the United

Kingdom 17.5 km. The dispersion of ports in central and eastern Europe is very low. Furthermore, these ports are in average 30 km more distant from the access points than those in Western Europe. For example, ports in Estonia are 31.1 km away from the main access points. In Latvia 29.5 km, in Lithuania 28.5 km, in Poland 27.5 km. The ports in north Europe have slightly different problem. This ports have 5 months of freezing conditions. The port in Finland is 29.5 km away from the main access points, in Sweden 24.5 and in Norway 30.5 km. The ports in Russia are 50.5 km away from the main access points. In southern Europe, the ports don't have a problem with the freezing of the sea. The average distance from the main access traffic points is 22 km. Ports in Portugal are 22.5 km away from the main traffic access points, in Spain 19.5 km, in south France 19.0 km, in Italy 17.5 km, in south Greece 18.5 km, in south Turkey 20.5 km. Better positions are those of ports on the islands across Europe. Ports in Greece, Cyprus, Great Britain, Malta, Ireland, Island, Denmark, Faroe Island, Turkey have better possibility for marine transport, despite the low connectivity with traffic access points (see Fig. 3).

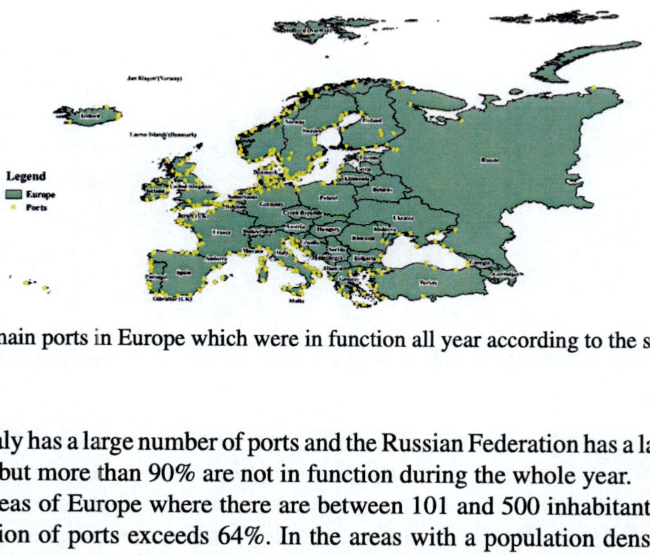

Fig. 3. The main ports in Europe which were in function all year according to the situation from 2019.

South Italy has a large number of ports and the Russian Federation has a large number of ports too but more than 90% are not in function during the whole year.

In the areas of Europe where there are between 101 and 500 inhabitants per 1 km^2 the distribution of ports exceeds 64%. In the areas with a population density between 101 and 500 inhabitants 20% of the ports are distributed. In the areas with the density of population between 10–51 inhabitants, there are 14% of the ports and in the areas with a density of population between 0 and 10 inhabitants there are 2% of the ports. 75% of the ports are at the distance of approximately 20 km away from the main access points. 10% of the ports are at the distance between 50 km and 100 km, 5% of the ports are between 100 km and 200 km away. Finally, 10% of the ports are more than 200 km away from the main access points in Europe. In 2100, more than 80% of inhabitants will be living in the urban area. 55% of inhabitants will be situated close to coastlines across Europe [20, 21]. In that case, the number of ports must grow rapidly. Even if the mobility and connectivity of marine traffic (ports) is low, this kind of traffic presents the safest and the cheapest traffic. This kind of traffic is well connected with road traffic, since 60% of ports in Europe have exits to the regional and international roads. 18% of ports are well connected with railways and 22% with local, regional, and international airports. Another problem of marine traffic is bad connection with other kinds of traffic. The problem occurs because of the fact that the timetable of marine traffic is not always in line with the timetable of other kinds of traffic.

4 Conclusion

GIS and the geographical analysis of the ports in Europe have shown their position and potential relation to other types of traffic. Major traffic features such as connectivity, mobility, availability, centrality, and clustering coefficient have shown all the advantages of marine traffic. The road traffic has a good possibility in connection with ports in 80% of territories. The connectivity of ports with other traffic networks is ranked as average – it is better with the road traffic system and less for 30% with railway traffic. Road traffic has high mobility, the highest connectivity, and availability. With the help of GIS and numerical analysis, one important part of the ports analysis was presented. In the end, this research provides a window into the systematic analyses of the ports and marine traffic in Europe. The research of the same topic in the future would include new results and analyses.

References

1. Zanutta, A., Lambertini, A., Vittuari, L.: UAV photogrammetry and ground surveys as a mapping tool for quickly monitoring shoreline and beach change. J. Marine. Sci. Eng. **8**(1), 52 (2020). https://doi.org/10.3390/jmse8010052
2. Palikaris, A., Mavraeidopoulos, K.A.: Electronic navigational charts: international standards and map projections. J. Marine Sci. Eng. **8**, 248 (2020). https://doi.org/10.3390/jmse7090316
3. Kujala, P., Hänninen, M., Arola Ylitalo, J.: Analysis of the marine traffic safety in the Gulf of Finland. Reliab. Eng. Syst. Saf. **94**(8), 1349–1357 (2009). https://doi.org/10.1016/j.ress.2009.02.028
4. Aydogdu, Y., Yurtoren, C., Park, J., Park, Y.A.: Study on local traffic management to improve marine traffic safety in the Istanbul Strait. J. Navig. **65**(1), 99–112 (2012). https://doi.org/10.1017/S0373463311000555
5. Hu, S., Zhang, J.: Risk assessment of marine traffic safety at coastal water area. Proc. Eng. **45**, 31–37 (2012). https://doi.org/10.1016/j.proeng.2012.08.116
6. Hashimoto, A., Okushima, T.: Evaluating marine traffic safety at channels. Accid. Anal. Prev. **22**(5), 421–442 (1990). https://doi.org/10.1016/0001-4575(90)90038-M
7. Silveira, P., Teixeira, A., Soares, C.: Use of AIS data to characterise marine traffic patterns and ship collision risk off the coast of Portugal. J. Navig. **66**(6), 879–898 (2013). https://doi.org/10.1017/S037346331300051
8. Yoo, Y., Kim, T.-G.: An improved ship collision risk evaluation method for Korea maritime safety audit considering traffic flow characteristics. J. Marine Sci. Eng. **7**(12), 448 (2019). https://doi.org/10.3390/jmse7120448
9. Blezard, R.H.: Calculated sea area of the New Zealand 200 nautical mile exclusive. NZ J. Mar. Freshwat. Res. **14**(2), 137–138 (1980). https://doi.org/10.1080/00288330.1980.9515854
10. Naturalearthdata.com. https://www.naturalearthdata.com/downloads/. Accessed 18 Apr 2020
11. Usgs.gov. https://earthexplorer.usgs.gov/. Accessed 18 Apr 2020
12. Earthdata. https://earthdata.nasa.gov/. Archived 19 Apr 2020

13. Sullivan, G.M., Artino, A.R.: Analyzing and interpreting data from likert-type scales. J. Grad. Med. Educ. **5**, 541–542 (2013)
14. Openfligths.org https://openflights.org/. Accessed on 20 July 2020
15. Valjarević, A., Djekić, T., Stevanović, V., Ivanović, R., Jandziković, B.: GIS numerical and remote sensing analyses of forest changes in the Toplica region for the period of 1953–2013. Appl. Geogr. **92**, 131–139 (2019). https://doi.org/10.1016/j.apgeog.2018.01.016
16. Wu, S.T., Chen, Y.S.: Examining eco-environmental changes at major recreational sites in Kenting National Park in Taiwan by integrating SPOT satellite images and NDVI. Tour. Manage. **57**, 23–36 (2016). https://doi.org/10.1016/j.tourman.2016.05.006
17. Valjarević, A., Filipović, D., Milanović, M., Valjarević, D.: New updated world maps of sea-surface salinity. Pure Appl. Geophys. **177**(6), 2977–2992 (2020). https://doi.org/10.1007/s00024-019-02404-z
18. Zanutta, A., Lambertini, A., Vittuari, L.: UAV photogrammetry and ground surveys as a mapping tool for quickly monitoring shoreline and beach change. Marine Sci. Eng. **8**(1), 52 (2020). https://doi.org/10.3390/jmse8010052
19. Clode, S., Rottensteiner, F., Kootsookos, P., Zelniker, E.: Detection and vectorization of roads from lidar data. Photogramm. Eng. Remote. Sens. **5**, 517–535 (2007)
20. Falkenmark, M., Rockstrom, M., Karlberg, L.: Present and future water requirements for feeding humanity. Food Secur. **1**, 59–69 (2009). https://doi.org/10.1007/s12571008-0003-x
21. Cullinane, K., Wang, T.F., Song, D.W., Ji. P.: The technical efficiency of container ports comparing data envelopment analysis and stochastic frontier analysis. Transport. Res. Part A. Policy Pract. **40**(4), 354–374 (2006). https://doi.org/10.1016/j.tra.2005.07.003

Observed Coastal Morphological Changes Associated with Coastal Engineering Works at Loc An Beach, South East Vietnam

Viola van Onselen[1], Tsung-Yi Lin[1(✉)], Vo-Le Phu[2,4], and Trong-Vinh Bui[3,4]

[1] Department of Geography, National Taiwan Normal University, 162 He Ping East Road, Section 1, Taipei 10610, Taiwan
aeolin@ntnu.edu.tw

[2] Faculty of Environment and Natural Resources, Ho Chi Minh City University of Technology (HCMUT), 268 Ly Thuong Kiet Street, Ho Chi Minh City, Vietnam

[3] Faculty of Geology and Petroleum Engineering, Ho Chi Minh City University of Technology (HCMUT), 268 Ly Thuong Kiet Street, Dist. 10, Ho Chi Minh City, Vietnam

[4] Vietnam National University Ho Chi Minh City, Linh Trung Ward, Thu Duc District, Ho Chi Minh City, Vietnam

Abstract. Coastal erosion is a common issue and has increasingly become one of the stark challenges along the Vietnamese coastline, in the age of global climate change. Large parts of the coast have already been eroded over the past decades and exacerbated by climate change effects, coastal erosion remains a serious threat to coastal habitats and residents. Many engineering structures have been constructed along the coast to combat erosion, such as in geotubes, sea-walls, jetties, groins, tetrapods, etc., but after construction adverse side effects become visible. This paper focusses on such an engineering project at Loc An Beach, Vung Tau City. At a natural land spit in front of the coast, stabilizing structures have been placed and currently two jetties have been constructed to create a new channel for the navigation of boats to the open sea. Coastal morphology changes are analyzed and described using both satellite images and site observations from before and after the constructions. From the data it is evident that the jetties block the longshore sediment transport, causing downdrift erosion and the jetties cannot fulfill their purpose as sediments need to be dredged to keep the channel open. The results of this analysis are important implications for the planning of future coastal management strategies to deal with coastal erosion; different methods might be considered in the framework of Integrated Coastal Zone Management and in terms of more natural (ecosystem-based) and sustainable approaches for better long-term results.

Keywords: Coastal erosion · Coastal engineering · Morphological changes · Sediment transportation

1 Introduction

The rate of socio-economic development of Vietnam is high, especially in the coastal zone, where people rely on marine resources and harbors for trade opportunities. This

D. V. K. Huynh et al. (Eds.): VSOE 2021, LNCE 208, pp. 537–544, 2022.
https://doi.org/10.1007/978-981-16-7735-9_60

triggers people to move towards the coastal zone, leading to rapid urbanization and high population densities [1]. A huge threat in the coastal zones of Vietnam is coastal erosion. Large stretches of beach have been eroded and erosion poses a threat to coastal ecosystems, residential and tourist areas [2] driven by underlying causes of natural forces or anthropogenic activities [3]. However, often human activities and interference in natural environments comes forwards as the most causal factor to coastal erosion [1].

To prevent erosion, engineering structures like sea-walls, jetties, groins or tetrapods are constructed along the coastline of Vietnam. These structures are designed for different purposes; from stabilizing beaches and dissipating waves to preventing erosion from waves and storm surges or sea-level rise. Jetties and breakwaters are also constructed for navigation purposes, by creating a channel from the harbor to the open sea. [4]. However, negative side effects of these structures become visible over time because structures like groins and jetties interrupt the sediment transport along the coastline and the sediment input on the downdrift site decreases, resulting in downdrift erosion [5]. This downdrift erosion requires further measures and it will become a recurring process, transforming sandy shorelines into an artificial coastline [6]. These so called 'hard structures' along the coastline will transform the geography of the area, disturb the natural environment and affect the sediment influxes, leading to changing morphologies. Without sufficient preliminary investigation, there is no knowledge sediment fluxes or on the disturbance in biodiversity. Moreover, these engineering structures are merely short-term solutions because they are often quickly constructed at low coast, causing incomplete designs or the use of unsustainable materials. In addition, many different stakeholders are involved in coastal planning and different responses to the similar issues can lead to fragmented approaches, limiting the effectiveness [4].

This study investigates a large sand body (sand spit) which evolved at the coast of Loc An. Because the sand spit was blocking a more direct route from the harbor to the open sea, 2 jetties were constructed at the end of the sand spit, opening a channel from the harbor to the sea. The presence of the jetties has altered the coastal morphodynamics and bottom profiles in this region, when sedimentation processes are interrupted and erosion is redistributed to the downdrift coast [5]. To evaluate the impacts of this construction and gather information about the shoreline dynamics in this area, satellite data were studied and field visits were conducted over several time periods. The results of this study aim to point out the long-term consequences of coastal engineering structures and find strategies for future engineering projects that are in line with the ICZM for a more sustainable development of coastal areas.

1.1 Study Area

Loc An coastal area is located in Ba Ria-Vung Tau province, approximately 30 km northeast from the city of Vung Tau at the south coast of Vietnam (Fig. 1). The shoreline orientation at the study area is northeast to southwest, different from other nearby areas, because of the curved coastline of Southern Vietnam. This can result in different environments and wave directions every few kilometers along the coastline. The area of Loc An is promoted by tourism bureaus as an area of 'white sandy beaches and clear blue water' and also includes sand dunes, several residential zones and tourist facilities. The climate in this region is considered tropical and humid, with rainy season from May -

October while dry periods are seen in December-April. Typhoons or tropical storms do occur in this area, but they are less common than in northern or central Vietnam [7].

The coastal area is characterized by narrow river outlet to the sea (Song Ray river) and a sand spit has developed parallel to the coastline, extending the river outlet to sea. The spit is mainly controlled by wave induced processes [8] and was slowly expending southwest, nourished by a continuous sediment supply from the longshore current along the coastline. Projects have been carried out in the last decade to stabilize the sand spit, by placing coastal revetments such as tetrapods and geotubes. But due to the dynamic nature of the spit, erosion is prevailing at the sea side of the northeast part and several resorts had to close down or relocate due to the severity of the erosion. Stabilizing projects are also found along the coastline where the river migrates to the southwest, here a fishing harbor and local settings were threatened by coastal erosion and tetrapods are placed along the coast.

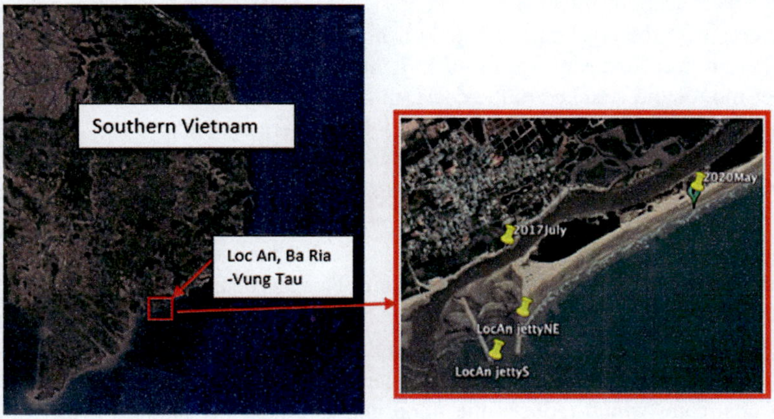

Fig. 1. Study Area pointing out the field study locations (July 2017 and May 2020) and the location of the two newly constructed jetties. (source: Google Earth Pro)

A pair of jetties has been constructed at the tip of the sand spit, to create a new channel from the fishing harbor to the sea, between April 2017 and December 2019. The opening width between the jetties is 500m and the northeastern jetty has a hooked construction, narrowing the channel towards the sea. The southern jetty has a length of approximately 530 m and is orientated perpendicular to the coastline and the northeastern jetty has a length of approximately 580 m.

2 Methods

Changes in sand spit movement and shoreline fluctuations are evaluated using satellite data and field surveys. Field investigations were carried out along the coastline near the river outlet (July 2017) to study the erosion sites and stabilizing constructions and the updrift side of the sand spit near the jetties (May 2020) to study the effects of the engineering works on the shoreline along the sand spit (see Fig. 1). Satellite images are retrieved from before and after the construction of the jetties and coastal morphology changes are analyzed using images from Google Earth Pro from 2000 until 2020.

3 Results

3.1 Natural Sand Spit Formation

The formation of a sand spit is controlled by waves and sediment input from the longshore drift and spits often form where the coastline changes direction. In this case the sand spit formed together with the alongshore migration of the Ray river mouth. The sand spit is characterized by a hooked end, caused by secondary wind or wave directions (see Fig. 2). When behaving naturally without human interference, the evolution of a sand spit will follow the pattern of gradual growth through sediment accretion, migration, followed by breaching and onshore collapse [9]. This process is seen throughout history in this area, with the last breaching and onshore migration of a sand spit a few decades ago. The old breached sand spit is still visible on the satellite images, when it closed the river mouth and attached to the coast, this resulted in a shoreline environment with unstable soft sediments and a wetland area with numerous ponds.

Huong (2020) studied the sand spit growth over time and pointed out that the sand spit growth rate is linear, as it is consistent with an unrestricted type of spit growth, where no (or hardly) any sediment leaves or enters the system. The sand spit has grown several kilometers in the last few decades and in 2012 revetments were constructed at the river side of the sand spit to prevent erosion. This has resulted in strengthening of the spit, but interrupted the natural process of breaching and onshore migration, causing the sand spit to keep growing parallel to the shoreline and causing the river mouth to migrate further southwest. The river became very narrow since the construction of a reservoir further upstream, which reduces the flood and sediment discharge downstream, also contributing to the elongation of the sand spit [10]. From 2000 till 2012 the spit was extending quickly, causing downdrift erosion along the coastline where the river mouth migrated southwest and it elongated further from 2012 till 2016. Due to the variation in dynamics of wave height, wave direction and sediment discharge, the spit is affected by the difference in seasons. Erosion is seen on the sea side of the northeast part of the spit and several resorts already closed down due to intense erosion.

The field investigation in 2017 studied the area west of the fishing harbor, where erosion persisted and clear signs where found in the field. At that time the constructions of the jetty already started at the northwest tip of the sand spit (see Fig. 3). The downdrift coastal erosion and the southwest migration of the river mouth caused problems for the fishing harbor and lead to difficulties in navigating out to the sea. Because of this, a pair of jetties was constructed to open a new navigation channel and redirect and stabilize the river mouth.

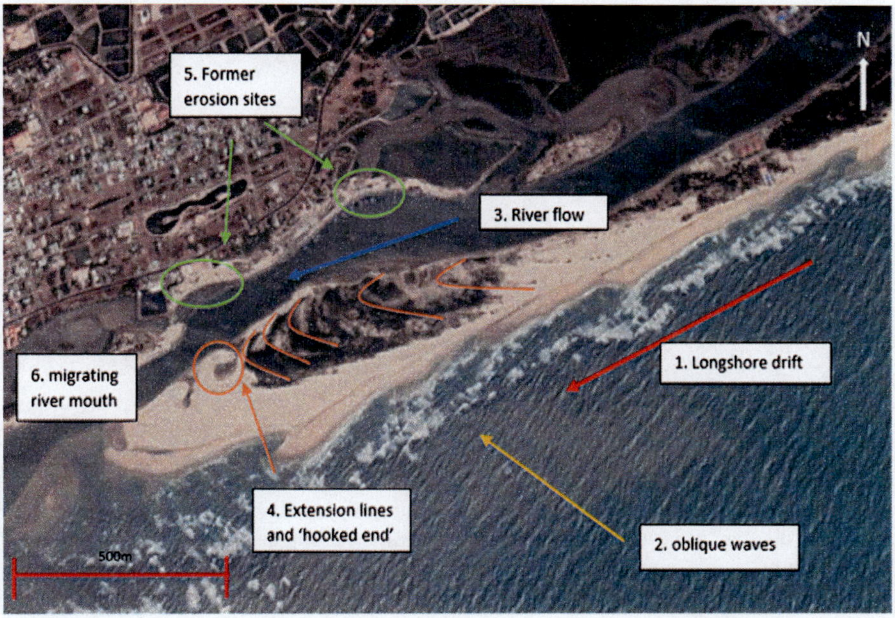

Fig. 2. Development of the sand spit at Loc An (image from 3/26/2016, Google Earth Pro). The longshore current (1) flows from northeast to southwest transporting sediments along the coastline and oblique wave direction (2) changes from season to season. Extension lines (4) show the growth direction of the sand spit and the river mouth (6) is migrating southwestwards due to the elongation of the sand spit. The former erosion sites (5) are a consequence of the growth of the sand spit and the migrating river mouth.

3.2 Observed Changes After Construction Engineering Works

By interrupting the natural process of sand spit formation, the sand spit is disturbed and there will be no longer be an equilibrium system, following the natural process of accretion and breaching. Since the engineering works are designed with stabilizing intensions, surrounding morphology starts to change and more erosion occurs at fixed points. Since the construction of the jetties, the NE jetty blocked the southwestward longshore drift, resulting in sand deposition northeast of the jetty forming a wide foreshore, which is already clearly visible on the satellite images (Fig. 5), while downdrift erosion will increase because of this sediment blockage. Moreover, after the construction of the jetties it turned out that the channel in between the jetties also traps sediment.

This can be traced back to the weak river flow, which is not strong enough to wash out the sediments. The jetties would function as a new gateway for ships to navigate to open sea, and shifted the river mouth, but now the channel needs to be dredged continually because of this sediment accumulation. At the southern beaches (point 3 in Fig. 5) land reclamation projects have started from the dredged sediments to counteract the downdrift erosion. This is only a short-term solution, since the process of downdrift erosion is a continuing process. Future building plans at this location have to take into account these processes and be careful with building close to the shoreline. During the field observation in May 2020, severe coastal erosion was observed at the sand spit, approximately 1 km away from the NE jetty at the seaward side, displayed in Fig. 4 and Fig. 5.

Fig. 3. Fieldwork in 2017 at the shoreline shows: (a) severe erosion at the downdrift beach (b) the start of the engineering works at the sand spit.

Fig. 4. Field visit in 2020: (a)Stabilizing project with geotubes (b) sand scarps (c) erosion causing the floor and fundaments to collapse (d) sand encroachment.

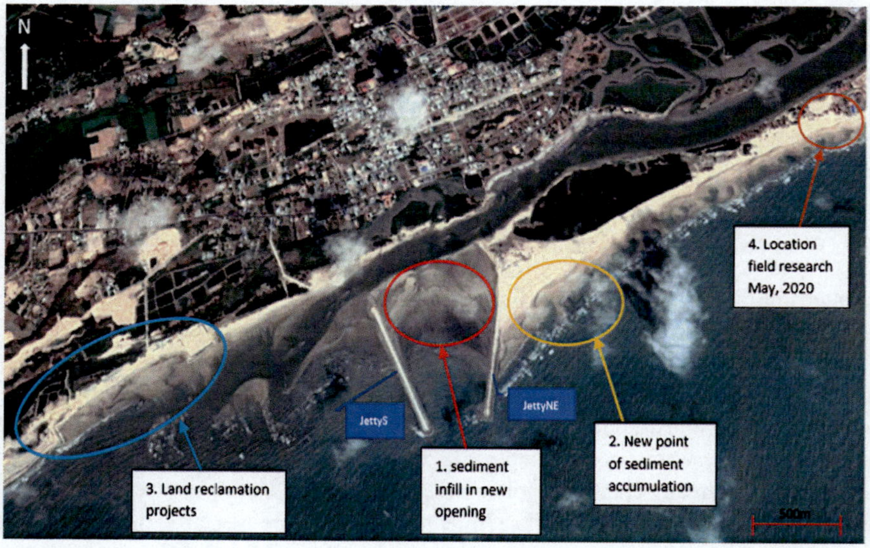

Fig. 5. Situation after construction of the jetties (image from 2/14/2020, GoogleEarth).

4 Discussion and Conclusion

There are many factors influencing the process of sand spit formation at the coast of Loc An. These factors show great variability as seasonal changes influence sediment discharge, river outflow, wave direction etc. Sand spit formation has been a natural process for many years or even decades at the Loc An coast, however human activities have disturbed this natural process. Due to a combination of activities such as the building of a reservoir upstream of the Ray river, stabilizing projects on the sand spit and ultimately the construction of the jetties at the tip of the sand spit, resulted in a disturbed environment. The longshore sediment drift is interrupted by the jetties, resulting in increased downdrift erosion and the jetties failed to open a deep channel for boats to navigate from the fishing harbor to open sea, because the channel is constantly filled up with sediment. This has resulted in many subsequent activities like dredging of sediments to keep the channel open between the jetties and land reclamation south of the jetties to prevent coastal erosion.

To improve coastal planning in the future, proper preliminary research is necessary. It is vital to first understand the coastal process and the mechanisms behind it over longer periods of time. Monitoring of the jetties or other coastal engineering works can contribute useful information to understand what will happen when humans interfere in these processes and other solutions for coastal erosion or sand spit breaching can be considered. Using models to predict outcomes after the construction of engineering works can be a useful tool to add to this research.

Different engineering measures can be applied to minimize coastal erosion. However, in this age of global climate change and an increase in awareness of the negative side effects of human activities, it is important for future activities to not only look at what is convenient for humans, but also think of the effects on the environment. The results of this study are important implications for the planning of future coastal management strategies to deal with coastal erosion, especially in the framework of Integrated Coastal Zone Management, which supports both coastal development and environmental protection. Many studies already pointed out that nature-based adaptations or natural engineering approaches in the long term often are more cost effective, which might be an important trigger for the Vietnamese government. By enhancing public understanding on the values of the coastal zone and the importance to keep an protect coastal resources and including all different stakeholders in the process of coastal development, future engineering activities can work more towards resilient and sustainable approaches like sand dune restoration, ecosystem enhancement or building further away from the shoreline to prevent buildings from collapsing. With this in mind, future research in this area could focus on alternative sustainable strategies compared to the hard engineering structures found along the coastline.

References

1. Lin, T.-Y., Vo, L.P., Nguyen, D.T., Bui, T.V., Nguyen, G.B., Dang, H.A.: Coastal processes in central coastline of Vietnam: a field study and its implications for coastal hazard management. In: 5th Vietnam-Japan Workshop on Estuaries, Coasts Rivers, pp. 209–220 (2017)
2. Hung, N.T., Vinh, B.T., Nam, S.Y., Lee, J.L.: Cause analysis of erosion-induced resort washout on Cua Dai Beach, Vietnam. J. Coast. Res. **79**, 214–218 (2017)
3. Ngo, N.C., Pham, H.T., Do, D.S., Nguyen, N.B.: Status of Coastal Erosion of Vietnam and Proposed Measures for protection (2005). (Report)

4. Vo, P.L., Lin, T.-Y., Nguyen, B.G.: Threat of coastal hazards to socio-economic development in the central coasts of Vietnam. In: International Symposium Lowland Technology (2018)
5. Noshi, Y., Uda, T., Kobayashi, A., Miyahara, S., Serizawa, M.: Beach changes observed in Phan Rang city in Southeast Vietnam. Procedia Eng. **116**, 163–170 (2015)
6. Uda, T., Yamamoto, K.: Beach erosion around a sand spit -an example of Mihono-Matsubara sand spit-. Coastal Eng. Chap. **197**, 2726–2740 (1994)
7. Nguyen, K., Liou, Y., Terry, J.P.: Science of the total environment vulnerability of Vietnam to Typhoons : a spatial assessment based on hazards, exposure and adaptive capacity. Sci. Total Environ. **682**, 31–46 (2019)
8. Allard, J., Bertin, X., Chaumillon, E., Pouget, F.: Sand spit rhythmic development: a potential record of wave climate variations? Arçay Spit, western coast of France. Mar Geol. **253**, 107–131 (2008)
9. Dabees, M.A.: Evolution of barrier island sand spits and effects on beach management practices. Case studies in central Florida Gulf coast, USA. In: Coastal Dynamics, pp. 870–888 (2017)
10. Hoang, V.C.: Long-term evolution of morphology at Loc An Estuary, Vung Tau, Vietnam. Mar Geod. **43**, 163–188 (2020)

Author Index

D. V. K. Huynh et al. (Eds.): VSOE 2021, LNCE 208, pp. 545–547, 2022.
https://doi.org/10.1007/978-981-16-7735-9